Lecture Notes in Computer Science   13423

More information about this series at https://link.springer.com/bookseries/558

Bohan Li · Lin Yue · Chuanqi Tao · Xuming Han ·
Diego Calvanese · Toshiyuki Amagasa (Eds.)

Web and Big Data

6th International Joint Conference, APWeb-WAIM 2022
Nanjing, China, November 25–27, 2022
Proceedings, Part III

Springer

Editors
Bohan Li
Nanjing University of Aeronautics
and Astronautics
Nanjing, China

Chuanqi Tao
Nanjing University of Aeronautics
and Astronautics
Nanjing, China

Diego Calvanese
Free University of Bozen-Bolzano
Bolzano, Italy

Lin Yue
Newcastle University
Callaghan, NSW, Australia

Xuming Han
Jinan University
Guangzhou, China

Toshiyuki Amagasa
University of Tsukuba
Tsukuba, Japan

ISSN 0302-9743   ISSN 1611-3349 (electronic)
Lecture Notes in Computer Science
ISBN 978-3-031-25200-6   ISBN 978-3-031-25201-3 (eBook)
https://doi.org/10.1007/978-3-031-25201-3

This Springer imprint is published by the registered company Springer Nature Switzerland AG
The registered company address is: Gewerbestrasse 11, 6330 Cham, Switzerland

Preface

These volumes (LNCS 13421–13423) contain the proceedings of the 6th Asia-Pacific Web (APWeb) and Web-Age Information Management (WAIM) Joint Conference on Web and Big Data (APWeb-WAIM). Researchers and practitioners from around the world came together at this leading international forum to share innovative ideas, original research findings, case study results, and experienced insights in the areas of the World Wide Web and big data, thus covering web technologies, database systems, information management, software engineering, knowledge graphs, recommender system and big data.

The 6th APWeb-WAIM conference was held in Nanjing during 25–27 November 2022. As an Asia-Pacific flagship conference focusing on research, development, and applications in relation to Web information management, APWeb-WAIM builds on the successes of APWeb and WAIM. Previous APWeb events were held in Beijing (1998), Hong Kong (1999), Xi'an (2000), Changsha (2001), Xi'an (2003), Hangzhou (2004), Shanghai (2005), Harbin (2006), Huangshan (2007), Shenyang (2008), Suzhou (2009), Busan (2010), Beijing (2011), Kunming (2012), Sydney (2013), Changsha (2014), Guangzhou (2015), and Suzhou (2016). And previous WAIM events were held in Shanghai (2000), Xi'an (2001), Beijing (2002), Chengdu (2003), Dalian (2004), Hangzhou (2005), Hong Kong (2006), Huangshan (2007), Zhangjiajie (2008), Suzhou (2009), Jiuzhaigou (2010), Wuhan (2011), Harbin (2012), Beidaihe (2013), Macau (2014), Qingdao (2015), and Nanchang (2016). The combined APWeb-WAIM conferences have been held in Beijing (2017), Macau (2018), Chengdu (2019), Tianjin (02020), and Guangzhou (2021). With the ever-growing importance of appropriate methods in these data-rich times and the fast development of web-related technologies, we believe APWeb-WAIM will become a flagship conference in this field.

The high-quality program documented in these proceedings would not have been possible without the authors who chose APWeb-WAIM for disseminating their findings. APWeb-WAIM 2022 received a total of 297 submissions and, after the double-blind review process (each paper received at least three review reports), the conference accepted 75 regular papers (including research and industry track) (acceptance rate 25.25%), 45 short research papers, and 5 demonstrations. The contributed papers address a wide range of topics, such as big data analytics, advanced database and web applications, data mining and applications, graph data and social networks, information extraction and retrieval, knowledge graphs, machine learning, recommender systems, security, privacy and trust, and spatial and multimedia data. The technical program also included keynotes by Ihab F. Ilyas Kaldas, Aamir Cheema, Chengzhong Xu, Lei Chen, and Haofen Wang. We are grateful to these distinguished scientists for their invaluable contributions to the conference program.

We would like to express our gratitude to all individuals, institutions, and sponsors that supported APWeb-WAIM 2022. We are deeply thankful to the Program Committee members for lending their time and expertise to the conference. We also would like to acknowledge the support of the other members of the organizing committee. All of them helped to make APWeb-WAIM 2022 a success. We are grateful for the guidance of the Honorary Chairs (Zhiqiu Huang), Steering Committee representative (Yanchun Zhang) and the General Co-chairs (Aoying Zhou, Wojciech Cellary and Bing Chen) for their guidance and support. Thanks also go to the Workshop Co-chairs (Shiyu Yang and Saiful Islam), Tutorial Co-chairs (Xiang Zhao, Wenqi Fan and Ji Zhang), Demo Co-chairs (Jianqiu Xu and Travers Nicolas), Industry Co-chairs (Chen Zhang Hosung Park), Publication Co-chairs (Chuanqi Tao, Lin Yue and Xuming Han), and Publicity Co-chairs (Yi Cai, Siqiang Luo and Weitong Chen).

We hope the attendees enjoyed the exciting program of APWeb-WAIM 2022 as documented in these proceedings.

November 2022

Toshiyuki Amagasa
Diego Calvanese
Xuming Han
Bohan Li
Chuanqi Tao
Lin Yue

Organization

General Chairs

Aoying Zhou	East China Normal University, China
Bing Chen	Nanjing University of Aeronautics and Astronautics, China
Wojciech Cellary	WSB University, Poland

Steering Committee

Aoying Zhou	East China Normal University, China
Divesh Srivastava	AT&T Research Institute, USA
Jian Pei	Simon Fraser University, Canada
Lei Chen	Hong Kong University of Science and Technology, China
Lizhu Zhou	Tsinghua University, China
Masaru Kitsuregawa	University of Tokyo, Japan
Mingjun Huang	National University of Singapore, Singapore
Tamor Özsu	University of Waterloo, Canada
Xiaofang Zhou	University of Queensland, Australia
Yanchun Zhang	Victoria University, Australia

Program Committee Chairs

Diego Calvanese	Free University of Bozen-Bolzano, Italy
Toshiyuki Amagasa	University of Tsukuba, Japan
Bohan Li	Nanjing University of Aeronautics and Astronautics, China

Publication Chairs

Chuanqi Tao	Nanjing University of Aeronautics and Astronautics, China
Lin Yue	University of Newcastle, Australia
Xuming Han	Jinan University, China

Program Committee

Alex Delis	University of Athens, Greece
An Liu	Soochow University, China
Aviv Segev	University of South Alabama, USA
Bangbang Ren	National University of Defense Technology, China
Baokang Zhao	NUDTCS, China
Baoning Niu	Taiyuan University of Technology, China
Bin Guo	Northwestern Polytechnical University, China
Bin Cui	Peking University, China
Bin Xia	Nanjing University of Posts and Telecommunications, China
Bin Zhao	Nanjing Normal University, China
Bolong Zheng	Huazhong University of Science and Technology, China
Byung Suk Lee	University of Vermont, USA
Carson Leung	University of Manitoba, Canada
Cheng Long	Nanyang Technological University, Singapore
Chengliang Chai	Tsinghua University, China
Cheqing Jin	East China Normal University, China
Chuanqi Tao	Nanjing University of Aeronautics and Astronautics, China
Cuiping Li	Renmin University of China, China
Dechang Pi	Nanjing University of Aeronautics and Astronautics, China
Dejun Teng	Shandong University, China
Demetrios Zeinalipour-Yazti	University of Cyprus, Cyprus
Derong Shen	Northeastern University, China
Dong Li	Liaoning University, China
Donghai Guan	Nanjing University of Aeronautics and Astronautics, China
Fabio Valdés	FernUniversität in Hagen, Germany
Fei Chen	Shenzhen University, China
Genoveva Vargas-Solar	CNRS, France
Giovanna Guerrini	University of Genoa, Italy
Guanfeng Liu	Macquarie University, Australia
Guodong Long	University of Technology Sydney, Australia
Guoqiong Liao	Jiangxi University of Finance & Economics, China
Haibo Hu	Hong Kong Polytechnic University, China
Hailong Liu	Northwestern Polytechnical University, China
Haipeng Dai	Nanjing University, China

Haiwei Pan   Harbin Engineering University, China
Haiwei Zhang   Nankai University, China
Hancheng Wang   Nanjing University, China
Hantao Zhao   Southeast University, China
Harry Kai-Ho Chan   University of Sheffield, UK
Hiroaki Ohshima   University of Hyogo, Japan
Hong Chen   Renmin University, China
Hongzhi Wang   Harbin Institute of Technology, China
Hongzhi Yin   University of Queensland, Australia
Hua Dai   Nanjing University of Posts and
                                        Telecommunications, China
Hua Wang   Victoria University, Australia
Hui Li   Xiamen University, China
Javier A. Espinosa-Oviedo   University of Lyon, France
Ji Zhang   University of Southern Queensland, Australia
Jia Yu   Washington State University, USA
Jiajie Xu   Soochow University, China
Jiali Mao   East China Normal University, China
Jian Yin   Sun Yat-Sen University, China
Jiangtao Cui   Xidian University, China
Jianqiu Xu   Nanjing University of Aeronautics and
                                        Astronautics, China
Jianxin Li   Deakin University, Australia
Jilin Hu   Aalborg University, Denmark
Jing Jiang   University of Technology Sydney, Australia
Jizhou Luo   Harbin Institute of Technology, China
Jun Gao   Peking University, China
Junhu Wang   Griffith University, Australia
Junjie Yao   East China Normal University, China
K. Selçuk Candan   Arizona State University, USA
Kai Zeng   Microsoft, USA
Kai Zheng   University of Electronic Science and Technology
                                        of China, China
Kazutoshi Umemoto   University of Tokyo, Japan
Krishna Reddy P.   International Institute of Information Technology,
                                        India
Ladjel Bellatreche   ISAE-ENSMA, France
Le Sun   Nanjing University of Information Science and
                                        Technology, China
Lei Duan   Sichuan University, China
Lei Zou   Peking University, China
Leong Hou U.   University of Macau, China

Liang Hong	Wuhan University, China
Lin Li	Wuhan University of Technology, China
Lin Yue	University of Queensland, Australia
Liyan Zhang	Yanshan University, China
Lizhen Cui	Shandong University, China
Lu Chen	Zhejiang University, China
Luyi Bai	Northeastern University, China
Makoto Onizuka	Osaka University, Japan
Maria Damiani	University of Milan, Italy
Maria Luisa Damiani	University of Milan, Italy
Markus Endres	University of Augsburg, Germany
Meng Wang	Southeast University, China
Miao Xu	University of Queensland, Australia
Mingzhe Zhang	University of Queensland, Australia
Min-Ling Zhang	Southeast University, China
Mirco Nanni	ISTI-CNR Pisa, Italy
Mizuho Iwaihara	Waseda University, Japan
My T. Thai	University of Florida, USA
Nicolas Travers	Léonard de Vinci Pôle Universitaire, France
Peer Kroger	Christian-Albrechts-Universität Kiel, Germany
Peiquan Jin	University of Science and Technology of China, China
Peisen Yuan	Nanjing Agricultural University, China
Peng Peng	Hunan University, China
Peng Wang	Fudan University, China
Qian Zhou	Nanjing University of Posts and Communications, China
Qilong Han	Harbin Engineering University, China
Qing Meng	Southeast University, China
Qing Xie	Wuhan University of Technology, China
Qiuyan Yan	China University of Mining and Technology, China
Qun Chen	Northwestern Polytechnical University, China
Quoc Viet Hung Nguyen	Griffith University, Australia
Reynold Cheng	University of Hong Kong, China
Rong-Hua Li	Beijing Institute of Technology, China
Saiful Islam	Griffith University, Australia
Sanghyun Park	Yonsei University, Korea
Sanjay Madria	Missouri University of Science & Technology, USA
Sara Comai	Politecnico di Milano
Sebastian Link	University of Auckland, New Zealand

Xin Cao	University of New South Wales, Australia
Xin Wang	Tianjin University, China
Xingquan Zhu	Florida Atlantic University, USA
Xuelin Zhu	Southeast University, China
Xujian Zhao	Southwest University of Science and Technology, China
Xuming Han	Jinan University, China
Xuyun Zhang	Macquarie University, Australia
Yaokai Feng	Kyushu University, Japan
Yajun Yang	Tianjin University, China
Yali Yuan	University of Göttingen, Germany
Yanda Wang	University of Bristol, UK
Yanfeng Zhang	Northeastern University, China
Yanghua Xiao	Fudan University, China
Yang-Sae Moon	Kangwon National University, Korea
Yanhui Gu	Nanjing Normal University, China
Yanjun Zhang	Deakin University, Australia
Yasuhiko Morimoto	Hiroshima University, Japan
Ye Liu	Nanjing Agricultural University, China
Yi Cai	South China University of Technology, China
Yingxia Shao	BUPT, China
Yong Tang	South China Normal University, China
Yong Zhang	Tsinghua University, China
Yongpan Sheng	Chongqing University, China
Yongqing Zhang	Chengdu University of Information Technology, China
Youwen Zhu	Nanjing University of Aeronautics and Astronautics, China
Yu Gu	Northeastern University, China
Yu Liu	Huazhong University of Science and Technology, China
Yu Hsuan Kuo	Amazon, USA
Yuanbo Xu	Jilin University, China
Yue Tan	University of Technology Sydney, Australia
Yunjun Gao	Zhejiang University, China
Yuwei Peng	Wuhan University, China
Yuxiang Zhang	Civil Aviation University of China, China
Zakaria Maamar	Zayed University, UAE
Zhaokang Wang	Nanjing University of Aeronautics and Astronautics, China
Zheng Zhang	Harbin Institute of Technology, China
Zhi Cai	Beijing University of Technology, China

Zhiqiang Zhang   Zhejiang University of Finance and Economics,
                        China
Zhixu Li   Soochow University, China
Zhuoming Xu   Hohai University, China
Zhuowei Wang   University of Technology Sydney, Australia
Ziqiang Yu   Yantai University, China
Zouhaier Brahmia   University of Sfax, Tunisia

Additional Reviewers

Bo Tang   Southern University of Science and Technology,
                        China
Fang Wang   Hong Kong Polytechnic University, China
Genggeng Liu   Fuzhou University, China
Guan Yuan   China University of Mining and Technology,
                        China
Jiahao Zhang   Hong Kong Polytechnic University, China
Jinguo You   Kunming University of Science and Technology,
                        China
Long Yuan   Nanjing University of Science and Technology,
                        China
Paul Bao   Nagoya University, Japan
Philippe Fournier-Viger Shenzhen University, China
Ruiyuan Li   Chongqing University, China
Shanshan Yao   Taiyuan University of Technology, China
Xiaofeng Ding   Huazhong University of Science and Technology,
                        China
Yaoshu Wang   Shenzhen University, China
Yunpeng Chai   Renmin University of China, China
Zhiwei Zhang   Beijing Institute of Technology, China

Contents – Part III

**Security, Privacy, and Trust and Blockchain Data Management and
Applications**

Spatial and Multi-media Data

Query Processing and Optimization

MSP: Learned Query Performance Prediction Using MetaInfo and Structure of Plans

Honghao Liu[1], Zhiyong Peng[1], Zhe Zhang[2], Huan Jiang[1], and Yuwei Peng[1(✉)]

[1] Wuhan University, Wuhan 430000, China
{liuhonghao,peng,xiaocao,ywpeng}@whu.edu.cn
[2] Huawei, Xi'an 710000, China

Abstract. Query performance prediction is important and challenging in database management systems. The traditional cost-based methods perform poorly predicting query performance due to inaccurate cost estimates. In recent years, research shows that learning-based query performance prediction without actual execution has outperformed traditional models. However, existing learning-based models still have limitations in feature encoding and model design. To address these limitations, we propose a method of query performance prediction based on the binary tree-structured model fully expressing the impact between plan tree nodes. We also present an efficient metadata encoding method, taking into account the data type and value distribution of the columns, which we call **metaInfo**. This encoding method can support various complex SQL queries on changing data. The experiments are conducted on real-world datasets, and the experimental results show that our approach outperforms the state-of-the-art method.

Keywords: Query performance prediction · Feature encoding · Tree-structured model

1 Introduction

Query optimizer is an essential part of the database management system, which aims to select the optimal execution plan for a SQL query. However, recent studies [9,12] have shown that traditional query optimizers based on cost model cannot guarantee to pick out the optimal plan every time due to their poor cost and cardinality estimation.

To solve this problem, learning-based query performance prediction (QPP) without actual execution has been proposed, which has evolved into an essential primitive included in a wide variety of database tasks, such as query monitoring [15], resource management [20], query scheduling [3] and admission control [22].

After QPP was proposed, database researchers have attempted to predict query plan performance with machine learning models [1,5,10]. With the development of deep learning, more neural network models have been applied to this

B. Li et al. (Eds.): APWeb-WAIM 2022, LNCS 13423, pp. 3–18, 2023.
https://doi.org/10.1007/978-3-031-25201-3_1

field [8,17]. Deep neural network with average pooling is ineffective to represent tree-structured query plan, thus tree-structured learning models are proposed [13,18].

The currently proposed learning-based model for query performance prediction still has some shortcomings. Firstly, encoding methods proposed recently adopt one-hot vectors of tables and columns. The dimensions of the encoding vectors will change with the number of tables and columns, which causes the neural network to be retrained, thus leading to the growth of time for training and evaluation of the network. Secondly, metadata significantly impacts on the execution time of a plan, so metadata information should be fed into the model as a priori knowledge. Finally, the learning models proposed so for treat two-child nodes equally. For example, a hash join node has two child nodes. The sum of the output vectors of the two child nodes is directly used as the input of the hash join node. However, if the left and right child nodes are interchanged, the obtained information is also the same. This is inconsistent with the actual situation. The execution time of the hash join is not the same as the exchange of the left and right nodes. Therefore we should fully capture the tree-structure information of the query plan, including not only the order and dependencies between physical operations in the query plan but also the different effects of operations on subsequent ones.

To address these challenges, we define the data type and value distribution of the columns in a table as **metaInfo**. In addition, we propose a feature encoding method that takes into account metaInfo and design a tree-structure model to capture more information about a query plan tree for predicting query performance.

In summary, our contributions are as follows:

(1) Considering **metaInfo** of the database, we propose an efficient feature extraction method that uses data types and value distribution of fields rather than the column and table names as metadata encoding, which can avoid frequent retraining of the model when underlying data changes.
(2) We propose a learning model based on tree structure which learns different influences on the parent node incurred by its child nodes through more fine-grained network parameters to capture the dependencies between nodes in the plan tree fully.
(3) Experiments are conducted on real-world datasets, and the experimental results show that our method outperforms the existing methods.

The paper is organized as followings. In Sect. 2, we summarize the related work. In Sect. 3, we propose a more adaptable encoding method. In Sect. 4, we design the tree-structured learning model. In Sect. 5, we evaluate our design through experiments and conclusions are made in Sect. 6.

2   Related Work

Query performance prediction (QPP) evaluates the query plan's performance by predicting the actual execution time. Learning-based QPP is comprised of three primary components: *Training Data Generator, Feature Extractor,* and *Learning-based Model*, as shown in Fig. 1.

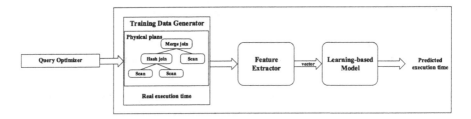

Fig. 1. Architecture of learning-based performance prediction

Training Data Generator first generates some queries according to the datasets' potential join graph and the workload predicates and the predicates in the workload, then for each query extracts a physical plan by the query optimizer and gets the real execution time. After getting the physical plans, valuable features of the plan are extracted through *Feature Extractor*. Each node in the query plan is encoded into feature vectors and each vector is organized into tensors, which are taken as the input of the training model. *Learning-based Model* can learn representations for the (sub)plans, which can be used in execution time estimation. The model is trained based on the training data, storing the updated parameters in the model to estimate execution time for new query plans.

In *Feature Extractor*, approaches for cardinality estimation [8,11,23] and query latency prediction [13] proposed their own methods of encoding query plans. All methods proposed so far one-hot encode tables present in a query plan or query and some methods also one-hot encode the columns across all tables. Runsheng [6] explored two methods for extending the 1-hot encoding to explicitly represent self-joins called *1-Hot+* and *Multiple 1-Hot*. For *Learning-based Model*, database researchers initially use machine learning models. Akdere et al. [1] proposed a prediction model based on support vector machines. Archana Ganapathi et al. [5] proposed kernel canonical correlation analysis to predict multiple performance indicators of SQL queries. Li et al. [10] proposed a method of merging database query knowledge and statistical models. The boosted regression tree is utilized as a benchmark statistical model and a linear scaling function to infer query performance. With the development of deep learning, more neural network models have been applied. MSCN [8] adopted the convolutional neural network to estimate cardinality. LSTM and ResNet [17] took full account of the serialized information and non-serialized information of the query plan and had more minor errors than machine learning approaches in QPP. To be generalized to support most SQL queries, tree-structured learning models are proposed according to the query plan tree's structure. TNN [13] is a tree structured neural network composed of many neural units according to the query plan tree, where each unit represents a specific plan tree node. Instead of standard neural units, TLSTM [18] uses LSTM units as a component of the tree structure model to overcome gradient vanishing and explosion for complicated queries.

3   Feature Encoding

In this section, we introduce the feature extraction and encoding method for the proposed MSP method. We first encode a query node as a node vector and then transform the node vectors into a tree-structured vector.

For each operation node in a query plan, three main factors may affect the execution time: the type of physical operation, metadata and query predicate. We'll introduce how to extract these features and encode them into vectors.

Physical Operation. According to the physical operation, all nodes in a query plan can be divided into the following categories:

- Scan Node, including sequential scan, index scan, etc.
- Join Node, including loop nest join, hash join, merge join.
- Other Node, such as materialization, result, etc.

The type of plan node will obviously affect the execution time of the plan. Since the types can be enumerated, they can be encoded as one-hot vectors.

MetaData. Metadata describes schema of tables, columns and indexes used in an operation node. Regardless of whether the column names are the same, the features of columns with the same data type and similar value distribution can be considered similar, which makes it unnecessary to use the one-hot encoding of names to distinguish the columns. We therefore define **metaInfo** to represent the characteristics of the metadata, mainly including data type and data statistics of each column in table.

Suppose the number of basic data types in the database is n, which is generally fixed. Columns and indexes on columns can be encoded as vectors with fixed-sized dimension, denoted by $V_{column} = [\alpha_1, \alpha_2, \cdots, \alpha_n, \alpha_{n+1}]$. $[\alpha_1, \alpha_2, \cdots, \alpha_n]$ is the one-hot encoding for the data type, where α_i corresponding to the data type of the column is set to 1 and the rest is set to 0. α_{n+1} is appended at the end to distinguish whether it is an index or a column. When the vector representation of each column and index is obtained, the encoding vector of the table is the average of the column vectors, denoted by $V_{table} = \frac{\sum V_{column}}{num\_column}$. num_column is the sum of the numbers of columns and indexes in the table. Taking the aka_name table in PostgreSQL as an example, the encoding vectors of its metadata are shown in Table 1.

Integer, floating-point, fixed-length characters, variable-length characters, time, and boolean are used as basic types in PostgreSQL, so the dimension of the vectors is 7.

The data statistics indicates the distribution of values in the column. The description of statistical information in each database is different and its maintenance is time-consuming. Therefore in the actual implementation, we encode the estimated number of tuples and the width of the tuples obtained by each plan operator into the encoding. Although there are errors in these estimates, as long as the underlying data don't change particularly sharply, the learning model can tolerate these errors, so the accuracy of the model will not be greatly affected (see Sect. 5.6). How to accurately estimate statistics will be further explored in future research and will not be mentioned in this article.

Table 1. MetaData encoding example

Column	DataType-one-hot
id	1000000
person_id	1000000
name	0010000
name_pcode_cf	0010000
name_pcode_nf	0010000
surname_pcode	0010000
md5sum	0010000
Index	**DataType-one-hot**
aka_name_pkey	1000001
Table	**DataType-one-hot**
aka_name	0.375 0 0.625 000 0.125

Compared with the one-hot vector encoded according to the table name and column name whose dimension is consistent with the number of tables and columns, vectors based on metaInfo overcome the problem of inconsistent dimensions caused by newly added or deleted columns, since the basic types of data are fixed so that the dimension of the encoded vector remains unchanged. So when the underlying data changes, the model does not need to be retrained. On the other hand, if the same model is now deployed for a different database, the data type is basically fixed, and the data statistics can be obtained from a new database, thus the representation is consistent for other databases(in contrast to one-hot encodings of columns).

Predicate. Query predicates are a collection of filtering and joining conditions in operating nodes such as mc.company_id = cn.id and season_nr> 6. Except such a single selection condition, there is a high probability that compound selection conditions may appear in a predicate, e.g.

(((season_nr>6) AND (season_nr <10)) OR ((season_nr<6) AND (episode_nr>32))) AND (Info like '%DVD%')

This compound selection condition can be expressed as a tree structure [18], as shown in Fig. 2.

We first generate vectors for each atomic predicate, and then splice these vectors into the final predicate vector according to the in-order traversal query tree sequence with Min-Max-Pooling (see Eq. 1).

Each atomic predicate mainly consists of three parts: *Column, Operator* and *Operand*. An *Operator* can be encoded as a one-hot vector. The encoding vector of *Column* has already been introduced above. For an *Operand*, if its value is numeric, we encode it by a normalized float; if it is a string, we use the skip-gram model [14] to generate a word vector as the vector of *Operand*. *Bool operator* is used to indicate whether it is an atomic predicate or a parallel conjunction. The encoding vectors of various predicate nodes are shown in Table 2 (padding means filling up the corresponding blocks with zeros).

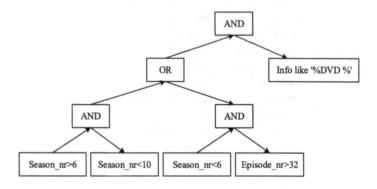

Fig. 2. The tree structure of a query predicate

Table 2. Predicate encoding

Node	Operator	Column	Operand	Bool operator
season_nr <12	0001	1000000	0.0046	00
episode_nr = 32	0010	1000000	0.0002	00
Info like '%DVD %'	0100	0010000	0.13,0.41,...,0.76	00
and	Padding	Padding	Padding	01
or	Padding	Padding	Padding	10

4   Proposed Tree-structured Model

As illustrated in Fig. 3, the proposed model is mainly divided into three parts, namely *embedding layer*, *representation layer* and *prediction layer*. Since the feature vectors obtained by feature extraction and encoding for each node in the plan tree are sparse, *embedding layer* is required to compress each input vector to obtain a high-dimensional feature vector as the input for *representation layer*. Since the plan is a tree-like structure, *representation layer* adopts a tree-structured model, where each node adopts the same representation model and the overall model structure is consistent with the plan tree. Each representation model (corresponding to each node of the physical plan in Fig. 1) can capture the characteristics of not only the plan node but also its sub-plans by considering the learned vectors of the child nodes to ensure the integrity of the learned plan features. Finally, *output layer* predicts the plan's actual execution time according to the output vector of the root node.

4.1   Embedding Layer

embedding layer is to transform a sparse vector into a dense one. As described in Sect. 3, the input feature vectors include three parts: *Operator Type, Metadata* and *Predicate*. Since the vectors of *Operation Type* and *Metadata* are sparse, they are embedded by a fully connected layer of the ReLU activation function. The structure of *Predicate* vector is complicated because it contains multiple AND/OR semantics, so we employ Min-Max-Pooling [18] to extract the high-dimensional feature vector of the predicate.

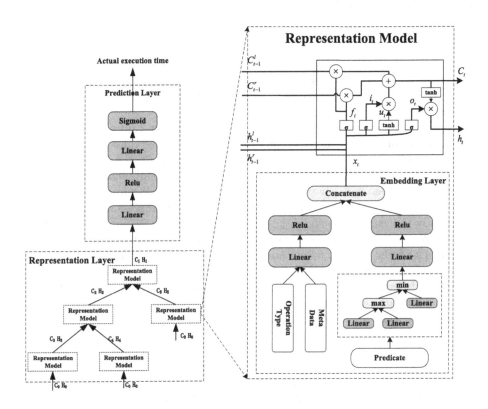

Fig. 3. Overall model framework

We use O_t, M_t, P_t to represent *Operation Type, Metadata* and *Predicate* of $node_t$, and denote each node of *Predicate* as P_t with P_t^l as its left child and P_t^r as its right child. Embedding Model can be formalized as Eq. 1. E is the output of the embedding layer, W is the weight of the fully connected layer, b is the bias of the fully connected layer.

$$E = [embed(O_t), embed(M_t), embed(P_t)]$$
$$embed(O_t) = ReLU(W_o O_t + b_o)$$
$$embed(M_t) = ReLU(W_m M_t + b_m)$$
$$embed(P_t) = \begin{cases} min(embed(P_t^l), embed(P_t^r)) & P_t = and \\ max(embed(P_t^l), embed(P_t^r)) & P_t = or \\ W_p * P_t + b_p & P_t = expr \end{cases} \quad (1)$$

4.2 Representation Layer

For the leaf nodes in the plan, such as an index scan on a single table, it is easy to estimate the cost of the node and further predict the execution time. But for the upper node, it is difficult because the correlation among nodes cannot be well captured, which results in information vanish. Therefore, the estimated execution time of a node is related to its own feature vector and the information from its child node(s).

Representation Layer is to solve the above problem by capturing the global information from the leaf node to the root node to avoiding information vanish. By recursively training the representation of the sub-plan, the representation vector learned from the model is used to represent the sub-plan, which is the input of the upper node. It is ensured that the lower-level information can be transmitted to the root node through the tree structure representation models, and finally the learning result of the global information is output to predict the execution time of the plan. As shown in Fig. 3, all units in this layer are neural networks with the same structure and share common parameters which we refer to as representation models. No matter how complex the plan tree structure is, it can be decomposed into basic plan nodes that use the same representation model. Therefore the trained model can support various complex query plans, ensuring the learning model's generalization.

Each representation model has three inputs, the embedding vector E, the vector $[C_{t-1}^l, h_{t-1}^l]$ of the left child, and the vector $[C_{t-1}^r, h_{t-1}^r]$ of the right child (for leaf nodes, the zero vector is used as the output vector of their child nodes). The output of the representation model is $[C_t, h_t]$.

The most important issue of the tree-structured model design is the choice of recursive neural network. A simple implementation is to use a fully connected neural network, which concatenates the locally transformed features with the output of its children as input. However, the lost information will make representation model with naive neural networks suffer from gradient vanishing and gradient explosion problems [2]. Compared with deep neural network, LSTM units can effectively solve these problems by using additional information channels. The structure of LSTM [7] unit is shown in Fig. 3. C_t is the channel of the long memory unit, and f_t controls which information should be forgotten in the long memory unit. i_t controls which information should be added to the long memory channel. o_t controls which information in the long memory channel should be taken as the representation of the sub-plans. Since C_t can be a path without any multiplication, LSTM avoids the problem of gradient disappearance. On the other hand, forget gate of Sigmoid helps LSTM solve the problem of gradient explosion.

How to connect the representation models of the parent node and the child node is also worth researching. The LSTM model based on the tree structure mainly has two connection forms, namely DTree-LSTM (Fig. 4a) and CTree-LSTM (Fig. 4b) [21]. DTree-LSTM was first proposed to be applied to the syntactic dependency tree structure, which is characterized by disordered sub-nodes and an indefinite number of sub-nodes. CTree-LSTM was first proposed to be applied to phrase structure trees, which is characterized by a fixed number of sub-nodes.

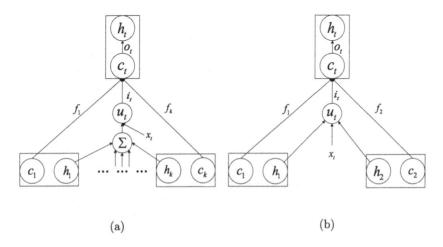

(a)   (b)

Fig. 4. Structure of DTree-LSTM and CTree-LSTM

The plan is a binary tree. For a node with only one child node, the other child of it can be represented with a zero vector, so the number of child nodes is fixed at 2. Because child nodes have different impacts on the parent node, directly taking the sum of the output vectors of the sub-nodes as the hidden state input of the parent node will ignore the differences of the influence from child nodes, resulting in insufficient information. To solve this problem, CTree-LSTM assigns a parameter matrix to each child node (see equation2). We choose CTree-LSTM as the representation model and the parameter N (the number of childs) in the model is set to 2. The representation layer can be formalized as Eq. 2. E is the embedding vector of the current node, $[C_{t-1}^l, h_{t-1}^l]$ is the output vector of the left child node, $[C_{t-1}^r, h_{t-1}^r]$ is the output vector of the right child node, and $[C_t, h_t]$ is the output vector.

$$x_t = E$$
$$i_t = \sigma \left(W^{(i)} x_t + U_l^{(i)} h_{t-1}^l + U_r^{(i)} h_{t-1}^r + b^{(i)} \right)$$
$$f_{tk} = \sigma \left(W^{(f)} x_t + U_{kl}^{(f)} h_{t-1}^l + U_{kr}^{(f)} h_{t-1}^r + b^{(f)} \right)$$
$$o_t = \sigma \left(W^{(o)} x_t + U_l^{(o)} h_{t-1}^l + U_r^{(o)} h_{t-1}^r + b^{(o)} \right) \tag{2}$$
$$u_t = \tanh \left(W^{(u)} x_t + U_l^{(u)} h_{t-1}^l + U_r^{(u)} h_{t-1}^r + b^{(u)} \right)$$
$$c_t = i_t \odot u_t + f_{t1} \odot c_{t-1}^l + f_{t2} \odot c_{t-1}^r$$
$$h_t = o_t \odot \tanh (c_t)$$

4.3   Prediction Layer

The output layer is a two-layer fully connected neural network that uses the ReLU function as its activator and finally uses a sigmoid function to predict the actual execution time of the plan. The output layer can predict the actual execution time of any sub-plan through the representation vector of the plan. In order to get the actual execution time of the entire plan, we take the representation h_t of the root node in the representation layer as input, and calculation equations are as follows:

$$time' = ReLU(W_{time'} h_t + b_{time'})$$
$$time = sigmoid(W_{time} time' + b_{time}) \tag{3}$$

For the regression issue of time prediction, mean absolute percentage error (MAPE) [4] and Q-error [8,16], are common loss functions. The mean absolute percentage error refers to the average value of the relative error between the true value and the predicted value, and the Q-error is the ratio between the true value and the predicted value. The loss functions are formalized as below:

$$MAPE = \frac{1}{n} \sum_{i=1}^{n} \frac{\hat{time}_i - time_i}{time_i}$$
$$Q - error = \frac{1}{n} \sum_{i=1}^{n} \frac{max(\hat{time}_i, time_i)}{min(\hat{time}_i, time_i)} \tag{4}$$

Among them, n is the number of query plans, $\hat{time}_i$ is the predicted query plan execution time, and $time_i$ is the actual query plan execution time.

In practice, it has been discovered that employing MAPE as the loss function causes the model to underestimate the time. Q-error can overcome this shortcoming, but when the error is close to 1, it may reduce the percentage of error [19]. Therefore, when the MAPE exceeds 1, the effect of Q-error is better, otherwise the effect of MAPE is better, thus these two loss functions will be comprehensively considered during the training process.

5   Experiments

5.1   Dataset

To verify the performance of the proposed method, real-world dataset IMDB is used for experiments. The IMDB dataset contains 21 tables covering more than

2.5 million movies produced by 234,997 companies over 133 years, involving more than 4 million actors and actresses. This dataset has more tables than TPC-H and more joins between tables, so there is more correlation between the tables. Therefore it is much more challenging to estimate query execution time than TPC-H.

5.2   Baselines

In the experiments, the proposed method is compared with popular deep learning models (as shown in Table 3). MSCN and TNN support time prediction for queries containing only numeric predicates. Dtree-LSTM and CTree-LSTM support time prediction for complex queries containing numbers and strings. We consider the performance of these networks with and without the metaInfo encoding.

Table 3. Baselines (MI means metaInfo encoding)

Methods	Represent network	MetaInfo encoding	Supported predicate
MSCN	NO	NO	NUMERIC
MSCN+MI	NO	YES	
TNN	NN	NO	
TNN+MI	NN	YES	
DTree-LSTM	LSTM	NO	NUMERIC &STRING
DTree-LSTM+MI	LSTM	YES	
CTree-LSTM	LSTM	NO	
CTree-LSTM+MI	LSTM	YES	

5.3   Experiment Setting

Table 4. Two types of query loads

Workload	Joins	Predicate
Synthetic	0–2	NUMERIC
Scale	0–4	
JOB-light	1–4	
JOB	3–16	NUMERIC &STRING

To meet the use scenarios of the models in the baselines, two types of workloads are mainly used, as shown in Table 4. For numeric workloads, we generate 10K queries with 0–4 joins containing only numeric predicates, using 90% of the generated queries as training data and 10% as verification data, and three workloads

Synthetic, Scale and *JOB-light* [8] are used to test the performance of all models in baselines. For numeric and string workloads, we generate 50,000 single-table queries, 50,000 queries with 1–4 joins, and 50,000 queries with more than 5 joins. We set 90% of the generated queries as training data, 10% as verification data, and use 113 JOB queries [9] as the test workload.

A machine with Intel(R) Xeon(R) CPU E5-2630 v4, 756GB Memory, GeForce GTX 1080 and PostgreSQL12.1 is used for the experiments. The model is trained with the batch size of 64 and a mini-batch stochastic gradient descent using Adaptive Moment Estimation (Adam) optimizer with a learning rate 0.001. Our method takes about 4 h to train 100K queries.

5.4   Experimental Results

Different methods are employed for training on two types of training sets. After the model converges, four query loads (see Table 4) are used for testing, and the final results are obtained. Figure 5 shows the verification error on JOB workload. Table 5 and Table 6 show the test error for various encodings and models on different workloads.

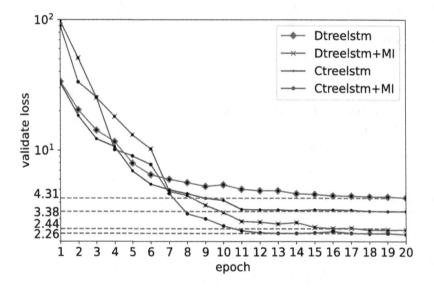

Fig. 5. validation error

Performance on Numeric Workloads. As shown in Table 6, Q-error of CTree-LSTM is 1/2, 1/3 and 1/5 of that of MCSN on three workloads. We can conclude that CTree-LSTM outperforms MSCN by 2 times, 3times and 5times on *Synthetic, Scale* and *JOB-light* workload respectively. We can observe that tree-structured model is good at representing complicated plans and predicates. CTree-LSTM has smaller prediction errors than TNN on all three workloads, especially on JOB-light. The result shows that LSTM uses the representation model to learn more robust representations for sub-plans. This is because LSTM has an extra channel to avoid information vanishing for complex queries. On three workloads, the test error of CTree-LSTM is also smaller than that of Dtree-LSTM on max and mean error.

Table 5. Test error (Q-error) on three numeric workloads

	Synthetic			Scale			JOB-light		
	Median	Max	Mean	Median	Max	Mean	Median	Max	Mean
MSCN	3.14	739	8.3	1.79	1027	12.22	4.75	987	27.4
MSCN+MI	**2.64**	**698**	**6.46**	**1.69**	**788**	**7.27**	**4.24**	**563**	**21.3**
TNN	1.49	718	4.35	1.61	714	5.53	2.06	401	19.1
TNN+MI	**1.45**	**642**	**3.54**	**1.59**	**672**	**4.61**	**1.55**	**293**	**14.5**
DTree-LSTM	1.48	532	3.99	1.58	254	4.39	1.85	123	5.76
DTree-LSTM+MI	**1.32**	**354**	**2.64**	**1.56**	**212**	**2.84**	**1.45**	**101**	**3.83**
CTree-LSTM	1.39	413	3.12	1.59	233	3.37	1.62	122	4.93
CTree-LSTM+MI	**1.16**	**289**	**2.38**	**1.54**	**168**	**2.46**	**1.46**	**95**	**2.47**

Performance on JOB Workload. We evaluated several methods supporting string predicate on JOB workload. CTree-LSTM+MI outperforms the best competitor by 40% on max error and 80% on mean error, as shown in Table 7. The reasons are two-fold. (1)MetaInfo captures information on metadata that is closely related to plan execution time, helping to improve prediction accuracy. (2)Assigning weights to child nodes can capture more information about the tree structure of plans.

Table 6. Test error (Q-error) on JOB workload

JOB	Median	Max	Mean
DTree-LSTM	4.07	67.3	6.98
DTree-LSTM+MI	**2.94**	**53.6**	**4.53**
CTree-LSTM	3.76	59.4	5.75
CTree-LSTM+MI	**2.47**	**47.8**	**3.86**

5.5 Ablation Analysis

We ablate MetaInfo encoding (MI) on the model of every neural networks and take test on two types of workload, as shown in Tables 5 and 6. Without MI at the embedding layer, we observe the accuracy of query performance prediction drops, while the prediction performance is significantly improved with MetaInfo encoding. In contrast, encoding based on table and column names does not capture the information of the metadata as well as MetaInfo encoding. It is apparent that MetaInfo encoding is more suitable for query performance prediction.

5.6   Model Adaptability

Table 7. Queries affected by database updates

Affected queries	Synthetic	Scale	JOB-light	JOB
total	5000	500	70	113
insert tuples to cast-info	1614	209	35	67
insert column role-id	508	57	15	26
insert table movie-keyword	1628	231	42	75

We evaluate our method on databases with modification. As table cast-info has the most join relations in the JOB join graph, (it has 3.6 million tuples and can join with 12 tables), we confine all tuple updates to this table. As table movie-keyword has lots of related queries in the test workload (over a half in Synthetic, Scale and JOB-light), we recreate this table to emulate table-level modification. As the attribute role-id is frequent in the predicate, we recreate this column to emulate column-level modification. Firstly, we remove table movie-keyword from the database and column role-id from table cast-info. Secondly, we extract 3,000,000 tuples (around 10%) from table cast-info as inserted tuples, and train the model on the rest tuples. For tuple insertions, we insert 1,000,000 tuples each time and insert 3 times to table cast-info. For column insertion, we insert column role-id to table cast-info. For table insertion, we insert table movie-keyword. Table 7 shows the amount of queries affected by the updates. After database and statistics update, we use the previously trained model to run the test workloads. Figure 6 shows the estimation errors. We can see that the errors

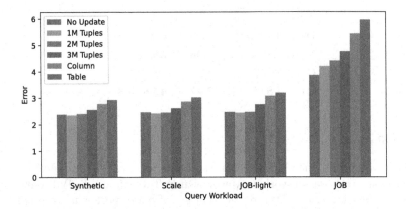

Fig. 6. Errors with Database Changes

are similar to those without updates. The main reason is that the MetaInfo-based encoding approach does not change the encoded dimensions when the underlying data changes, allowing the trained model to be used directly. This suggests that our approach is more appropriate for application to real databases.

6   Conclusion

This paper proposes a deep learning model based on tree structure to predict the execution time of query plans. Using metadata encoding based on metaInfo instead of traditional one-hot encoding can better capture the characteristics of metadata in the database and avoid frequent retraining when changing under-lying data. At the same time, a more detailed CTree-LSTM model is used to capture the different effects between nodes in a plan tree. Experiments on real-world datasets demonstrate that the proposed method is superior to the currently existing technologies.

References

1. Akdere, M., Çetintemel, U., Riondato, M., Upfal, E., Zdonik, S.B.: Learning-based query performance modeling and prediction. In: IEEE (2012)
2. Bengio, Y., Simard, P., Frasconi, P.: Learning long-term dependencies with gradient descent is difficult. IEEE Trans. Neural Netw. **5**(2), 157–166 (1994)
3. Chi, Y., Moon, H.J., Hacigümüs, H.: ICBS: Incremental cost based scheduling under piecewise linear SLAS. Proc. VLDB Endow. **4**(9), 563–574 (2011)
4. De Myttenaere, A., Golden, B., Le Grand, B., Rossi, F.: Mean absolute percentage error for regression models. Neurocomputing **192**, 38–48 (2016)
5. Ganapathi, A., Kuno, H.A., Dayal, U., Wiener, J.L., Patterson, D.A.: Predicting multiple metrics for queries: better decisions enabled by machine learning. In: IEEE International Conference on Data Engineering (2009)
6. Guo, R.B., Daudjee, K.: Research challenges in deep reinforcement learning-based join query optimization. In: Proceedings of the Third International Workshop on Exploiting Artificial Intelligence Techniques for Data Management, pp. 1–6 (2020)
7. Hochreiter, S., Schmidhuber, J.: Long short-term memory. Neural Comput. **9**(8), 1735–1780 (1997)
8. Kipf, A., Kipf, T., Radke, B., Leis, V., Boncz, P.A., Kemper, A.: Learned cardi-nalities: estimating correlated joins with deep learning. In: CIDR (2019)
9. Leis, V., Gubichev, A., Mirchev, A., Boncz, P., Kemper, A., Neumann, T.: How good are query optimizers, really? Proc. VLDB Endow. **9**(3), 204–215 (2015)
10. Li, J., König, A., Narasayya, V., Chaudhuri, S.: Robust estimation of resource consumption for SQL queries using statistical techniques. Proc. VLDB Endow. **5**(11), 1555–1566 (2012)
11. Liu, H., Xu, M., Yu, Z., Corvinelli, V., Zuzarte, C.: Cardinality estimation using neural networks. IBM Corp. (2015)
12. Lohman, G.: Is query optimization a "solved" problem. In: Proceedings of Work-shop on Database Query Optimization, vol. 13, p. 10. Oregon Graduate Center Computer Science and Technical Report (2014)

13. Marcus, R., Papaemmanouil, O.: Plan-structured deep neural network models for query performance prediction. In: Proceedings of the VLDB Endowment (2019)
14. Mikolov, T., Chen, K., Corrado, G., Dean, J.: Efficient estimation of word representations in vector space. Computer Science (2013)
15. Mishra, C., Koudas, N.: The design of a query monitoring system. ACM Trans. Database Syst. **34**(1), 1–51 (2009)
16. Moerkotte, G., Neumann, T., Steidl, G.: Preventing bad plans by bounding the impact of cardinality estimation errors. Proc. VLDB Endow. **2**(1), 982–993 (2009)
17. Ni, J., Zhao, Y., Zeng, K., Su, H., Zheng, K.: DeepQT?: learning sequential context for query execution time prediction. In: Nah, Y., Cui, B., Lee, S.-W., Yu, J.X., Moon, Y.-S., Whang, S.E. (eds.) DASFAA 2020. LNCS, vol. 12114, pp. 188–203. Springer, Cham (2020). https://doi.org/10.1007/978-3-030-59419-0_12
18. Sun, J., Li, G.: An end-to-end learning-based cost estimator. Proc. VLDB Endow. **13**(3), 307–319 (2019)
19. Sun, J., Li, G., Tang, N.: Learned cardinality estimation for similarity queries. In: Proceedings of the 2021 International Conference on Management of Data, pp. 1745–1757 (2021)
20. Taft, R., Lang, W., Duggan, J., Elmore, A.J., Dewitt, D.: Step:scalable tenant placement for managing database-as-a-service deployments. In: ACM Symposium on Cloud Computing (2016)
21. Tai, K.S., Socher, R., Manning, C.: Improved semantic representations from tree-structured long short-term memory networks. Comput. Sci. **5**(1), 36 (2015)
22. Tozer, S., Brecht, T., Aboulnaga, A.: Q-cop: avoiding bad query mixes to minimize client timeouts under heavy loads. In: 2010 IEEE 26th International Conference on Data Engineering (ICDE 2010) (2010)
23. Wu, C., Jindal, A., Amizadeh, S., Patel, H., Rao, S.: Towards a learning optimizer for shared clouds. Proc. VLDB Endow. **12**(3), 210–222 (2018)

Computing Online Average Happiness Maximization Sets over Data Streams

Zhiyang Hao[1] and Jiping Zheng[1,2(✉)] 🆔

[1] College of Computer Science and Technology, Nanjing University of Aeronautics and
Astronautics, Nanjing, China
{haozhiyang,jzh}@nuaa.edu.cn
[2] State Key Laboratory for Novel Software Technology, Nanjing University, Nanjing, China

Abstract. Finding a small subset representing a large dataset is an important
functionality in many real applications such as data mining, recommendation and
web search. The average happiness maximization set problem also known as the
average regret minimization set problem was recently proposed to fulfill this task
and it can additionally satisfy users on average with the representative subset.
In this paper, we study the online average happiness maximization set (Online-
AHMS) problem over data streams where each data point should be decided
to be accepted or discarded when it arrives, and the discarded data points will
never be considered. We provide an efficient online algorithm named GreedyAT
with theoretical guarantees for the Online-AHMS problem which greedily selects
data points based on the adaptive thresholds strategy. Experimental results on the
synthetic and real datasets demonstrate the efficiency and effectiveness of our
GreedyAT algorithm.

Keywords: Happiness maximization set · Online algorithm · Adaptive
threshold

1 Introduction

In many real applications such as data mining [27], recommendation [18,30] and web
search [29], an important functionality is to select a succinct subset from a large dataset
to meet the requirements of various users. To fulfill this task, three popular tools are
proposed in the last three decades, namely the top-k query [17], the skyline query [5,11]
and the happiness maximization set (also known as the regret minimization set, or the
k-regret) query [21,35]. The top-k query is to use the concept of utility function to
quantify a user's preference on different attributes and the top-k data points with the
largest utilities are returned. However, the weakness of the top-k query is that the utility
function is often unclear or only vaguely known which limits the applicability of this
tool. By utilizing the concept of *domination*: a point p dominates a point q iff p is as
good as q on all attributes and strictly better than q on at least one attribute, the skyline
query returns all data points not dominated by other data points. Though no specific
utility functions are needed, the skyline query does not effectively reduce the solution
size over high-dimensional datasets [5].

To avoid the limitations of the top-k and skyline queries, the Happiness Maximization Set (HMS) query (called the k-regret query when first proposed [21]) was proposed

B. Li et al. (Eds.): APWeb-WAIM 2022, LNCS 13423, pp. 19–33, 2023.
https://doi.org/10.1007/978-3-031-25201-3_2

to simultaneously have the merits of both top-k and skyline queries, resulting in many studies in the database community [1,3,7,10,13,19,22,24,26,28,32,35–39]. Specifically, the happiness ratio is defined to quantify the happiness level of a user with a utility function for a set of data points, compared with when s/he has seen the entire dataset. In HMS, a subset of size k is chosen from the dataset such that the minimum happiness ratio of any possible utility function between the best data point in the selected subset and the best point in the whole dataset is maximized. The happiness maximization set query is to select a size-k subset maximizing the minimum happiness ratio. We can see that the minimum happiness ratio may unfairly prioritize the least satisfied users [19,26,38,39]. Instead, the Average Happiness Maximization Set (AHMS) query [38,39] is provided to maximize the users' happiness ratios on average, which is more proper to satisfy the majority of users. Note that the Average Happiness Maximization Set query is identical to the Average Regret Minimization Set query, but the average happiness ratio function of the AHMS query shows the property of submodularity which allows for the deviation of stronger theoretical results. For the AHMS query [19,26,38,39], the users' utility functions are not just uniformly distributed but follow a probability distribution which can be obtained from users' historical preferences and feedback provided in real applications [12,16,27,31]. Since the average happiness ratio over the utility function distribution can be approximated by sampling, the AHMS query selects a subset to maximize the average happiness ratio on a sample of M utility functions instead of on the distribution of the utility functions.

The HMS query and its variants show their NP-hardness on any dataset when the dimensionality d is larger than 2, i.e., $d \geq 3$, and many promising algorithms for the happiness maximization set query (or the regret minimization set query) exhibit their advantages to solve the related problems [1,3,7,10,13,19,22,24,26,36–39]. However, instead of considering the streaming setting where the data points arrive one by one, the existing algorithms almost aim at the static setting of the dataset where full access to the entire dataset is always available. Wang et al. [32] and Zheng et al. [42] considered the dynamic environment of datasets where data points are inserted and deleted dynamically and efficient algorithms, namely FD-RMS and DynCore were proposed to solve this problem, respectively. Ma et al. [20] assumed the data points are only valid in a sliding window and an efficient coreset-based algorithm was proposed to answer the regret minimization set query. However, these researches do not consider the scenario that an immediate decision should be made for each on-arriving data point in a data stream, i.e., whether the current data point should be included in the result set or not, and the discarded data points will never be considered. Our model is also distinct from the existing streaming setting [30] where a small portion of the data points can be buffered and added to the solution later. Moreover, the scenario in the online setting occurs in many real applications, such as news recommendation, football player recruitment, job seeking, etc. when a news is published by a news agency, a news portal (e.g., Sina News), should make an immediate decision on whether the news is to be included in the fixed-size headlines; otherwise, other news portals will have the chance to release the news at the earliest time. The same scenario happens for a football team recruiting a player, or a company employing a staff. If they miss the chance to recruit the player or employ the job-seeker, they may have no chance to consider her/him again, because the player or job-seeker may be recruited or employed by other football teams or companies.

In this paper, we study the Online Average Happiness Maximization Set (Online-AHMS) problem. Since we have no foreknowledge of future data points in the online setting, we define the online average happiness ratio function as a reasonable measure. The goal of the Online-AHMS query is to return a subset with k data points while maximizing the online average happiness ratio at any time in the online setting. To solve this problem, we provide an efficient online-selection algorithm named GreedyAT where *adaptive thresholds* are set to filter the data points. In addition, by utilizing the property that our online average happiness ratio function is submodular, we can guarantee a constant competitive ratio of the GreedyAT algorithm. To sum up, the main contributions of this paper are listed as follows:

– We study the online average happiness maximization set problem which selects the data points irrevocably on the arriving of them, and an online algorithm GreedyAT is proposed based on the adaptive thresholds strategy.
– We perform theoretical analysis of our GreedyAT algorithm and the competitive ratio of the GreedyAT algorithm is provided based on the submodularity property of our online average happiness ratio function.
– Extensive experiments on the synthetic and real datasets are conducted to verify the efficiency, effectiveness and applicability of the GreedyAT algorithm in the online setting.

The rest of the paper is organized as follows. We introduce the related work in Sect. 2. Related concepts are provided and we formally define our Online-AHMS problem in Sect. 3. In Sect. 4, we propose our GreedyAT algorithm based on the adaptive thresholds strategy and further provide the theoretical analysis to obtain the competitive ratio of the GreedyAT algorithm. We conduct experiments on the synthetic and real datasets to evaluate our GreedyAT algorithm in Sect. 5. Lastly, we conclude our work in Sect. 6.

2   Related Work

To avoid the drawbacks of the top-k query (needs users to provide exact utility functions) and the skyline query (the output size is uncontrollable), the happiness maximization set query has been investigated in the last decade which was called the k-regret query first proposed by Nanongkai et al. [21]. Due to the NP-hardness of the k-regret query [10,35], many promising methods are proposed and useful variants are introduced. Geometry-based methods, such as GeoGreedy [24] and Sphere [36] are proposed to improve the efficiency of the k-regret query. To achieve the same goal, various techniques, e.g., ϵ-kernel [1,7], hitting set [1], discretized matrix [3] are borrowed to answer the k-regret query efficiently. As useful variants of the k-regret query, the kRMS query [10], the interactive regret minimization query [22,34,40], the regret minimization query on nonlinear utility functions [13,25], the rank-based regret minimization query [4,33] and the regret minimization query with approximation guarantees [41] are proposed to meet different situations. As a completely identical concept to the k-regret query, the happiness maximization set query was recently studied in [19,26,37] where if the happiness ratio is defined as hr then the regret ratio rr is one minus hr, i.e., $rr = 1 - hr$.

The researches above related to the happiness maximization set query (or the k-regret query) focused on maximizing the minimum happiness ratio (or minimizing the maximum regret ratio) and optimizing the worst-case scenario. To address the issue that the happiness maximization set query may prioritize the least satisfied users, the average regret minimization query was proposed by Zeighami et al. [38]. They introduced the concept of the average regret ratio to measure the user's regret ratio in the average case which is more reasonable since it considers the expectations of different users. Moreover, Zeighami et al. [39] proved the average regret minimization set problem is NP-hard and an efficient algorithm Greedy-Shrink was proposed. Luenam et al. [19] showed that the happiness maximization set query can admit stronger approximation guarantees than the regret minimization set query.

Although the massive volume and the real-time generation of data points imply a growing need for non-static algorithms, few researches considered the this setting of the happiness maximization set query or the k-regret query. Wang et al. [32] studied the k-regret query on dynamic datasets where data points were arbitrarily inserted or deleted. They provided the FD-RMS algorithm which transformed the fully-dynamic k-regret query to a dynamic set cover problem and constantly maintained the result with theoretical guarantees. Further, Zheng et al. [42] also considered the same problem and a more efficient method named DynCore was proposed to achieve a better regret ratio bound with a lower time complexity. Ma et al. [20] assumed there was a sliding window on a data stream and the data points inside the window were valid. They provided a coreset-based method to continuously maintain the result efficiently. In the online setting, however, upon the arrival of a data point, we need to decide whether to accept it or not immediately as an online selection which usually occurs in many real applications. The above researches do not consider this scenario because the data points are buffered without immediate decisions. In this paper, we study the Online-AHMS problem which is very promising in real-world applications.

3 Problem Definition

In this section, we formally define the online average happiness maximization set (Online-AHMS) problem. Some useful concepts such as utility function, happiness ratio, average happiness ratio and their online versions are introduced before we state our problem. Let D be a d-dimensional dataset containing n data points where each data point $p =< p[1], p[2], \ldots, p[d] >\in D$ is described by d numerical attributes which are normalized in the range $[0, 1]$. We assume that a larger value in each dimension is preferable to all users.

Definition 1 (Utility Function). *A utility function f is a mapping $f: \mathbb{R}_+^d \rightarrow \mathbb{R}_+$ that assigns a non-negative utility $f(p)$ to each data point $p \in D$ which shows how satisfied the user is with the data point p.*

Following [10,21], we assume the form of the utility functions to be linear, i.e., f is represented as a d-dimensional vector $u =< u[1], u[2], \ldots, u[d] >\in \mathbb{R}_+^d$ where $u[i]$ denotes the importance of the i-th dimension in user's happiness. W.l.o.g., we assume that u is normalized such that $||u||_1 = \sum_{i=1}^{d} u[i] = 1$. Thus, the utility of a data point $p \in D$ w.r.t. f can be expressed as $f(p) = u \cdot p = \sum_{i=1}^{d} u[i]p[i]$.

Definition 2 (Happiness Ratio). *Given a subset $S \subseteq D$ and a user's utility function f, the happiness ratio of S over D for f, denoted by $hr_{D,f}(S)$, is defined to be $\frac{\max_{p \in S} f(p)}{\max_{p \in D} f(p)}$.*

Intuitively, we have $\max_{p \in S} f(p) \leq \max_{p \in D} f(p)$, so the happiness ratio of a user ranges from 0 to 1. The happiness ratio measures how happy a user is if s/he sees the subset S instead of the entire dataset D. The happiness ratio of a user is closer to 1 which indicates that the user feels happier with the selected subset S.

However, the users are generally difficult or not willing to provide their utility functions explicitly. We thus assume that all users' utility functions belong to a utility function class, denoted by $\mathcal{FC}$. Therefore, $\mathcal{FC}$ is the set of all possible linear utility functions, i.e., $\mathcal{FC} = \{f | f(p) = u \cdot p\}$. We focus on that the users' utility functions in $\mathcal{FC}$ follow a probability distribution Θ. Unless specified otherwise, we consider arbitrary type of distribution. Let $\eta(f)$ be the pobability density function for utility functions f in $\mathcal{FC}$ corresponding to Θ. Next we formally define the average happiness ratio.

Definition 3 (Average Happiness Ratio). *Given a subset $S \subseteq D$ and a utility function class $\mathcal{FC}$ with the probability density function $\eta(.)$ corresponding to a probability distribution Θ, the average happiness ratio of S is defined as*

$$ahr_{D,\mathcal{FC}}(S) = \int_{f \in \mathcal{FC}} hr_{D,f}(S) \cdot \eta(f) df. \qquad (1)$$

Unfortunately, the set of all possible linear utility functions $\mathcal{FC}$ is uncountable, so evaluating the above average happiness ratio needs to compute an integral over $\mathcal{FC}$, which is very time-consuming. Hence, we utilize a sampling technique from [39] to compute the average happiness ratio in Definition 3 with a theoretical bound. Specifically, we sample M utility functions according to the distribution Θ and obtain a sampled utility function class $\mathcal{FC}_M = \{f_1, f_2, \ldots, f_M\}$ where $|\mathcal{FC}_M| = M$. With $\mathcal{FC}_M$, we compute the estimated average happiness ratio by averaging the happiness ratio of the M sampled utility functions, as follows,

$$ahr_{D,\mathcal{FC}_M}(S) = \frac{1}{M} \sum_{f \in \mathcal{FC}_M} \frac{\max_{p \in S} f(p)}{\max_{p \in D} f(p)}. \qquad (2)$$

It holds that $ahr_{D,\mathcal{FC}_M}(S) \in [0, 1]$, and when $ahr_{D,\mathcal{FC}_M}(S)$ is close to 1, it indicates that the subset S satisfies the majority of users with utility functions in $\mathcal{FC}_M$. We claim that this estimated average happiness ratio $ahr_{D,\mathcal{FC}_M}(S)$ differs from the exact average happiness ratio $ahr_{D,\mathcal{FC}}(S)$ by an error $\epsilon \in [0, 1]$ (Theorem 4 in [39]). In the following, we ignore the error ϵ and only focus on maximizing $ahr_{D,\mathcal{FC}_M}(S)$ instead of $ahr_{D,\mathcal{FC}}(S)$ and further provide the average happiness maximization set problem based on the sampled utility function class $\mathcal{FC}_M$.

Definition 4 (Average Happiness Maximization Set, AHMS). *Given a dataset $D \in \mathbb{R}_+^d$ with n data points, a user-specified positive integer k, and the sampled utility function class $\mathcal{FC}_M$, we want to find a set $S \subseteq D$ containing at most k data points such that $ahr_{D,\mathcal{FC}_M}(S)$ is maximized, i.e.,*

$$S = \underset{S' \subseteq D : |S'| \leq k}{\arg\max} \ ahr_{D,\mathcal{FC}_M}(S'). \qquad (3)$$

In this paper, we focus on solving the Online-AHMS problem. We first present some descriptions of our online model, then we give the formal definition of the Online-AHMS problem based on the model.

In the online setting, the data points of D are revealed one by one in an online fashion. At each timestamp $t \in \{1, 2, \ldots, n\}$, a data point $p_t \in D$ is revealed, the algorithm must immediately decide whether to accept p_t into the solution set S_{t-1}, which is initially empty.

Moreover, we relax the model by allowing the algorithm to preempt previously accepted data points. That is, when the algorithm adds the newly revealed data point p_t into the current solution S_{t-1}, a data point may be removed from S_{t-1}, as long as this preemption can improve the solution quality. We cannot reconsider those data points that have been rejected from the solution or have been discarded because of preemption.

In the online setting, since at any timestamp $t \in \{1, 2, \ldots, n\}$, there are only a portion of the data points $D_t = \{p_1, p_2, \ldots, p_t\}$ revealed. We have no foreknowledge of the future data points and only select a subset S from D_t, so we would like to find a practical method to compute $\max_{p \in D_t} f(p)$ for each utility function $f \in \mathcal{FC}_M$. Note that this actually calculates the maximum utility among all data points in D_t for f, the natural idea is to maintain an auxiliary variable $m_{f,t}$ which holds the current maximum utility for each utility function f after the point p_t is revealed. Further, we utilize $m_{f,t}$ to calculate the happiness ratio of S over D_t for f, i.e., $hr_{D_t,f}(S) = \frac{\max_{p \in S} f(p)}{m_{f,t}}$.

With the above analysis, we next reformalize the average happiness ratio in the online setting and define the online average happiness ratio.

Definition 5 (Online Average Happiness Ratio). *Given a portion of the revealed data points $D_t = \{p_1, p_2, \ldots, p_t\}$ at timestamp $t \in \{1, 2, \ldots, n\}$, a sampled utility function class $\mathcal{FC}_M = \{f_1, f_2, \ldots, f_M\}$ and a subset $S \subseteq D_t$, the online average happiness ratio is defined as*

$$\widehat{ahr}_{\mathcal{FC}_M,t}(S) = \frac{1}{M} \sum_{f \in \mathcal{FC}_M} \frac{\max_{p \in S} f(p)}{m_{f,t}}, \tag{4}$$

where $m_{f,t}$ is an auxiliary variable which holds the current maximum utility for the utility function $f \in \mathcal{FC}_M$ over D_t at timestamp t.

Obviously, the value of $\widehat{ahr}_{\mathcal{FC}_M,t}(S)$ does not depend on the entire unknown data stream D, but on the current maximum utility $m_{f,t}$ for each f and the subset S, which makes it an ideal objective function for selecting representative data points in the online setting. In fact, the online average happiness ratio defined above still satisfies submodularity which we will prove in Sect. 4.1, thus our model is reasonable in practice.

Definition 6 (Online Average Happiness Maximization Set, Online-AHMS). *Given an unknown data stream D with n data points, a user-specified positive integer k, and the sampled utility function class $\mathcal{FC}_M = \{f_1, f_2, \ldots, f_M\}$. The Online-AHMS problem returns a feasible selected set S_t with $|S_t| \leq k$ over the currently revealed data points $D_t \subseteq D$ such that the online average happiness ratio $\widehat{ahr}_{\mathcal{FC}_M,t}(S_t)$ is maximized at any timestamp $t \in \{1, 2, \ldots, n\}$, i.e.,*

$$S_t = \underset{S' \subseteq D_t : |S'| \leq k}{\arg \max} \, \widehat{ahr}_{\mathcal{FC}_M,t}(S'). \tag{5}$$

Unfortunately, according to the existing results in [39], the Online-AHMS problem is NP-hard for any $d \geq 3$. Thus, it is unlikely to have a polynomial-time algorithm that solves the Online-AHMS problem optimally, unless P=NP. Hence, we focus on designing the approximate algorithm for the Online-AHMS problem in this paper.

4   The GreedyAT Algorithm

In this section, we will first introduce *monotonicity* and *submodularity* properties of the online average happiness ratio function and show how these properties can be used in designing our approximation algorithm. Then, we present our GreedyAT algorithm for the Online-AHMS problem.

4.1   Properties

Given a dataset D, a non-negative set function $g: 2^D \rightarrow \mathbb{R}_+$ and a subset $S \subseteq D$, a set function g is naturally associated with a marginal gain $\Delta g(p|S) := g(S \cup \{p\}) - g(S)$, which represents the increase of $g(S)$ when adding a data point $p \in D \backslash S$ to S.

Definition 7 (Monotonicity). *A set function g is monotone if and only if for any $S \subseteq D$ and $p \in D \backslash S$, it holds that $\Delta g(p|S) \geq 0$.*

Definition 8 (Submodularity). *A set function g is submodular if and only if for any $S \subseteq T \subseteq D$ and $p \in D \backslash S$, it holds that $\Delta g(p|S) \geq \Delta g(p|T)$.*

A set function is submodular if the gain of adding a data point p to a set S is always no less than the gain of adding the same data point to a superset of S. In [38], it was shown that the estimated average regret ratio $arr_{\mathcal{FC}_M}(S)$ is monotonically decreasing and supermodular. Since $ahr_{\mathcal{FC}_M}(S) = 1 - arr_{\mathcal{FC}_M}(S)$, it trivially follows that $ahr_{\mathcal{FC}_M}(S)$ is monotonically increasing and submodular. Next, we show the online average happiness ratio function $\widehat{ahr}_{\mathcal{FC}_M,t}(S)$ satisfies above two properties.

Lemma 1. $\widehat{ahr}_{\mathcal{FC}_M,t}(S)$ *is a monotonically increasing function.*

Proof. At timestamp $t \in \{1, 2, \ldots, n\}$, the data point p_t is revealed, $D_t \subseteq D$ is the set containing all currently revealed data points. $\widehat{ahr}_{\mathcal{FC}_M,t}(S) = \frac{1}{M} \sum_{f \in \mathcal{FC}_M} \frac{\max_{p \in S} f(p)}{m_{f,t}}$. For each utility function $f \in \mathcal{FC}_M$, the denominator $m_{f,t}$ is a fixed value, because it represents the maximum utility of all data points in D_t on f at this timestamp t. Let S and T be subsets of D_t, where T is a superset of S. It clearly yields that $\max_{p \in S} f(p) \leq \max_{p \in T} f(p)$ as the additional data points in $T \backslash S$ can only contribute larger utility for each f. So, we can easily have $\widehat{ahr}_{\mathcal{FC}_M,t}(S) \leq \widehat{ahr}_{\mathcal{FC}_M,t}(T)$.

Lemma 2. $\widehat{ahr}_{\mathcal{FC}_M,t}(S)$ *is a submodular function.*

Proof. Similar to Lemma 1, at timestamp $t \in \{1, 2, \ldots, n\}$, the data point p_t is revealed, and $D_t \subseteq D$ is the set that contains all currently revealed data points. We need to show that for all $S \subseteq T \subseteq D$ and for any data point $p \in D_t \backslash T$,

$\Delta\widehat{ahr}_{\mathcal{FC}_M,t}(p|S) \geq \Delta\widehat{ahr}_{\mathcal{FC}_M,t}(p|T)$. To do so, consider a data point $p \in D_t\backslash T$. There are two possibilities depending on whether p is the best data point in S for any user with utility function $f \in \mathcal{FC}_M$ or not. If p is not the best data point in $S \cup \{p\}$ (and consequently, since $S \subseteq T$, p is not the best data point in $T \cup \{p\}$ either) for any utility function, $\Delta\widehat{ahr}_{\mathcal{FC}_M,t}(p|S)$ and $\Delta\widehat{ahr}_{\mathcal{FC}_M,t}(p|T)$ are both zero which proves the result in this case.

Otherwise, if p is the best data point in $S \cup \{p\}$ for some utility functions, then by the definition of the online average happiness ratio, we easily have $\Delta\widehat{ahr}_{\mathcal{FC}_M,t}(p|S) = \frac{1}{M}\sum_{f \in F} \frac{\max_{p \in S \cup \{p\}} f(p) - \max_{p \in S} f(p)}{m_{f,t}}$, where F is the set of utility functions whose best data points change when p is added to S. Similarly, we have $\Delta\widehat{ahr}_{\mathcal{FC}_M,t}(p|T) = \frac{1}{M}\sum_{f \in F} \frac{\max_{p \in T \cup \{p\}} f(p) - \max_{p \in T} f(p)}{m_{f,t}}$ holds for the same reason. And if the best data point of a user changes when p is added to T, the utility function must be in F since S is a subset of T. We can show that $\Delta\widehat{ahr}_{\mathcal{FC}_M,t}(p|S)$ is larger than or equal to $\Delta\widehat{ahr}_{\mathcal{FC}_M,t}(p|T)$, which implies that $\widehat{ahr}_{\mathcal{FC}_M,t}(S)$ is a submodular function.

4.2   The Algorithm

In this section, we present our greedy online-selection algorithm with adaptive thresholds, i.e., GreedyAT for the Online-AHMS problem.

According to the lemmas in Sect. 4.1, $\widehat{ahr}_{\mathcal{FC}_M,t}(S)$ is a monotonically increasing and submodular function. Thus, we can transform the Online-AHMS problem to an online submodular maximization problem under a cardinality constraint, which has been recently studied extensively [6,8,9]. The current state-of-the-art solution is the online algorithm *Preemption* [6] with a competitive ratio (approximation ratio for online algorithms) of at least $1/4$. Initially, Preemption accepts the first k revealed data points sequentially and gets the solution set S_k. When the $(k+1)$-th data point is revealed, the Preemption algorithm needs to find a swap which replaces some data point in S_k. However, the swap happens only when the increased value of the solution, i.e., $\widehat{ahr}_{\mathcal{FC}_M,k+1}(S_k \cup \{p_{k+1}\} - \{p_i\}) - \widehat{ahr}_{\mathcal{FC}_M,k}(S_k)$ is large enough to pass a given threshold $c \cdot \widehat{ahr}_{\mathcal{FC}_M,k}(S_k)/k$ where $p_i \in S_k$ and c is a given positive constant, i.e., $c > 0$. According to Corollary 4.3 in [6], it is known that Preemption provides a $\frac{c}{(c+1)^2}$-competitive ratio, and the best competitive ratio is $1/4$ when $c = 1$. However, once the parameter c is fixed, it can be obviously observed that after several swaps, the value of the solution set will reach a high value and consequent swaps have less chance to increase this value. Thus, the solution quality is not satisfactory.

To improve the quality of the solution set when swapping the on-arriving data point p_t with the data point in the current solution set S_{t-1}, we adaptively set the thresholds based on the fact that n is known because it is obvious that in the scenarios in Sect. 1, the numbers of the candidate players, employees and coming news can be predicated or are known in advance. Our adaptive thresholds strategy is that we partition the stream into two parts and the first part is set with a larger threshold according to the parameter c_1 and the second part with a smaller threshold corresponding to the parameter c_2 which is smaller than c_1, i.e., $c_2 < c_1$. The strategy is simple but empirically effective. To do this, we introduce a balancing parameter $\beta \in (0,1)$, such that the first part is with the points $\{p_{k+1}, \ldots, p_{\lfloor \beta n \rfloor}\}$ while the second part contains the points $\{p_{\lceil \beta n \rceil}, \ldots, p_n\}$ (or $\{p_{\lceil \beta n \rceil + 1}, \ldots, p_n\}$ when $\lceil \beta n \rceil$ is an integer). When the data points in the first part $\{p_{k+1}, \ldots, p_{\lfloor \beta n \rfloor}\}$ are revealed, we use a higher threshold with parameter c_1 to determine the happening of the swap while when the data points in the second part $\{p_{\lceil \beta n \rceil}, \ldots, p_n\}$ are revealed, we use a lower threshold with parameter c_2 to decide whether to swap or not. When the swap happens, we greedily select the point p_i in S_{t-1} to be replaced to maximize the increased value and p_i will be discarded forever. Based on the above idea, we provide our greedy online-selection algorithm with adaptive thresholds, GreedyAT as shown in Algorithm 1.

Algorithm 1: GreedyAT

Input: Data stream $D = \{p_1, p_2, \ldots, p_n\}$, sampled utility function class
$\mathcal{FC}_M = \{f_1, f_2, \ldots, f_M\}$, user-specified positive integer k, balancing parameter
$\beta \in (0, 1)$, and threshold parameters $c_1 \geq c_2 > 0$
Output: The solution set S_t at timestamp t

1 $S_0 \leftarrow \emptyset$;
2 **for** $t \leftarrow 1, \ldots, n$ **do**
3   **if** $t \leq k$ **then**
4          $S_t \leftarrow S_{t-1} \cup \{p_t\}$;
5   **else**
6   **if** $t \leq \lfloor \beta n \rfloor$ **then**
7              $c \leftarrow c_1$;
8   **else if** $t \geq \lceil \beta n \rceil$ **then**
9              $c \leftarrow c_2$;
10   Let p_i be the point in S_{t-1} maximizing $\widehat{ahr}_{\mathcal{FC}_M, t}(S_{t-1} \cup \{p_t\} \setminus \{p_i\})$;
11   **if** $\widehat{ahr}_{\mathcal{FC}_M, t}(S_{t-1} \cup \{p_t\} \setminus \{p_i\}) - \widehat{ahr}_{\mathcal{FC}_M, t}(S_{t-1}) \geq c \cdot \widehat{ahr}_{\mathcal{FC}_M, t}(S_{t-1})/k$ **then**
12              $S_t \leftarrow S_{t-1} \cup \{p_t\} \setminus \{p_i\}$;
13   **else**
14              $S_t \leftarrow S_{t-1}$;

15 **return** S_t;

In Algorithm 1, when $t \leq k$, it is obvious that the data points revealed are all selected into the solution set (Lines 3–4). Otherwise, two thresholds along with parameters c_1, c_2 are set for the data points in $\{p_{k+1}, \ldots, p_{\lfloor \beta n \rfloor}\}$ and $\{p_{\lceil \beta n \rceil}, \ldots, p_n\}$, respectively (Lines 6–10). We greedily online select the point p_i in S_{t-1} with the maximum increased value after the swap (Line 10). If the increased value overpasses the threshold, we replace p_i with the current revealed data point p_t (Lines 11–12). Note that when the data point p_t is revealed, we calculate the utility of p_t for f, i.e., $f(p_t)$, and update the current maximum utility $m_{f,t}$ by $m_{f,t} \leftarrow \max\{m_{f,t}, f(p_t)\}$ which is used to compute $\widehat{ahr}_{\mathcal{FC}_M, t}(S)$.

Our GreedyAT algorithm improves the Preemption algorithm due to the fact that after first several swaps, the online average happiness ratio will reach a high value (almost equal to 1 in most cases) and a smaller threshold will improve the solution quality inevitably.

4.3 Theoretical Analysis

In this section, we provide the competitive ratio of the GreedyAT algorithm.

Theorem 1. *The competitive ratio of GreedyAT is at least* $\min\{\frac{c_1}{(c_1+1)^2}, \frac{c_2}{(c_2+1)^2}\}$.

Proof. Let S_t^* denote the subset of size at most k that maximizes the online average happiness ratio at timestamp $t \in \{1, 2, \ldots, n\}$, i.e., the optimal solution with the value $OPT = \widehat{ahr}_{\mathcal{F}C_M,t}(S_t^*)$. Buchbinder et al. [6] proved that their algorithm which compares each preemption against a threshold determined by a fixed parameter $c > 0$ without the adaptive thresholds strategy, yields a $\frac{c}{(c+1)^2}$-competitive ratio. At any timestamp $t \in \{1, \ldots, p_{\lfloor \beta n \rfloor}\}$, the GreedyAT algorithm outputs a selected subset S_t such that $|S_t| \leq k$ and $\widehat{ahr}_{\mathcal{F}C_M,t}(S_t) \geq \frac{c_1}{(c_1+1)^2} \cdot OPT$. And at any timestamp $t \in \{p_{\lceil \beta n \rceil}, \ldots, n\}$, the GreedyAT algorithm outputs a selected subset S_t such that $|S_t| \leq k$ and $\widehat{ahr}_{\mathcal{F}C_M,t}(S_t) \geq \min\{\frac{c_1}{(c_1+1)^2}, \frac{c_2}{(c_2+1)^2}\} \cdot OPT$. Hence, the competitive ratio of the GreedyAT algorithm is at least $\min\{\frac{c_1}{(c_1+1)^2}, \frac{c_2}{(c_2+1)^2}\}$.

Besides the competitive ratio derived (Theorem 1), our GreedyAT algorithm also has the advantages of current graceful online submodular maximization algorithms [6, 8,9]. GreedyAT only needs one pass over the data stream and the space complexity is $O(k)$ ($O(1)$ if k is a constant).

5 Experimental Evaluation

In this section, we experimentally evaluate the performance of our proposed algorithm GreedyAT for solving the Online-AHMS problem on both synthetic and real datasets. We first introduce the experimental setup in Sect. 5.1. Then, we present the experimental results in Sect. 5.2.

5.1 Setup

All algorithms were implemented in C++. The experiments were conducted on a machine running Ubuntu 16.04 with an Intel Core i7-5500 CPU and an 8GB RAM.

Datasets. Due to space limitation, we run our experiments only on 1 synthetic dataset and 1 real dataset which are both popular in the literature [1,2,21,36,37]. The datasets used in our experiments are listed as follows:

- **Anti-correlated.** The Anti-correlated dataset is a synthetic dataset generated by the synthetic dataset generator [5]. The dataset contains 10,000 random data points with 4 anti-correlated attributes.
- **Tweet.** The Tweet dataset is a real dataset for streaming applications and we adapted it for our comparison. The dataset was obtained from an archive website[1]. In the Tweet dataset, the data points describe tweets delivered during a certain period and the goal is to select the most popular tweets of a fixed size. After pre-processing, we

---

[1] https://web.archive.org/.

selected 12,197 tweets described by 7 attributes, namely *TweetID*, *UserID*, *FollowerCount*, *FollowingCount*, *ReplyCount*, *LikeCount* and *RetweetCount*. Among them, the first two attributes identify a specific tweet, while the last five attributes characterize the popularity of the tweet and are used to conduct our experiments. The detailed meaning of each attribute is: (1) *TweetID* is the tweet identifier, (2) *UserID* is the anonymized user identifier, (3) *FollowerCount* is the number of accounts following the user, (4) *FollowingCount* is the number of accounts followed by the user, (5) *ReplyCount* is the number of tweets replying to this tweet, (6) *LikeCount* is the number of likes that this tweet received, (7) *RetweetCount* is the number of retweets that this tweet received.

For these two datasets, all attributes of the data points are normalized into [0,1] after pre-processing. We considered the entire dataset as a data stream and processed the data points in the order.

Algorithms. The algorithms compared are listed as follows:

- **GREEDY**: The GREEDY algorithm is a classical algorithm for the offline submodular maximization problem under a cardinality constraint which achieves a $(1-1/e)$ approximation factor [23]. GREEDY iterates k times over the entire dataset and greedily selects the data point with the largest marginal gain in each iteration. Although GREEDY is not an online algorithm, we used GREEDY as the benchmark algorithm providing typically the best solution quality, which allows us to compare the relative performance of other algorithms.
- **Random**: A randomized algorithm with replacement via Reservoir Sampling [14]. Originally, Random is a random sampling of k data points as a solution. We adapted it in the online setting by unconditionally accepting the first k data points and using Reservoir Sampling [14] to randomly swap a data point in the current solution. For constraint maximization problems, it cannot return a solution with theoretical guarantees. In spite of this, we still considered the empirical performance of Random as a simple baseline.
- **StreamGreedy**: A variation of GREEDY for the streaming cardinality-constraint submodular maximization problem in [15]. In the online setting, StreamGreedy unconditionally accepts the first k data points and replaces the newly revealed data point with any data point in the solution if the replacement can improve the current solution by a constant threshold $\eta > 0$.
- **Preemption**: Current state-of-the-art algorithm for the online submodular maximization problem in [6]. Preemption achieves a competitive ratio of $\frac{c}{(c+1)^2}$, where $c > 0$ is a fixed parameter. It works similarly to StreamGreedy, but instead of using a constant threshold η, it uses a more suitable threshold, which depends on the current solution. Unless stated explicitly, we set the value of parameter c in Preemption to 1, such that Preemption can achieve the best competitive ratio.
- **GreedyAT**: Our Online-AHMS algorithm based on the preemption and adaptive thresholds strategy proposed in Sect. 4.

To evaluate the online average happiness ratio function of a solution for each algorithm above, we only considered that utility functions used are linear and the learned probability distribution of the utility functions is uniform. By default, the sample size

M of the utility functions is set to $1,000$, i.e., $M = 1,000$. The algorithms are evaluated from two perspectives, namely the average happiness ratio (AHR) and the CPU time.

5.2 Experimental Results

We first evaluate the performance of the GreedyAT algorithm when varying the parameters, by changing two values of the balanced parameter β and the threshold parameters c_1 and c_2. Since the change of parameter values in GreedyAT only affects the quality of the solution, we ignore the results for the CPU time of GreedyAT. Moreover, we only show the experimental results on the real dataset Tweet due to space limitation, and fix the solution size k to 20.

Effect of Parameters β, c_1, and c_2. Figure 1(a) shows the effect of parameters β and c_1 on GreedyAT. We report the average happiness ratios (AHR) of GreedyAT for $c_2 = 0.4$ by varying $\beta \in \{0.1, 0.3, 0.5, 0.7, 0.9\}$ and $c_1 \in \{0.6, 0.8, 1.0, 1.2, 1.4\}$. First of all, for any c_1, the quality of solutions is better when the value of β is 0.3 or 0.5, but degrades if β is too small or too large. This shows that the larger or smaller value of β will lead to the malfunction of our adaptive thresholds strategy, because the effectiveness of GreedyAT with the thresholds under larger or smaller β value is same as that there is only one threshold c_1 or c_2. On the other hand, when β is fixed, the AHR decreases as c_1 becomes larger. This is due to the fact that we have discarded some important data points when c_1 is large.

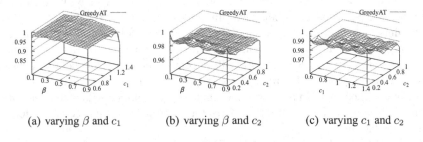

(a) varying β and c_1 (b) varying β and c_2 (c) varying c_1 and c_2

Fig. 1. Performance of GreedyAT under different adaptive thresholds ($k = 20$)

Figure 1(b) shows the effect of parameters β and c_2 on GreedyAT. We report the AHR of GreedyAT for $c_1 = 1.0$ by varying $\beta \in \{0.1, 0.3, 0.5, 0.7, 0.9\}$ and $c_2 \in \{0.2, 0.4, 0.6, 0.8, 1.0\}$. Similarly, we can observe that the AHR decreases when c_2 is larger. This illustrates the marginal gains of the remaining data points could be small when a portion of the data stream is accessed. Moreover, if the value of β is moderate, the solution quality will increase, which shows that it is pathological for β to be too large or too small.

Figure 1(c) shows the effect of parameters c_1 and c_2 on GreedyAT. We report the AHR of GreedyAT for $\beta = 0.5$ by varying $c_1 \in \{0.6, 0.8, 1.0, 1.2, 1.4\}$ and $c_2 \in \{0.2, 0.4, 0.6, 0.8, 1.0\}$. When fixing c_2, for larger c_1, the solution quality of GreedyAT remains relatively stable. But if c_1 is fixed, the AHR becomes smaller by increasing c_2.

Thus, considering the theoretical guarantee and practical performance, the values of three parameters (β, c_1, and c_2) in GreedyAT will be properly decided in the remaining experiments.

Effect of Solution Size k. We report the performance of all the algorithms on synthetic and real datasets by varying $k \in \{10, 15, 20, 25, 30\}$. We present two groups of experiments with different parameter settings. In the first group of the experiments, we set the parameter $\eta = 0.1$ in StreamGreedy, $c = 1.0$ in Preemption, and $\beta = 0.5, c_1 = 1.0, c_2 = 0.4$ in GreedyAT. While in the second group of the experiments, we set the parameter $\eta = 0.05$ in StreamGreedy, $c = 0.8$ in Preemption, and $\beta = 0.7, c_1 = 1.2, c_2 = 0.2$ in GreedyAT.

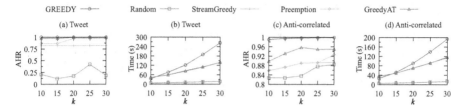

Fig. 2. The performance of all algorithms with varying k ($\eta = 0.1$, $c = 1.0$ and $\beta = 0.5, c_1 = 1.0, c_2 = 0.4$)

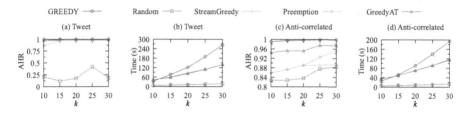

Fig. 3. The performance of all algorithms by varying k ($\eta = 0.05$, $c = 0.8$ and $\beta = 0.7, c_1 = 1.2, c_2 = 0.2$)

As shown in Figs. 2 and 3, generally the average happiness ratio and the query time of each algorithm grow with the increase of k. We can see that Random is the fastest algorithm, but Random shows the weakest quality, especially on the real dataset, Tweet. Only the AHR of Random decreases when k increases (i.e., $k = 30$). Compared with GREEDY, GreedyAT takes less CPU time except in the case of $k = 10$ on the Tweet and Anti-correlated datasets, and the superiority is more obvious when k is larger. Therefore, GreedyAT shows higher efficiency than GREEDY, especially for larger k. On the other hand, the only difference between the GreedyAT algorithm and the Preemption and StreamGreedy algorithms is the use of the adaptive thresholds strategy, so the CPU time is almost the same for these algorithms.

In terms of the solution quality (i.e., AHR) as depicted in Figs. 2 and 3, the quality of GREEDY is the best among all algorithms. This is obvious for GREEDY has accessed the whole stream many times. Also, this also can be seen as the cost of the online algorithms, i.e., StreamGreedy, Preemption and GreedyAT, which only access the stream once. However, compared with Preemption and StreamGreedy, the solution quality of GreedyAT is generally closer to that of GREEDY for different k. GreedyAT almost has the same solution quality as GREEDY on the Tweet dataset (Figs. 2(a) and 3(a)) and provides solutions of at least 90% quality of GREEDY on the Anti-correlated dataset (Figs. 2(c) and 3(c)). This confirms the effectiveness of the adaptive thresholds strategy we used in GreedyAT. Hence, GreedyAT generally outperforms Preemption and StreamGreedy.

6   Conclusion

In this paper, we studied the online average happiness maximization set (Online-AHMS) problem and formulated our problem as an online submodular maximization problem under the cardinality constraint. Then we provided an efficient online algorithm called GreedyAT based on the adaptive thresholds strategy to solve the Online-AHMS problem. Our proposed GreedyAT has been proved with a constant competitive ratio. Extensive experimental results on the synthetic and real datasets confirmed the efficiency, effectiveness, and applicability of our proposed algorithm.

References

1. Agarwal, P.K., Kumar, N., Sintos, S., Suri, S.: Efficient algorithms for k-regret minimizing sets. In: SEA, pp. 7:1–7:23 (2017)
2. Alami, K., Maabout, S.: A framework for multidimensional skyline queries over streaming data. Data Knowl. Eng. **127**, 101792 (2020)
3. Asudeh, A., Nazi, A., Zhang, N., Das, G.: Efficient computation of regret-ratio minimizing set: a compact maxima representative. In: SIGMOD, pp. 821–834 (2017)
4. Asudeh, A., Nazi, A., Zhang, N., Das, G., Jagadish, H.V.: RRR: rank-regret representative. In: SIGMOD, pp. 263–280 (2019)
5. Börzsöny, S., Kossmann, D., Stocker, K.: The skyline operator. In: ICDE, pp. 421–430 (2001)
6. Buchbinder, N., Feldman, M., Schwartz, R.: Online submodular maximization with preemption. ACM Trans. Algorithms. **15**(3), 30:1–30:31 (2019)
7. Cao, W., et al.: k-regret minimizing set: Efficient algorithms and hardness. In: ICDT, pp. 11:1–11:19 (2017)
8. Chakrabarti, A., Kale, S.: Submodular maximization meets streaming: matchings, matroids, and more. Math. Program. **154**(1–2), 225–247 (2015)
9. Chan, T.H.H., Huang, Z., Jiang, S.H.C., Kang, N., Tang, Z.G.: Online submodular maximization with free disposal. ACM Trans. Algorithms. **14**(4), 56:1–56:29 (2018)
10. Chester, S., Thomo, A., Venkatesh, S., Whitesides, S.: Computing k-regret minimizing sets. In: VLDB, pp. 389–400 (2014)
11. Chomicki, J., Ciaccia, P., Meneghetti, N.: Skyline queries, front and back. SIGMOD Rec. **42**(3), 6–18 (2013)
12. Chu, W., Ghahramani, Z.: Preference learning with gaussian processes. In: ICML, pp. 137–144 (2005)
13. Faulkner, T.K., Brackenbury, W., Lall, A.: k-regret queries with nonlinear utilities. In: VLDB, pp. 2098–2109 (2015)
14. Feige, U., Mirrokni, V.S., Vondrak, J.: Maximizing non-monotone submodular functions. SIAM J. Comput. (SICOMP) **40**(4), 1133–1153 (2011)
15. Gomes, R., Krause, A.: Budgeted nonparametric learning from data streams. In: ICML, pp. 391–398 (2010)
16. Houlsby, N., Huszar, F., Z. Ghahramani, Z., Hernández-lobato, J.: Collaborative gaussian processes for preference learning. In: NIPS, pp. 2096–2104 (2012)

17. Ilyas, I.F., Beskales, G., Soliman, M.A.: A survey of top-k query processing techniques in relational database systems. CSUR. **40**(4), 11:1–11:58 (2008)
18. Li, Y., et al.: Hyperbolic hypergraphs for sequential recommendation. In: CIKM, pp. 988–997 (2021)
19. Luenam, P., Chen, Y.P., Wong, R.C.: Approximating happiness maximizing set problems. CoRR abs/2102.03578, pp. 1–13 (2021)
20. Ma, W., Zheng, J., Hao, Z.: A coreset based approach for continuous k-regret minimization set queries over sliding windows. In: WISA, pp. 49–61 (2021)
21. Nanongkai, D., Sarma, A., Lall, A., Lipton, R., Xu, J.: Regret-minimizing representative databases. In: VLDB, pp. 1114–1124 (2010)
22. Nanongkai, D., Lall, A., Das Sarma, A., Makino, K.: Interactive regret minimization. In: SIGMOD, pp. 109–120 (2012)
23. Nemhauser, G.L., Wolsey, L.A., Fisher, M.L.: An analysis of approximations for maximizing submodular set functions-i. Math. Program. **14**(1), 265–294 (1978)
24. Peng, P., Wong, R.C.W.: Geometry approach for k-regret query. In: ICDE, pp. 772–783 (2014)
25. Qi, J., Zuo, F., Samet, H., Yao, J.C.: k-regret queries using multiplicative utility functions. TODS. **43**(2), 10:1–10:41 (2018)
26. Qiu, X., Zheng, J., Dong, Q., Huang, X.: Speed-up algorithms for happiness-maximizing representative databases. In: APWebWAIM DS Workshop, pp. 321–335 (2018)
27. Qu, M., Ren, X., Han, J.: Automatic synonym discovery with knowledge bases. In: KDD, pp. 997–1005 (2017)
28. Storandt, S., Funke, S.: Algorithms for average regret minimization. In: AAAI, pp. 1600–1607 (2019)
29. Stoyanovich, J., Yang, K., Jagadish, H.: Online set selection with fairness and diversity constraints. In: EDBT, pp. 241–252 (2018)
30. Wang, Y., Li, Y., Tan, K.: Efficient representative subset selection over sliding windows. TKDE **31**(7), 1327–1340 (2019)
31. Wang, Y., Mathioudakis, M., Li, Y., Tan, K.: Minimum coresets for maxima representation of multidimensional data. In: PODS, pp. 138–152 (2021)
32. Wang, Y., Li, Y., Wong, R.C.W., Tan, K.L.: A fully dynamic algorithm for k-regret minimizing sets. In: ICDE, pp. 1631–1642 (2021)
33. Xiao, X., Li, J.: Rank-regret minimization. CoRR abs/2111.08563, pp. 1–15 (2021)
34. Xie, M., Wong, R.C.W., Lall, A.: Strongly truthful interactive regret minimization. In: SIGMOD, pp. 281–298 (2019)
35. Xie, M., Wong, R.C.W., Lall, A.: An experimental survey of regret minimization query and variants: bridging the best worlds between top-k query and skyline query. VLDB J. **29**, 147–175 (2020)
36. Xie, M., Wong, R.C.W., Li, J., Long, C., Lall, A.: Efficient k-regret query algorithm with restriction-free bound for any dimensionality. In: SIGMOD, pp. 959–974 (2018)
37. Xie, M., Wong, R.C.W., Peng, P., Tsotras, V.J.: Being happy with the least: achieving α-happiness with minimum number of tuples. In: ICDE, pp. 1009–1020 (2020)
38. Zeighami, S., Wong, R.C.W.: Minimizing average regret ratio in database. In: SIGMOD, pp. 2265–2266 (2016)
39. Zeighami, S., Wong, R.C.: Finding average regret ratio minimizing set in database. CoRR abs/1810.08047, pp. 1–18 (2018)
40. Zheng, J., Chen, C.: Sorting-based interactive regret minimization. In: APWeb-WAIM, pp. 473–490 (2020)
41. Zheng, J., Dong, Q., Wang, X., Zhang, Y., Ma, W., Ma, Y.: Efficient processing of k-regret minimization queries with theoretical guarantees. Inf. Sci. **586**, 99–118 (2022)
42. Zheng, J., Wang, Y., Wang, X., Ma, W.: Continuous k-regret minimization queries: a dynamic coreset approach. TKDE (2022). https://doi.org/10.1109/TKDE.2022.3166835

A Penetration Path Planning Algorithm for UAV Based on Laguerre Diagram

Dan Li$^{(\boxtimes)}$, Xu Chen, Panpan Ding, and Jiani Huang

Nanjing University of Aeronautics and Astronautics, Nanjing 211106, China
lidansusu007@163.com

Abstract. Due to the special requirements of modern military, unmanned combat aircraft (UAVs) attracts more and more attention from the researchers all over the world. It possesses many advantages, such as zero life risk, stronger combat capability ,and better adaptability to harsh combat environments than manual-controlled aircraft. The wide application of UAVs on the battlefield determines their important position in the war. With the rapid development of military science and technology. Nevertheless, there is a conflict between large amount of computing data in complex battlefield environments and limited computing capability of onboard computers. Therefore, a stable and efficient path planning algorithm is critical. This paper aims to dramatically improve the efficiency of path planning by using Laguerre diagram method. Through combing the Laguerre diagram method and the Dijkstra algorithm, the path planning algorithm we proposed in this paper could obtain feasible solutions with lower computational cost by decreasing path nodes and reducing the scale of the problem further. Furthermore, the current research mainly focuses on 2D scenes, which is inconsistent with the actual UAV flight situation. Therefore, this paper proposes a Laguerre diagram based path planning algorithm in 3D scenes. Simulation shows that the execution time of the stage of path planning reduced to one-third compare to the Dijkstra algorithm on the grid.

Keywords: Path planning · Laguerre diagram · Low altitude penetration

1 Introduction

With the development and application of high-tech military equipment, UAV has been widely used in modern air combat. Intelligent technologies such as cluster technology, autonomous technology, and collaborative technology are promoting the application of small UAV [7,11]. UAV survival ability and combat ability plays a vital role in combat, but the military drone technology still exists in complex battlefield environment, the communication link is not stable, calculation of large amount of data, the onboard computer computing ability is limited, therefore, route planning problem as the key to improve the UAV autonomy

© The Author(s), under exclusive license to Springer Nature Switzerland AG 2023
B. Li et al. (Eds.): APWeb-WAIM 2022, LNCS 13423, pp. 34–41, 2023.
https://doi.org/10.1007/978-3-031-25201-3_3

technology is becoming a hot spot of research at home and abroad, It is very important to study efficient and reliable path planning algorithm [2,8].

The map modeling method will have a significant impact on the efficiency of the path planning algorithm. Beard et al. [1] used Voronoi diagram for the first time to generate the initial route set of UAV, and connected UAV and target point respectively with the vertices of the three Voronoi diagrams nearest to them to form a complete state transition sequence. However, Voronoi diagram does not consider the influence of the coverage of the two threat sources, so the composition will generate some unfeasible paths. Dai et al. [3] proposed a method for local correction of Voronoi diagram in view of the above deficiencies. Laguerre diagram [9]is a derivative of Voronoi diagram, which is especially suitable for solving the geometry problem of circle, and requires small storage space, low algorithm complexity, especially suitable for military UAV penetration path planning algorithm. However, the current research is still in the 2D scene, which is inconsistent with the actual route.

This paper proposes a 3D route chart construction method based on Laguerre diagram, and combines Dijkstra algorithm to solve the optimal path. The results show that the algorithm proposed in this paper saves a large number of feasible paths and nodes, and on the basis of ensuring that the optimal solution is close to the theoretical one, it can effectively save the algorithm execution time.

This paper is organized as follows. In Sect. 2, the problem description is presented. In Sect. 3, we introduce the composition method of Laguerre diagram. On this basis, a route chart construction algorithm is proposed and optimized for sparse generating sources. In the fourth section, we make a simulation and give the performance analysis results of the algorithm. Finally, in the fifth part, we draw the conclusion of this paper.

2   Problem Description

We assume that UAV is carrying out reconnaissance missions in battlefield environment, so UAV route planning is a kind of path planning in essence. The path planning method is based on the space search theory of geometry and generally does not consider the kinematics and dynamics model of UAV. The problem of UAV penetration path planning is to avoid UAV being detected by military radar. In order to study the control of flight track points, the influence of wind force and other external conditions is ignored. It is assumed that UAV moves in three-dimensional space. The radar reflection area of the UAV is denoted as σ.Coordinate is (x, y, z).There are K radars are distributed in space. Specify the position of the radar on plane $z = 0$ (on the ground) by coordinates $(x_k, y_k, 0)$. The threat range of a single radar $R_{max}$is given by the radar equation [6,10]:

$$R_{\max} = \left[\frac{P_t A^2 \sigma}{4\pi \lambda^2 \, S_{\min}} \right]^{\frac{1}{4}} \tag{1}$$

where P_t represents the transmitting power of the radar transmitter and A represents the effective receiving area of the antenna. The wavelength of the electromagnetic wave is denoted by λ. Then S_{min} represents Minimum detectable power of receiver.

It is assumed that the starting point is S_p, the target point is G_p, and the all nodes of Laguerre diagram form viable node set V together, and the edges of all Laguerre diagram form viable path set E together. The diagram is expressed as $G = (V, E)$, then all nodes and paths in diagram G are safe. [4] Our goal is to find the shortest sequence of state transitions from the initial state to the target state in the diagram. And since the resolution of UAV is affected by altitude, the flight height of UAV should also be concerned. The evaluation function of the whole solution path is given by the following formula 2:

$$f(x) = w_1 Dis(x) + w_2 H(x) \tag{2}$$

where w_1 and w_2 represents the coefficient. The total distance of the path is denoted by $Dis(x)$ and $H(x)$ represents average altitude.

3   A Penetration Path Planning Algorithm Based on Laguerre Diagram

We propose a route planning method based on Laguerre diagram in 3D environment, which can be used to construct the route set of UAV, aiming to save nodes and paths by mathematical means and save time for the algorithm under the condition that it is as close to the theoretical optimal solution as possible. When UAV penetrates at low altitude, the change of altitude is also crucial to the success of penetration. Therefore, we propose a construction method in 3D environment, which can be directly used to construct the route set of UAV.

3.1   Route Chart Construction

Step1: First of all, we need to arrange all known radar threats in descending order according to the maximum threat radius. Here, we specify the flying altitude range of UAV as 2–4 km, and disperse the flying altitude of UAV into three planes according to the unit of one kilometer. For this reason, we divide all known radars into three groups (R > 4 km, R > 3 km, R > 2 km), initialize a Delaunay triangle set named $DT(S)$, and a Laguerre vertex set named $L(S)$.

Step2: The Laguerre diagram took a group of circles on the plane as the generating element. Starting from the highest layer, the section (R > 4 km) was made for the threat range of radar, and the section circle was taken as the generating element. Write down the set of n radar cross section circles at this height as $S = C_1(p_1, r_1), C_2(P_2, r_2)...Cn(p_n, r_n)$, where p is the coordinates of the center of the circle, r is the radius of the cross section circle. Next according to the set S generating Delaunay triangle set $DT(S)$, Generates a triangle large enough to contain the centers of all generators, then add a circle p to the $DT(S)$, Find the circumferences of the triangles that contain the center p and delete their common edges, around the center of the circle produced a convex polygon, connect p to all vertices of a convex polygon, and so on, until the center of all radar sections affecting that height is added.

Step3: According to the constructed Delaunay diagram, construct the Laguerre diagram.

Calculate the vertices of Laguerre diagram according to Delaunay triangle, and then connect the vertices calculated by adjacent Delaunay triangles to form Laguerre diagram. Let the three vertices of the triangle be C_i, C_j, C_k, and the coordinates Q of Laguerre vertex calculated conform to:

$$d_L(Q, C_i) = d_L(Q, C_j) = d_L(Q, C_k) \tag{3}$$

Connect all vertices and the resulting Laguerre graph is shown in the Fig. 1:

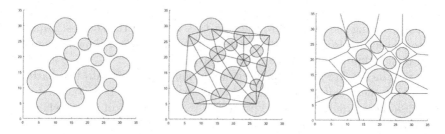

Fig. 1. Generators, corresponding Delaunay triangles, and corresponding Laguerre diagram

Step4: The only feasible route of UAV under a specified height is not enough, also need to build a feasible path of UAV under different altitude flight connection mode, Start building the Laguerre at the next level, in the process of composition, we need to always maintain the Laguerre diagram of edge and vertex information, Laguerre diagram of each edge is generated by the threat of two adjacent source, in the build connection relations, We continue to build from the top down. Assuming that the vertices of the Laguerre diagram formed by two adjacent heights H_i and $H_j(H_i > H_j)$ are $L(SH_i)$ and $L(SH_j)$ respectively, then all the vertices in $L(SH_i)$ are equivalent to the previous state of $L(SH_j)$.

When a new generator is added, a new Delaunay triangle will be formed with the surrounding threat sources, some vertices will be deleted and new Laguerre vertices will be generated. Our connection rule is to connect the newly generated vertices in $L(SH_j)$ with corresponding nodes in the deleted $L(SH_i)$. Nodes (X_1, Y_1, H_i) on the same horizontal and vertical coordinates of $L(SH_i)$ and $L(SH_j)$ are connected with (X_2, Y_2, H_j), and (X_1, Y_1, H_j) are connected with (X_2, Y_2, H_i). And so on until the entire diagram is constructed. The route chart is shown in the Fig. 2.

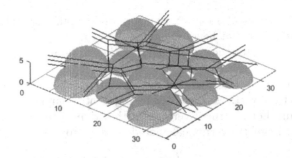

Fig. 2. 3D route chart generated based on Laguerre diagram.

3.2 Path Planing

After constructing the route map, the next task is to find the shortest sequence of state transitions from the initial state to the target state. Here we combine Dijkstra algorithm [5] to find the optimal path.

Step1: The set $L(S)$ of all points in the graph is divided into two parts, V and U. V is the set of points where the optimal path has been taken. In the initial case there is only the source point s in V. U is the set of points where the optimal path has not been taken, and in the initial case is all the points except s.

Step2: We should not only focus on the shortest path, but also set a certain limit on the altitude of the flight, and set a penalty value for the path that stays aloft for a long time, The evaluation value of two points $a(x_1, y_1, z_1)$ and $b(x_2, y_2, z_2)$ in three-dimensional space is given by the formula 4:

$$d_{a,b} = \sqrt{(x_1 - x_2)^2 + (y_1 - y_2)^2 + (z_1 - z_2)^2} + \frac{z_1 - z_2}{h_{average}} \qquad (4)$$

where, $h_{average}$ represents the average of upper and lower limits of UAV.

Because each iteration needs to specify a point in the V set that is currently being iterated, that point is set to the intermediate point. Naturally, first set s to the intermediate point k, and then start the iteration.

Step3: In the process of each iteration, the point k with the lowest evaluation value from U to the intermediate point is obtained, k is added to V set. Then point k is deleted from U set, and the process is repeated until U set is empty. All of the points in V now constitute the optimal path that we need.

4 Simulation and Analysis

In order to illustrate the effectiveness and practicability of the algorithm, we compared the algorithm with the grid method in the virtual radar coverage area. The UAV height was limited to 2 km–5 km, unit height $h = 1$ km, and the UAV movement range was limited to 35 km $*$ 35 km, the number of threat sources was N, and the coverage range was 2 km–5 km.

We compared our approach to the grid method, which is divided into grids per kilometer of the XYZ axis and allows the drone to move in 26 surrounding directions per step. The starting point and end point of the Laguerre diagram are firstly located in the Laguerre region, and the starting point and end point can be directly connected to all vertices in the region. Constitute a complete pathway. Finally, Dijkstra algorithm is used to accurately solve the shortest path. Assuming that the starting point is $(0, 0, 2)$ and the terminal coordinate is $(35, 35, 2)$, we compare the two methods to solve the shortest path and the length of the shortest path on the basis of ensuring that all paths do not pass through the threat area, as shown in the Fig. 3. The red solid line is the theoretical optimal solution obtained by the grid method. The black dotted line is the optimal solution of our algorithm.

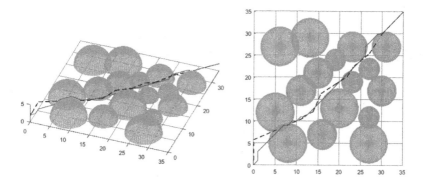

Fig. 3. The optimal path obtained by two algorithms.

As shown in the Fig. 4, The shortest path length solved by our algorithm is 61.67 km and the average flight height is 2.21 km. The shortest path length solved by the grid method is 55.38 km, and the average flight height is 2.18 km. However, compared with the grid method, our algorithm saves a large number of paths and nodes, and the shortest path reaches more than 90% of the solving quality of the grid method.

It can be seen from the Figs. 3 and 4 that our method is very close to the theoretical optimal solution. However, compared with the grid method, our method saves a large number of paths and nodes, and the shortest path reaches more than 90% of the solving quality of the grid method. The scale of nodes and path sets generated by our method and the grid method is shown in the following Table 1:

The most direct impact brought by saving nodes and path sets is the impact on the efficiency of Dijkstra algorithm. Therefore, we compared the execution time of the two algorithms. In order to make the data more scientific and reliable, we randomly selected 100 groups of starting points and ending points with the average Euclidean distance of 40 km to calculate the time. Our algorithm consists of Laguerre graph construction time and solving time, while the grid method

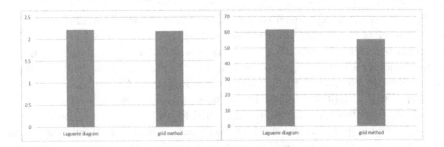

Fig. 4. Average flight height and the shortest path length of the two algorithms

Table 1. The number of nodes and path sets generated by the two algorithms.

Method	Node	Path
Laguerre diagram	187	1097
Grid method	3264	50358

only consists of solving time. The numerical results of through-route planning executed by different algorithms are shown in the Fig. 5:

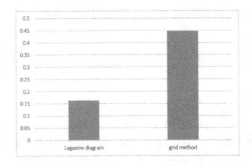

Fig. 5. The execution time of two algorithms.

As can be seen in the Fig. 5, the time to calculate a single path is 0.163s, and the time to calculate a single path by grid method is 0.447 s. Therefore, our algorithm is better than the grid method in terms of execution time, and with the increase of the problem scale, our algorithm will obtain more obvious advantages in terms of time.

5   Conclusion

This paper has designed a new 3D route chart construction algorithm is proposed and applied to UAV low-altitude penetration path planning to improve the search efficiency of the path planning algorithm. In a virtual battlefield environment, we carried out many experiments, the experimental results show that the optimal path obtained by our algorithm is almost equal to that obtained by the grid method in terms of path length and average flight altitude. Our algorithm greatly reduces the scale of the problem, reducing the node set and path set to 5% of the grid method, and shortening the time of path planning to one-third of the original.

The model and algorithm proposed in this paper can quickly generate the penetration path according to the known environment and have certain military application value. Future work will focus on fast penetration route planning for UAVs in combat scenarios including terrain concealment and some uncertain moving targets.

References

1. Beard, R.W., et al.: Autonomous vehicle technologies for small fixed-wing UAVs. J. Aerosp. Comput. Inf. Commun. **2**(1), 2003–6559 (2005)
2. Contreras-Cruz, M.A., Ayala-Ramirez, V., Hernandez-Belmonte, U.H.: Mobile robot path planning using artificial bee colony and evolutionary programming. Appl. Soft Comput. J. **30**, 319–328 (2015)
3. Dai, T., Li, B.: PORP: parallel optimization strategy of route planning for self-driving vehicles. J. Zhejiang Univ. (Eng. Sci.). **56**, 329–337 (2022)
4. Erlandsson, T., Niklasson, L.: Automatic evaluation of air mission routes with respect to combat survival. Inf. Fusion **20**, 88–98 (2014)
5. Guo, Q., Zheng, Z., Yue, X.: Path-planning of automated guided vehicle based on improved Dijkstra algorithm. In: Control & Decision Conference (2017)
6. Inanc, T., Muezzinoglu, M.K., Misovec, K., Murray, R.M.: Framework for low-observable trajectory generation in presence of multiple radars. J. Guid. Control. Dyn. **31**(6), 1740–1749 (2008)
7. Jin, H., Cui, W., Fu, H.: Improved RRT-connect algorithm for urban low-altitude UAV route planning. J. Phys. Conf. Ser. **1948**(1), 012048 (2021)
8. Liu, P., Wang, M., Cui, J., Li, H.: Top-k competitive location selection over moving objects. Data Sci. Eng. **6**(4), 392–401 (2021)
9. de Berg, M.: Computational Geometry: Algorithms and Applications (2005)
10. Tina, E.: Route planning for air missions in hostile environments. J. Defense Model. Simul. App. Methodol. Technol. JDMS **12**(3), 289–303 (2015)
11. Zhan, W., Wei, W., Chen, N., Chao, W.: Efficient UAV path planning with multiconstraints in a 3d large battlefield environment. Math. Probl. Eng. **2014**(pt.3), 1–12 (2014)

Scheduling Strategy for Specialized Vehicles Based on Digital Twin in Airport

Hongying Zhang[1], Minglong Liu[1], Chang Liu[2(✉)], Qian Luo[2], and Zhaoxin Chen[2]

[1] School of Electronic Information and Automation, Civil Aviation University of China, Tianjin 300300, China
[2] Engineering Technology Research Center, Second Research Institute of Civil Aviation Administration of China, Chengdu 610041, China
carole_zhang@vip.163.com

Abstract. Flight schedule fluctuations are very common in many airports, which means the arrival of flights tend to be different from the plan. It has caused huge challenges for airports to real-time schedule the specialized vehicles to finish the ground-service operations. However, the existing scheduling methods, such as optimization algorithms and off-line simulations, rarely utilize the real-time information and could not adjust the specialized vehicles dispatching according to the flight schedule fluctuations. As a result, precise operation control for specialized vehicles couldn't be achieved. Therefore, this paper proposes a real-time scheduling strategy for specialized vehicles based on digital-twin, which is constructed by cloud-and-edge computing architecture. This method uses real-time arrival of flights and the current status of vehicles to generate and adjust the schedules of specialized vehicles. In this method, specialized vehicle priority and flight one are taken into account, and both of them are calculated based on the real-time information in the digital twin model. A number of experiments were processed and experimental results show that the proposed method have a better performance in shortening ground-service duration of three kinds of flights and improving the number of flights departed on time.

Keywords: Flight schedule fluctuation · Digital twin · Cloud-and-edge computing · Real-time scheduling

1 Introduction

Ground-service operations of turnaround flights, which are mainly undertaken by specialized vehicles, plays a pivotal role in determining level of flights departure on time. And it is also one of the most essential factors to reflect airport operating efficiency. So scheduling strategies of specialized vehicles in airports directly and positively influence both the efficiency of airports and flights' departure time. However, the phenomenon that flights' real arriving time fluctuates from its estimated arriving time are very common in many airports, increasing the pressure of scheduling specialized vehicles. In fact, once

Supported by Sichuan Science and Technology Program (No. 2021003).

fluctuations in flight schedule happen, it means the previous plan of ground-service has to be adjusted. Therefore, in order to deal with the flight schedule fluctuations and improve the efficiency of ground-service process, it's very urgent for airport to dynamically adjust the schedule of specialized vehicles in real-time.

Actually, with modern communication and automatic driving technologies [1] applied in future smart airport [2], such as wireless sensor network, wireless local area network, the futuristic mobile communication network 5G, automated driven specialized vehicles and blockchain technology [4, 5], all of them provide a basis to achieve real-time information transmission and real-time control of specialized vehicles [6, 7]. However; there is still lack of a model that organizes them organically as a whole.

Recent years, the concept of digital twin has attracted many researchers' attention, which is a digital representation of physical system in cyber space, integrating multiple disciplines, scales and physical quantities [8]. The typical of digital twin are virtual and real-time integration and real-time interaction, iterative operation a characteristics and optimization, etc. [9]. Generally speaking, not only can digital twin compensate the deficiencies of traditional simulating method which cannot achieve online simulation based on real-time information, but achieve dynamical adjustment according to changes in the physical environment so as to accurately and precisely control [10]. Therefore, digital twin provides a new approach to build a model, which integrates specialized vehicle operations as a whole, offering the real-time information to formulate new scheduling schemes of specialized vehicles and achieve and precise control and real-time adjustment.

As a consequence, the contributions of this paper have been concluded as below:

1) This paper first proposed a digital twin model of specialized vehicles based on cloud-and-edge computing, which are responsible to provide a model as basis to achieve real-time control of four kinds of specialized vehicles.
2) And we also proposed a scheduling strategy about specialized vehicles, which can achieve self-adjustment in real-time according to flights schedule fluctuation in order to improve the level of turnaround flights departure on time.

This paper is organized as follows. First the related works review the past studies about specialized vehicle scheduling strategies in Sect. 2. Then the problem statement describes the ground-service process of turnaround flights in Sect. 3. The method has been divided into three parts in Sect. 4. Section 4.1 describes overall architecture of digital twin scheduling model based on edge-and-cloud computing. Section 4.2 illustrates the real-time scheduling strategy based-on digital twin model. Finally, the experiments and conclusions are shown up in the Sect. 5.

2 Related Works

At present, researches about how to allocate the ground-service resources are mainly divided into twofold aspects, which are optimization algorithms and off-line simulation respectively. And they have been illustrated as below.

2.1 Optimized Algorithms

The first one is optimization algorithm study based on the characteristics of specialized vehicles during working, which include genetic algorithm, heuristic algorithm, integer programming and time-window based algorithm. Angus Cheung et al. [11] took the total working time of vehicles as the goal, and designed separate scheduling models for different types of vehicles, using genetic algorithms to solve scheduling problem of water trucks, cleaning trucks and tractors. He Danni [12] used heuristic algorithm to establish an optimization model aiming at minimizing dispatched vehicles and the total distance of vehicles as well as balancing vehicle task allocation according to the service process's time window constraints of ferry vehicles. Kuhn [13] put forward simple integer programming model based on reducing shift delays and running costs to study the airport ground support service vehicles scheduling. Yin Long and Heng Hongjun [14] constructed a specialized vehicle scheduling model with time window, solved it with the nearest neighbor algorithm. However, all of them can hardly dynamically adjust the plan according to the change of flight plan, once the scheduling strategy has been made.

2.2 Off-Line Simulation

On the other hand, some researchers carried out simulation researches about the problem of scheduling specialized vehicles in airport. Huang Lishi [15] proposed a vehicle scheduling optimization model based on SIMIO, aiming at reasonably allocating equipment tasks and reducing flight delays, simulating ferry vehicles and refueling vehicles to optimize resources. Tao Jingjing [16] used the discrete-time simulation method to study the scheduling problem of ferry vehicles in remote positions, and simulated it on the platform of software SIMIO to reduce the flight delay caused by improper scheduling of aircraft ground support services. Although these methods have taken some uncertainties into account during simulation, they rarely considered the real-time information or fluctuations happening when the strategy was come into service. Therefore; these solutions might not achieve precise and accurate control of the ground-service process in real-time.

3 Problem Statement

Ground-service of turnaround flights is composed of a series of subunit task. And each task is mainly undertaken by one type of SV. Meanwhile since the duration of different flights tends to be different, meaning that ground-service operation have to be finished in different limited time window before flights departure. Meanwhile the duration of ground-service operation is determined by longest path of the whole process which is also called critical path as Fig. 1; as a result, we take the critical path as study object which contains passenger-stairs vehicles, passenger bus, fuel dispenser and tractor. Therefore, this paper intends to build a digital twin model based on cloud-and-edge computing to schedule specialized vehicles, which are passenger-stairs vehicles, fuel dispenser passenger bus and tractor, respectively.

4   Method

4.1   Digital Twin Scheduling System Based on Edge-Computing

The overall architecture of digital twin scheduling model, which is based on cloud-edge about specialized vehicles, designed to achieve dynamically scheduling [17] and precisely control to eliminate the fluctuations in real situation is illustrated as Fig. 1 which adopts a three-layer hierarchy, cloud center, edge layer, and end nodes of vehicles respectively [18]. In this section, the functions of each layer and the interaction among them will be introduced in detail.

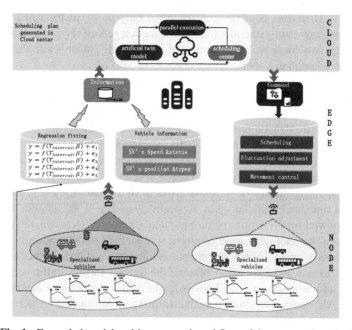

Fig. 1. Expanded model architecture and workflow of the proposed method

Firstly, as the most fundamental layer, end nodes are four different types of specialized automatic guided vehicles. These four kinds of specialized vehicles are directly connected with the edge layer, which can be regarded as executive agents of the model. There are two kinds of functions included in it. One is mainly responsible for collecting information and sending it to edge layer, such as the time duration of SV's operation to finish the ground-service task, the position and speed of SVs and the status of SVs. The other is to receive commands and guidance published by edge to control its operation in real-time, such as guiding route, speed control and scheduling commands. Along with other objects and such as roads, flights, stations and buildings, SVs of end nodes are composed of the physical entity of digital twin model.

Edge layer consists of a series of intelligent road-side units (RSD) which are distributed at different point of location on the ground to cover the whole area of SVs' operation. It is also combined with two function modulars, information modular and command modular respectively. The information modular is in charge of preliminarily processing of raw data gathered by SV, such as the uncomplicated computation about them, and then send them to the center cloud. Moreover, the information modular also conveys and transmits information, such as SV's position, status and speed etc., between the end nodes and cloud center. Secondly the command modular is to schedule and control the SV's operation in real time according to the scheduling plan generated in cloud center. Every road-side unit take responsibility of collecting and processing SV' information within its coverage which reduces the excessive or unbalanced computation burden greatly. Therefore, edge devices could not only polish distributed intelligence during real-time raw data gathering and processing but reduce the computing loads of the cloud center to achieve real-time control of SV. The edge layer is also the data layer of digital twin scheduling system.

Thirdly, the cloud center, can be divided into threefold function modulars, artificial twin model, computing and scheduling center, and parallel execution namely. Firstly, with the help of software-defined objects (SDOs), software-defined relationships (SDRs), and software-defined processes (SDPs), artificial elements and the relationship among them in real world, such as vehicle to vehicle and vehicle to infrastructure, are described and designed in artificial twin model. And it's achieved by using the bottom-up multiagent method. In this part, we build seven species of intelligent agents, artificial specialized vehicle, artificial road, artificial RSU, artificial base stations, artificial buildings, artificial time, and artificial flight, respectively as Fig. 2. Meantime; all of these agents are endowed with functions of simple calculation and interaction, through which we can improve the intelligent level of physical objectives without additional costs to update their hard-ware to finish some simple intelligent functions. Through the redefinition of the actions and interaction rules of the agents, various traffic scenarios can be simulated and assessed, and the knowledge of different situations can be acquired by means of AI, data mining, and machine learning. Additionally, on account of the fact that they are connected with physical entities in real-time through information layer of edge, it can inspect the evolvement of physical scene situation as well as provide real-time information to computing and scheduling center which can compute and make decision to formulate the scheduling plans of the specialized vehicles. Then the function of computing and scheduling center is to design the quantitative grouping strategy and sequential interaction rules for all kinds of intelligent vehicles with the help of the artificial twin model in order to generate specializes vehicles' scheduling plan. Finally, the modular of parallel execution will simulate and verify the scheduling and the outcome will send to the command layer of edge to schedule and control the specialized vehicles operation and moving. It also the virtual twin of digital twin model.

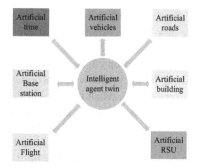

Fig. 2. Elements of the digital twin model

In a conclusion, the end nodes of vehicles are responsible for collecting and transmitting information to edge layer. Next edge layer processes and submit them to cloud center, and cloud center can make use of the real-time information to simulate and generate scheduling schemes. Then they will be conveyed to the command layer which can control the operation of specialized vehicles. Finally, a digital twin scheduling system based on edge-computing is constructed, which can achieve close-loop control [19, 20].

4.2 Real-Time Scheduling Strategy Based-On Digital Twin

After the digital twin model has been constructed, the process of scheduling strategy is completed in the cloud center of the digital twin. Next it will be described in details as follow.

At the moment when a flight arrives at random, the scheduling function in scheduling center in digital twin will be triggered in real time. There are twofold processes comprised scheduling function. The first is decompose ground-service task the into sequence of unit subtasks according to the characteristic of flight and get the priority of the ground-service task as a whole. The second is to compare the priorities of all flights in ground and make a list of them. The highest priority flight among all flights will get the privilege to schedule ground-service resources. Then the scheduling rights will be allocated to the rest flights in the light of flights priority list.

The concept of priority in this paper can be divided into two aspects, which are priority of specialized vehicles and priority of flights respectively. And they are introduced in detail as follow.

1) **Specialized vehicle priority**

The specialized vehicle priority is measured by two parts, which is the time that a SV spends to finish the task of target flight as well as the resource consumption.

First the $Priority_1$ has been shown as Eq. (1) and (2) below:

$$Priority_1 = -(T_1 + T_2) \tag{1}$$

$$T_1 = T_m + T_n \tag{2}$$

T_1 represents the rest of time that a SV finishes its current task. T_2 Represents the time to finish ground-service task of target flight. And it can be divided into two parts T_m and T_n as Eq. (2). T_m is the time that SV spends to reach the point of target flight. Because of the digital twin model based-on edge-cloud, it can be calculated by its near edge device and sent to cloud model to update model information in real-time. T_n is the time that SV finishes ground-service task when it arrives at the position of target flight. On account of that every single SV's operation time interval data has been recorded by the edge device, road-side unit, therefore in the cloud model, these time interval data can be fitted by machine learning algorithm such as interval-valued data regression algorithm [20]. In the end, the function of Tm can be provided as Eq. (3) below where X is time interval in a day, ε_i:

$$T_m = f(X, \varepsilon_i) + \theta_i \tag{3}$$

Then in the cloud digital twin model, intelligent twin SV will plus the $-T_1$ and $-T_2$. The higher the value is, the lower the priority is. Secondly, the priority2 can be calculated by Eq. (4).

$$Priority_2 = -P(v) * t(v) = -P(v) * \frac{s}{v} \tag{4}$$

$P(v)$ represents the power when SV's speed is average speed v; s denotes the path length. And $t(v)$ can be represented by the Tm. In the same case, the higher the consumption is the lower the priority2 is. Then a SV's priority can be calculated by Eq. (5) as below where the γ, ρ are indicator weights;

$$Priority_{vehicle} = \gamma \cdot Priority_1 + \rho \cdot Priority_2 \tag{5}$$

When all SV' priorities have been calculated by the intelligent twin SV in cloud model, they will be conveyed to the scheduling center where they will be compared and the highest priority SV will be chosen. Through calculating the specialized vehicle (SV) priority, it will help to match the highest priority vehicle to each subunit task of ground-service at the minimum time and resource consumption.

2) Flight priority

However, some flights will arrive in a same time interval and there are some other flights on ground waited to be served; meanwhile, their estimated taking-off time is different. So, their requirements for ground-service duration are different too. As a result, it's necessary to calculate flight priority to allocate the privilege of scheduling SVs. And it denotes as Eq. (6)

$$Priority_{flight} = -\beta * k * (t_{Eend} - t_{end}) + \varepsilon * (1 - k) * (t_{end} - t_{Eend}) \tag{6}$$

As mentioned above, when all the SVs with highest priority are chosen for the target flight, the whole time of ground-service operation can be obtained in the end. So t_{end} denotes the end time of flight's ground-service operation which can be obtained from

3). t_{Eend} is the expected ending time of the ground-service which is given by airport. β, ε are indicator weights. And the value of k has been set as Eq. (7).

$$k = \begin{cases} 1 & t_{EOBT} > t_{end} \\ 0 & t_{EOBT} < t_{end} \end{cases} \tag{7}$$

Therefore; when the priorities for all flights on ground are calculated, then the scheduling center will comparel them and the highest priority flight will be chosen and allocated the preferential privilege of SV resources. And other flights will enter the next round of SV scheduling. The whole process will be expressed in details as 3) and Fig. 3.

3) Scheduling plan generation

a) Scheduling center will decompose ground-service operations of arriving flights in a same time interval and send them to the intelligent twin SVs.

b) Intelligent SVs will compute vehicle priorities as described in 1) and feed back to scheduling center and the highest priority SV will be chosen for each ground-service operation of each flight.

c) Scheduling center will calculate the priorities of flights as described in 2) according to b) and Eq. 5 and compare their values.

d) Flight with highest priority will be chosen and the matched highest priority SV will be locked. Then the flight will withdraw the cycle.

e) Other flights will go on the cycle until all flights are allocated the SV for ground-service operation.

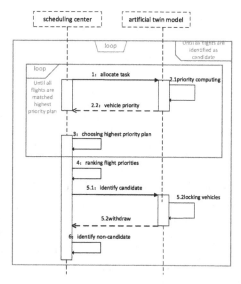

Fig. 3. Sequence diagram of the highest priority flights selection

5 Experiment

This section will provide a case to verify the effectiveness of the proposed approach on lowering the ground-service duration while improving the level of flight's taking-off on time when flight schedule fluctuates.

A simulation model has been constructed at the simulating platform of Anylogic 8.7.9 Professional as Fig. 4. To verify the real-time scheduling strategy based-on digital twin, three sets of comparative simulations has been implemented. And we take the 'First arrive, First serve' (FAFS) as the comparison. The relative parameter of the model has been shown as Table 1.

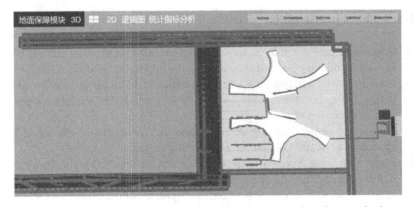

Fig. 4. A digital twin platform for ground-service process based on Anylogic

Table 1. Simulation parameters

Parameter	Number	Speed
Passenger-stair vehicle	10	In interval [1.39 8.3] m/s
Passenger bus	7	In interval [1.39 5.6] m/s
Fuel dispenser	10	In interval [1.39 8.3] m/s
Tractor	10	In interval [1.39 8.3] m/s

In the model, there are three types of flights which are A320 A300 A380 respectively as examples. They are typical representatives of small passenger planes, medium-sized passenger planes, and large passenger air bus and very common in airports. At mean time, each of them contains 15 flights which will arrive at random time. The outcomes of the experiment have been shown as Fig. 5 where the DT means the proposed method and FAFS means the mothed of First-Arrive-First-Serve.

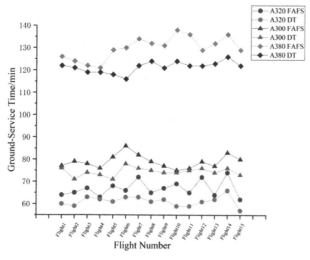

Fig. 5. Comparison of ground-service duration of three types of flights in DT and FAFS

We take the duration of ground-service as the metric to compare the effectiveness of the two kinds of method. The performance of proposed method is apparently better than the First-arrive-First-serve in all of these three kinds of flights. And the average time of these two methods have been shown as Table 2.

Table 2. Average time comparison of FAFS and DT

Flight types	FAFS	DT
A320	66.9 min	61.2 min
A300	79 min	74.4 min
A380	130 min	121.4 min

Then we compare the improvements of the proposed method on these three types of flights and the outcome has been shown as Fig. 6. As can be seen from Fig. 6. The improvement for large-sized air bus is the most obvious but the for medium-sized passenger planes the improvement is lower than that of other two kinds of flights. In a conclusion, the real-time scheduling strategy based-on digital twin has a better performance of scheduling specialized vehicles compared to FAFS such as the ground-service duration and the level of flights' taking-off on time when the flight schedule fluctuates.

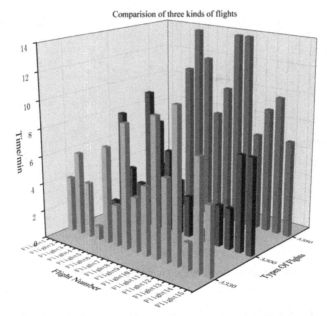

Fig. 6. Improvements of the proposed method on these three types of flights

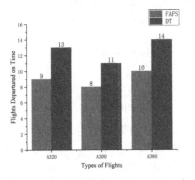

Fig. 7. Numbers of flights departure on time   **Fig. 8.** Comparison of the total ground-service duration

And meanwhile according to Fig. 7 and Fig. 8 it shows the numbers of flights departed on time under the proposed scheduling strategy based on digital twin (DT) have achieved a significant raise compared to the traditional FAFS in all of these three kinds of flights; moreover, the total time of flights' ground-service process has also realized a moderate improvement. In a conclusion, the proposed real-time scheduling strategy based-on digital twin (DT) has a better performance in indicators such as efficiency of ground-service and level of flights departure on time in face of the arrival of flights in real time.

6 Conclusion

In this paper, we firstly proposed a real-time scheduling strategy of specialized vehicles based on digital twin in face of the real-time arrival of flights and the fluctuations happened in the actual implementations of the specialized vehicle operation. First, we adopt cloud-and-edge computing architecture to construct our digital twin model in order to achieve precise and real-time control of the specialized vehicles. Based on the digital twin model, we proposed a new scheduling strategy from the perspective of vehicle priority and flight priority by considering their real-time information. A lot of experiments have shown the proposed method achieves a satisfactory performance in shortening the duration of ground-service and improving the level of flights departed on time in face of the real-time arrival of flights.

References

1. Ulm, G., Smith, S., Nilsson, A., et al.: OODIDA: on-board/off-board distributed real-time data analytics for connected vehicles. Data Sci. Eng. **6**(1), 102–117 (2021)
2. AlMashari, R., AlJurbua, G., AlHoshan, L., et al.: IoT-based smart airport solution. In: 2018 International Conference on Smart Communications and Networking (SmartNets), p. 1. IEEE (2018)
3. Li, B., Liang, R., Zhou, W., et al.: LBS meets blockchain: an efficient method with security preserving trust in SAGIN. IEEE Internet Things J. **9**(8), 5932–5942 (2021)
4. Li, B., Liang, R., Zhu, D., et al.: Blockchain-based trust management model for location privacy preserving in VANET. IEEE Trans. Intell. Transp. Syst. **22**(6), 3765–3775 (2020)
5. Salis, A., Jensen, J.: A smart fog-to-cloud system in airport: challenges and lessons learnt. In: 21st IEEE International Conference on Mobile Data Management (MDM), pp. 359–364. IEEE (2020)
6. Koroniotis, N., Moustafa, N., Schiliro, F., et al.: A holistic review of cybersecurity and reliability perspectives in smart airports. IEEE Access **8**, 209802–209834 (2020)
7. Zhang, M., Tao, F., Nee, A.Y.C.: Digital twin enhanced dynamic job-shop scheduling. J. Manuf. Syst. **58**, 146–156 (2021)
8. Saifutdinov, F., Jackson, I., Tolujevs, J., et al.: Digital twin as a decision support tool for airport traffic control. In: 61st International Scientific Conference on Information Technology and Management Science of Riga Technical University (ITMS), pp. 1–5. IEEE (2020)
9. Guo, H., Chen, M., Mohamed, K., et al.: A digital twin-based flexible cellular manufacturing for optimization of air conditioner line. J. Manuf. Syst. **58**, 65–78 (2021)
10. Cheung, A., Ip, W.H., Lu, D., et al.: An aircraft service scheduling model using genetic algorithms. J. Manuf. Technol. Manag. **16**, 109–119 (2005)
11. He, D.: Research on ground service vehicle scheduling problem of large airport flights. Master's thesis, Beijing Jiaotong University, Beijing (2018)
12. Kuhn, K., Loth, S.: Airport service vehicle scheduling (2009)
13. Yin, L., Heng, H.: Research on the application of airport special vehicle scheduling based on nearest neighbor algorithm. Comput. Technol. Devt. **26**(7), 151–155 (2016)
14. Huang, L.: Simulation study on apron vehicle scheduling based on SIMIO. Master's thesis, Nanjing University of Aeronautics and Astronautics, Nanjing (2013)
15. Tao, J.: Simulation and optimization of apron support service equipment scheduling. Master's Thesis, Nanjing University of Aeronautics and Astronautics, Nanjing (2011)
16. Chen, H., Zhu, X., Liu, G., et al.: Uncertainty-aware online scheduling for real-time workflows in cloud service environment. IEEE Trans. Serv. Comput. **14**(4), 1167–1178 (2018)

17. Jia, P., Wang, X., Shen, X.: Digital-twin-enabled intelligent distributed clock synchronization in industrial IoT systems. IEEE Internet Things J. **8**(6), 4548–4559 (2020)
18. Wang, X., Han, S., Yang, L., et al.: Parallel internet of vehicles: ACP-based system architecture and behavioral modeling. IEEE Internet Things J. **7**(5), 3735–3746 (2020)
19. Min, Q., Lu, Y., Liu, Z., et al.: Machine learning based digital twin framework for production optimization in petrochemical industry. Int. J. Inf. Manag. **49**, 502–519 (2019)
20. de Carvalho, F.A.T., Neto, E.A.L., da Silva, K.C.F.: A clusterwise nonlinear regression algorithm for interval-valued data. Inf. Sci. **555**, 357–385 (2021)

Recommender Systems

Graph-Based Sequential Interpolation Recommender for Cold-Start Users

Aoran Li[1], Jiajun Wu[1], Shuai Xu[1,2(✉)], Yalei Zang[1], Yi Liu[1], Jiayi Lu[1], Yanchao Zhao[1], Gaoxu Wang[3], Qing Meng[4], and Xiaoming Fu[5]

[1] Nanjing University of Aeronautics and Astronautics, Nanjing, China
`xushuai7@nuaa.edu.cn`
[2] State Key Laboratory for Novel Software Technology, Nanjing University, Nanjing, China
[3] Nanjing Hydraulic Research Institute, Nanjing, China
[4] Singapore University of Technology and Design, Singapore, Singapore
[5] University of Goettingen, Goettingen, Germany

Abstract. Sequential recommendation systems aim to capture the users' dynamic preferences from users' historical interactions to provide more accurate recommendations. However, in many scenarios, there are a large number of cold-start users with limited user-item interactions. To address this challenge, some studies utilize auxiliary information to infer users' interests. But with the increasing awareness of personal privacy protection, it is difficult to obtain detailed descriptions of users. Therefore, we propose a model called GitRec (**G**raph-based sequential **i**nterpolation **Rec**ommender for cold-start users) to address this issue. Our proposed model GitRec captures users' latent preferences and provides more proper recommendations for cold-start users without any help from external information. Specifically, (i) GitRec constructs the global graph based on all sequences to obtain the core interests of the warm-start users. (ii) Then, GitRec selects content-rich sequences that are most similar to the short sequences and constructs the local graph to find out the potential preferences of cold-start users. (iii) Finally, GitRec can find suitable candidates to enrich the information of the cold-start users by using graph neural networks, and provide better recommendations with the help of core interests and potential preferences. We conducted extensive experiments on three real-world datasets. The experimental results demonstrate that GitRec has significant improvement over state-of-the-art methods.

Keywords: Recommendation systems · Cold-start · Graph neural network

1 Introduction

Recommendation systems (RS) attempt to help users select interesting content in various online scenarios to address the problem of information overload. Traditional recommendation systems infer the users' preferences by modeling users'

B. Li et al. (Eds.): APWeb-WAIM 2022, LNCS 13423, pp. 57–71, 2023.
https://doi.org/10.1007/978-3-031-25201-3_5

historical interactions in a static way. While sequential recommendation systems (SRS) dynamically predict users' next behaviors from chronological user-item interactions to give more accurate and personalized recommendations. Consequently, sequential recommenders attract more attention and become a new paradigm of RS.

Some early works on sequential recommendation are based on Markov chains, which are hard to model the high-order dependency [21]. Then, with the rapid development of deep learning, some neural networks are applied to recommendation tasks, such as recurrent neural network (RNN) [7,16,25], convolutional neural network (CNN) [19,22,23], graph neural network (GNN) [27–29], and attention mechanism [9,15]. These methods model the complex sequential patterns based on historical interactions and achieve remarkable performance.

However, in many real scenarios, recommenders have difficulty making accurate recommendations for users who have limited interactions with items [1]. For E-commerce platforms, many new users are introduced to the platform every day. If the recommender can effectively provide recommendations for cold-start users while considering long-time users, it will greatly increase customer stickiness. Due to the dependence on the number of users' records, most existing methods can not characterize users' preferences with limited data. Therefore, for sequential recommender, capturing the preferences of cold-start users is significant and urgently needs to be solved.

To handle the user cold-start problem, most research works use external information to draw the portrait of users and infer their preferences, such as user attribute information [11,17], knowledge graph [3,8,13] and social network information [14]. However, with the increasing awareness of privacy protection, valid personal information becomes more difficult to get. A few methods choose to interact with users directly to collect information about users' interests. One of the challenges is that people may not be willing to spend time and energy on the query process, such as filling out the questionnaires or rating some given items. Some other methods apply meta-learning to model users' transition patterns with only limited interactions [24]. But the model learned in this way is only for cold-start users which did not comprehensively consider all sequences from the global perspective. Therefore, it is valuable to design a brand-new sequential recommendation model, which can address the user cold-start problem and adapt to long-time users well without using any auxiliary information.

In this paper, we propose an effective and accurate sequential recommender GitRec to provide appropriate recommendations for cold-start users. Compared with the previous work only focuses on the current sequence, we use all sequences to construct the global graph and extract the global core interests by employing the PageRank algorithm. We construct the local graphs with the selected similar sequences to capture users' potential preferences. By employing graph neural networks, we can learn the item embeddings from two aspects. Then, we can enrich the information of the cold-start users by utilizing the appropriate candidate items to provide more accurate and personalized recommendations. The primary contributions of this paper are summarized as follows:

- We define a set of rules to measure the similarity between sequences from three aspects: number, order, and tightness (NOT). In this way, we can effectively use similar users' information to uncover the potential preferences of cold-start users.
- We construct the global graph through all sequences and local graphs through the selected similar sequences. By employing graph neural networks, we can find suitable candidates to enrich the information of the cold-start sequence without using any auxiliary information.
- Extensive experiments on three real-world datasets verify the effectiveness of our sequential recommender for cold-start users. The results demonstrate that GitRec outperforms the state-of-the-art methods.

2   Preliminaries

In this section, we first describe the formal definition of the problem, then we propose two graph models, the global graph and local graph, and finally define a set of rules to measure the similarity between sequences.

2.1   Problem Formulation

In sequential recommendation, let $U = \{u_1, u_2, \ldots, u_m\}$ and $V = \{v_1, v_2, \ldots, v_n\}$ denote the set of users and items. For user u, the interactions can be listed as $S_u = \{v_{u,1}, v_{u,2}, \ldots, v_{u,l}\}$, where $v_{u,i}$ represents the i-th item interacted with the user. In our task, the goal is to recommend the top-N items for the cold-start user who has limited historical records, generally considered to be less than 5.

Input: Each user's historical interactions $S_u = \{v_{u,1}, v_{u,2}, \ldots, v_{u,l}\}$, where $l \leq 5$.

Output: The top-N item list ranking by probabilities estimated by the proposed model at the *(l+1)-th* step.

2.2   Graph Models

In this subsection, we introduce the global graph and local graph (see Fig. 1) to fully utilize all sequences to enrich the information of the cold-start user.

Global Graph Model. Previous work shows that the sequence graph can effectively capture the complex potential relationships of sequences [26,28]. Inspired by [26], we construct the global graph to gain the core interests of all users. Considering that high-order relationships are weak, for each item v_i, we build the edge $\xi_i = (v_i, v_j)$, where v_j is the first- or second-order neighbor of v_i. In this way, we construct the global graph $G_g = \{V_g, \xi_g\}$, V_g and ξ_g represent the set of items and edges.

To extract the core interests, we use the PageRank algorithm to obtain the weight of each item from a global perspective. PageRank is an algorithm that

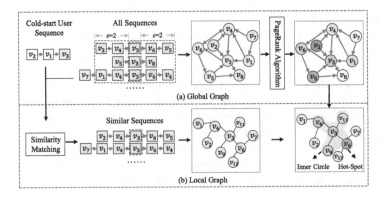

Fig. 1. Illustration of the global graph and local graph.

estimates the importance of a node in the network by calculating the number and quality of nodes linked to it. PageRank value (PR) can be calculated as follows,

$$PR(v_i) = \alpha \sum_{v_j \in In_{v_i}} \frac{PR(v_j)}{Out(v_j)} + \frac{(1-\alpha)}{N}, \tag{1}$$

where In_{v_i} and Out_{v_j} denote the items that point to v_i and the items that v_i points to. N is the number of all items in the global graph. α is the probability that the user randomly clicks on an item, generally set to 0.85. We set the initial PR value of each node to 1. After a continuous iterative process, we can obtain the PR value of each item.

Local Graph Model. We utilize the information of users who have similar behavior patterns as the cold-start user to help us better infer the user's current interests. But how to match similar sequences is a challenging task. Edit Distance (ED) evaluates the similarity by calculating the minimum number of times $d(S, S')$ required to turn sequence S into S'. Since it can effectively solve the sequence of displacement, disturbance, and deformation [20], we improve it to remeasure the similarity between sequences from three aspects **Number**, **Order**, and **Tightness** (NOT) by defining the following rules.

Rule 1. Number of the same items. If the sequence S has the same item as the query sequence S', we believe that the similarity is proportional to the number of the same items $Num(S, S')$ defined as follows:

$$Num(S, S') = \{v_i \mid v_i = v_j; v_i \in S; v_j \in S'\}. \tag{2}$$

Rule 2. Order of the items. If sequences S and S' have the same items, we believe that the more consistent their order is, the more similar they are. The order is defined as follows:

$$Order(S, S') = 1 - \frac{d(S, S')}{d_{max}(S, S')}, \tag{3}$$

where $d_{max}(S, S') = \max(d(S, S'))$.

Rule 3. Tightness of the relation. Under the condition that the same items are in the same order, it's reasonable to assume that the tightness of the relation positively affects the similarity. We directly use the distance between items to show the tightness,

$$\text{Tightness}(S, S') = \frac{1}{\sum_i^n \left(\text{Loc}_{v_{i+1}} - \text{Loc}_{v_i}\right)}, v_i \in \text{Num}(S, S'), \tag{4}$$

where Loc_{v_i} is the position of v_i in the sequence S. The rules rank 1, 2 and 3 in importance. For each item v_i in the cold-start sequence Seq, we first select the sequences containing v_i. Then, we filter the sequences most similar to Seq according to the importance of the rules. Finally, we use them to construct the local graph $G_l = \{V_l, \xi_l\}$. Next, we give two definitions and then elaborate on the proposed method.

Definition 1. *Inner Circle* $(I_c(v_i))$. *For each item* v_i, *the inner circle of* v_i *indicates the first-order neighbors defined as follows:*

$$I_c(v_i) = \{v_j \mid \exists \xi = (v_i, v_j); \xi \in G_l\}. \tag{5}$$

Definition 2. *Hot Spot* $(H_s(v_i))$. *For the nodes in* $I_c(v_i)$, *we divide them into the left and right parts of* v_i. *For each part, we define the node with the highest PR value as the hot spot:*

$$H_S(v_i) = \{v_h \mid \forall v_k, PR(v_k) < PR(v_h); v_k, v_h \in I_c(v_i); v_k \neq v_h\}. \tag{6}$$

3 The Proposed Method

We propose an effective sequence recommender named GitRec, which makes recommendations for cold-start users without using any auxiliary information. Figure 2 shows the architecture of GitRec, which comprises four components. Global graph layer utilizes all sequences to construct the global graph and uses the PageRank algorithm to extract the users' core interests. Local graph layer selects sequences most similar to the cold-start sequence to construct the local graph to mine users' potential interests. After employing graph neural networks, we can learn the item representations at two levels. Interpolation layer selects appropriate items to enrich the information of the cold-start sequence. Prediction layer scores each candidate item and outputs the probability that the user will click them at the next step.

3.1 Global Graph Layer

Since graph neural networks (GNN) can utilize the complex underlying relationships between nodes [28], we model the global graph through GNN to get the latent vector representation of each item. We employ the graph attention mechanism to assign attention weights. For each item v_i, its first-order neighbor's features are linearly combined based on the attention score,

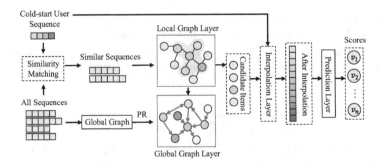

Fig. 2. Overview of the proposed GitRec model.

$$h^g_{I_c(v_i)} = \sum_{v_j \in I_c(v_i)} h_{v_j} \cdot w(v_i, v_j), \tag{7}$$

where $h_{I_c(v_i)}$ is the set of first-order neighbor of v_i, $w(v_i, v_j)$ is the attention weight. As some items may appear more than once in all sequences, we assign weights as follows,

$$w(v_i, v_j) = \frac{n(v_i, v_j)}{\sum \text{Out}(v_i)} \cdot \frac{n(v_i, v_j)}{\sum \text{In}(v_j)}, \tag{8}$$

where $n(v_i, v_j)$ is the occurrence of the edge $\xi = (v_i, v_j)$. $\sum \text{Out}(v_i)$ is the out-degree of the item v_i, and $\sum \text{In}(v_j)$ is the in-degree of the item v_j. We apply a softmax function to normalize the attention weight.

Then, item v_i can be represented by aggregating its representation h_{v_i} and its neighbors' representations $h^g_{I_c(v_i)}$,

$$h^g_{v_i} = \text{relu}\left(W_1\left(h_{v_i} \| h^g_{I_c(v_i)}\right)\right), \tag{9}$$

where $relu$ is the activation function, and W_1 is the trainable parameter.

3.2 Local Graph Layer

For the cold-start sequence Seq = $\{v_1, v_2, \ldots, v_n\}$, we build a local graph for each item v_m. We still choose GNN to get the latent vector representations of the items, and use the attention mechanism to learn the weights between nodes,

$$h^l_{I_c(v_m)} = \sum_{v_n \in I_c(v_m)} h_{v_n} \cdot w(v_m, v_n), \tag{10}$$

where $w(v_m, v_n)$ represents the attention coefficient, which can be calculated by Eq.(8). Then, we use the softmax function to normalize the weight for easy processing and comparison:

$$w(v_m, v_n) = \frac{\exp(w(v_m, v_n))}{\sum_{v_k \in I_c(v_m)} \exp(w(v_m, v_k))}. \tag{11}$$

Then, we can obtain the representation of each item by integrating its neighbor's features in a linear combination,

$$h^l_{v_i} = \text{relu}\left(W_2\left(h_{v_i}\|h^l_{I_c(v_i)}\right)\right),\tag{12}$$

where we choose $relu$ as the activation function and W_2 is the trainable parameter.

3.3   Interpolation Layer

Given the global and local graph, we select appropriate items to enrich the information of the cold-start user. From the perspective of statistics and probability, we assume that an item accepted by most users has a high probability of being accepted by new users. Therefore, for each item v_i in the cold-start sequence, the candidate item is defined as follows:

$$v_{can} = \{v_k \mid v_k \in I_c\left(H_s\left(v_i\right)\right)\}.\tag{13}$$

We score the candidate item by multiplying its embedding by v_i's latent vector:

$$\text{score}\ _{v_k} = h^l_{v_i} \cdot h_{v_k}.\tag{14}$$

By sorting the scores in descending order, we select the Top-N items for interpolation. To be noticed, for the order between the inserted item and v_i, we follow the order of the original sequence data. If more than one candidate item is on the same side of the v_i, we assume that the item with a higher score should be closer to v_i.

After the interpolation, for the new sequence S $= \{v_1, v_2, \ldots, v_m\}$, we represent it by linearly combining the representations of all items,

$$S = \sum_{i=1}^{m} w_i h'_{v_i},\tag{15}$$

where the corresponding weights w_i are learned through the soft attention mechanism.

3.4   Prediction Layer

We take the local-level representation of the local graph layer $h^l_{v_i}$ and the graph-level representation of the global graph layer $h^g_{v_i}$ as the user's final representation:

$$h'_{v_i} = h^g_{v_i} + h^l_{v_i}.\tag{16}$$

We define the output $\hat{y}_i$ as the dot product of the initial representation h_{v_i} with the new sequence representation S:

$$\hat{\mathbf{y}}_i = \text{softmax}\left(\mathbf{h}_{v_i} \cdot \mathbf{S}\right),\tag{17}$$

where $\hat{\mathbf{y}}_i$ represents the probability that the item v_i will be clicked by the user next step. We apply the cross-entropy of the prediction results $\hat{\mathbf{y}}$ as the loss function written as follows,

$$\mathcal{L}(\hat{\mathbf{y}}) = -\sum_{i=1}^{m} \mathbf{y}_i \log(\hat{\mathbf{y}}_i) + (1 - \mathbf{y}_i) \log(1 - \hat{\mathbf{y}}_i), \qquad (18)$$

where y is the one-hot encoding vector of the ground truth item.

4   Experiments

In this section, we conduct extensive experiments on three real-world datasets to evaluate the performance of our proposed method GitRec by answering the following four questions:

- RQ1: Compared with state-of-the-art sequential recommenders, how does GitRec perform?
- RQ2: Can the performance of GitRec be improved by learning item embeddings at the two levels of the local graph and global graph?
- RQ3: Is it effective to handle the cold-start user problem by enriching the information of sequences?
- RQ4: What is the effect of different hyper-parameter settings on GitRec's performance?

4.1   Datasets

We evaluate the recommender on four datasets, ml-$1m$[1], $Amazon$[2], and $Tmall$[3]. The basic statistics of the datasets are summarized in Table 1. For all datasets, we filter out the sequences of length 1, and the items with fewer than 5 interactions. Then, we sort items according to the timestamp that the user clicks. Due to the large amount of data in $Amazon$ dataset, we select the most frequent clicked 5,000 items. In the experiment, we divide each dataset into the training set, test set, and verification set in a ratio of 7:2:1.

4.2   Baselines

To demonstrate the performance of our model, we compare it with seven classic and state-of-the-art sequential recommenders.

- POP: It directly recommends the top-N frequent items in the training set.
- FPMC [21]: It is a sequential recommender combining the matrix factorization and the Markov chain.

---

[1] http://files.grouplens.org/datasets/movielens/.
[2] http://jmcauley.ucsd.edu/data/amazon/.
[3] https://tianchi.aliyun.com/dataset/dataDetail?dataId=42.

Table 1. Statistics of the datasets.

Dataset	ml-1m	Amazon	Tmall
#users	6,039	70,570	65,286
#items	3,702	4,999	40,727
#actions	999,868	581,546	416,554
#sparsity	4.47%	0.16%	0.02%
#cold-start user	0	24,662	34,745
#proportion of cold-start user	0%	34.95%	53.22%

- GRU4REC [7]: It applies Gated Recurrent Units to model the users' sequences based on RNN.
- NARM [16]: It is an RNN-based model with an attention mechanism, which can provide more accurate recommendations.
- STAMP [18]: It employs a short-term memory priority model to capture the users' general interests.
- SR-GNN [28]: It utilizes GNN to model sequences and predicts user behavior by combining short-term and long-term preferences.
- Meta-TL [24]: It exploits meta-learning to learn transition patterns to provide cold-start users with more accurate recommendations.
- GitRec: Our proposed model.

4.3   Experimental Setup

Evaluation Metrics. We use three ranking metrics MRR@N(Mean Reciprocal Rank), NDCG@N(Normalized Discounted Cumulative Gain), and HR@N(Hit Rate) to evaluate the model. In the experiment, we set N for 10 and 20.

Parameter Setting. Following previous work [16, 26, 28], the batch size is set to 100, and the dimension of the latent vectors is fixed to 100. For our model, the initial learning rate is 0.001 and decays 0.1 after every 3 epochs, and the dropout rate is in [0, 0.2, 0.4, 0.6, 0.8]. The L2 penalty is set to 10^{-5}. For the datasets *Amazon*, *Tmall* and *ml-1m*, we set the maximum sequence length to 10, 10 and 40 respectively. Moreover, we set the maximum number of rounds of the PageRank algorithm to 10.

4.4   Overall Performance (RQ1)

Table 2 shows the results of the three datasets. The best model in each column is bolded, and the second best is underlined. We note that:

- **Our proposed model GitRec achieves the best performance in most cases.** It performs better on datasets with more cold-start users. It proves that GitRec is effective in dealing with cold-start users and has no negative effect on warm-start users (according to the results on the *ml-1m* dataset).

Table 2. Performance comparisons of all models on the three datasets.

Datasets	Metric	N	POP	FPMC	GRU4REC	NARM	STAMP	SR-GNN	Meta-TL	GitRec	Improvement
ml-1m	MRR	10	0.0104	0.0498	0.0441	0.0721	0.0856	<u>0.0892</u>	0.0854	**0.0911**	2.13%
		20	0.0120	0.0544	0.0487	0.0794	0.0924	<u>0.0956</u>	0.0906	**0.0991**	3.66%
	NDCG	10	0.0114	0.0700	0.0618	0.1010	0.1141	<u>0.1185</u>	0.1103	**0.1201**	1.35%
		20	0.0159	0.0867	0.0789	0.1278	0.1390	<u>0.1421</u>	0.1294	**0.1490**	4.86%
	HR	10	0.0307	0.1366	0.0789	0.1962	0.2081	<u>0.2151</u>	0.1922	**0.2201**	2.32%
		20	0.0551	0.2025	0.1884	0.2525	0.3070	<u>0.3087</u>	0.2680	**0.3180**	3.01%
Amazon	MRR	10	0.0095	0.0594	0.0585	0.0664	0.0704	0.0716	<u>0.0772</u>	**0.0803**	4.02%
		20	0.0111	0.0629	0.0630	0.0711	0.0747	0.0758	<u>0.0822</u>	**0.0857**	4.26%
	NDCG	10	0.0109	0.0766	0.0779	0.0883	0.0899	0.0909	<u>0.1011</u>	**0.1068**	5.64%
		20	0.0151	0.0897	0.0945	0.1058	0.1060	0.1064	<u>0.1194</u>	**0.1301**	8.96%
	HR	10	0.0291	0.1332	0.1419	0.1605	0.1542	0.1544	<u>0.1796</u>	**0.1974**	9.91%
		20	0.0521	0.1852	0.2079	0.2298	0.2181	0.2159	<u>0.2526</u>	**0.2699**	6.85%
Tmall	MRR	10	0.0042	0.0370	0.0413	0.0498	0.0538	0.0710	<u>0.0743</u>	**0.0808**	8.75%
		20	0.0045	0.0404	0.0433	0.0551	0.0581	0.0759	<u>0.0804</u>	**0.0867**	7.84%
	NDCG	10	0.0050	0.0495	0.0523	0.0719	0.0642	0.0938	<u>0.1011</u>	**0.1093**	8.11%
		20	0.0061	0.0622	0.0597	0.0912	0.0734	0.1118	<u>0.1233</u>	**0.1310**	6.24%
	HR	10	0.0109	0.0911	0.0885	0.1452	0.0892	0.1688	<u>0.1893</u>	**0.2040**	7.77%
		20	0.0157	0.1418	0.1176	0.2222	0.1379	0.2402	<u>0.2775</u>	**0.2898**	4.43%

- **Traditional methods cannot handle the cold-start problem.** POP only considers the frequency of items and recommends popular items for users. FPMC performs better than POP verifies that the Markov chain is meaningful. However, it is hard to model the high-order dependence, which leads to its poor performance.
- **Neural network-based methods outperform the traditional methods.** RNN-based models can mine chronological information in data, resulting in better results than conventional methods. Compared with them, SR-GNN employs GNN to further improve performance, which indicates that it is more suitable to model sequences as graphs.

Besides, we find Meta-TL shows a certain superiority in handling cold-start users by using meta-learning. But they don't utilize the global information, or only focus on the last clicked item. There's still a gap between these methods and our model.

4.5   Ablation Study

Impact of Graph Models (RQ2). We propose the global graph to obtain the core interests of warm-start users and propose the local graph to capture the potential preferences of cold-start users. Now we study whether the two graph models are necessary. The results in Table 3 show that the performance improves under each condition, which demonstrates the effectiveness of our proposed graph models. It proves that learning item embedding at the two levels of the local graph and global graph is significant.

Impact of Interpolation Module(RQ3). We enrich the information of the cold-start users by interpolating the sequences. Table 4 shows the performance comparison of models using different interpolation methods. We notice that the result of the model with the similarity sequence is slightly poor. That is because although it can filter irrelevant information, the similarity between users is likely to be 0 due to data sparsity. The performance of the model with the PageRank algorithm is improved, which shows the effectiveness of mining the core interests of users. The model with the interpolation module works best. That means that enriching the information of the cold-start user is meaningful.

Table 3. Performance comparison of the models using different embedding methods.

Dataset	Model	MRR	NDCG	HR
Amazon	w/o Global Graph	0.0685	0.0921	0.1699
	w/o Local Graph	0.0723(+5.55%)	0.0963(+4.56%)	0.175(+3.00%)
	w/ Graph Models	0.0793(+15.77%)	0.1058(+14.88%)	0.1875(+10.36%)
Tmall	w/o Global Graph	0.0673	0.0883	0.1677
	w/o Local Graph	0.0776(+13.82%)	0.1052(+10.31%)	0.187(+5.55%)
	w/ Graph Models	0.0805(+19.61%)	0.1097(+24.24%)	0.1959(+16.82%)

Table 4. Performance comparison of the models using different interpolation methods.

Dataset	Model	MRR	NDCG	HR
Amazon	w/o Interpolation	0.071	0.0966	0.1761
	w/ Similarity sequence	0.0724(+1.97%)	0.0951(−1.55%)	0.1759(−0.11%)
	w/ PageRank	0.0749(+5.49%)	0.1009(+4.45%)	0.1872(+6.30%)
	w/ Interpolation	0.0794(+11.83%)	0.1058(+9.52%)	0.1925(+9.31%)
Tmall	w/o Interpolation	0.079	0.1044	0.197
	w/ Similarity sequence	0.0752(−4.81%)	0.1052(+0.77%)	0.1974(+0.20%)
	w/ PageRank	0.0799(+1.14%)	0.1082(+3.64%)	0.2003(+1.68%)
	w/ Interpolation	0.0810(+2.53%)	0.1101(+5.46%)	0.2059(+4.52%)

4.6 Hyper-Parameter Study (RQ4)

Effectiveness of Embedding Size. Embedding size is related to the specific dataset, and it generally needs to be evaluated by specific tasks. Figure 3(a) shows the results. We can observe that the low dimension doesn't perform well due to its poor representation while the high dimension may lead to overfitting problems. The performance is best when the embedding size is 100.

Effectiveness of the Number of Similar Sequences. We select sequences similar to cold-start users to mine their potential preferences. To see if it is necessary, we conduct experiments to explore the effect of the number of similar sequences. As shown in Fig. 3(b), the number is proportional to the model's performance. It demonstrates that users' interests can be well discovered by selecting similar sequences.

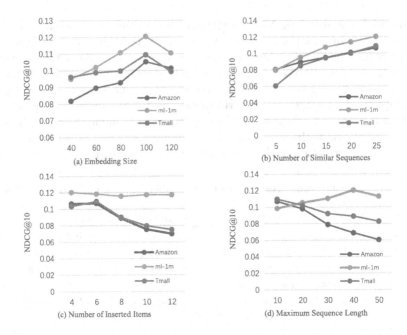

Fig. 3. Performance comparison of the proposed GitRec model using different parameter settings.

Effectiveness of the Number of Inserted Items. Since we add items to the item's left and right sides simultaneously to enrich the information of the cold-start users, the number of inserted items grows by two. Figure 3(c) shows that the number has little impact on the performance of *ml-1m* datasets. The results of *Amazon* and *Tmall* datasets demonstrate that performance improves with increasing items at first, but too many items may introduce noise and affect the model.

Effectiveness of the Maximum Sequence Length. We limit the length of the sequence to better process sequence data. The results are shown in Fig. 3(d). Each dataset has different data characteristics, so we have to treat it differently. In the comparison experiment, we choose the appropriate length for each dataset to obtain the best performance.

5   Related Work

5.1   Sequential Recommendation

Traditional sequential recommendation methods are mostly based on Markov Chains (MC). MC-based methods regard recommendation as a Markov decision problem [5]. FPMC combines MF and MC to model sequence data but ignores the users' latent representations [21]. Lately, the deep neural network is introduced to the sequential recommendation. Due to its intrinsic dominant position in modeling sequential dependency, many RNN-based methods are proposed. GRU4REC [7] first applied RNN for the sequential recommendation, which is built on Gate Recurrent Units. KPRN [25]models longer and more complex dependencies by using Long Short-Term Memory. CNN has been applied in sequential recommendation because of its success in other fields [10]. [23] proposes a 3D CNN model to model the sequential patterns in sequences. In recent years, the GNN-based methods learn item representation by employing GNN on sequence graphs. SR-GNN [28] first introduces GNN to the sequential recommendation and achieves good performance. GCE-GNN [26] uses item transitions overall sessions to infer the users' interests. We employ GNN to model sequences for better capturing the core interests and potential preferences of users for recommendations.

5.2   Cold-Start Recommendation

The main challenge of the cold-start problem is the lack of information required for recommendations. Based on the way of collecting information, solutions can be classified as explicit and implicit. Explicit methods collect information by interacting with the user directly, such as asking users to fill out questionnaires or rate several given items. CPR [12] regards conversational recommendation as an interactive path reasoning problem on a graph by following user feedback. [2] gains the users' preferences by asking several questions prepared in advance. But users are usually unwilling to get involved because of the time and effort required. Implicit methods attempt to understand users' real intentions using auxiliary information like social media. [4] utilizes social network information to recommend items across multiple domains. [6] creates a behavioral profile to classify the user through social media data. However, with the increasing awareness of privacy protection, valid personal information becomes more difficult to get. MetaTL models users' transition patterns through meta-learning [24]. But the model is designed to accommodate cold-start users and has certain limitations. In this paper, we propose a sequential recommender, which can address the user cold-start problem without using any auxiliary information.

6   Conclusion

In this work, we study the user cold-start problem in sequential recommendation systems. We propose a recommender that can provide appropriate recommendations for cold-start users without using any external information. The model uses the remarkable ability of graph neural networks to extract users' core interests and potential preferences from historical records. By utilizing all sequences' information, our model enriches the information of the cold-start user for better prediction. Extensive experiments on three public datasets confirm that the proposed method can significantly outperform the state-of-art methods.

Acknowledgements. This work is supported by the Natural Science Foundation of Jiangsu Province (No. BK20210280), the Fundamental Research Funds for the Central Universities (NO. NS2022089), the Jiangsu Provincial Innovation and Entrepreneurship Doctor Program under Grants No. JSSCBS20210185.

References

1. Chen, J., et al.: Co-purchaser recommendation for online group buying. Data Sci. Eng. **5**(3), 280–292 (2020)
2. Christakopoulou, K., Radlinski, F., Hofmann, K.: Towards conversational recommender systems. In: Proceedings of the 22nd ACM SIGKDD International Conference on Knowledge Discovery and Data Mining, pp. 815–824 (2016)
3. Cui, Z., et al.: Reinforced KGS reasoning for explainable sequential recommendation. World Wide Web **25**(2), 631–654 (2022)
4. Fernández-Tobías, I., Braunhofer, M., Elahi, M., Ricci, F., Cantador, I.: Alleviating the new user problem in collaborative filtering by exploiting personality information. User Model. User-Adapt. Interact. **26**(2), 221–255 (2016)
5. He, R., McAuley, J.: Fusing similarity models with Markov chains for sparse sequential recommendation. In: 2016 IEEE 16th International Conference on Data Mining (ICDM), pp. 191–200. IEEE (2016)
6. Herce-Zelaya, J., Porcel, C., Bernabé-Moreno, J.: New technique to alleviate the cold start problem in recommender systems using information from social media and random decision forests. Inf. Sci. **536**, 156–170 (2020)
7. Hidasi, B., Karatzoglou, A., Baltrunas, L., Tikk, D.: Session-based recommendations with recurrent neural networks. Computer Science (2015)
8. Huang, Y., Zhao, F., Gui, X., Jin, H.: Path-enhanced explainable recommendation with knowledge graphs. World Wide Web **24**(5), 1769–1789 (2021). https://doi.org/10.1007/s11280-021-00912-4
9. Kang, W.C., McAuley, J.: Self-attentive sequential recommendation. In: 2018 IEEE International Conference on Data Mining (ICDM), pp. 197–206. IEEE (2018)
10. Karpathy, A., Toderici, G., Shetty, S., Leung, T., Sukthankar, R.: Large-scale video classification with convolutional neural networks. In: Proceedings of the IEEE conference on Computer Vision and Pattern Recognition, pp. 1725–1732 (2014)
11. Lee, H., Im, J., Jang, S., Cho, H., Chung, S.: MELU: meta-learned user preference estimator for cold-start recommendation. In: Proceedings of the 25th ACM SIGKDD International Conference on Knowledge Discovery and Data Mining, pp. 1073–1082 (2019)
12. Lei, W., et al.: Interactive path reasoning on graph for conversational recommendation. In: Proceedings of the 26th ACM SIGKDD International Conference on Knowledge Discovery and Data Mining, pp. 2073–2083 (2020)
13. Li, B.H., Liu, Y., Zhang, A.M., Wang, W.H., Wan, S.: A survey on blocking technology of entity resolution. J. Comput. Sci. Technol. **35**(4), 769–793 (2020)
14. Li, C., Wang, F., Yang, Y., Li, Z., Zhang, X.: Exploring social network information for solving cold start in product recommendation. In: Wang, J., et al. (eds.) WISE 2015. LNCS, vol. 9419, pp. 276–283. Springer, Cham (2015). https://doi.org/10.1007/978-3-319-26187-4_24
15. Li, J., Wang, Y., McAuley, J.: Time interval aware self-attention for sequential recommendation. In: Proceedings of the 13th International Conference on Web Search and Data Mining, pp. 322–330 (2020)

16. Li, J., Ren, P., Chen, Z., Ren, Z., Lian, T., Ma, J.: Neural attentive session-based recommendation. In: Proceedings of the 2017 ACM on Conference on Information and Knowledge Management, pp. 1419–1428 (2017)
17. Li, J., Jing, M., Lu, K., Zhu, L., Yang, Y., Huang, Z.: From zero-shot learning to cold-start recommendation. In: Proceedings of the AAAI Conference on Artificial Intelligence, vol. 33, pp. 4189–4196 (2019)
18. Liu, Q., Zeng, Y., Mokhosi, R., Zhang, H.: Stamp: short-term attention/memory priority model for session-based recommendation. In: Proceedings of the 24th ACM SIGKDD International Conference on Knowledge Discovery and Data Mining, pp. 1831–1839 (2018)
19. Liu, Y., Li, B., Zang, Y., Li, A., Yin, H.: A knowledge-aware recommender with attention-enhanced dynamic convolutional network. In: Proceedings of the 30th ACM International Conference on Information and Knowledge Management, pp. 1079–1088 (2021)
20. Qian, X., Li, M.: E-commerce user type recognition based on access sequence similarity. J. Organ. Comput. Electron. Commer. **30**(3), 209–223 (2020)
21. Rendle, S., Freudenthaler, C., Schmidt-Thieme, L.: Factorizing personalized Markov chains for next-basket recommendation. In: Proceedings of the 19th International Conference on World Wide Web, pp. 811–820 (2010)
22. Tang, J., Wang, K.: Personalized top-n sequential recommendation via convolutional sequence embedding. In: Proceedings of the eleventh ACM International Conference on Web Search and Data Mining, pp. 565–573 (2018)
23. Tuan, T.X., Phuong, T.M.: 3d convolutional networks for session-based recommendation with content features. In: Proceedings of the Eleventh ACM Conference on Recommender Systems, pp. 138–146 (2017)
24. Wang, J., Ding, K., Caverlee, J.: Sequential recommendation for cold-start users with meta transitional learning. In: Proceedings of the 44th International ACM SIGIR Conference on Research and Development in Information Retrieval, pp. 1783–1787 (2021)
25. Wang, X., Wang, D., Xu, C., He, X., Cao, Y., Chua, T.S.: Explainable reasoning over knowledge graphs for recommendation. In: Proceedings of the AAAI Conference on Artificial Intelligence, vol. 33, pp. 5329–5336 (2019)
26. Wang, Z., Wei, W., Cong, G., Li, X.L., Mao, X.L., Qiu, M.: Global context enhanced graph neural networks for session-based recommendation. In: Proceedings of the 43rd International ACM SIGIR Conference on Research and Development in Information Retrieval, pp. 169–178 (2020)
27. Wu, S., Zhang, Y., Gao, C., Bian, K., Cui, B.: Garg: anonymous recommendation of point-of-interest in mobile networks by graph convolution network. Data Sci. Eng. **5**(2), 433–447 (2020)
28. Wu, S., Tang, Y., Zhu, Y., Wang, L., Xie, X., Tan, T.: Session-based recommendation with graph neural networks. In: Proceedings of the AAAI Conference on Artificial Intelligence, vol. 33, pp. 346–353 (2019)
29. Zang, Y., et al.: GISDCN: a graph-based interpolation sequential recommender with deformable convolutional network. In: Zang, Y., et al. (eds.) Database Systems for Advanced Applications. DASFAA 2022. LNCS, vol. 13246, pp. 289–297. Springer, Cham (2022). https://doi.org/10.1007/978-3-031-00126-0_21

Self-guided Contrastive Learning
for Sequential Recommendation

Hui Shi[1], Hanwen Du[1], Yongjing Hao[1], Victor S. Sheng[2], Zhiming Cui[3], and Pengpeng Zhao[1(✉)]

[1] School of Computer Science and Technology, Soochow University, Suzhou, China
{hshi1,hwdu,yjhaozb}@stu.suda.edu.cn, ppzhao@suda.edu.cn
[2] Department of Computer Science, Texas Tech University, Lubbock, USA
Victor.Sheng@ttu.edu
[3] School of Electronic and Information Engineering, Suzhou University of Science and Technology, Suzhou, China
zmcui@mail.usts.edu.cn

Abstract. Sequential recommendation has injected plenty of vitality into online marketing and retail industry. Existing contrastive learning-based models usually resolve data sparsity issue of sequential recommendation with data augmentations. However, the semantic structure of sequences is typically corrupted by data augmentations, resulting in low-quality views. To tackle this issue, we propose **Self**-guided contrastive learning enhanced **BERT** for sequential recommendation (**Self-BERT**). We devise a self-guided mechanism to conduct contrastive learning under the guidance of BERT encoder itself. We utilize two identically initialized BERT encoders as view generators to pass bi-directional messages. One of the BERT encoders is parameter-fixed, and we use the all Transformer layers' output as a series of views. We employ these views to guide the training of the other trainable BERT encoder. Moreover, we modify the contrastive learning objective function to accommodate one-to-many positive views constraints. Experiments on four real-world datasets demonstrate the effectiveness and robustness of **Self-BERT**.

Keywords: Sequential recommendation · Contrastive learning · BERT

1 Introduction

Recommender systems, owing to their ability to give suggestions for users effectively, can alleviate the information overload issue and help users derive valuable insights from big data. Sequential Recommendation aims to model user behaviors dynamically by taking the sequential patterns of user interaction history into consideration [9,11,16–18]. Given recent observations of users, sequential recommendation models are built to capture sequential item relationships.

Due to the high practical value of sequential recommendation, various works are proposed to resolve it. Early works [16] on sequential recommendation are based on the Markov Chain (MC) assumption to model users' pair-wise behavior

B. Li et al. (Eds.): APWeb-WAIM 2022, LNCS 13423, pp. 72–86, 2023.
https://doi.org/10.1007/978-3-031-25201-3_6

transition. Due to the limitation of MC-based models, Recurrent Neural Network (RNN) [3] is adopted in sequential recommendation to model sequence-wise relationships. Recently, Transformer [19] adopts the self-attention mechanism to encode sequences, which has proved significantly powerful in various fields [2,20, 21]. Thus, many researchers design Transformer-based models in sequential recommendation [11,17]. For example, BERT4Rec [17] employs Bi-directional Encoder Representations from Transformers (BERT) [4], stacked by multiple Transformer encoder layers, to reveal the correlations of sequential items.

Although existing approaches have achieved great recommendation results, the issue of data sparsity is still not well explored [13,22], leading to suboptimal performance. In detail, the data sparsity problem includes inadequate data amount and short data length, causing insufficient model training. More recently, a contrastive learning paradigm is introduced to alleviate the above issue [13,22]. In detail, this paradigm usually constructs both positive and negative views of an original sequence through data augmentations. The goal is to push positive views close to the original sample by optimizing the value of contrastive learning loss, while negative views are the opposite. Such a paradigm can enhance the discrimination ability of the encoders and improve the robustness of the model.

However, we consider that there are two points in existing contrastive learning models that could be improved. First, most contrastive learning sequential recommendation models are modified based on unidirectional Transformers. In fact, bi-directional Transformers perform better than unidirectional Transformers, which means we can consider integrate bi-directional Transformers (such as BERT) with contrastive learning to obtain better recommendation performance. Second, the data augmentations adopted in the existing contrastive learning-based models have two shortcomings, i.e., (1) finding and realizing the optimal data augmentation method for different datasets is very time-consuming, (2) the data augmentation process of generating views has a certain degree of randomness, leading to the destruction of important original semantic information. Therefore, instead of data augmentation-based solutions, a more efficient and stable contrastive learning scheme is highly demanded.

To this end, we propose a new framework, **Self**-guided contrastive learning enhanced **BERT** for sequential recommendation (**Self-BERT**) to cope with the above issues. It consists of three essential parts: (1) a traditional BERT-based sequential recommendation task; (2) a self-guided contrastive learning paradigm to take advantage of all the information captured in BERT encoder; (3) a joint-learning training framework to optimize two loss functions simultaneously in contrastive learning and recommendation tasks. Specifically, Self-BERT uses two BERT encoders. One BERT encoder is fixed after initialization and its outputs of all hidden layers are regarded as positive views. The other is used to obtain general sequence representations. In this way, multiple pair of positive views are automatically generated. It is equivalent to using the information inside the BERT encoder to guide its training, and we call it a self-guided mechanism. In addition, we modify NT-Xent loss [1] by extending its one-to-one view constraints into one-to-many view constraints, which allows our model to train multiple pairs of positive and negative views simultaneously. We conduct experiments on four real-world datasets to verify the effectiveness and robustness of our **Self-BERT**. To distinguish our work from other sequential recommendation solutions, main contributions of this paper are listed as follows:

- To the best of our knowledge, this is the first work to apply self-guided contrastive learning-based BERT to sequential recommendation.
- We propose a novel data augmentation-free contrastive learning paradigm to tackle the unstable and time-consuming challenges in contrastive learning. It exploits self-guided BERT encoders and extends one-to-many view constraints to preserve the view-wise semantic correlations.
- We conduct extensive experiments on four real-world benchmark datasets. Our experimental results improve competitive baselines with a large margin under the challenges of data sparsity.

2   Related Work

2.1   Sequential Recommendation

Sequential recommendation predicts future items that users may be interested in by capturing item correlations in history interaction sequences. Early proposed models, such as FPMC [16], adopt the MC assumption to capture pairwise item correlations. Besides, FPMC incorporates Matrix Factorization (MF) to simulate users' personal preferences. With the advancement of neural networks in many other research domains [27,28], RNN [3] and its variants are widely used in sequential recommendation. For example, Hidasi et al. [9] propose GRU4Rec, which exploits Gated Recurrent Unit (GRU) [5], and it aims to dynamically model long-term users' preferences from their historical interaction sequences. In addition, other deep neural networks, like Convolutional Neural Network (CNN), also made remarkable achievements in sequential recommendation. Tang et al. design a CNN-based model named Caser [18], which views history interaction sequences as "images" and captures local patterns via convolutional filters. Yuan et al. [25] devise a generative model (NextItNet) by stacking holed convolutional layers to increase the receptive fields of convolutional filters. Recently, inspired by the application of self-attention network [19] in Natural Language Processing (NLP), various attention-based models are proposed in sequential recommendation [23,26]. Kang et al. [11] develop SASRec to characterize advanced item transition correlations by adapting Transformer layer [19]. Sun et al. [17] propose BERT4Rec by applying BERT [4] to obtain better sequence representations. BERT4Rec essentially consists of a stack of Transformers that can fuse contextual information from two directions. Nowadays, multiple contrastive learning-based models are proposed, trying to obtain characteristic signals within the sequence. Both S^3-Rec [13] and CL4SRec [22] are very competitive models. However, the encoders in these models only consider unidirectional information transfer without the ability to fuse contextual information in the sequence.

2.2 Self-supervised Learning

The goal of Self-Supervised Learning (SSL) is to utilize the unlabeled data to obtain valuable signals for downstream tasks. Generally speaking, SSL models can mainly be divided into generative and contrastive approaches [10]. Representative models of generative SSL-based models includes Generative Adversarial Networks (GAN) [7] and Variational Auto Encoders (VAE) [12]. These models are designed to make the generated data close to the original data. Different from generative models, contrastive SSL-based models are built to maximize the consistency between pairs of positive views and the opposite for pairs of negative views. Positive views are obtained by performing a series of augmentation operations on original sequences. Recently, Contrastive SSL-based methods have achieved remarkable success in various research domains [6,8,29]. For different data types, views are generated by applying reasonable augmentation methods. Chen et al. [1] propose SimCLR and employ random crop, color distortions, and Gaussian blur to obtain augmented visual representations. GraphCL [24] devises four data augmentations to generate views, including node dropping, edge perturbation, attribute masking, and subgraph selection. Existing contrastive SSL-based sequential recommendation models fully consider the sequence characteristics of user interaction history and design effective view generators. Zhou et al. [13] propose S^3-Rec to learn the item transition correlations among attribute, item, subsequence and sequence by optimizing four auxiliary self-supervised objectives. CL4SRec [22] devises three data augmentation operations (crop, mask, reorder) to randomly generate views of original sequences. However, we hold the opinion that the contrastive learning views generated by data augmentations cannot preserve the complete semantic information of the original sequences.

3 Method: Self-BERT

In this section, we first provide the problem definition of sequential recommendation. Then, we introduce the two tasks of Self-BERT, the main recommendation task and the auxiliary self-guided contrastive learning task. Finally, we train the above two tasks together through a joint learning framework.

3.1 Problem Definition

In this paper, we symbolize user and item sets as $\mathcal{U}$ and $\mathcal{V}$, respectively. $|\mathcal{U}|$ and $|\mathcal{V}|$ represent the number of users and items. Each user has a chronological interaction sequence consisting of a series of items. For user u, we denote the corresponding sequence as $S_u = [v_1, v_2, ..., v_{|S_u|}]$, where v_t represents user u interacted with this item at time t. Given the interaction sequence S_u of user u, the target of sequential recommendation is to predict the most possible item that user u will interact with at time $|S_u|+1$, which can be formulated as follows:

$$\arg\max_{v_i \in \mathcal{V}} P(v_{|S_u|+1} = v_i | S_u). \tag{1}$$

3.2   Recommendation Task

In this subsection, we introduce the recommendation task implemented by a BERT encoder. As shown in Fig. 1, the BERT encoder consists of stacked Transformer encoder layers. Unlike the left-to-right unidirectional Transformer, the BERT encoder can fuse bi-directional contextual information, which gives the model a global receptive field to capture the dependencies in any distance.

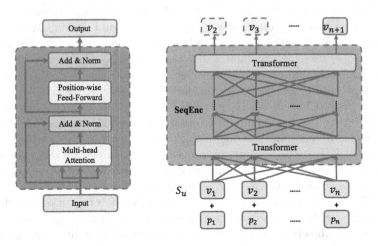

Fig. 1. The architecture of Transformer encoder (left) and BERT (right).

Embedding Layer. To take full advantage of the sequential information of the input, for a given item v_i of sequence S_i, its embedding representation h_i^0 considers both original item embedding v_i and corresponding positional embedding p_i. The formula is as follows:

$$h_i^0 = v_i + p_i, \tag{2}$$

where h_i^0, v_i, p_i are all d-dimensional embedding vectors (d represents the dimension of embedding). By this way, a positional embedding representation $H^0 \in \mathbb{R}^{N \times d}$ of sequence S_i is obtained, where N denotes the maximum length of all sequences[1].

Transformer Layer. Given a sequence of length N, after passing through the embedding layer, we iteratively pass it through L layers of Transformers to get the hidden layer representation h_i^l of each layer l at position i. We stack the hidden layer representations of all positions together to get $H^l \in \mathbb{R}^{N \times d}$. Next, we pass H^l through a multi-head self-attention network to capture the dependencies

---

[1] We pad the sequence length with zeros on the left when the length is less than N.

between representation pairs at arbitrary distances in the sequence. We linearly projects H^l into h subspaces, and the formula is as follows:

$$\mathrm{MH}(H^l) = \mathrm{concat}(\mathrm{head}_1; \mathrm{head}_2; \cdots ; \mathrm{head}_h)W^O,$$
$$\mathrm{head}_i = \mathrm{Attention}(H^l W_i^Q, H^l W_i^K, H^l W_i^V), \tag{3}$$

where W^O, W_i^Q, W_i^K, and W_i^V are trainable parameters. Q, K, and V represent queries, keys, and values, respectively. The Attention function is implemented by the Scaled Dot-Product Attention:

$$\mathrm{Attention}(Q, K, V) = \mathrm{softmax}(QK^\top/\sqrt{d/h})V, \tag{4}$$

After linear projection through multi-head self-attention, we utilize a position-wise feed-forward network (PFFN) to endow the interactions between different dimensions at each position h_i:

$$\mathrm{PFFN}(H^l) = [\mathrm{FFN}(h_1^l)^\top; ...; \mathrm{FFN}(h_N^l)^\top]^\top, \tag{5}$$

where $\mathrm{FFN}(\cdot)$ represents a two-layer feedforward network with GELU as an activation function. We also use residual connections, dropout, and normalization to connect the two sub-layers of multi-head self-attention and position-wise feed-forward network. The process of stacking Transformer layers is as follows:

$$H^l = \mathrm{Trm}(H^{l-1}) = \mathrm{LayerNorm}(F^{l-1} + \mathrm{Dropout}(\mathrm{PFFN}(F^{l-1}))),$$
$$F^{l-1} = \mathrm{LayerNorm}(H^{l-1} + \mathrm{Dropout}(\mathrm{MH}(H^{l-1}))). \tag{6}$$

Finally, we pass H^L through a linear prediction layer and a softmax operation to get an output distribution $P(v)$ over all candidate items.

Training with the Cloze Task. In order to realize the bi-directional delivery of information in the Transformer encoder, the Cloze task is selected to train the model. The input sequence is randomly masked according to the ratio ρ during the training process, and the masked items are predicted based on the two-way context information. The objective function of the recommendation task is:

$$\mathcal{L}_{Rec} = \frac{1}{|S_{u,m}|} \sum_{v_m \in S_{u,m}} -\log P(v_m = v_m^* | S_u'), \tag{7}$$

where S_u' represents the masked user behavior sequence S_u. $S_{u,m}$ is a collection of random masked items. v_m and v_m^* represent masked item and true item, respectively. In the testing process, we append a "[mask]" token at the end of the input sequence to predict the next item.

3.3   Self-guided Contrastive Learning Task

Unlike existing models that construct views through data augmentation, we propose Self-BERT and use the hidden layer representations obtained by the

BERT encoder itself as positive views. The main architecture of Self-BERT is shown in Fig. 2. Essentially, these representations are all yielded from the same input sequence and thus can be treated as positive views. We call it self-guided contrastive learning since the encoder uses its internal signals to guide the contrastive learning.

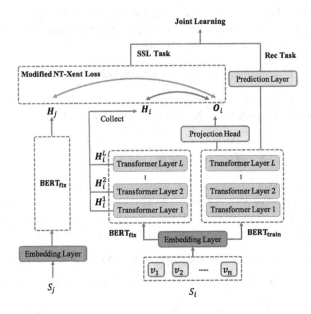

Fig. 2. The overall framework of Self-BERT. The positive views H_i (a series of hidden layer outputs) and O_i are obtained through the BERT_{fix} and $\text{BERT}_{\text{train}}$ encoders, respectively. The negative view H_j is obtained through the BERT_{fix}. And we calculate the contrastive learning loss through this ternary relationship.

Specifically, we first pre-train the BERT encoder for z epochs[2] on the Cloze task. Then, we use two BERT encoders to obtain positive views in the contrastive learning task. One of the encoders, called $\text{BERT}_{\text{train}}$, is initialized according to the parameters in the trained BERT, and the parameters are updated synchronously in the subsequent training process. The other encoder, called BERT_{fix}, is also a copy of the trained BERT, but its parameters are fixed after initialization. Self-BERT uses all hidden layer outputs obtained by BERT_{fix} as positive views for contrastive learning. Therefore, the information captured by each Transformer layer in the BERT encoder can be fully utilized.

Given an interaction sequence S_i of user i, we first pass it through the same embedding layer as in the recommendation task to get the embedded input S_i^E of the encoder. A contrastive learning view O_i is obtained directly by passing

---

[2] We found in experiments that the performance of Self-BERT does not change much with different z, and we choose $z = 50$.

S_i^E through $\text{BERT}_{\text{train}}$ and a projection head. At the same time, we feed S_i^E into BERT_{fix} and obtain the output of each layer of Transformers:

$$H_i = \text{BERT}_{\text{fix}}(S_i^E) = [H_i^1, ..., H_i^k, ..., H_i^L] \quad (1 \leq k \leq L). \tag{8}$$

Since each Transformer layer in the BERT encoder can capture different level information, we regard O_i and each element in H_i as a pair of positive views. Thus, there are L pairs of positive views in total.

As for negative views, we directly choose other sequences in the batch where S_i is located. For example, Fig. 2 shows that we feed the negative sample S_j into BERT_{fix} and acquire a series of negative views H_j.

After computing these vectors, we compute the NT-Xent loss [1] for each positive sample, which is commonly used in contrastive learning.

$$\phi(u, v) = \exp(\text{sim}(f(u), f(v))/\tau), \tag{9}$$

$$\mathcal{L}_{Cl} = -\log \frac{\phi(O_i, H_i^L)}{\phi(O_i, H_i^L) + \Sigma_{j=1, j \neq i}^b \phi(O_i, H_j^L)}, \tag{10}$$

where b denotes the current batch size, τ denotes the temperature parameter, $\text{sim}(\cdot)$ and $f(\cdot)$ symbolize the cosine similarity function and a projection head, H_i^L and H_j^L represent the last layer output of BERT_{fix}.

However, as shown in Eq. (10), the commonly used NT-Xent loss only computes a pair of positive and negative views. Therefore, we modify Eq. (10) so that it could take into account the output of intermediate layers and calculate the loss of L pairs of positive and negative views simultaneously:

$$\mathcal{L}_{Cl} = \frac{1}{L} \sum_{k=1}^L -\log \frac{\phi(O_i, H_i^k)}{\phi(O_i, H_i^k) + \Sigma_{j=1, j \neq i}^b \Sigma_{l=1}^L \phi(O_i, H_j^l)}. \tag{11}$$

3.4 Joint Learning

Finally, we train the recommendation task and the self-guided contrastive learning task simultaneously through a joint learning framework to improve the performance of sequential recommendation. The overall loss function is as follows:

$$\mathcal{L} = \mathcal{L}_{Rec} + \lambda \mathcal{L}_{Cl}, \tag{12}$$

where λ is a hyperparameter that controls the proportion of the auxiliary contrastive learning task in the two tasks.

4 Experiments

In this section, we conduct extensive experiments to answer the following Research Questions (**RQs**):

- **RQ1:** How does Self-BERT perform compared to the existing methods?
- **RQ2:** How is the sensitivity of the hyper-parameters in Self-BERT?
- **RQ3:** How does the self-guided mechanism contribute to the whole model?
- **RQ4:** How does Self-BERT perform under the issue of data sparsity?

4.1   Experimental Settings

Datasets. We conduct experiments on four real-world public datasets to answer the four **RQs**. The Movielens dataset includes movies, users, and corresponding rating data, commonly used in evaluating recommendation systems. We employ Movielens 1 m (**ML-1M**)[3] and Movielens 20 m (**ML-20M**)[4] to carry out our experiments. McAuley et al. [14] developed a series of datasets based on the Amazon web store, which is split according to the products categories. We select two of the datasets[5], i.e., Beauty and Toys, for experiments.

As for data preprocessing, we follow the standard practice in [13,17,22]. First, we regard the cases with ratings or reviews as positive samples and the rest as negative samples. Then for each user, the corresponding items are sorted chronologically. We follow the "5-core" principle, which means we discard users with less than 5 interactions and items related with less than 5 users. The statistics of the four datasets after data preprocessing are shown in Table 1.

Table 1. Statistics of four datasets after preprocessing.

Dataset	Users	Items	Actions	Avg. length	Sparsity
ML-1M	6,040	3,416	999,611	165.5	95.21%
ML-20M	138,493	18,345	19,984,024	144.3	99.21%
Beauty	22,363	12,101	198,502	8.9	99.93%
Toys	19,412	11,924	167,597	8.6	99.93%

Baselines. We use the following baseline methods for comparison to demonstrate the effectiveness of Self-BERT.

- **Pop:** It is a non-personalized recommendation method based on item popularity. The items that occur most frequently in the interaction sequence are recommended for all users.
- **BPR-MF** [15]: It is a matrix factorization model that captures pairwise item correlations and optimizes a Bayesian Personalized Ranking (BPR) loss.
- **GRU4Rec** [9]: This method employs GRU, which is a variant of RNN, to obtain better representations of user interaction sequences.
- **Caser** [18]: It is a CNN-based model that utilizes convolution kernels in horizontal and vertical orientations to capture local patterns.
- **SASRec** [11]: It adopts a left-to-right Transformer encoder to model users' interests dynamically.
- **BERT4Rec** [17]: This model regards the recommendation task as the Cloze task to train model so that the Transformer encoder could fuse contextual information for better sequence representations.

---

[3] https://grouplens.org/datasets/movielens/1m/.
[4] https://grouplens.org/datasets/movielens/20m/.
[5] http://jmcauley.ucsd.edu/data/amazon/.

- **S^3-Rec** [13]: It devises four auxiliary SSL objectives to gain the relations among attribute, item, subsequence and sequence. However, for the fairness of the comparison, we remove the modules related to attribute.
- **CL4SRec** [22]: It applies contrastive learning to construct self-supervision signals from the original sequences by three data augmentations.

Metrics. To compare the next-item prediction effectiveness among Self-BERT and baseline methods, we follow [11,17] to choose Hit Ratio (HR) and Normalized Discounted Cumulative Gain (NDCG) as evaluation metrics. It is worth noting that we rank the predictions on each whole dataset without negative sampling for a fair comparison. We report HR and NDCG with $k = 5, 10$. For all evaluation metrics, the higher values indicate better model performance.

Implementation Details. We use code provided by the authors for GRU4Rec[6], Caser[7], SASRec[8], and S^3-Rec[9]. We implement BPR, BERT4Rec, and CL4SRec on public resources. All hyperparameters are set following the original papers and tuned on the performance of the validation set. We report the results of each baseline at its optimal hyperparameter setting. For ML-20M, the batch size is 64 due to insufficient GPU memory, while for other datasets, the batch size is 256. We tune the max sequence length on different datasets and choose the corresponding optimal parameters. The contrastive loss proportion is tuned (see Sect. 4.3 for more details) and is finally decided as 0.1. As for other hyperparameters, we follow the guidance from BERT4Rec. We train our model using Adam optimizer with learning rate of 0.0001, $\beta_1 = 0.9$, $\beta_2 = 0.999$.

4.2 Overall Performance Comparison (RQ1)

To answer **RQ1**, we compare the performance of Self-BERT with the baseline methods. These baselines can be divided into two categories, non-SSL-based models and SSL-based models. Table 2 and Table 3 present the best results of all models on four real-world datasets individually. The best score and the second best score in each column are bolded and underlined, respectively. Improvements over the best baseline method are indicated in the last row. From the experimental results, we obtain the following observations:

Comparison with Non-SSL-Based Baselines. Non-personalized methods, such as PopRec and BPR-MF, exhibit worse recommendation performance on all datasets, which indicates the necessity of capturing sequential features in next item recommendation. In sequential recommendation, the Transformer-based models (e.g., SASRec and BERT4Rec) perform better than other models. This phenomenon shows that the self-attention mechanism can more effectively mine the information in the sequence than RNN and CNN. BERT4Rec performs relatively well on datasets with longer average sequence lengths, such as Movielens. We speculate that BERT4Rec can capture more contextual information

---

[6] https://github.com/hidasib/GRU4Rec.
[7] https://github.com/graytowne/caser_pytorch.
[8] https://github.com/kang205/SASRec.
[9] https://github.com/RUCAIBox/CIKM2020-S3Rec.

as sequence gets longer, hence obtain better performance. However, Self-BERT achieves better performance than BERT4Rec, which verifies the effectiveness of the self-supervised contrastive learning task under a joint learning framework.

Comparison with SSL-Based Baselines. Both S^3-Rec and CL4SRec are SSL-based models, but S^3-Rec performs worse than CL4SRec. One possible reason is that the two-stage training mode may lead to catastrophic forgetting. We also observe that Self-BERT performs better than CL4SRec. It is likely that data augmentations might corrupt the semantic information in sequence. In contrast, Self-BERT uses the complete sequence representation obtained in the encoder without data augmentations, improving the validity of contrastive learning.

Table 2. Overall performance of different methods on Movielens datasets. Bold scores are the best in method group, while underlined scores are the second best. The last row is the relative improvements compared with the best baseline results. (H is short for HR, N is short for NDCG)

Dataset	ML-1M				ML-20M			
Metric	H@5	H@10	N@5	N@10	H@5	H@10	N@5	N@10
PopRec	0.0078	0.0162	0.0052	0.0079	0.0165	0.0311	0.0094	0.0140
BPR-MF	0.0366	0.0603	0.0226	0.0302	0.0280	0.0482	0.0168	0.0233
GRU4Rec	0.0806	0.1344	0.0475	0.0649	0.0691	0.1187	0.0436	0.0595
Caser	0.0912	0.1442	0.0565	0.0734	0.0637	0.1051	0.0398	0.0531
SASRec	0.1071	0.1727	0.0634	0.0845	0.1276	0.1895	0.0842	0.1041
S^3-Rec	0.1020	0.1724	0.0612	0.0839	0.1187	0.1807	0.0775	0.0974
CL4SRec	0.1142	0.1810	0.0705	0.0920	0.1108	0.1782	0.0707	0.0924
BERT4Rec	<u>0.1308</u>	<u>0.2219</u>	<u>0.0804</u>	<u>0.1097</u>	<u>0.1380</u>	<u>0.2092</u>	<u>0.0928</u>	<u>0.1157</u>
Self-BERT	**0.1834**	**0.2620**	**0.1227**	**0.1480**	**0.1732**	**0.2468**	**0.1199**	**0.1436**
Improv	40.21%	18.07%	52.61%	34.91%	25.51%	17.97%	29.20%	24.11%

Table 3. Overall performance of different methods on Amazon datasets. Bold scores are the best in method group, while underlined scores are the second best. The last row is the relative improvements compared with the best baseline results. (H is short for HR, N is short for NDCG)

Dataset	Beauty				Toys			
Metric	H@5	H@10	N@5	N@10	H@5	H@10	N@5	N@10
PopRec	0.0075	0.0143	0.0041	0.0063	0.0066	0.0094	0.0046	0.0055
BPR-MF	0.0143	0.0253	0.0091	0.0127	0.0124	0.0178	0.0087	0.0105
GRU4Rec	0.0206	0.0332	0.0132	0.0172	0.0121	0.0184	0.0077	0.0097
Caser	0.0254	0.0436	0.0154	0.0212	0.0205	0.0333	0.0125	0.0166
SASRec	0.0371	0.0592	0.0233	0.0305	0.0429	<u>0.0652</u>	0.0248	0.0320
S^3-Rec	0.0365	0.0610	0.0228	0.0306	0.0405	0.0644	0.0258	<u>0.0335</u>
CL4SRec	<u>0.0396</u>	<u>0.0630</u>	0.0232	<u>0.0307</u>	<u>0.0434</u>	0.0635	0.0249	0.0314
BERT4Rec	0.0370	0.0598	<u>0.0233</u>	0.0306	0.0371	0.0524	<u>0.0259</u>	0.0309
Self-BERT	**0.0516**	**0.0740**	**0.0350**	**0.0421**	**0.0540**	**0.0759**	**0.0381**	**0.0452**
Improv	30.30%	17.46%	50.21%	37.13%	24.42%	16.41%	47.10%	34.93%

4.3 Parameter Sensitivity (RQ2)

To answer **RQ2**, we investigate the influence of essential hyperparameters in Self-BERT, including the weight of $\mathcal{L}_{Cl}$ and the batch size b.

Impact of Contrastive Learning Loss. In this section, we study the effect of contrastive learning proportion λ in the model. We select two datasets (ML-1M and Beauty) and conduct experiments with different λ values (0, 0.05, 0.1, 0.2, 0.5, 1, 2, 4) while keeping other parameters optimal. The obtained experimental results (NDCG@5) are shown in Fig. 3 (a). We observe that appropriately increasing the value of λ can improve the performance. However, the performance become worse when λ exceeds a certain threshold. The above observation shows that if λ is too large, the contrastive learning dominates the training process, which may influence the performance of sequential recommendation. We finally choose $\lambda = 0.1$ to achieve a balance between the contrastive learning task and the recommendation task for better recommendation performance.

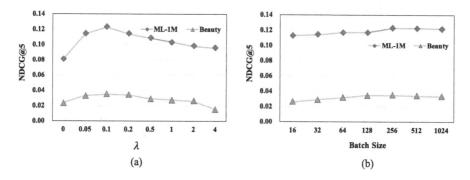

Fig. 3. Performance comparison (in NDCG@5) on Self-BERT w.r.t. different λ (a) and batch size (b) on ML-1M and Beauty datasets.

Impact of Batch Size. In this section, we study the effect of different batch size b on the model. We set the batch size as 16, 32, 64, 128, 256, 512, and 1024 for experiments. The results (NDCG@5) on the ML-1M and Beauty datasets are shown in Fig. 3(b). We observe that a large batch size has an advantage over the smaller ones, but the influence tends to be tiny as the batch size increases. This finding is similar to the experimental results in SimCLR [1]. A larger batch size can provide more negative samples for contrastive learning and promote the model to convergence. According to the experimental results, when the batch size is larger than 256, the effect of the model does not improve much, thus we select batch size $b = 256$.

4.4 Ablation Study (RQ3)

To answer **RQ3**, we conduct ablation study on Self-BERT to analyze the impact of the components in Self-BERT. To verify the effectiveness of these components, we propose the following three variants of Self-BERT: (B) delete the auxiliary contrastive learning task of Self-BERT, (C) only use the last layer output of BERT_{fix} as positive view, (D) cancel the parameter-fix mechanism in BERT_{fix} and use its hidden layer representations as positive views. The experimental results on four datasets are shown in Table 4. From the results, we observe that:

- Self-BERT under the joint learning framework (A) performs better than a single recommendation task model (B). It can be concluded that the auxiliary contrastive learning task can help obtain high quality representations for sequential recommendation.
- In our model, the auxiliary contrastive learning task with one-to-many view constraints (A) performs better than one-to-one view constraints (C). Self-BERT makes full use of the output of all Transformer layer as positive views, which improves the performance of the contrastive learning task by increasing the number of positive views.
- Self-BERT (A) performs better than the model which cancel the parameter-fix mechanism in BERT_{fix} (D). We think that the self-guided mechanism in Self-BERT prevents training signal of BERT_{fix} from being degenerated.

Table 4. Ablation study of Self-BERT.

Model	ML-1M		ML-20M		Beauty		Toys	
	HR@5	NDCG@5	HR@5	NDCG@5	HR@5	NDCG@5	HR@5	NDCG@5
(A) Self-BERT	**0.1834**	**0.1227**	**0.1732**	**0.1199**	**0.0516**	**0.0350**	**0.0540**	**0.0381**
(B) w/o CL	0.1308	0.0804	0.1380	0.0928	0.0370	0.0233	0.0371	0.0259
(C) only H^L	0.1715	0.1106	0.1592	0.1091	0.0479	0.0325	0.0488	0.0353
(D) w/o fix	0.1498	0.0956	0.1464	0.1014	0.0392	0.0279	0.0427	0.0284

4.5 Robustness Analysis (RQ4)

To answer **RQ4**, we conduct experiments to illustrate the robustness of Self-BERT when facing the data sparsity issue. To simulate the data sparsity problem, we only use part of the data (25%, 50%, 75%, 100%) for training and keep the test data unchanged. We compare the proposed model with CL4SRec and BERT4Rec, and the results are shown in Fig. 4. We observe that performance drops when using less training data, but Self-BERT consistently outperforms the other two baseline methods. Although the data sparsity issue effects vary on different datasets, our model can alleviate the influence of this issue for sequential recommendation to some extent.

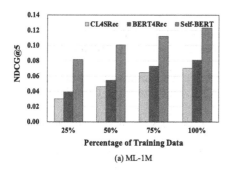
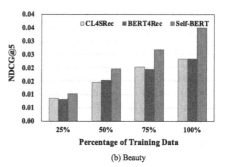

Fig. 4. Model performance (in NDCG@5) comparison w.r.t. sparsity ratio on ML-1M (a) and Beauty (b) datasets.

5   Conclusion

In this paper, we propose self-guided contrastive learning enhanced BERT for sequential recommendation (**Self-BERT**). High-quality contrastive views can be stably generated by introducing the self-guided mechanism, which means the hidden layer representations produced by a fixed BERT encoder are used to guide the training of another trainable BERT encoder. Moreover, we also improve the commonly used contrastive learning loss function (NT-Xent) to make it more suitable for **Self-BERT**. Experimental results on four real-world datasets demonstrate the effectiveness of our **Self-BERT**. We also experimentally verify the data sparsity robustness of our **Self-BERT**.

Acknowledgements. This research was supported by the NSFC(61876117, 61876217, 62176175), the major project of natural science research in Universities of Jiangsu Province (21KJA520004), Project Funded by the Priority Academic Program Development of Jiangsu Higher Education Institutions and Exploratory Self-selected Project of the State Key Laboratory of Software Development Environment.

References

1. Chen, T., Kornblith, S., Norouzi, M., Hinton, G.E.: A simple framework for contrastive learning of visual representations. In: ICML. Proceedings of Machine Learning Research, vol. 119, pp. 1597–1607. PMLR (2020)
2. Chen, W., et al.: Probing simile knowledge from pre-trained language models. In: ACL (2022)
3. Cho, K., et al.: Learning phrase representations using RNN encoder-decoder for statistical machine translation. In: EMNLP, pp. 1724–1734. ACL (2014)
4. Devlin, J., Chang, M., Lee, K., Toutanova, K.: BERT: pre-training of deep bidirectional transformers for language understanding. In: NAACL-HLT (1), pp. 4171–4186. Association for Computational Linguistics (2019)
5. Donkers, T., Loepp, B., Ziegler, J.: Sequential user-based recurrent neural network recommendations. In: RecSys, pp. 152–160. ACM (2017)

6. Gao, T., Yao, X., Chen, D.: Simcse: simple contrastive learning of sentence embeddings. In: EMNLP. Association for Computational Linguistics (2021)
7. Goodfellow, I.J., et al.: Generative adversarial networks. CoRR abs/1406.2661 (2014)
8. He, K., Fan, H., Wu, Y., Xie, S., Girshick, R.B.: Momentum contrast for unsupervised visual representation learning. In: CVPR, pp. 9726–9735. Computer Vision Foundation/IEEE (2020)
9. Hidasi, B., Karatzoglou, A., Baltrunas, L., Tikk, D.: Session-based recommendations with recurrent neural networks. In: ICLR (Poster) (2016)
10. Jaiswal, A., Babu, A.R., Zadeh, M.Z., Banerjee, D., Makedon, F.: A survey on contrastive self-supervised learning. CoRR abs/2011.00362 (2020)
11. Kang, W., McAuley, J.J.: Self-attentive sequential recommendation. In: ICDM, pp. 197–206. IEEE Computer Society (2018)
12. Kingma, D.P., Welling, M.: Auto-encoding variational bayes. In: ICLR (2014)
13. Liu, Y., Li, B., Zang, Y., Li, A., Yin, H.: A knowledge-aware recommender with attention-enhanced dynamic convolutional network. In: CIKM, pp. 1079–1088 (2021)
14. McAuley, J.J., Targett, C., Shi, Q., van den Hengel, A.: Image-based recommendations on styles and substitutes. In: SIGIR, pp. 43–52. ACM (2015)
15. Rendle, S., Freudenthaler, C., Gantner, Z., Schmidt-Thieme, L.: BPR: Bayesian personalized ranking from implicit feedback. CoRR abs/1205.2618 (2012)
16. Rendle, S., Freudenthaler, C., Schmidt-Thieme, L.: Factorizing personalized Markov chains for next-basket recommendation. In: WWW. ACM (2010)
17. Sun, F., Liu, J., Wu, J., Pei, C., Lin, X., Ou, W., Jiang, P.: BERT4Rec: sequential recommendation with bidirectional encoder representations from transformer. In: CIKM, pp. 1441–1450. ACM (2019)
18. Tang, J., Wang, K.: Personalized top-n sequential recommendation via convolutional sequence embedding. In: WSDM, pp. 565–573. ACM (2018)
19. Vaswani, A., et al.: Attention is all you need. In: NIPS, pp. 5998–6008 (2017)
20. Wang, J., et al.: Knowledge enhanced sports game summarization. In: WSDM (2022)
21. Wang, J., et al.: A survey on cross-lingual summarization. arXiv abs/2203.12515 (2022)
22. Xie, X., et al.: Contrastive learning for sequential recommendation. arXiv preprint arXiv:2010.14395 (2020)
23. Xu, C., et al.: Long- and short-term self-attention network for sequential recommendation. Neurocomputing **423**, 580–589 (2021)
24. You, Y., Chen, T., Sui, Y., Chen, T., Wang, Z., Shen, Y.: Graph contrastive learning with augmentations. In: NeurIPS (2020)
25. Yuan, F., Karatzoglou, A., Arapakis, I., Jose, J.M., He, X.: A simple convolutional generative network for next item recommendation. In: WSDM. ACM (2019)
26. Zhao, J., Zhao, P., Zhao, L., Liu, Y., Sheng, V.S., Zhou, X.: Variational self-attention network for sequential recommendation. In: ICDE. IEEE (2021)
27. Zhou, Z., Wang, Y., Xie, X., Chen, L., Liu, H.: Riskoracle: a minute-level citywide traffic accident forecasting framework. In: AAAI, vol. 34, pp. 1258–1265 (2020)
28. Zhou, Z., Wang, Y., Xie, X., Chen, L., Zhu, C.: Foresee urban sparse traffic accidents: a spatiotemporal multi-granularity perspective. IEEE Trans. Knowl. Data Eng. (2020)
29. Zhu, Y., Xu, Y., Yu, F., Liu, Q., Wu, S., Wang, L.: Graph contrastive learning with adaptive augmentation. In: WWW, pp. 2069–2080. ACM/IW3C2 (2021)

Time Interval Aware Collaborative Sequential Recommendation with Self-supervised Learning

Chenrui Ma[1], Li Li[2(✉)], Rui Chen[1(✉)], Xi Li[3], and Yichen Wang[4]

[1] Harbin Engineering University, Heilongjiang, China
{cherry1892,ruichen}@hrbeu.edu.cn
[2] University of Delaware, Delaware, USA
lilee@udel.edu
[3] Hospital of Chengdu University of Traditional Chinese Medicine, Sichuan, China
[4] Hunan University, Hunan, China

Abstract. Over the last few years, sequential recommender systems have achieved a great success in different applications. In the literature, it is generally believed that items farther away from the recommendation time have a weaker impact on the recommendation results. However, simply considering the distance between the interaction time and the recommendation time would prevent effective user representations. To solve this issue, we propose a **T**ime **I**nterval **A**ware **C**ollaborative (TIAC) model with self-supervised learning for sequential recommendation. We propose to adjust the attention score learned from the time interval between an interaction time and the recommendation time using a time kernel function to achieve a better user representation. We also introduce self-supervised learning to combine the collaborative information obtained from a graph convolutional network and the sequential information learned from gated recurrent units to further enrich user representation. Extensive experiments on four real-world benchmark datasets show that our proposed TIAC model consistently outperforms state-of-the-art models under various evaluation metrics.

Keywords: Sequential recommendation · Attention mechanism · Self-supervised learning · Graph convolutional network

1 Introduction

Recommender systems (RSs) have become an integral part of many real-world applications, such as e-commerce websites, online video sharing platforms, and social media services. Among various types of recommender systems, sequential recommendation has received great attention recently. The general idea of sequential recommendation is to explicitly exploit users' historical interactions to predict the next items that users are likely to interact with. There have been

Supported by the National Key R&D Program of China under Grant No. 2020YFB1710200.

B. Li et al. (Eds.): APWeb-WAIM 2022, LNCS 13423, pp. 87–101, 2023.
https://doi.org/10.1007/978-3-031-25201-3_7

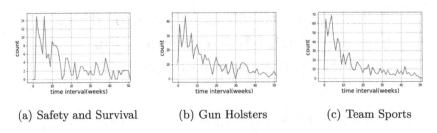

(a) Safety and Survival   (b) Gun Holsters   (c) Team Sports

Fig. 1. The distribution of re-purchase time intervals, where the x-axis represents the time gap between purchasing the same or two similar items from the same user, and the y-axis counts the number of users with the same time gap.

a plethora of studies on sequential recommendation based on different sequence models [3,5,6,14,19,25,26]. In these methods, a user's current representation is normally determined by the most recent items he/she interacted with. However, in practice this assumption may not always hold. For example, users exhibit repeated purchase tendency, and may buy items again when they approach the end of their lifetimes [17,18].

Although the model in [17] has taken into consideration different lifetimes of different items in repeated purchases, we argue that even the same item can have different lifetimes for different users due to different usage habits. As shown in Fig. 1, we can observe from, for example, the Amazon-Sports dataset that users usually purchase the same or similar products again within a certain time window. Figure 1(a) shows that most of the re-purchase events under the category of "Safety and Survival" happen between 4 and 10 weeks, while Fig. 1(b) and Fig. 1(c) show that the peak values of repeated purchases under the categories of "Gun Holsters" and "Team Sports" happen between 1 and 10 weeks. Therefore, we need to consider the fact that, in addition to different items, the same items can also have different re-purchase cycles to generate better user representations by adjusting items' attention scores accordingly.

The above idea helps to capture richer sequential information. However, it generates user representations from a single perspective, which is not comprehensive enough to fully depict users. Some previous research [22,24] has shown that the use of self-supervised learning could be an effective means to fuse multiple user representations in recommender systems. Our insight is that combining information from different perspectives will substantially enrich the resultant user representations.

Based on these observations, in this paper we propose a novel **Time Interval Aware Collaborative** (TIAC) model with self-supervised learning to obtain better user representations for sequential recommendation. The improvements come from two aspects: (1) based on the time interval between an interaction time and the recommendation time, attention scores are adjusted by considering different re-purchase cycles with a time kernel function to learn a better user representation from sequential information; (2) using the collaborative information obtained by graph convolutional networks (GCNs) to further improve user representations through self-supervised learning. The main contributions of this paper are summarized as follows:

- We propose a new attention module to adjust attention coefficients in sequential recommendation to learn better user representations based on time perception;
- We make use of self-supervised learning to combine users' sequential information and collaborative information to further enhance user representations;
- We have conducted extensive experiments on multiple real-world datasets to show the superiority of the TIAC model and the effectiveness of each proposed module.

2   Related Work

2.1   Sequential Recommendation

Early research on sequential recommendation makes use of Markov chains to mine sequential patterns from historical data. FMPC [14] models users' long-term preferences by combining first-order Markov chains and matrix factorization for next-basket recommendation. In the era of deep learning, different deep neural networks have been proposed for sequential recommendation. GRU4Rec [4] is the first model that applies recurrent neural networks (RNNs) to session-based recommendation. Caser [16] considers convolutional neural networks (CNNs) as the backbone network to embed recent items into an "image" so as to learn sequential patterns as local features of the image using convolutional filters. HGN [10] models long-term and short-term interests via feature gating and instance gating modules. SASRec [5] captures sequential dynamics based on self-attention. CTA [18] proposes to use multiple kernel functions to learn time dynamics, and then deploys different scored kernels in the context information. IMfOU [2] uses relative recommendation time intervals to model user representations. By introducing knowledge graphs, Chrous [18] constructs the relationship between two different items, and considers the distance between the interaction time and the recommendation time to make recommendations. There are also other latest results based on self-attention [7,9]. The above models generally assume that the next item to recommend heavily relies on the most recent interactions, which often does not hold in practice. SLRC [17] is the closest to our work. It uses the Hawkes process to model temporal dynamics of repeat consumption. However, it does not recognize the fact that many items can have different lifetimes for different users.

2.2   Self-supervised Learning in RS

Self-supervised learning [8] is an emerging paradigm to learn and obtain supervisory signals from data itself. Early research on self-supervised learning is mainly driven by graph learning, where the common types of stochastic augmentations include uniform node/edge dropout, random feature/attribute shuffling, and subgraph sampling using random walk. Inspired by the success of graph self-supervised learning, there have been some recent works [11, 21] applying the self-supervised learning idea to recommendation applications. A self-supervision method is introduced in [27] by randomly masking attributes of items and skipping items of a given sequence. Yao *et al.* [23] propose a uniform feature masking and dropout dual-tower DNN structure for self-supervision. Ma *et al.* [11] use feature masks to reconstruct future sequences for self-supervision. SGL [21] uses the random enhancement method on graphs, and introduces a general self-supervised graph learning framework. GroupIM [15] uses self-supervision among a group of specific members to improve user representations and make recommendations. Most of the above methods use user information from homogeneous sources, for example, either sequential information or collaborative (graph) information. However, for sequential recommendation, a user's sequential information and collaborative information can be combined to further enrich his/her representation.

3 Problem Formulation

Let $\mathcal{U}$, $\mathcal{I}$ and $\mathcal{T}$ represent the set of users, items and timestamps, respectively. For a user $u \in \mathcal{U}$, the associated sequence of interacted items is denoted by $S_u = [(s_1, t_1), (s_2, t_2), \cdots, (s_n, t_n)]$, where $s_i \in \mathcal{I}$, $t_i \in \mathcal{T}$ and $t_i < t_{i+1}$. Each pair (s_i, t_i) indicates that user u interacted with item s_i at timestamp t_i. Let $\mathcal{O}^+ = \{y_{ui} | u \in \mathcal{U}, i \in \mathcal{I}\}$ be the set of observed interactions, where y_{ui} indicates that the user u has interacted with item i before. We can construct a bipartite graph $\mathcal{G} = \{\mathcal{V}, \mathcal{E}\}$, where the node set $\mathcal{V} = \mathcal{U} \cup \mathcal{I}$ includes all users and items, and the edge set $\mathcal{E} = \mathcal{O}^+$ represents all historical interactions. Based on graph $\mathcal{G}$, the adjacency matrix between users and items is defined as $\mathcal{R} \in \mathbb{R}^{m \times n}$, where m is the number of users and n is the number of items. For each (u, i) in $\mathcal{R}$, $r_{ui} = 0$ if user u has not interacted with item i; otherwise, r_{ui} is the number of interactions between user u and item i. Let $E^I \in \mathbb{R}^{n \times d}$ be the item embedding matrix generated by an item embedding layer, e_i^I the embedding vector of item i, $E^U \in \mathbb{R}^{m \times d}$ the user embedding matrix generated by a user embedding layer, and e_j^U the embedding vector of user j, where $e_i^I \in E^I$, $e_j^U \in E^U$, and d is the embedding size.

Given the historical interaction sequence S_u of user u, our goal is to predict the next item that user u is likely to interact with. In particular, we generate a list of top-K interesting items for user u at the recommendation time.

4 Proposed Method

Similar to previous research [6], if a sequence's length is greater than a predefined value L, we truncate it by selecting the L most recent items; otherwise, we pad the sequence to be of length L. In this section, we explain the TIAC model in detail. The overall architecture of the TIAC model is illustrated in Fig. 2.

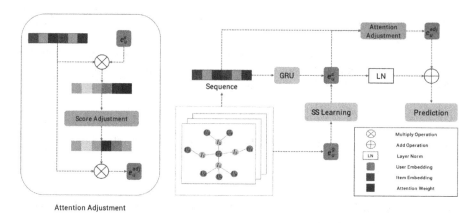

Fig. 2. The overall architecture of the proposed TIAC model.

4.1 Time Interval Aware Collaborative Model with Attention Adjustment

User's Collaborative Information from GCNs. As suggested in previous works [20,24], we use graph convolutional networks (GCNs) to obtain a user's collaborative information. We first build the adjacency matrix from the user-item interactions.

Here we take a user u as an example. For any interacted user-item pair (u, i), we define the process of information transmission in the l-th layer as:

$$m_{u \leftarrow i}^{(l)} = p_{ui}(W_1^{(l)} e_i^{(l-1)} + W_2^{(l)}(e_i^{(l-1)} \odot e_u^{(l-1)})),$$
$$m_{u \leftarrow u}^{(l)} = W_1^{(l)} e_u^{(l-1)}, \tag{1}$$

where $W_1, W_2 \in \mathbb{R}^{d \times d}$ are the trainable weight matrices to distill useful information during propagation, and p_{ui} is a coefficient to control the decay factor. $e_u^{(l-1)}$ is the user representation generated from the previous layer, which memorizes the messages from its $(l-1)$-hop neighbors. At the same time, the user's self-loop is added to ensure the retention of the original information. Then, we can obtain the representation of user u at the layer l. Note that $e_u^{(0)}$ and $e_i^{(0)}$ are initialized as e_u^U and e_i^I, respectively.

Similarly, we can define the information transmission process of an item. As shown in Eq. 1, a user or an item is capable of receiving the messages propagated from its l-hop neighbors. In the l-th layer, the representation of user u is updated via:

$$e_u^{(l)} = \text{LeakyReLu}(m_{u \leftarrow u}^{(l)} + \sum_{i \in \mathcal{N}_u} m_{u \leftarrow i}^{(l)}), \tag{2}$$

where $\mathcal{N}_u$ denotes the l-hop neighbors of user u. We perform the same operation on items. After propagating with l layers, we obtain multiple representations for user u and item i. Since the representations obtained in different layers emphasize the information transmitted by different connections, they can reflect different characteristics of users (or items). Therefore, we concatenate them to constitute the final embedding of each user (or each item), and obtain the final user embedding and item embedding as follows:

$$e_u^* = e_u^{(0)}|| \cdots ||e_u^{(l)}, \quad e_i^* = e_i^{(0)}|| \cdots ||e_i^{(l)}, \tag{3}$$

where $||$ is the concatenation operation. By doing this, on the one hand, we can enrich the initial embeddings with the embedding propagation layer; on the other hand, we can control the range of information propagation by tuning l.

User's Sequential Information and Attention Adjustment. To capture the sequentiality among the items in a sequence, we inject a learnable positional embedding to each item in the sequence:

$$e_i^s = e_i^* + e_i^P, \tag{4}$$

where e_i^* is obtained from Eq. 3, and $e_i^P \in \mathbb{R}^d$ is the embedding of position i in the sequence.

In particular, we define $X^u = [e_1^s, e_2^s, \cdots, e_L^s]$ as the user's sequence without timestamps, aiming to learn the influence of the user's historical interaction sequence. There has been a lot of research on capturing sequential information using different sequence models. Here we choose gated recurrent units (GRUs) to capture the sequential information:

$$e_u^S = \text{GRUs}(X_u). \tag{5}$$

In general, most existing methods assume that the closer an item is to the recommended time, the greater impact it has on the recommendation result. However, based on our observation, the impact of certain items on the final recommendation needs to be reinforced after a period of time. To solve this issue, after obtaining the user representation from the sequential information extracted by the GRUs, we first calculate the attention score of each item in the sequence as:

$$a_{u,i} = e_u^{S^\top} e_i^s, \tag{6}$$

where e_i^s is the i-th item of X^u. Then we use a kernel function to construct the effect of time intervals on attention scores:

$$a_{u,i}^f = a_{u,i} + \pi \mathcal{N}(\triangle t | \mu_i^u, \sigma_i^u), \tag{7}$$

where $\triangle t$ is the time interval between the item interaction time and the recommendation time, $\mathcal{N}(\triangle t | \mu_i^u, \sigma_i^u)$ is a Gaussian distribution with mean μ_i^u and standard deviation σ_i^u, and π is used to control the change ratio. Here μ and σ are learned according to the items purchased by users, and μ_i^u and σ_i^u are extracted from learnable parameter matrices $E^\mu \in \mathbb{R}^{m \times L}$ and $E^\sigma \in \mathbb{R}^{m \times L}$, respectively. Therefore, we can adjust the corresponding attention score according to the changes in the learning intensity function of different items for different users.

Model Optimization and Recommendation Generation. Next, we leverage the above learned attention scores to delineate the user representation:

$$e_u^{adj} = \sum_{i=1}^{L} a_{u,i}^f e_i^s. \tag{8}$$

Then, to obtain the final user representation, we use a linear strategy to fuse the user representation after attention adjustment (e_u^{adj}) and the user representation obtained by GRUs (e_u^S)[1]:

$$e_u = \mathrm{LN}(e_u^{adj}) + e_u^S, \tag{9}$$

where $\mathrm{LN}(\cdot)$ is layer normalization. We add layer normalization to ensure stable learning.

Here we adopt the binary cross entropy loss function:

$$\mathcal{L}_r = - \sum_{(u,t,i,j)\in D} \log(\sigma(r_{ui}^t)) + \log(1 - \sigma(r_{uj}^t)) + \lambda||\Theta||_F^2, \tag{10}$$

where $\Theta = \{E^I, E^U, E^P, E^\mu, E^\sigma\}$ is a subset of model parameters, $||\cdot||$ denotes the Frobenius norm, λ is the regularization parameter, and $\sigma(\cdot)$ is the sigmoid activation function. D is the training dataset that includes both ground-truth interaction pairs (u, t, i) and negative sample items j that u did not interact with at time t.

4.2   Enhancing User Representation with Self-Supervised Learning

Inspired by the success of self-supervised learning on graphs, we integrate self-supervised learning into our proposed network to combine user representations from sequential information and collaborative information to further enhance user representations. In particular, we design an auxiliary task to benefit our recommendation task with the following two steps.

Creation of Self-supervision Signals. In TIAC, we generate user representations in two ways, namely GRUs and GCNs. Since these two methods only use either sequential or collaborative information, the two types of user embeddings may effectively complement each other, resulting in richer user representations. For each mini-batch including n users during the training process, there is a bijective mapping between the two types of user embeddings. Naturally, the two types can be the ground truth of each other for self-supervised learning. If the two user embeddings both represent the same user, we label this pair as a positive ground truth, otherwise we label it as negative.

---

[1] Our experimental results show that the linear fusing strategy achieves a reasonable balance between model complexity and performance.

Self-supervised Learning. We contrast the two types of user embeddings, and adopt InforNCE [12] with a standard binary cross-entropy loss between the samples from the ground truth (positive) and the corrupted samples (negative) as our learning objective, which is defined as follows:

$$\mathcal{L}_s = -\log(\sigma(f_D(e_u^S, e_u^*))) - \log(\sigma(1 - f_D(e_u^S, \hat{e}_u^*))), \tag{11}$$

where $\hat{e}_u^*$ is the negative sample obtained by corrupting e_u^* with row-wise shuffling, e_u^* is the user representation obtained by GCNs, e_u^S is the user representation obtained by GRUs, and $f_D(\cdot) : \mathbb{R}^d \times \mathbb{R}^d \to \mathbb{R}$ is the discriminator function that takes two vectors as the input and then scores the agreement between them. We implement the discriminator as the dot product between two vectors. This learning objective is used to maximize the mutual information between the user embeddings learned in different ways. By doing so, they can acquire information from each other to improve their performance.

We unify the recommendation task and this self-supervised learning task into a primary and auxiliary learning framework, with the recommendation task as the primary task and the self-supervised learning task as the auxiliary task. The joint learning objective is formalized as:

$$\mathcal{L} = \mathcal{L}_r + \beta \mathcal{L}_s, \tag{12}$$

where β controls the influence of the self-supervised learning task.

Finally, the recommendation list is generated by computing the similarity score between a user's embedding e_u and each candidate item's embedding e_o with $r_o = e_u e_o^\top$, and choosing the top-K candidate items with the highest similarity scores.

5   Experiments

In this section, we perform an extensive experimental evaluation of the proposed TIAC model on four real-world benchmark datasets.

5.1   Datesets

In the experiments, we use four real-world benchmark datasets that are widely used in the literature, including:

- **Tafeng**[2] is a dataset that contains 4-month shopping transactions (from November 2000 to February 2001) at the Tafeng grocery store.
- **Amazon**[3] is a collection of datasets that contain item reviews and metadata in different product categories from Amazon. We use two datasets of **Beauty** and **Sports** in the experiments.

---

[2] https://www.kaggle.com/chiranjivdas09/ta-feng-grocery-dataset.
[3] http://jmcauley.ucsd.edu/data/amazon/links.html.

Table 1. The statistics of the datasets.

Dataset	#Users	#Items	#Interactions	Avg. length	Time range
Tafeng	16,034	8,906	644,960	40.22	2000.11.01–2001.02.28
Sports	35,571	18,349	293,644	8.25	2009.01.01–2014.07.23
Beauty	22,352	12,101	197,514	8.83	2008.01.01–2014.07.22
MovieLens-1M	6,041	3,707	1,000,209	163.60	2000.04.26–2003.03.01

- **MovieLens-1M (ML1M)**[4] is a dataset that contains rating data of multiple movies submitted by multiple users, as well as movie metadata and user attribute information.

To generate representative experimental results, we choose datasets of different time lengths to show that our model can work well in different time lengths. We follow the same procedure in [6] to preprocess the datasets. In particular, we treat a review or rating as implicit feedback and order the items of a user by timestamps. We only keep the users with at least five ratings/reviews and the items with at least five ratings/reviews. For each sequence, we use the penultimate item as the validation set for hyperparameter tuning and the last item as the test set. The statistics of the datasets are summarized in Table 1.

5.2   Baselines and Evaluation Metrics

We compare TIAC with the following nine representative methods:

- **POP** is a widely used baseline which recommends items based on item popularity in the training dataset.
- **BPR** [13] is a classical recommendation algorithm which optimizes the matrix factorization (MF) model with a pairwise ranking loss function via stochastic gradient descent.
- **FPMC** [14] combines MF and first-order Markov chains to capture long-term preferences and dynamic transitions, respectively.
- **GRU4Rec** [4] employs GRUs to model item sequences for session-based recommendation.
- **SASRec** [5] leverages self-attention to model sequentiality embedded in interaction sequences. It does not consider the impact of more detailed time information in purchase sequences on the recommendation results.
- **SLRC** [17] combines the Hawkes process and collaborative filtering to model temporal dynamics. We choose Bayesian personalized ranking (BPR) to derive the base intensity.
- **TiSASRec** [6] is a state-of-the-art sequential recommender that incorporates item time interval information using self-attention.
- **TimelyRec** [1] learns two specific types of temporal patterns, namely periodic patterns and evolving patterns, for sequential recommendation.

---

[4] http://files.grouplens.org/datasets/movielens/.

- **Chorus** [18] is a state-of-the-art method with item relationship and time evolution. It requires a large amount of prior knowledge, such as the categories of items and the relationship between categories.
- S^3-**Rec** [27] uses the correlation of the original data to construct self-supervised signals and enhances the data representation through a pre-training method.

To make a fair comparison, we do not consider models whose code is not publicly available.

Similar to previous studies, we adopt HR@10 and NDCG@10 to evaluate different models. We train and evaluate each model five times and report the average results. In particular, following the existing works, for each user, we randomly sample 100 negative items and rank these items with the ground-truth item. We calculate HR@10 and NDCG@10 based on the rankings of these 101 items.

5.3 Experimental Settings

For all datasets, the L_2 regularization parameter λ is 0.0001, the embedding size d is 50, the dropout rate is 0.5, the learning rate is 0.0001, and the number of layers in GCNs is 4. We determine the maximum length L of input sequences based on the average sequence length, and the maximum length L of Tafeng, Beauty, Sports, and ML1M is set to 50, 20, 15, and 50, respectively. The batch size of Tafeng, Beauty, Sports, and ML1M is set to 128, 1024, 1024, and 256, respectively. The self-supervised learning weight parameter β of Tafeng, Beauty, Sports, and ML1M is set to 0.001, 0.002, 0.02, and 0.02, respectively. The hyper-parameters of all baseline algorithms are carefully tuned by grid search, and the best performances are reported. We implemented our model in PyTorch 1.8 and Python 3.7. All experiments were run on a workstation with an Intel Xeon Platinum 2.40 GHz CPU, a NVIDIA Quadro RTX 8000 GPU, and 500 GB RAM.

5.4 Experimental Results

We present the main experimental results in Table 2. We can make a few key observations.

- First, our proposed TIAC model achieves the best performance on all datasets in terms of all evaluation metrics. TIAC's improvements are significant with p-value < 0.05 and consistent.
- Second, without utilizing the sequential information, the models cannot achieve a reasonable performance for sequential recommendation, as demonstrated by POP and BPR.
- Third, only using sequential information without paying attention to the impact of the time interval between items' interaction time and recommendation time will also lead to suboptimal performance, as demonstrated by GRU4Rec and SASRec.

Table 2. Performances of different models. The best and second-best results are bold-faced and underlined, respectively. The symbol * indicates that the improvement is significant with p-value < 0.05.

Method	Beauty		Sports		Tafeng		MovieLens-1M	
	NDCG@10	HR@10	NDCG@10	HR@10	NDCG@10	HR@10	NDCG@10	HR@10
POP	0.1472	0.2801	0.1642	0.2968	0.2963	0.4775	0.2513	0.4575
BPR	0.2492	0.4073	0.2511	0.4180	0.3639	0.5584	0.4063	0.6849
FRMC	0.2775	0.4205	0.259	0.4223	0.3555	0.55	0.537	0.7816
GRU4Rec	<u>0.3207</u>	0.4725	0.257	0.4282	0.3819	0.5817	0.5068	0.7656
SASRec	0.2994	<u>0.4738</u>	0.2961	0.4775	0.3886	<u>0.5963</u>	0.5463	0.8021
TiSASRec	0.2995	0.4655	<u>0.2986</u>	<u>0.4794</u>	0.3931	0.5938	0.5404	0.8057
TimelyRec	0.1964	0.3409	0.1598	0.3078	0.3931	0.4926	0.3228	0.5312
SLRC	0.2603	0.4253	0.2597	0.4308	<u>0.3972</u>	0.5802	0.3211	0.5788
Chorus	0.2939	0.4623	0.2986	0.4710	-	-	0.4097	0.6886
S^3-Rec	0.2978	0.4572	0.2744	0.4497	0.3503	0.5466	<u>0.5723</u>	<u>0.8125</u>
TIAC	**0.3516***	**0.5357***	**0.3371***	**0.5458***	**0.4121***	**0.6038***	**0.5811***	**0.8187***
Improvement	9.63%	13.06%	12.89%	12.75%	3.75%	1.26%	1.54%	0.76%

- Fourth, the experimental results confirm that considering the fact that different items have different lifetimes for different users in TIAC is beneficial. This explains why TIAC can outperform TiSASRec and TimelyRec, which already consider the impact of the time interval between items' interaction time and the recommendation time, and SLRC and Chorus, which already use time kernel functions to update correlation scores. In addition, compared with Chorus, our TIAC model does not require item category information to generate complementary and substitutive relationships, and thus enjoys better applicability. In fact, Chorus is not able to generate recommendations for Tafeng because it does not have obvious category information.
- Fifth, compared with S^3-Rec that does not take into account the periodic and collaborative information of items, our method can significantly outperform it on the datasets (e.g., Beauty and Sports), which have strong periodic and collaborative information.

5.5   Ablation Study

The previous section has shown the benefits of having both attention adjustment and the fusion of sequential and collaborative information on the final recommendation results. In this section, we show the effectiveness of each of the two parts. Due to the space limitation, we only present the results on Beauty and Sports in Table 3. Similar trends can be observed on the other datasets. In Table 3, "w/o" means "without", "w/o AA" means that we do not apply attention adjustment,

Table 3. Benefits of each proposed component.

Method	Beauty		Sports	
	NDCG@10	HR@10	NDCG@10	HR@10
w/o AA	0.3424	0.5216	0.3238	0.5212
w/o SSL	0.3350	0.5242	0.3280	0.5292
TIAC	**0.3516**	**0.5357**	**0.3371**	**0.5458**

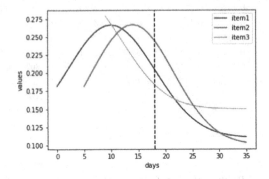

Fig. 3. An illustration of how attention scores of different items in a sequence change over time through the time kernel function.

and "w/o SSL" means that the two information sources are concatenated directly instead of using self-supervised learning. It can be seen that every part plays an important role in improving performance. We can observe that both the attention adjustment module and the self-supervised learning module are beneficial to the final TIAC model, which confirms the importance of considering a variety of information sources and adjusting attention scores. It well justifies our motivation.

5.6   Case Study

We provide a case study on the Sports dataset to demonstrate that our attention adjustment design can indeed recommend more relevant items according to the time interval between recommendation time and interaction time. In Fig. 3, the lines of different colors represent the last three items previously purchased by a user. Here, we take the interaction time of the antepenultimate item as the starting point of the x-axis, which represents the time interval from the starting point. The dotted line in Fig. 3 represents the recommendation time. The y-axis represents the attention score. We draw the attention scores of different items over time. It can be observed that the time kernel function in Eq. 7 can model more complicated temporal patterns, allowing an item interacted a long time ago to exhibit a large attention score.

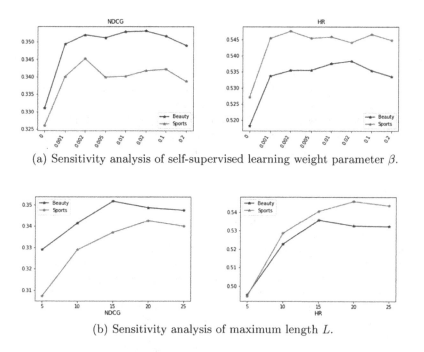

(a) Sensitivity analysis of self-supervised learning weight parameter β.

(b) Sensitivity analysis of maximum length L.

Fig. 4. Experimental results under different hyperparameter settings.

5.7 Hyperparameter Sensitivity

In the last set of experiments, we study our model's performance with respect to different key hyperparameters.

Impact of Self-supervised Learning Weight β. In Fig. 4(a), we vary the value of the self-supervised learning weight β in {0, 0.001, 0.002, 0.005, 0.01, 0.02, 0.1, 0.2} while keeping other hyperparameters unchanged. It can be observed that initially increasing β leads to performance improvement in terms of both HR@10 and NDCG@10, confirming the benefits of introducing the self-supervised learning task. However, the performance will eventually drop as we further increase β. We deem that it is due to the gradient conflicts between the two learning tasks (i.e., the primary recommendation task and the auxiliary self-supervised learning task).

Impact of Maximum Length L. In Fig. 4(b), we vary the number of the maximum length L in {5, 10, 15, 20, 25} while keeping other hyperparameters unchanged. We can see that a sufficiently large L is important as it allows to capture enough sequential information for effective sequential recommendation. However, when L becomes overly large, it will introduce excessive noise to short sequences, which in turn jeopardizes the quality of learned user representations. In practice, setting L based on the average sequence length of a dataset normally achieves reasonable performance.

6   Conclusion

In this paper, we proposed a novel TIAC model for sequential recommendation. We first obtained user representations generated from collaborative information through GCNs and sequential information through GRUs and then learned better user representations through the attention adjustment mechanism. Then, self-supervised learning was introduced to combine user representations learned from collaborative and sequential information. Extensive experimental results show that our model outperforms state-of-the-art models.

References

1. Cho, J., Hyun, D., Kang, S., Yu, H.: Learning heterogeneous temporal patterns of user preference for timely recommendation. In: Proceedings of the 30th International Conference on World Wide Web (WWW) (2021)
2. Guo, X., Shi, C., Liu, C.: Intention modeling from ordered and unordered facets for sequential recommendation. In: Proceedings of the 29th International Conference on World Wide Web (WWW) (2020)
3. Hidasi, B., Karatzoglou, A.: Recurrent neural networks with top-k gains for session-based recommendations. In: Proceedings of the 27th ACM International Conference on Information and Knowledge Management (CIKM) (2018)
4. Hidasi, B., Karatzoglou, A., Baltrunas, L., Tikk, D.: Session-based recommendations with recurrent neural networks. In: Proceedings of the 4th International Conference on Learning Representations (ICLR) (2016)
5. Kang, W., McAuley, J.J.: Self-attentive sequential recommendation. In: Proceeding of the 18th IEEE International Conference on Data Mining (ICDM) (2018)
6. Li, J., Wang, Y., McAuley, J.J.: Time interval aware self-attention for sequential recommendation. In: Proceedings of the 13th International Conference on Web Search and Data Mining (WSDM) (2020)
7. Liu, C., Li, X., Cai, G., Dong, Z., Zhu, H., Shang, L.: Noninvasive self-attention for side information fusion in sequential recommendation. In: Proceedings of the 35th AAAI Conference on Artificial Intelligence (AAAI) (2021)
8. Liu, X., et al.: Self-supervised learning: generative or contrastive. CoRR abs/2006.08218 (2020)
9. Liu, Y., Li, B., Zang, Y., Li, A., Yin, H.: A knowledge-aware recommender with attention-enhanced dynamic convolutional network. In: CIKM, pp. 1079–1088 (2021)
10. Ma, C., Kang, P., Liu, X.: Hierarchical gating networks for sequential recommendation. In: Proceedings of the 25th ACM SIGKDD International Conference on Knowledge Discovery and Data Mining (KDD) (2019)
11. Ma, J., Zhou, C., Yang, H., Cui, P., Wang, X., Zhu, W.: Disentangled self-supervision in sequential recommenders. In: Proceedings of the 26th ACM SIGKDD International Conference on Knowledge Discovery and Data Mining (KDD) (2020)
12. van den Oord, A., Li, Y., Vinyals, O.: Representation learning with contrastive predictive coding. CoRR abs/1807.03748 (2018)
13. Rendle, S., Freudenthaler, C., Gantner, Z., Schmidt-Thieme, L.: BPR: Bayesian personalized ranking from implicit feedback. In: Proceedings of the 25th Conference on Uncertainty in Artificial Intelligence (UAI) (2009)

14. Rendle, S., Freudenthaler, C., Schmidt-Thieme, L.: Factorizing personalized Markov chains for next-basket recommendation. In: Proceedings of the 19th International Conference on World Wide Web (WWW), pp. 811–820 (2010)
15. Sankar, A., Wu, Y., Wu, Y., Zhang, W., Yang, H., Sundaram, H.: Groupim: a mutual information maximization framework for neural group recommendation. In: Proceedings of the 43nd International ACM SIGIR Conference on Research and Development in Information Retrieval (SIGIR) (2020)
16. Tang, J., Wang, K.: Personalized top-n sequential recommendation via convolutional sequence embedding. In: Proceedings of the 11th International Conference on Web Search and Data Mining (WSDM) (2018)
17. Wang, C., Zhang, M., Ma, W., Liu, Y., Ma, S.: Modeling item-specific temporal dynamics of repeat consumption for recommender systems. In: Proceedings of the 28th International Conference on World Wide Web (WWW), pp. 1977–1987 (2019)
18. Wang, C., Zhang, M., Ma, W., Liu, Y., Ma, S.: Make it a chorus: knowledge- and time-aware item modeling for sequential recommendation. In: Proceedings of the 43nd International ACM SIGIR Conference on Research and Development in Information Retrieval (SIGIR) (2020)
19. Wang, J., Louca, R., Hu, D., Cellier, C., Caverlee, J., Hong, L.: Time to shop for valentine's day: shopping occasions and sequential recommendation in e-commerce. In: Proceedings of the 13th International Conference on Web Search and Data Mining (WSDM) (2020)
20. Wang, X., He, X., Wang, M., Feng, F., Chua, T.: Neural graph collaborative filtering. In: Proceedings of the 42nd International ACM SIGIR Conference on Research and Development in Information Retrieval (SIGIR) (2020)
21. Wu, J., et al.: Self-supervised graph learning for recommendation. In: Proceedings of the 44nd International ACM SIGIR Conference on Research and Development in Information Retrieval (SIGIR) (2021)
22. Xia, X., Yin, H., Yu, J., Wang, Q., Cui, L., Zhang, X.: Self-supervised hypergraph convolutional networks for session-based recommendation. In: Proceedings of the 35th AAAI Conference on Artificial Intelligence (AAAI) (2021)
23. Yao, T., et al.: Self-supervised learning for deep models in recommendations. CoRR abs/2007.12865 (2020)
24. Yu, J., Yin, H., Gao, M., Xia, X., Zhang, X., Hung, N.Q.V.: Socially-aware self-supervised tri-training for recommendation. In: Proceedings of the 23th ACM SIGKDD International Conference on Knowledge Discovery and Data Mining (KDD) (2021)
25. Yu, Z., Lian, J., Mahmoody, A., Liu, G., Xie, X.: Adaptive user modeling with long and short-term preferences for personalized recommendation. In: Kraus, S. (ed.) Proceedings of the Twenty-Eighth International Joint Conference on Artificial Intelligence (IJCAI) (2019)
26. Zang, Y., et al.: GISDCN: a graph-based interpolation sequential recommender with deformable convolutional network. In: International Conference on Database Systems for Advanced Applications, pp. 289–297 (2022)
27. Zhou, K., et al.: S3-Rec: self-supervised learning for sequential recommendation with mutual information maximization. In: Proceedings of the 29th ACM International Conference on Information and Knowledge Management (CIKM) (2020)

A2TN: Aesthetic-Based Adversarial Transfer Network for Cross-Domain Recommendation

Chenghua Wang[✉] and Yu Sang

School of Computer Science and Technology, Soochow University, Suzhou, China
20205227042@stu.suda.edu.cn, sangyu@suda.edu.cn

Abstract. To address the long-standing data sparsity problem in recommender systems, Cross-Domain Recommendation (CDR) has been proposed to leverage the relatively richer information from a richer domain to improve the recommendation performance in a sparser domain. Therefore, enhancing the transferability of features in different domains is crucial for improving the recommendation performance. However, existing methods are usually confronted with negative transfer due to the data sparsity problem. To this end, we propose an Aesthetic-based Adversarial Transfer Network (A2TN) for CDR, which takes advantage of the transferability of aesthetic preferences to ensure the effect of feature transfer. Specifically, we first utilize an aesthetic network to extract aesthetic features and a general feature layer to embed general features, which can collaboratively capture user's comprehensive preferences. Then, we adopt the adversarial transfer network to generate domain-dependent features for avoiding negative transfer and domain-independent features for effective knowledge transfer. Moreover, an attention mechanism is used to fuse different preferences with different weights to reflect their importance in better portraying user's preferences. Finally, the experimental results on real-world datasets demonstrate that our proposed model outperforms the benchmark recommendation methods.

Keywords: Cross-domain recommendation · Aesthetic preferences · Adversarial transfer network · Attention mechanism

1 Introduction

In most real-world application scenarios, users cannot provide ratings or reviews for plenty of items, thereby leading to the long-standing data sparsity problem in recommender systems [7,22]. In order to address this problem that seriously reduces the recommendation accuracy, Cross-Domain Recommendation (CDR) [1] appears as a new trend by leveraging the relatively richer information from the richer (source) domain, e.g., ratings [1], tags [3], reviews [19] and latent features [7,26], to improve the recommendation performance in the sparser (target) domain [25].

B. Li et al. (Eds.): APWeb-WAIM 2022, LNCS 13423, pp. 102–116, 2023.
https://doi.org/10.1007/978-3-031-25201-3_8

The existing CDR approaches can be generally classified into two categories, i.e., transfer-based CDR and content-based CDR [11, 26]. Transfer-based CDR approaches pay attention to transfer the latent factors from the source domain to the target domain via common or similar users/items, which is achieved by employing classical machine learning methods, such as transfer learning [7] and deep neural networks [26]. In contrast, content-based CDR approaches typically tend to link different domains by identifying similar user/item attributes, e.g., ratings [1], tags [3] and reviews [19]. Different from the transfer-based CDR, the idea of conventional content-based CDR approaches is to treat the similar user/item attributes between two domains as a bridge, so that the similar users/items in different domains could share their knowledge across domains [25]. Consequently, as a bridge for knowledge transfer across domains, the bridge needs to be stable. Although the user/item attributes (e.g., ratings, tags and reviews) used in the existing content-based CDR approaches can improve the recommendation performance to some extent [9, 13, 22], they are unstable. For example, user's ratings and item's tags can be various in different situations. In other words, user's ratings and item's tags in different domains are difficult to align, causing it hard to find shared knowledge for transfer.

As we all know, user's aesthetic preferences as personality traits will not change easily, then they are alignable and could be used as a bridge between different domains for knowledge transfer [11]. When shopping on the Internet with tons of products, especially appearance-first products, e.g., clothing and furniture, we usually look through product images to determine whether their appearances meet our tastes firstly, and then make the decision. Product images provide abundant visual information [5], including design, color schemes, decorative patterns, texture, and so on [21]. We can even estimate the quality and the price of a product from its images. As a consequence, aesthetic features could play a vital role in improving the cross-domain recommendation performance of appearance-first products. [11] is the first work to consider aesthetic preferences as a bridge between different domains for cross-domain recommendation, the result demonstrated that the introduction of aesthetic features into the cross-domain recommendation does have a significant effect on improving the recommendation performance [21]. However, [11] simply assumes all aesthetic preferences are transferable among different domains, and fails to investigate the transferability of features in different domains.

Unfortunately, aesthetic preferences in different domains can not be all served as a bridge for knowledge transfer across domains effectively. On the one hand, this is because although aesthetic preferences vary significantly from person to person, the same user's aesthetic preferences in different domains could be consistent [9, 11], which are called domain-independent aesthetic preferences. On the other hand, precisely due to different domains, we argue that users will also have inconsistent aesthetic preferences in different domains, which are called domain-dependent aesthetic preferences. Therefore, the direct transfer of all aesthetic preferences between different domains will introduce noise from the source domain into the target domain, leading to negative transfer [9]. If domain-dependent aesthetic features are transferred across domains as a bridge, it will not help to improve the recommendation performance. In addition, despite aesthetic is indeed an opportunity for CDR, the performance improvement is limited. The main reason for this problem is that only considering aesthetic preferences modeling is not satisfactory to portray user's comprehensive preferences accurately. Therefore, general preferences [6,7,13,17], which reflect the features of interaction behaviors between users and items and are as important as aesthetic preferences, should be effectively fused with aesthetic preferences with different weights to reflect their importance in better modeling user's preferences.

To address the above challenges, we propose an Aesthetic-based Adversarial Transfer Network for Cross-Domain Recommendation, termed as A2TN, which employs adversarial transfer learning [9] to find transferable knowledge applicable to both the source domain and the target domain. Specifically, in A2TN, we first utilize an aesthetic feature extraction network to extract the holistic features which represent the aesthetic elements of a product photo (e.g., the aesthetic elements can be color, structure, proportion, style, etc.). Also, a general feature embedding layer is used to embed general features, which reflect the features of interaction behaviors between users and items. After that, we input aesthetic features and general features into the adversarial transfer network to generate the representations of domain-dependent features and domain-independent features for each domain. In this way, the domain-dependent features are used for avoiding negative transfer, while the domain-independent features that the domain discriminator cannot identify which domain they belong to are used to transfer knowledge across domains. At last, for the purpose of better portraying the user's preferences, we apply an element-wise attention mechanism to fuse aesthetic preferences and general preferences with different weights to balance their importance. In conclusion, the main contributions of this paper are summarized as follows:

- We propose a novel cross-domain recommendation model A2TN, which can effectively enhance the performance of knowledge transfer in different domains. Specifically, it takes advantage of the transferability of aesthetic features to ensure the effect of feature transfer. To avoid negative transfer, it employs adversarial transfer network to find domain-independent features as intrinsic preferences to transfer across domains, and domain-dependent features are learned simultaneously to avoid being transferred.
- We further adopt an element-wise attention mechanism to balance the importance of aesthetic preferences and general preferences in modeling user's preferences, and fuse them for the purpose of better portraying user's comprehensive preferences accurately.
- We conduct extensive experiments on two real-world cross-domain datasets. Compared with some state-of-the-art single-domain and cross-domain recommendation methods, our model A2TN is demonstrated to be effective via comprehensive analysis.

2   Related Work

With the development of the Internet, recommender systems have attracted a lot of research interests in recent years for their ability to alleviate the growing problem of information overload, and more and more recommender systems [12,18,22] have achieved significant success. For example, some works on POI recommendation [20] and trip recommendation [2,10] can make personalized recommendations for users accurately and timely. After extensive researches on graph embedding representation [23,24], recommender systems [20] employing graph have also achieved extraordinary recommendation performance.

2.1   Cross-Domain Recommendation

Transfer-based CDR approaches tend to employ classical machine learning methods (e.g., transfer learning and deep neural networks) to transfer user/item latent factors across domains. The novel transfer learning method proposed in [7] is called collaborative cross networks (CoNet), which could learn complex user-item interaction relationships and enable dual knowledge transfer across domains. The authors suggested in [26] using deep neural networks to learn the nonlinear mapping function between the source domain and the target domain.

Content-based CDR approaches focus on establishing domain links by identifying similar user/item attributes, e.g., ratings, tags and reviews, and then transfer user's preferences or item's details across domains. The idea proposed in [1] targeted at solving the data sparsity problem by importing and aggregating vectors of users' ratings in different application domains. The work in [3] enriched both user and item profiles with independent sets of tag factors, better capturing the effects of tags on ratings, so that it can generate more accurate recommendations. The authors in [19] uncovered the correlations between different domains by making full use of text information.

2.2   Adversarial Transfer Network

Recently, adversarial neural networks have extraordinary performance in computer vision (CV) area and natural language processing (NLP) area. In cross-domain recommendation, the idea of adversarial neural networks has been introduced into domain adaptation. Adversarial transfer learning is a technique that incorporates adversarial learning inspired by generative adversarial nets (GAN) into transfer learning. However, few studies have explored how to use adversarial transfer network for cross-domain recommendation. The work in [22] proposed a deep domain adaptation model that is capable of extracting and transferring patterns from rating matrices only. The work in [9] proposed an adversarial transfer learning based model which effectively captures domain-shared and domain-specific features for cross-domain recommendation.

3   Notations and Problem Definition

In this section, we will introduce the important notations used in this paper and formalize the problem definition. Given a source domain $\mathcal{S}$ and a target domain $\mathcal{T}$, where the same set of users are shared, denoted by $\mathcal{U}=\{u_1,u_2,...,u_m\}$ (its size $m = |\mathcal{U}|$). The item sets from the source domain and the target domain are denoted by $\mathcal{I}_{\mathcal{S}}=\{i_1,i_2,...,i_{n_S}\}$ and $\mathcal{I}_{\mathcal{T}}=\{j_1,j_2,...,j_{n_T}\}$ (their item size $n_S=|\mathcal{I}_{\mathcal{S}}|$ and $n_T=|\mathcal{I}_{\mathcal{T}}|$) respectively. Then, we use matrix $\mathcal{R}_{\mathcal{S}} \in \mathbb{R}^{m \times n_S}$ to represent the user-item interaction matrix in the source domain, where the entry $r_{ui} \in \{0,1\}$ is 1 (observed) if the user u has purchased item i and 0 (unobserved) otherwise. Similarly, another matrix $\mathcal{R}_{\mathcal{T}} \in \mathbb{R}^{m \times n_T}$ is used to denote the user-item interactions in the target domain, where the entry $r_{uj} \in \{0,1\}$ is 1 (observed) if the user u has an interaction with item j and 0 (unobserved) otherwise. We denote the set of observed item interactions given by user u as $\mathcal{I}_u^S$ ($\mathcal{I}_u^T$), and the unobserved item interactions as $\overline{\mathcal{I}}_u^S$ ($\overline{\mathcal{I}}_u^T$) for the source (target) domain.

Problem Definition. The task of cross-domain recommendation can be specified as selecting a subset from $\overline{\mathcal{I}}_u^T$ (items set in the target domains) for user u based on his/her history records in both domains ($\mathcal{I}_u^S$ and $\mathcal{I}_u^T$), with a goal of maximizing his/her satisfaction in the target domain. In other words, our task aims at improving the recommendation performance in the target domain with the help of aesthetic features and general features from the source domain. In A2TN, each domain is jointly trained to improve the recommendation performance through mutual knowledge transfer.

4   The Proposed Model

In this section, we will give an overview of the proposed aesthetic-based adversarial transfer network (A2TN) for cross-domain recommendation. As is shown in Fig. 1, A2TN consists of four components: an Aesthetic Feature Extraction Network, a General Feature Embedding Layer, an Adversarial Transfer Network and an Element-Wise Attention Mechanism. In the following subsections, the details of each model component will be introduced, and finally the corresponding network learning strategy will be discussed in detail.

4.1   Aesthetic Feature Extraction Network

The pre-trained novel deep convolutional neural network named ILGNet [8] is used to extract high-level aesthetic features from item images. The ILGNet (I: Inception, L: Local, G: Global) combines both the inception modules and an connected layer of both local features and global features. In this paper, we apply the pre-trained ILGNet to extract the aesthetic features of each item in the source domain and the target domain. For each item in the source domain, we use $x_i^{Aes} \in \mathbb{R}^{1 \times 1024}$ to represent the aesthetic features of the corresponding image in advance. Similarly, for each item in the target domain, $x_j^{Aes} \in \mathbb{R}^{1 \times 1024}$ is obtained as its aesthetic features.

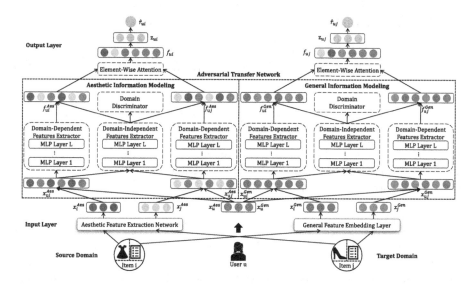

Fig. 1. Framework of A2TN for cross-domain recommendation.

4.2 General Feature Embedding Layer

For the purpose of representing the input of the model, we adopt the one-hot encoding to encode user-item interaction indices. For user u, item i in the source domain and item j in the target domain, we map them into one-hot encoding $\mathbb{X}_u \in \{0,1\}^m$, $\mathbb{X}_i \in \{0,1\}^{n_S}$, $\mathbb{X}_j \in \{0,1\}^{n_T}$ where only the element corresponding to that index is 1 and all others are 0. Then we embed one-hot encodings into continuous representations: $x_u^{Aes} = P_{Aes}^T \mathbb{X}_u$, $x_u^{Gen} = P_{Gen}^T \mathbb{X}_u$, $x_i^{Gen} = Q_i^T \mathbb{X}_i$ and $x_j^{Gen} = Q_j^T \mathbb{X}_j$ via four embedding matrices P_{Aes}, P_{Gen}, Q_i^T and Q_j^T respectively. Then we concatenate them (including x_i^{Aes} and x_j^{Aes}) and represent them as $x_{ui}^{Aes} = [x_u^{Aes}, x_i^{Aes}]$, $x_{uj}^{Aes} = [x_u^{Aes}, x_j^{Aes}]$, $x_{ui}^{Gen} = [x_u^{Gen}, x_i^{Gen}]$ and $x_{uj}^{Gen} = [x_u^{Gen}, x_j^{Gen}]$ to be the input of following adversarial transfer network, where x_i^{Gen} is the general features of item i in the source domain and x_j^{Gen} is the general features of item j in the target domain respectively.

4.3 Adversarial Transfer Network

To guarantee that domain-dependent features not to cause negative transfer and domain-independent features to be effectively transferred, we propose the adversarial transfer network which consists of domain-dependent features extractor, domain-independent features extractor, domain discriminator and cross-domain transfer layer. Next we will introduce them and how to implement them separately.

Domain-Dependent Features Extractor. For the domain-dependent features extractor, we utilize private Multi-layer Perception Layers (MLP) to extract domain-dependent features which are not supposed to be transferred across domains. Formally, the MLP layers can be defined as:

$$MLP(\boldsymbol{x}|\theta_{dependent}) = \phi_L(\phi_{L-1}(...(\phi_1(\boldsymbol{x}))...)) \tag{1}$$

$$\phi_L(\boldsymbol{x}) = \sigma_L(\boldsymbol{W}_L^T\boldsymbol{x} + \boldsymbol{b}_L) \tag{2}$$

where $\boldsymbol{W}_L^T$, $\boldsymbol{b}_L$ and σ_L denote the weight matrix, the bias vector and the activation function for the L-th layer respectively, the input $\boldsymbol{x}$ can be $\boldsymbol{x}_{ui}^{Aes}$ or $\boldsymbol{x}_{ui}^{Gen}$ for the source domain and $\boldsymbol{x}_{uj}^{Aes}$ or $\boldsymbol{x}_{uj}^{Gen}$ for the target domain, and $\theta_{dependent}$ denotes all the parameters for the whole MLP layers. After going through this process, we can get domain-dependent features: $\boldsymbol{f}_{dependent}^S$ from the source domain and $\boldsymbol{f}_{dependent}^T$ from the target domain.

Domain-Independent Features Extractor. For the domain-independent features extractor, we utilize shared Multi-layer Perception Layers (MLP) and domain discriminator to extract domain-independent features which are supposed to be transferred across domains. Similar to domain-dependent features extractor, the MLP layers can be defined as:

$$MLP(\boldsymbol{x}|\theta_{independent}) = \phi_L(\phi_{L-1}(...(\phi_1(\boldsymbol{x}))...)) \tag{3}$$

$$\phi_L(\boldsymbol{x}) = \sigma_L(\boldsymbol{W}_L^T\boldsymbol{x} + \boldsymbol{b}_L) \tag{4}$$

where $\boldsymbol{W}_L^T$, $\boldsymbol{b}_L$ and σ_L denote the weight matrix, the bias vector and the activation function for the L-th layer respectively, the input $\boldsymbol{x}$ can be $\boldsymbol{x}_{ui}^{Aes}$ or $\boldsymbol{x}_{ui}^{Gen}$ for the source domain and $\boldsymbol{x}_{uj}^{Aes}$ or $\boldsymbol{x}_{uj}^{Gen}$ for the target domain, and $\theta_{independent}$ denotes all the parameters for the whole MLP layers. After going through this process, we can get domain-independent features: $\boldsymbol{f}_{independent}^S$ from the source domain and $\boldsymbol{f}_{independent}^T$ from the target domain.

Domain Discriminator. To guarantee that only domain-independent features are transferred, the domain discriminator is used to determine which domain the features belong to. The domain discriminator can be expressed as:

$$D(\boldsymbol{f}_{independent}|\theta_D) = \sigma(MLP(\boldsymbol{f}_{independent}^c)) \tag{5}$$

where θ_D denotes the parameters of the domain discriminator, which predicts the corresponding domain label $c \in \{S, T\}$, while S denotes the input vector belongs to the source domain and T the target domain (domain classifier). By this way, the domain-independent features extractor can learn features to mislead the domain discriminator, so that the domain discriminator tries its best to predict the corresponding domain label correctly. Finally, the domain discriminator can not distinguish which domain label these features belong to, and these features are called domain-independent features.

Cross-Domain Transfer Layer. After going through the above operations, we can obtain domain-dependent features and domain-independent features from aesthetic information and general information: $f_{dependent}^{Aes}$, $f_{dependent}^{Gen}$, $f_{independent}^{Aes}$ and $f_{independent}^{Gen}$ in the target domain. Then, we get f^{Aes} and f^{Gen} respectively by combining them:

$$f^{Aes} = \sigma(W_{Aes}^T f_{dependent}^{Aes} + H_{Aes}^T f_{independent}^{Aes}) \tag{6}$$

$$f^{Gen} = \sigma(W_{Gen}^T f_{dependent}^{Gen} + H_{Gen}^T f_{independent}^{Gen}) \tag{7}$$

where W_{Aes}^T and W_{Gen}^T denote the transform matrix of domain-dependent features, and H_{Aes}^T and H_{Gen}^T denote the transfer matrix of domain-independent features in the target domain. And σ is the activation function.

4.4   Element-Wise Attention

For the purpose of better modeling user's preferences, we fuse aesthetic features (f^{Aes}) and general features (f^{Gen}) by an element-wise attention mechanism. By this way, the combined embedding (f) of user's comprehensive preferences can remain both preferences with different proportions and our element-wise attention tends to pay more attention to the more informative preferences from f^{Aes} and f^{Gen}. The combined embedding f can be represented as:

$$f = \left[W^A \odot f^{Aes}, (1 - W^A) \odot f^{Gen} \right] \tag{8}$$

where $\odot$ is the element-wise multiplication and W^A is the weight matrix for the attention network in the target domain.

Finally, we let it go through the dimension reduction operation (MLP) to get z_{uj}:

$$z_{uj} = MLP(\theta_f | f) \tag{9}$$

where θ_f denotes all the parameters for the MLP layers in the output layer. Then the items in the target domain are ranked by their predicted scores:

$$\widehat{r}_{uj} = \sigma(z_{uj}) \tag{10}$$

where σ is the activation function.

4.5   Model Learning

Due to the task of item recommendation and the nature of implicit feedback, we adopt cross-entropy loss as our loss function for model optimization, which is defined as follows:

$$\mathcal{L}_0 = - \sum_{(u,i) \in R^+ \cup R^-} r_{ui} \log \widehat{r}_{ui} + (1 - r_{ui}) \log(1 - \widehat{r}_{ui}) \tag{11}$$

where $\mathbf{R}^+$ and $\mathbf{R}^-$ are the observed interaction matrix and randomly sampled negative examples [15] respectively. This objective function has probabilistic interpretation and is the negative logarithm likelihood of the following likelihood function:

$$\mathcal{L}(\Theta|\mathbf{R}^+ \cup \mathbf{R}^-) = \prod_{(u,i)\in\mathbf{R}^+} \widehat{r}_{ui} \prod_{(u,i)\in\mathbf{R}^-} (1 - \widehat{r}_{ui}) \tag{12}$$

where Θ are model parameters. Now we define the final joint loss function of the proposed model (A2TN):

$$\mathcal{L}(\Theta) = \mathcal{L}_S(\Theta_S) + \mathcal{L}_T(\Theta_T) + \lambda \mathcal{L}_{Adv}(\Theta_{Adv}) \tag{13}$$

where the model parameters $\Theta = \Theta_S \cup \Theta_T \cup \Theta_{Adv}$ and λ is a hyper-parameter which controls the portion of loss contributed by the domain classifier. Instantiating the base loss $\mathcal{L}_0$ by the loss of the source domain ($\mathcal{L}_S$), loss of the target domain ($\mathcal{L}_T$) and loss of the domain classification $\mathcal{L}_{Adv}(\Theta_{Adv}$ including Θ_D). Then the optimization objective can be expressed as follows:

$$(\widehat{\Theta}_S, \widehat{\Theta}_T) = \underset{\Theta_S,\Theta_T}{\arg\min} \mathcal{L}(\Theta_S, \Theta_T, \Theta_{Adv}) \tag{14}$$

$$\widehat{\Theta}_{Adv} = \underset{\Theta_{Adv}}{\arg\max} \mathcal{L}(\Theta_S, \Theta_T, \Theta_{Adv}) \tag{15}$$

To address the above minimax optimization problem, we introduce a stochastic gradient descent (SGD)-like algorithm: the gradient reversal layer (GRL) [4]. Unlike stochastic gradient descent algorithm, during the forward propagation, GRL acts as an identity transform, while during the back propagation, GRL takes the gradient from the subsequent level, multiplies it by $-\lambda$ and passes it to the preceding layer. Formally, GRL can be treated as a "pseudo-function" defined by two equations:

$$\Phi(\boldsymbol{x}) = \boldsymbol{x} \quad (forward\ propagation) \tag{16}$$

$$\frac{d\Phi(\boldsymbol{x})}{d\boldsymbol{x}} = -\lambda \boldsymbol{I} \quad (back\ propagation) \tag{17}$$

where $\boldsymbol{I}$ is the identity matrix. Finally, we can update the parameters as follows:

$$\Theta_S = \Theta_S - \eta \frac{\partial \mathcal{L}_S}{\partial \Theta_S} \tag{18}$$

$$\Theta_T = \Theta_T - \eta \frac{\partial \mathcal{L}_T}{\partial \Theta_T} \tag{19}$$

$$\Theta_{Adv} = \Theta_{Adv} - \eta \cdot \lambda \frac{\partial \mathcal{L}_{Adv}}{\partial \Theta_{Adv}} \tag{20}$$

where η is the learning rate which can vary over time.

5   Experiments

In this section, we will systematically evaluate the proposed A2TN model on multiple real-world datasets with shared users in different domains. Then we will show its outstanding performance over the state-of-the-art recommendation models based on a wide range of baselines according to the broadly accepted evaluation protocols and make comprehensive analysis on experimental results.

5.1   Experimental Settings

Datasets. We evaluate our proposed model on a real-world cross-domain datasets *Amazon*[1]. It contains reviews and metadata from different kinds of domains, which have been widely used to evaluate the performance of recommendation tasks. In this paper, we use three domains: (1) Clothing, Shoes and Jewelry (2) Sports and Outdoors (3) Tools and Home Improvement, and conduct cross-domain recommendation experiments based on their pairwise combinations. We convert the ratings of 4–5 as positive samples, which means that users like these samples. The basic statistics of datasets are summarized in Table 1. Note that our dataset was crawled in March 2022 and is different from the dataset in [11] (different number of users, items, interactions and density).

Table 1. Datasets and statistics.

Dataset	Domain	#users	#items	#interactions	#density (%)
Dataset 1	Source Domain: Clothing, Shoes and Jewelry	3846	12684	20107	0.041
	Target Domain: Tools and Home Improvement	3846	11123	26807	0.063
Dataset 2	Source Domain: Sports and Outdoors	8009	24596	58088	0.029
	Target Domain: Clothing, Shoes and Jewelry	8009	33798	57336	0.021

Evaluation Metrics. For item recommendation task, the leave-one-out (LOO) evaluation is widely used and we follow the protocol in [6]. It suggests that we reserve one interaction as the test item for each user. We determine hyperparameters by randomly sampling another interaction per user as the validation set. We follow the common strategy which randomly samples 99 items (negative) that are not interacted by the user and then evaluate how well the recommender can rank the test item against these negative ones. Since we aim at Top-N item recommendation, the typical evaluation metrics are hit ratio (HR) and normalized discounted cumulative gain (NDCG), where the ranked list is cut off at Top-N = {5, 10, 20}. The higher the value, the better the performance.

---

[1] http://jmcauley.ucsd.edu/data/amazon/links.html.

Parameter Setup. We set the dimension of aesthetic features as 1024, the dimension of general features as 32 and the embedding dimension of users as 32. For the adversarial transfer network component, the number of MLP layers can be set to 3 to enhance feature representation. In the following experiments, we use Adam to optimize all trainable parameters of our model and then we set the initial learning rate η as 0.0001.

Baselines

- **BPRMF** [16]: Bayesian personalized ranking is a typical collaborate filtering approach, which learns the user and item latent factors via matrix factorization and pairwise rank loss.
- **VBPR** [5]: VBPR is a visual-based recommendation method, which incorporates visual signals into predictors of people's opinions.
- **CMF** [17]: Collective matrix factorization is a multiple relations learning approach, which can jointly factorize relation matrices of domains by sharing user latent factors between two domains and vice versa.
- **CDCF** [13]: Cross-domain collaborate filtering recommendation is the first work to exploit an extension of factorization machines on cross domain recommendation, which incorporates domain-specific user interactions patterns.
- **MLP** [6]: Multi-layer perception is a typical neural collaborate filtering approach, which can effectively learn a nonlinear user-item interaction function by neural networks.
- **MLP++**: MLP++ is degenerated method without cross transfer units, which is composed of two MLP model by sharing the user embedding matrix.
- **CSN** [14]: The cross-stitch network is a deep multi-task learning model, which can learn two networks separately via a sharing cross-stitch unit. It can learn an optimal combination of shared and task-specific representations.
- **CoNet** [7]: CoNet is a deep cross-domain recommendation model, which enables dual knowledge transfer across domains based on the cross-stitch networks by introducing cross connections from one base network to another.
- **ATLRec** [9]: ATLRec is an adversarial transfer learning network for cross-domain recommendation, which aims to address the problem of not being able to prevent domain-specific features to be transferred.
- **ACDN** [11]: ACDN is the latest collaborate networks for cross-domain recommendation, which leverages an aesthetic network to extract aesthetic features and integrates them into a cross-domain network to transfer users' domain independent aesthetic preferences.

5.2 Performance Comparison

The performance comparison results of different baselines are illustrated in Table 2. Here the improvement is computed as the difference of A2TN and the best state-of-the-are method on that metric, shown in percentage. From the table we can observe that:

Table 2. Performance comparison results of different methods on two datasets.

Dataset	Dataset 1						Dataset 2					
Method	HR@5	NDCG@5	HR@10	NDCG@10	HR@20	NDCG@20	HR@5	NDCG@5	HR@10	NDCG@10	HR@20	NDCG@20
BPRMF	0.1862	0.1159	0.2964	0.1517	0.4049	0.1681	0.1833	0.1267	0.2701	0.1623	0.3508	0.1664
VBPR	0.1903	0.1204	0.2980	0.1553	0.4091	0.1712	0.1884	0.1310	0.2735	0.1647	0.3531	0.1692
CMF	0.1984	0.1270	0.3032	0.1601	0.4163	0.1793	0.1946	0.1366	0.2799	0.1705	0.3578	0.1728
CDCF	0.1937	0.1250	0.3008	0.1579	0.4116	0.1756	0.1904	0.1348	0.2761	0.1682	0.3553	0.1716
MLP	0.2101	0.1396	0.3141	0.1743	0.4318	0.1980	0.2055	0.1461	0.2952	0.1814	0.3716	0.1843
MLP++	0.2125	0.1422	0.3187	0.1763	0.4349	0.2005	0.2086	0.1493	0.2985	0.1846	0.3712	0.1870
CSN	0.2168	0.1455	0.3223	0.1790	0.4379	0.2066	0.2115	0.1514	0.3022	0.1887	0.3748	0.1903
CoNet	0.2200	0.1487	0.3271	0.1834	0.4412	0.2127	0.2164	0.1543	0.3082	0.1903	0.3789	0.1946
ATLRec	0.2270	0.1521	0.3315	0.1860	0.4529	0.2182	0.2206	0.1574	0.3116	0.1926	0.3819	0.2014
ACDN	0.2306	0.1543	0.3328	0.1890	0.4615	0.2216	0.2302	0.1651	0.3161	0.1955	0.4053	0.2186
A2TN	**0.2379**	**0.1604**	**0.3440**	**0.1944**	**0.4704**	**0.2262**	**0.2405**	**0.1723**	**0.3208**	**0.1998**	**0.4188**	**0.2210**
Improvement	3.17%	3.95%	3.37%	2.86%	1.93%	2.08%	4.47%	4.36%	1.49%	2.20%	3.33%	1.10%

Firstly, we can conclude that deep methods perform better than shallow methods in either single-domain or cross-domain. Specifically, deep single-domain methods (i.e., MLP) outperform shallow single-domain methods (i.e., BPRMF and VBPR), and deep cross-domain methods (i.e., MLP++, CSN, CoNet, ATLRec and ACDN) outperform shallow cross-domain methods (i.e., CMF and CDCF) in all cases on two datasets. This indicates that non-linear relationships and more parameters learned by deep neural networks benefit not only single-domain but also cross-domain recommendation.

Secondly, we can see that compared with single-domain methods (i.e., BPRMF, VBPR and MLP), cross-domain methods (i.e., CMF, CDCF, MLP++, CSN, CoNet, ATLRec and ACDN) show better performance no matter in the case of shallow methods or deep methods in all cases on two datasets. This shows that cross-domain methods benefit from the knowledge transferred from the auxiliary domains, especially encountered the problem of data sparsity. In addition, VBPR outperforms BPRMF, which suggests that visual features extracted from product images can really improve recommendation performance.

Thirdly, we can notice that our proposed model called A2TN performs better than all baselines at all settings on both datasets, including shallow cross-domain methods (i.e., CMF and CDCF) and deep cross-domain methods (i.e., MLP++, CSN, CoNet, ATLRec and ACDN). The reason why A2TN outperforms the state-of-the-art method ACDN is that the fusion of aesthetic features and general features for cross-domain recommendation is extremely effective. Also, what features to transfer and what features not to transfer is a matter deserving great concern.

Finally, we can summarize that the empirical comparison results reveal the superiority of the proposed model for cross-domain recommendation called A2TN, which transfers domain-independent features to guarantee the effect of feature transfer and learns domain-dependent features to avoid negative transfer. Furthermore, the element-wise attention mechanism is employed to balance the different importance of aesthetic preferences and general preferences in modeling user's preferences.

5.3 Impacts of the Adversarial Transfer Network and Information Modeling

In order to verify the impacts of the adversarial transfer network and information modeling, we keep the architecture of our proposed model and then design another three models: A2TN-MLP, A2TN-G and A2TN-A.

- **MLP** is a simplified model without adversarial transfer learning network.
- **A2TN-MLP** replaces adversarial transfer network with MLP layers.
- **A2TN-G** only consider aesthetic features to capture the preferences of a user, ignoring the general features.
- **A2TN-A** only consider general features to capture the preferences of a user, ignoring the aesthetic features.

Table 3. Performance of A2TN compared with variants of A2TN on two datasets.

Dataset	Dataset 1						Dataset 2					
Method	HR@5	NDCG@5	HR@10	NDCG@10	HR@20	NDCG@20	HR@5	NDCG@5	HR@10	NDCG@10	HR@20	NDCG@20
MLP	0.2101	0.1396	0.3141	0.1743	0.4318	0.1980	0.2055	0.1461	0.2952	0.1814	0.3716	0.1843
A2TN-MLP	0.2246	0.1506	0.3326	0.1872	0.4559	0.2135	0.2237	0.1542	0.3069	0.1880	0.3953	0.1946
A2TN-G	0.1251	0.1126	0.1851	0.1291	0.2813	0.1376	0.1201	0.0763	0.1864	0.0927	0.3166	0.1271
A2TN-A	0.2275	0.1551	0.3287	0.1860	0.4566	0.2181	0.2261	0.1625	0.2994	0.1869	0.3986	0.2006
A2TN	**0.2379**	**0.1604**	**0.3440**	**0.1944**	**0.4704**	**0.2262**	**0.2405**	**0.1723**	**0.3208**	**0.1998**	**0.4188**	**0.2210**

As is illustrated in Table 3, we can find that our proposed model A2TN always performs the better HR and NDCG results than the variants of A2TN at all settings on both datasets. Specifically, the adversarial transfer network can learn domain-dependent features and domain-independent features separately, and only transfers domain-independent features across domains without transferring domain-dependent features. This explains that our model A2TN can ensure effective knowledge transfer instead of negative transfer. A2TN-G only considers aesthetic features while A2TN-A only considers general features, which is unsatisfactory to consider them separately for cross domain recommendation. In contract, the reason why our proposed model performs best is that A2TN considers not only aesthetic features but also general features and fuse them well with different importance. This phenomenon exactly proves that aesthetic can be treated as a bridge between different domains for knowledge transfer, and the addition of general features can further improve recommendation performance.

5.4 Effects of the Hyper-Parameter λ

We analyse the sensitivity of the hyper-parameter λ which controls the portion of loss contributed by the domain classifier. We keep the structure of our proposed model A2TN and then optimize the performance of A2TN varying with $\lambda \in [0.01, 0.1]$ on both datasets. From Fig. 2, we can conclude that Dataset 1 and Dataset 2 achieve their best performance when λ is equal to 0.05 and 0.07

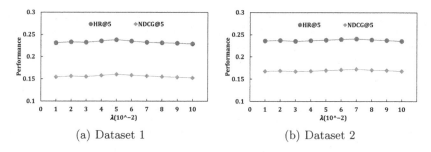

(a) Dataset 1   (b) Dataset 2

Fig. 2. Effects of the hyper-parameter λ.

respectively. The hyper-parameter λ represents the trade-off between extracting domain-dependent features and domain-independent features. As a result, the larger the hyper-parameter λ, the more domain-independent features will be transferred across domains.

6 Conclusion

In this paper, we proposed an Aesthetic-based Adversarial Transfer Network (A2TN), which can take advantage of aesthetic and adversarial transfer network to find transferable knowledge. Specifically, it can capture domain-dependent features for avoiding negative transfer and domain-independent features for effective knowledge transfer. Moreover, for the purpose of better learning user's comprehensive preferences, we fuse aesthetic features and general features with different importance. Finally, we conduct extensive experiments on various datasets and the results demonstrate that our model has the best performance by comparing with the state-of-the-art methods. In the future work, we will investigate the effectiveness of our model based on larger appearance-first product datasets.

References

1. Berkovsky, S., Kuflik, T., Ricci, F.: Cross-domain mediation in collaborative filtering. In: User Modeling, pp. 355–359 (2007)
2. Chen, X., et al.: S^2r-tree: a pivot-based indexing structure for semantic-aware spatial keyword search. GeoInformatica **24**(1), 3–25 (2020)
3. Fernández-Tobías, I., Cantador, I.: Exploiting social tags in matrix factorization models for cross-domain collaborative filtering. In: CBRecSys@RecSys, pp. 34–41 (2014)
4. Ganin, Y., Lempitsky, V.S.: Unsupervised domain adaptation by backpropagation. In: ICML, pp. 1180–1189 (2015)
5. He, R., McAuley, J.J.: VBPR: visual Bayesian personalized ranking from implicit feedback. In: AAAI, pp. 144–150 (2016)
6. He, X., Liao, L., Zhang, H., Nie, L., Hu, X., Chua, T.: Neural collaborative filtering. In: WWW, pp. 173–182 (2017)

7. Hu, G., Zhang, Y., Yang, Q.: Conet: collaborative cross networks for cross-domain recommendation. In: CIKM, pp. 667–676 (2018)
8. Jin, X., et al.: ILGNet: inception modules with connected local and global features for efficient image aesthetic quality classification using domain adaptation. IET Comput. Vis. **13**, 206–212 (2019)
9. Li, Y., Xu, J., Zhao, P., Fang, J., Chen, W., Zhao, L.: ATLRec: an attentional adversarial transfer learning network for cross-domain recommendation. J. Comput. Sci. Technol. **35**, 794–808 (2020)
10. Liu, H., Xu, J., Zheng, K., Liu, C., Du, L., Wu, X.: Semantic-aware query processing for activity trajectories. In: WSDM, pp. 283–292. ACM (2017)
11. Liu, J., et al.: Exploiting aesthetic preference in deep cross networks for cross-domain recommendation. In: WWW, pp. 2768–2774 (2020)
12. Liu, Y., Li, B., Zang, Y., Li, A., Yin, H.: A knowledge-aware recommender with attention-enhanced dynamic convolutional network. In: CIKM, pp. 1079–1088. ACM (2021)
13. Loni, B., Shi, Y., Larson, M.A., Hanjalic, A.: Cross-domain collaborative filtering with factorization machines. In: ECIR, pp. 656–661 (2014)
14. Misra, I., Shrivastava, A., Gupta, A., Hebert, M.: Cross-stitch networks for multi-task learning. In: CVPR, pp. 3994–4003 (2016)
15. Pan, R., et al.: One-class collaborative filtering. In: ICDM, pp. 502–511 (2008)
16. Rendle, S., Freudenthaler, C., Gantner, Z., Schmidt-Thieme, L.: BPR: Bayesian personalized ranking from implicit feedback. In: UAI, pp. 452–461 (2009)
17. Singh, A.P., Gordon, G.J.: Relational learning via collective matrix factorization. In: KDD, pp. 650–658 (2008)
18. Sun, H., Xu, J., Zheng, K., Zhao, P., Chao, P., Zhou, X.: MFNP: a meta-optimized model for few-shot next POI recommendation. In: IJCAI, pp. 3017–3023. ijcai.org (2021)
19. Tan, S., Bu, J., Qin, X., Chen, C., Cai, D.: Cross domain recommendation based on multi-type media fusion. Neurocomputing **127**, 124–134 (2014)
20. Wu, S., Zhang, Y., Gao, C., Bian, K., Cui, B.: GARG: anonymous recommendation of point-of-interest in mobile networks by graph convolution network. Data Sci. Eng. **5**(4), 433–447 (2020)
21. Yu, W., Zhang, H., He, X., Chen, X., Xiong, L., Qin, Z.: Aesthetic-based clothing recommendation. In: WWW, pp. 649–658 (2018)
22. Yuan, F., Yao, L., Benatallah, B.: Darec: deep domain adaptation for cross-domain recommendation via transferring rating patterns. In: IJCAI, pp. 4227–4233 (2019)
23. Zhou, A., Wang, Y., Chen, L.: Finding large diverse communities on networks: the edge maximum k*-partite clique. Proc. VLDB Endow. **13**(11), 2576–2589 (2020)
24. Zhou, A., Wang, Y., Chen, L.: Butterfly counting on uncertain bipartite networks. Proc. VLDB Endow. **15**(2), 211–223 (2021)
25. Zhu, F., Chen, C., Wang, Y., Liu, G., Zheng, X.: DTCDR: a framework for dual-target cross-domain recommendation. In: CIKM, pp. 1533–1542 (2019)
26. Zhu, F., Wang, Y., Chen, C., Liu, G., Orgun, M.A., Wu, J.: A deep framework for cross-domain and cross-system recommendations. In: IJCAI, pp. 3711–3717 (2018)

MORO: A Multi-behavior Graph Contrast Network for Recommendation

Weipeng Jiang[1], Lei Duan[1,2]($\boxtimes$), Xuefeng Ding[1], and Xiaocong Chen[1]

[1] School of Computer Science, Sichuan University, Chengdu, China
{jiangweipeng,chenxiaocong}@stu.scu.edu.cn,
{leiduan,dingxf}@scu.edu.cn
[2] Med-X Center for Informatics, Sichuan University, Chengdu, China

Abstract. Multi-behavior recommendation models exploit diverse behaviors of users (*e.g.*, page view, add-to-cart, and purchase) and successfully alleviate the data sparsity and cold-start problems faced by classical recommendation methods. In real-world scenarios, the interactive behaviors between users and items are often complex and highly dependent. Existing multi-behavior recommendation models do not fully utilize multi-behavior information in the following two aspects: (1) The diversity of user behavior resulting from the individualization of users' intents. (2) The loss of user multi-behavior information due to inappropriate information fusion. To fill this gap, we hereby propose a multi-behavior graph contrast network (MORO). Firstly, MORO constructs multiple behavior representations of users from different behavior graphs and aggregate these representations based on behavior intents of each user. Secondly, MORO develops a contrast enhancement module to capture information of high-order heterogeneous paths and reduce information loss. Extensive experiments on three real-world datasets show that MORO outperforms state-of-the-art baselines. Furthermore, the preference analysis implies that MORO can accurately model user multi-behavior preferences.

Keywords: Multi-behavior recommendation · Graph neural network · Multi-task learning · Representation learning

1 Introduction

Recommender systems are widely used in online retail platforms and review sites as techniques to alleviate information overload. How to exploit user behavior data to learn effective user/item representations is the key problem of effective recommendations [5,7,20]. Practically, users will perform different behaviors to items under different intents. As shown in Fig. 1, there are three behaviors (*i.e.*, "page view", "add-to-cart", and "purchase") between users and items, which indicate three kinds of user intents. However, the classical methods [2,5,12,13,20]

This work was supported by the National Natural Science Foundation of China (61972268), the National Key Research and Development Program of China (2018YFB0704301-1), Med-X Center for Informatics Funding Project (YGJC001).

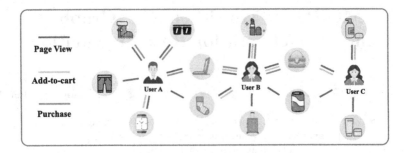

Fig. 1. A toy example of users' multi-behavior data. Best viewed in color. (Color figure online)

consider only single behavior (*e.g.*, purchase) instead of the diversity of multiple behaviors, making it difficult to provide a good list of recommendations to users who have no purchase behavior (*e.g.*, user C in Fig. 1).

To utilize the diversity of user behaviors, several efforts on multi-behavior recommender systems have been made. These works can be roughly classified into two categories. *Deep collaborate filtering-based methods* [3,4,17] use neural network techniques to enhance model representation learning. For instance, MATN [17] employs a transformer module to uniformly capture the dependencies among user behaviors. However, these methods ignore high-order information between users and items [1,16]. Then, *graph neural network-based methods* [1,6,19,21–23] are proposed recently, which model user multi-behavior in two different ways: (1) constructing a unified graph of multi-behavior data and learning user representations on the unified graph [1,6]; (2) constructing subgraph for each user behavior type, learning the representations on different subgraphs, and finally aggregating them [16,19]. However, these methods are still insufficient in their use of multi-behavioral data in two main ways.

- **The lack of modeling diverse user behavior intents.** As shown in Fig. 1, there are various user multi-behavior patterns. Specifically, user A likes to "purchase" items directly after "page view", while user B will "add-to-cart" and then "purchase" items. Uniform modeling the user behaviors as MATN [17] and GNMR [16] will lose this customized feature which is important for modeling user preference. Therefore, modeling users' personalized multi-behavior patterns is one of the goals in this paper.
- **The loss of fusion multi-behavior information.** Due to the influence of other behaviors, modeling on a unified graph cannot fully explore users' preferences of a specific behavior [6]. Moreover, modeling on different behavior subgraphs is difficult to capture the information of high-order heterogeneous paths between users and items (*e.g.*, user→page view→item→purchase →user) [19]. Thus, another goal of this paper is to reduce the information loss in multi-behavior information fusion.

To address these limitations, we propose *multi-behavior graph contrast networks* (MORO) to model complex user behaviors effectively. First, MORO uses graph convolution networks to construct user/item representations under different behavior graphs. Then, we mine the user's personalized multi-behavior patterns with a behavior perceptron module, which is our proposed novel idea enabling MORO to leverage personalized information about user behavior intents. In addition, we propose a contrast enhancement module to reduce the information loss of the multi-behavior information aggregation.

The main contributions of this work are summarized as follows:

- We emphasize the importance of modeling user personalized multi-behavior patterns and the loss of fusion multi-behavior information.
- We propose a novel recommendation model, MORO, which exploits users' personalized behavior patterns and tries to reduce the information loss in the aggregation process of different behavior representations.
- We conduct extensive experiments on three real-world datasets whose results demonstrate that our model outperforms baselines. Further studies on user preferences validate the interpretability of our model.

2   Related Work

In this section, we review works on multi-behavior recommendation. We roughly divide these works into two categories: *deep collaborate filtering-based methods* and *graph neural network-based methods*.

Deep collaborate filtering-based methods leveraged neural networks to learn effective representations of users and items from the interaction data. For example, NCF [5] employed a multilayer perceptron to replace the inner product to calculate the user's acceptance probability of the item. DMF [20] projected representations into the same semantic space, which reduced the noise of inner product computation. Moreover, AutoRec [12] and CDAE [15] introduced autoencoders into recommendation systems by minimizing the representation reconstruction loss. Recently, NMTR [3] proposed a multi-task framework for performing the cascade prediction of different behaviors. DIPN [4] leveraged attention mechanism to model user representations from user multi-behaviors and predict the purchase intent of users. MATN [17] introduced transformer to the multi-behavior recommendation. However, these methods are difficult to capture higher-order neighbor information which makes improvement limited.

Graph neural network-based methods employed graph neural network (GNN) [14,18,21] to aggregate the high-order neighbor information by message propagation mechanism. Wang et al. [13] proposed NGCF to aggregate neighbor information from the user-item bipartite graph. GNMR [16] leveraged GNN to learn representations from multi-behavior interactions between users and items. MBGCN [6] learned user preferences through multi-behavior interaction and learned item representations by item-relevance aware propagation. Moreover, MGNN [23] constructed a multiplex graph and proposed a multiplex graph neural network for multi-behavior recommendation. Recently, GHCF [1] employed efficient multi-task learning without sampling in multi-behavior recommendation. MB-GMN [19] combined meta-learning with GNN to learn the meta-knowledge of user

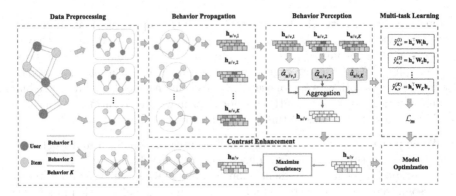

Fig. 2. Illustration of the proposed MORO. Best viewed in color. (Color figure online)

behaviors and improved accuracy of recommendation results. However, due to the differences in building methods, these methods will have some information loss in multi-behavior information fusion.

Different from these methods, we propose MORO to exploit user personalized behavior patterns from multi-behavior data. Moreover, we design a contrastive enhancement module to reduce the information loss on representation aggregation from different behaviors.

3   Preliminary

In multi-behavior recommendation, we need to define a behavior (*e.g.*, purchase) as *target behavior* which we aim to predict. Other relevant behaviors (*e.g.*, page view, add-to-cart, and add-to-favorite) will be defined as *source behaviors*. We begin with introducing key notations and considering a multi-behavior recommendation scenario with users and items:

Definition 1. *Behavior Graph.* *A behavior graph* $\mathcal{G}_k = (\mathcal{U}_k, \mathcal{V}_k, \mathcal{E}_k)$ *represents the behavior* k *from* $|\mathcal{U}_k|$ *users over* $|\mathcal{V}_k|$ *items. Specifically, each* $e_{u,v,k} \in \mathcal{E}_k$ *denotes an observed behavior* k *between user* $u \in \mathcal{U}_k$ *and item* $v \in \mathcal{V}_k$.

Definition 2. *Multi-behavior Graph.* *A multi-behavior graph* $\mathcal{G} = (\mathcal{U}, \mathcal{V}, \mathcal{E})$ *represents all kinds of behaviors* $\mathcal{E} = \bigcup_{k=1}^{K} \mathcal{E}_k$ *from users* $\mathcal{U} = \bigcup_{k=1}^{K} \mathcal{U}_k$ *over items* $\mathcal{V} = \bigcup_{k=1}^{K} \mathcal{V}_k$. *Particularly,* K *denotes the number of behavior categories. In the multi-behavior recommendation scenario,* K *is fixed.*

Task Formulation. Given a multi-behavior graph $\mathcal{G} = (\mathcal{U}, \mathcal{V}, \mathcal{E})$ and a target behavior k, the task of multi-behavior recommendation is to learn a predictive model $\mathcal{F}$ which outputs the probability $\hat{y}_{uv}^{(k)}$ that user $u \in \mathcal{U}$ performs target behavior k to item $v \in \mathcal{V}$.

4   Methodology

We now present our proposed MORO, which exploits multi-behavior data to learn users' preferences. Figure 2 illustrates the framework of MORO, which consists of four key components: (1) behavior propagation module, which models the representations of users and items in different behavior graphs (Sect. 4.1); (2) behavior perception module, which exploits user personalized multi-behavior preferences and aggregates representations from different behavior graphs (Sect. 4.2); (3) contrast enhancement module, which maximizes consistency between multi-behavior aggregated representations and global behavior representations to reduce information loss (Sect. 4.3); (4) multi-task learning module, which makes full use of the information of multi-behavior data (Sect. 4.4).

4.1   Behavior Propagation

In the E-commerce scenario, different behaviors between users and items contain different user intents. For example, "purchase" performs more obvious user preference than "page view". However, the "purchase" behavior is often sparse in practical application scenarios. To take full advantage of multi-behavior, MORO constructs the representations of users and items from different behavior graphs.

As shown in Fig. 2, MORO splits the multi-behavior graph $\mathcal{G}$ into several behavior graphs $\mathcal{G}_k \subset \mathcal{G}$ according to the behavior types. For each behavior graph $\mathcal{G}_k$, MORO performs a behavior propagation to engage the neighborhood information and obtain user/item representations:

$$\mathbf{h}_{u,k}^l = \sum_{v \in \mathcal{N}_k(u)} \beta_{u,v}^{(k)} \mathbf{h}_{v,k}^{l-1} \odot \mathbf{h}_k \tag{1}$$

$$\mathbf{h}_{v,k}^l = \sum_{u \in \mathcal{N}_k(v)} \beta_{v,u}^{(k)} \mathbf{h}_{u,k}^{l-1} \odot \mathbf{h}_k \tag{2}$$

where $\mathbf{h}_{u,k}^l \in \mathbb{R}^d$ and $\mathbf{h}_{v,k}^l \in \mathbb{R}^d$ denote the representations of user u and item v collecting the information from l-hops in behavior graph $\mathcal{G}_k$, respectively; $\mathbf{h}_k$ denotes the ID embedding of behavior k; and $\odot$ is the Hadamard product. Particularly, $\mathbf{h}_{u,k}^0$ and $\mathbf{h}_{v,k}^0$ denote the ID embeddings of user u and item v, respectively.

Inspired by UltraGCN [8], we calculate propagation weight $\beta_{v,u}^{(k)}$ of user u to item v in the behavior graph $\mathcal{G}_k$ in the following way:

$$\beta_{u,v}^{(k)} = \frac{1}{f(u,\mathcal{G}_k)} \sqrt{\frac{f(u,\mathcal{G}_k)+1}{f(v,\mathcal{G}_k)+1}} \tag{3}$$

where $f(u,\mathcal{G}_k)$ denotes the degrees of user u in behavior graph $\mathcal{G}_k$; $f(v,\mathcal{G}_k)$ denotes the degrees of item v in $\mathcal{G}_k$.

After L-layer propagations, we obtain the user/item representations at different layers. The user/item representations about behavior k are obtained:

$$\mathbf{h}_{u,k} = \frac{1}{L+1} \sum_{l=0}^{L} \mathbf{h}_{u,k}^l, \quad \mathbf{h}_{v,k} = \frac{1}{L+1} \sum_{l=0}^{L} \mathbf{h}_{v,k}^l \tag{4}$$

4.2   Behavior Perception

In the multi-behavior recommendation, user behaviors tend to show personalized. Taking the E-commerce scenario as an example, some users "purchase" the products after "page view", while others will "purchase" the products they prefer after they have "page view" and "add-to-cart". Therefore, it is necessary to model users' multi-behavior preferences individually.

In order to fuse the different behavior information, we define behavior perception weight for a particular behavior k for user u denoted as $\hat{\alpha}_{u,k}$:

$$\alpha_{u,k} = \mathbf{W}_1 \sigma_1(\mathbf{W}_2 \mathbf{h}_{u,k} + \mathbf{b}_2) + b_1 \tag{5}$$

$$\hat{\alpha}_{u,k} = \frac{exp(\alpha_{u,k})}{\sum_{k'=1}^{K} exp(\alpha_{u,k'})} \tag{6}$$

where $\mathbf{W}_1 \in \mathbb{R}^{1 \times d'}$ and $\mathbf{W}_2 \in \mathbb{R}^{d' \times d}$ denote two project matrices; $b_1 \in \mathbb{R}$ and $\mathbf{b}_2 \in \mathbb{R}^{d'}$ are bias terms; and the activate function $\sigma_1(\cdot)$ is LeakyReLU. d' and d denote the embedding dimensions ($d' < d$). Symmetrically, the behavior-aware weight of a specific item v can also be calculated by Eqs. (5–6).

MORO aggregates the representations from multiple behavior graphs, and obtains the multi-behavior aggregated representations about user u and item v:

$$\mathbf{h}_u = \sum_{k=1}^{K} \hat{\alpha}_{u,k} \mathbf{h}_{u,k}, \quad \mathbf{h}_v = \sum_{k=1}^{K} \hat{\alpha}_{v,k} \mathbf{h}_{v,k} \tag{7}$$

Therefore, MORO is able to aggregate the information of different behavior graph and generate multi-behavior aggregated representations of users and items. The perception weights also explain the multi-behavior preferences of users.

4.3   Contrast Enhancement

Since the multi-behavior aggregated representations (e.g., $\mathbf{h}_u$ and $\mathbf{h}_v$) are aggregated from different behavior graphs, there will be information loss about high-order heterogeneous path. As shown in Fig. 2, we design a contrast enhancement module to reduce the information loss of high-order heterogeneous path (e.g., user→purchase→item→page view →user). Here, we use R-GCN [11] to encode global behavior representations from multi-behavior graph:

$$\hat{\mathbf{h}}_u^l = \sigma\left(\sum_{k=1}^{K} \sum_{v \in \mathcal{N}_k(u)} \frac{1}{|\mathcal{N}_k(u)|} \mathbf{W}_k^{l-1} \hat{\mathbf{h}}_v^{l-1} + \mathbf{W}_0^{l-1} \hat{\mathbf{h}}_u^{l-1}\right) \tag{8}$$

$$\hat{\mathbf{h}}_v^l = \sigma\left(\sum_{k=1}^{K} \sum_{u \in \mathcal{N}_k(v)} \frac{1}{|\mathcal{N}_k(v)|} \mathbf{W}_k^{l-1} \hat{\mathbf{h}}_u^{l-1} + \mathbf{W}_0^{l-1} \hat{\mathbf{h}}_v^{l-1}\right) \tag{9}$$

where $\mathbf{W}_k^{l-1} \in \mathbb{R}^{d \times d}$ denotes the transition matrix for $(l-1)$-hop neighbors in the semantic space of behavior k; $\mathbf{W}_0^{l-1} \in \mathbb{R}^{d \times d}$ is the transition matrix for self-loop in all behaviors; and the activate function $\sigma(\cdot)$ is LeakyReLU. Particularly, $\hat{\mathbf{h}}_u^0$ and $\hat{\mathbf{h}}_v^0$ denote the ID embeddings of user u and item v, respectively.

Thus, the multi-behavior aggregated representation $\mathbf{h}_u$ and the global behavior representation $\hat{\mathbf{h}}_u$ to form two representations for user u from different perspectives. With these representations, we can efficiently model the relation between $\mathbf{h}_u$ and $\hat{\mathbf{h}}_u$. Based on InfoNCE [9], we design the graph-contrastive learning objective to maximize the consistency between $\mathbf{h}_u$ and $\hat{\mathbf{h}}_u$ as follows:

$$\mathcal{L}_{user} = \sum_{i \in \mathcal{U}} -log \frac{exp(s(\mathbf{h}_i, \hat{\mathbf{h}}_i)/\tau)}{\sum_{j \in \mathcal{U}} exp(s(\mathbf{h}_i, \hat{\mathbf{h}}_j)/\tau)} \tag{10}$$

where τ is the temperature hyperparameter of softmax; and $s(\cdot)$ denotes the cosine similarity function. Similarly, the graph-contrastive loss of the item side $\mathcal{L}_{item}$ can be obtained. And the complete graph-contrastive loss is the weighted sum of the above two losses:

$$\mathcal{L}_c = \mathcal{L}_{user} + \gamma \mathcal{L}_{item} \tag{11}$$

where γ is a hyperparameter to balance the weight of the two terms in graph-contrastive loss.

4.4   Multi-task Learning

In the multi-behavior recommendation scenario, the target behavior presents more obvious sparsity and cold-start problems than the source behaviors. As shown in Fig. 1, user C only performs "page view" and "add-to-cart" behaviors, and no purchase records. Therefore, using only "purchase" behavior data as target behavior to provide supervision signals is insufficient. We design a multi-task learning module to make full use of multi-behavior data to learn accurate user/item representations. Specifically, the probability of user u and item v having behavior k is calculated as follows:

$$\hat{y}_{u,v}^{(k)} = \mathbf{h}_u^\top \mathbf{W}_k \mathbf{h}_v \tag{12}$$

where $\mathbf{W}_k \in \mathbb{R}^{d \times d}$ denotes the project matrix from global semantic space to the semantic space of target behavior k. We optimize BRP loss [10] for each target behavior and sum them up to obtain the final loss:

$$\mathcal{L}_m = \sum_{u \in \mathcal{U}} \sum_{k=1}^{K} \sum_{(s,s') \in \mathcal{S}_{u,k}} -log(\sigma_2(\hat{y}_{u,s}^{(k)} - \hat{y}_{u,s'}^{(k)})) \tag{13}$$

where $\mathcal{S}_{u,k}$ denotes the set of positive and negative item sample pair of user u under behavior k. For each pair $(s, s') \in \mathcal{S}_{u,k}$, s denotes the positive item sample, s' denotes the negative item sample. The activate function $\sigma_2(\cdot)$ is softmax.

By combining $\mathcal{L}_c$ and $\mathcal{L}_m$, the following objective function for training model can be obtained:

$$\mathcal{L} = \mathcal{L}_m + \lambda_1 \mathcal{L}_c + \lambda_2 ||\Theta||^2 \tag{14}$$

where λ_1 and λ_2 are the hyper-parameters to control the weights of contrastive object and the regularization term, respectively, and Θ is the model parameters.

Table 1. Statistics of datasets.

Dataset	#User	#Item	#Interaction	Interaction behavior type
Taobao	147,894	99,037	7,658,926	{Page View, Favorite, Cart, Purchase}
Beibei	21,716	7,977	3,338,068	{Page View, Cart, Purchase}
Yelp	19,800	22,734	1,400,036	{Tip, Dislike, Neutral, Like}

5   Experiments

In this section, we evaluate the proposed MORO on three real-world datasets. The experiments are designed to answer the following research questions:

- **RQ1:** How does MORO perform compared with other baselines?
- **RQ2:** What is the impact of module designs on the improvement of MORO?
- **RQ3:** Can MORO successfully capture users' personalized behavior patterns?

5.1   Experimental Setup

We first introduce the datasets, evaluation metrics, baseline methods, and parameter settings involved in the experiments.

Data Description. To evaluate the effectiveness of MORO, we utilize three real-world datasets: Taobao[1], Beibei[2], and Yelp[3], which are publicly accessible and vary in terms of domain, size, and sparsity. The statistical information of them is shown in Table 1.

- *Taobao Dataset.* It is a benchmark dataset for the performance evaluation of multi-behavior recommendations. There are four types of user behaviors contained in this dataset, *i.e.*, page view (pv), add-to-cart (cart), tag-as-favorite (fav), and purchase (buy). We use "purchase (buy)" as the target behavior to evaluate the effectiveness of MORO.
- *Beibei Dataset.* This benchmark dataset is collected from infant product online retailing site Beibei. It involves three types of user behaviors, including page view (pv), add-to-cart (cart), and purchase (buy). We use "purchase (buy)" as the target behavior to evaluate the effectiveness of MORO.
- *Yelp Dataset.* This dataset is collected from the public data repository from Yelp platform. We differentiate user's behaviors over items in terms of the rating scores r, *i.e.*, negative behavior ($r \leqslant 2$), neutral behavior ($2 < r < 4$), and positive behavior ($r \geqslant 4$). Besides the users' rating behaviors, this data also contains the tip behavior if a user gives a tip on his/her visited venues. We use "positive behavior" as the target behavior.

---

[1] https://tianchi.aliyun.com/dataset/dataDetail?dataId=649.
[2] https://www.beibei.com/.
[3] https://www.yelp.com/dataset/download.

Evaluation Metrics. We adopt two widely-used evaluation metrics: *Hit Ratio* (Hit@N) and *Normalized Discounted Cumulative Gain* (NDCG@N) [16,17]. The higher Hit@N and NDCG@N, the better the model performance. Following the same experimental settings in [16,19], the leave-one-out evaluation is leveraged for training and test set partition. For efficient and fair model evaluation, we pair each positive item instance with 99 randomly sampled no-interactive items for each user, which shares the same settings in [16,19].

Baselines. In order to comprehensively verify the performance of MORO, we consider following baselines:

- **BiasFM** [7] is a classical matrix factorization model that considers the biased information from users and items.
- **DMF** [20] introduces a neural network to matrix factorization and leverages explicit interactions and implicit feedback to refine the representations.
- **NCF** [5] is a collaborative filtering-based method that replaces inner product computation with multilayer perceptrons.
- **AutoRec** [12] stacks multilayer auto-encoder to transfer user-item interaction into a low-dimensional space and fetch user/item representations.
- **NGCF** [13] uses message passing architecture to aggregate information over the user-item interaction and exploits high-order relationships.
- **NMTR** [3] proposes a multi-task framework for performing the cascade prediction of different types of behaviors.
- **DIPN** [4] is a classical multi-behavior recommendation method that leverages an attention mechanism to predict users' purchase intent.
- **MATN** [17] employs a transformer module to capture the relationships among user behaviors and refine the representations learning.
- **GNMR** [16] leverages GNN to learn representations from multi-behavior interactions between users and items.
- **R-GCN** [11] is a graph neural network-based method that leverages relation type information for knowledge graph completion.
- **GHCF** [1] is a stat-of-the-art method that proposes efficient multi-task learning without sampling for parameter optimization.
- **MBGCN** [6] is one of the state-of-the-art methods which uses a graph convolution network to perform behavior-aware embedding propagation.
- **MB-GMN** [19] is another state-of-the-art method that combines GNN with meta-learning to learn the meta-knowledge of user behaviors.

Parameter Settings. We implement the proposed MORO using Pytorch and release our implementation[4] (including the codes, datasets, parameter settings, and training logs) to facilitate reproducibility. MORO is optimized using Adam Optimizer during the training phase. We set the dimension of MORO d as 64 and the number of propagation layer L as 2. The batch size and the learning rate in MORO is set as 256 and 10^{-2}. In addition, we turn the hyper-parameters λ_1 and λ_2 in $[10^{-8}, 10^{-4}]$, τ in $[0.1, 1]$ with grid search.

---

[4] https://github.com/1310374310/MORO.

Table 2. Overall performance comparison.

Dataset	Taobao		Beibei		Yelp	
Model	Hit@10	NDCG@10	Hit@10	NDCG@10	Hit@10	NDCG@10
BiasMF [7]	0.262	0.153	0.588	0.331	0.775	0.481
DMF [20]	0.305	0.189	0.597	0.336	0.756	0.485
NCF [5]	0.319	0.191	0.595	0.332	0.714	0.429
AutoRec [12]	0.313	0.190	0.607	0.341	0.765	0.472
NGCF [13]	0.302	0.185	0.611	0.375	0.789	0.501
NMTR [3]	0.332	0.179	0.613	0.349	0.790	0.478
DIPN [4]	0.317	0.178	0.631	0.394	0.811	0.540
MATN [17]	0.463	0.271	0.626	0.385	0.822	0.523
GNMR [16]	0.424	0.249	0.604	0.367	0.848	0.559
R-GCN [11]	0.338	0.191	0.605	0.344	0.826	0.520
MBGCN [6]	0.369	0.222	0.642	0.376	0.779	0.465
GHCF [1]	0.377	0.218	0.693	0.411	0.791	0.485
MB-GMN [19]	0.491	0.301	0.691	0.410	0.852	0.567
MORO	**0.619**	**0.403**	**0.754**	**0.455**	**0.877**	**0.583**

5.2 Overall Performance (RQ1)

The performance comparison results are presented in Table 2. From the results, We have the following observations.

Compared to state-of-the-art baseline methods, MORO achieves the best performance on all datasets. Specifically, MORO achieves relatively 26%, 9%, and 3% improvements in terms of Hit@10, and 33%, 10%, and 3% improvements in terms of NDCG@10 on Taobao, Beibei, and Yelp datasets, respectively. It reflects the effectiveness of MORO on multi-behavior recommendation tasks. The improvements can be attributed to three reasons: (1) the advantages of behavior propagation and perception modules which effectively exploit user multi-behavior preferences; (2) the contrastive enhancement, which maximizes the consistencies of representations and reduces the information loss of multi-behavior aggregated representations; (3) multi-task learning, which fully utilizes the signals of multi-behavior data to refine representations.

Further analysis reveals that the methods in which injection multi-behavior information could boost the performance (*e.g.*, MATN, GNMR, GHCF, MBGCN, and MB-GMN). It illustrates the importance of considering multiple behaviors. Moreover, better performance is achieved due to the GNN-based methods to encode the neighbor information in the graph. Furthermore, the GNN-based methods considering the heterogeneity of edges (*e.g.*, MBGCN) perform better than the GNN-based algorithms on homogeneous graphs (*e.g.*, NGCF).

In addition, we also work with different N to validate the effectiveness of top-N recommendations. The experimental results on the Beibei dataset are given in Table 3. We can observe that MORO achieves the best performance under different values of N. It indicates the consistent superiority of MORO as compared to other baselines in assigning higher scores to the user's interested item in the top-N list. We attribute this to the fact that MORO exploits the

Table 3. Comparison results on Beibei dataset with varying N value in terms of Hit@N and NDCG@N.

Model	$N = 1$		$N = 3$		$N = 5$		$N = 7$	
	Hit	NDCG	Hit	NDCG	Hit	NDCG	Hit	NDCG
BiaMF [7]	0.118	0.118	0.310	0.228	0.453	0.287	0.537	0.316
NCF [5]	0.123	0.123	0.317	0.232	0.447	0.283	0.530	0.315
AutoRec [12]	0.128	0.128	0.321	0.236	0.456	0.291	0.540	0.322
MATN [17]	0.184	0.184	0.361	0.286	0.467	0.330	0.543	0.356
GNMR [16]	0.168	0.168	0.336	0.265	0.436	0.307	0.504	0.328
R-GCN [11]	0.134	0.134	0.323	0.242	0.453	0.295	0.535	0.323
MBGCN [6]	0.167	0.167	0.374	0.284	0.498	0.337	0.541	0.322
GHCF [1]	0.179	0.179	0.390	0.300	0.525	0.356	0.611	0.385
MB-GMN [19]	0.183	0.183	0.411	0.306	0.527	0.359	0.608	0.389
MORO	**0.201**	**0.201**	**0.451**	**0.344**	**0.591**	**0.402**	**0.676**	**0.431**

information of multi-behavior and achieves less information loss in representation aggregation by contrast enhancement.

5.3 Study of MORO (RQ2)

Impact of Module. To evaluate the rationality of designed modules in MORO, we consider four model variants as follows:

- **MORO-GCN**: To verify the effectiveness of the behavior propagation module, we replace the behavior propagation module with GCN, *i.e.*, the β of Eq. (1) is replaced by $1/\sqrt{|\mathcal{N}_u||\mathcal{N}_v|}$.
- **MORO-Rel**: We replace the behavior perception module with MLP, and aggregate the information from concatenated representations, *i.e.*, $\mathbf{h}_u = MLP(\mathbf{h}_{u,1}||\cdots||\mathbf{h}_{u,K})$, where $||$ denotes the concatenation operation.
- **MORO-Con**: We remove the contrast enhancement loss $\mathcal{L}_c$ from $\mathcal{L}$ to evaluate the effect of the contrast enhancement module.
- **MORO-Task**: We only sample positive and negative examples under the target behavior (*e.g.*, "purchase") and use them to train the model.

The ablation study results are shown in Table 4. From the evaluation results, we have the following observations.

- MORO outperforms all variants on all datasets. It shows the validation of each module in MORO. Specifically, in the e-commerce scenario (*i.e.*, Taobao and Beibei), the improvement of recommendation accuracy by each module is more pronounced. It demonstrates the application value of our proposed model in complex scenarios such as e-commerce.
- The performance gap between MORO and MORO-Rel indicates the advantage of the behavior perception module, which exploits user multi-behavior preferences and aggregates multiple graph representations. It also shows that capturing personalized user multi-behavior preferences is more important than modeling them uniformly.

Table 4. Ablation study on key components of MORO.

Dataset	Taobao		Beibei		Yelp	
Model	Hit@10	NDCG@10	Hit@10	NDCG@10	Hit@10	NDCG@10
MORO-GCN	0.584	0.370	0.694	0.427	0.869	0.580
MORO-Rel	0.577	0.364	0.702	0.418	0.871	0.574
MORO-Con	0.605	0.392	0.746	0.446	0.872	0.579
MORO-Task	0.578	0.368	0.603	0.351	0.843	0.550
MORO	**0.619**	**0.403**	**0.754**	**0.455**	**0.878**	**0.584**

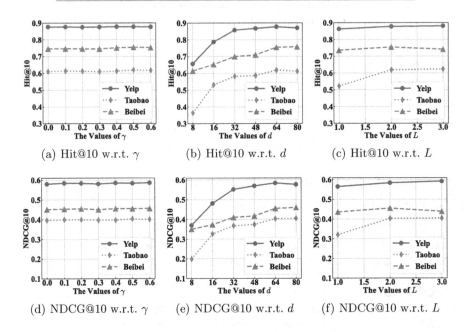

(a) Hit@10 w.r.t. γ (b) Hit@10 w.r.t. d (c) Hit@10 w.r.t. L

(d) NDCG@10 w.r.t. γ (e) NDCG@10 w.r.t. d (f) NDCG@10 w.r.t. L

Fig. 3. Hyperparameter study of MORO

- Moreover, the contrast enhancement module improves the performance of MORO. We attribute this to the ability of the contrast enhancement module to reduce the information loss in multi-behavior aggregated representations, making the representations more effective.
- The evaluation results illustrate the limitations of single-task learning in multi-behavior recommendation tasks (MORO-task). We believe that it is caused by data sparsity for a single target behavior. That multi-task learning can help alleviate this problem.

Hyperparameter Study. To analyze the effect of different parameter settings, we perform experiments to evaluate the performance of MORO with different hyperparameter configurations (*i.e.*, coefficient γ, embedding size d, and number of propagation layers L). The results are shown in Fig. 3.

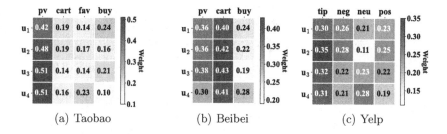

Fig. 4. Weights of user's behaviors

In Eq. (11), the coefficient γ can balance the two terms of $\mathcal{L}_c$ for optimization. To analyze the influence of coefficient γ, we turn γ in $[0, 0.6]$. The model's performance improves slightly as gamma increases, and the model overall achieves the best performance when $\gamma = 0.5$. With the further increase of γ, the performance remains stable, which shows that MORO is robust to γ.

We turn the embedding size d from 8 to 80. With the increasing embedding size d from 8 to 64, the performance improves due to a stronger representation space. After d reaches 64, better performance is not always obtained as d continues to increase. The reason is that a larger representation dimension reduces the ability of MORO to learn representations.

Finally, we analyze the number of propagation layers L. MORO achieves the best performance on all three data sets when stacking two propagation modules (*e.g.*, $L = 2$). Increasing the number of propagation layers L brings the noise to the representations, which affects the performance of MORO (*e.g.*, $L = 3$).

5.4 Study of User Preferences (RQ3)

To analyze user preferences, we visualize user attention weights calculated by Eqs. (5–6). As shown in Fig. 1, we can observe that for each user, MORO successfully captures his/her personalized multi-behavior patterns. The darker the color of the grid, the greater the contribution of that behavior to modeling user preferences. Moreover, for different datasets, the contributions of user behaviors show different distributions. Specifically, the "page view (pv)" behavior of users in the Taobao dataset contributes more to the preference modeling of users. Users' "add-to-cart (cart)" behavior in the Beibei dataset contributes more than other behaviors.

To verify the accuracy of user multi-behavior preferences in Fig. 4, we conduct data ablation experiments, removing specific source behavior data (*e.g.*, -pv) or only using the target behavior data (*e.g.*, +buy) for MORO. Figure 5 shows the results of data ablation experiment. We can find that after removing the "page view" (-pv) behavior data, the model's performance on the Taboao dataset has dropped significantly. For the Beibei dataset, removing the "add-to-cart" (-cart) behavior data impacts the performance of the recommendation. It shows that MORO can accurately model users' multi-behavior preferences and improve

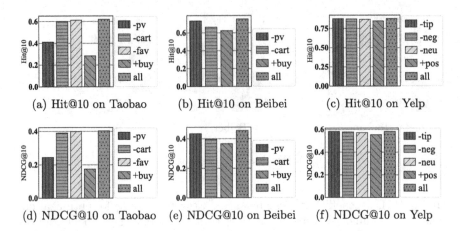

Fig. 5. Results of data ablation. Here, "-pv", "-cart", "-fav", "-tip", "-neg", and "-neu" represent MORO without incorporation of "page view", "add-to-cart", "add-to-favorite", "tip", "negative", and "neutral" behavior data, respectively. "+buy" and "+pos" denote the variants which only relies on the target behavior data.

model performance with all behaviors. Further analysis shows that the more behavior data eliminated, the worse the performance of MORO. For example, in the Beibei and Taobao datasets, if only "purchase" behavior data is kept (+buy), the model performance will decline because of the sparsity of the "purchase" behavior data. It also indicates that the source behaviors are vital in multi-behavior recommendation tasks.

6   Conclusion and Future Work

In this paper, we propose MORO for multi-behavior recommendation tasks. Considering the different information of user behaviors, we use the behavior propagation module and the behavior perception module to aggregate the representations of different behavior graphs. Then we employ the contrast enhancement module to enhance the multi-behavior aggregated representations. Extensive experiments on three real-world datasets demonstrate the superiority of MORO over other methods. In the future, we plan to capture the dynamic multi-behavior preferences of users from their multi-behavior time series data.

References

1. Chen, C., et al.: Graph heterogeneous multi-relational recommendation. In: AAAI, pp. 3958–3966 (2021)
2. Du, C., Li, C., Zheng, Y., Zhu, J., Zhang, B.: Collaborative filtering with user-item co-autoregressive models. In: AAAI, pp. 2175–2182 (2018)
3. Gao, C., et al.: Neural multi-task recommendation from multi-behavior data. In: ICDE, pp. 1554–1557 (2019)

4. Guo, L., Hua, L., Jia, R., Zhao, B., Wang, X., Cui, B.: Buying or browsing?: predicting real-time purchasing intent using attention-based deep network with multiple behavior. In: KDD, pp. 1984–1992 (2019)

5. He, X., Liao, L., Zhang, H., Nie, L., Hu, X., Chua, T.: Neural collaborative filtering. In: WWW, pp. 173–182 (2017)

6. Jin, B., Gao, C., He, X., Jin, D., Li, Y.: Multi-behavior recommendation with graph convolutional networks. In: SIGIR, pp. 659–668 (2020)

7. Koren, Y., Bell, R.M., Volinsky, C.: Matrix factorization techniques for recommender systems. Computer 42(8), 30–37 (2009)

8. Mao, K., Zhu, J., Xiao, X., Lu, B., Wang, Z., He, X.: Ultragcn: ultra simplification of graph convolutional networks for recommendation. In: CIKM, pp. 1253–1262 (2021)

9. van den Oord, A., Li, Y., Vinyals, O.: Representation learning with contrastive predictive coding. CoRR abs/1807.03748 (2018)

10. Rendle, S., Freudenthaler, C., Gantner, Z., Schmidt-Thieme, L.: BPR: Bayesian personalized ranking from implicit feedback. In: UAI, pp. 452–461 (2009)

11. Schlichtkrull, M.S., Kipf, T.N., Bloem, P., van den Berg, R., Titov, I., Welling, M.: Modeling relational data with graph convolutional networks. In: ESWC, vol. 10843, pp. 593–607 (2018)

12. Sedhain, S., Menon, A.K., Sanner, S., Xie, L.: AutoRec: autoencoders meet collaborative filtering. In: WWW, pp. 111–112 (2015)

13. Wang, X., He, X., Wang, M., Feng, F., Chua, T.: Neural graph collaborative filtering. In: SIGIR, pp. 165–174 (2019)

14. Wu, S., Zhang, Y., Gao, C., Bian, K., Cui, B.: GARG: anonymous recommendation of point-of-interest in mobile networks by graph convolution network. Data Sci. Eng. 5(4), 433–447 (2020)

15. Wu, Y., DuBois, C., Zheng, A.X., Ester, M.: Collaborative denoising auto-encoders for top-n recommender systems. In: WSDM, pp. 153–162 (2016)

16. Xia, L., Huang, C., Xu, Y., Dai, P.: Multi-behavior enhanced recommendation with cross-interaction collaborative relation modeling. In: ICDE, pp. 659–668 (2021)

17. Xia, L., Huang, C., Xu, Y., Dai, P., Zhang, B., Bo, L.: Multiplex behavioral relation learning for recommendation via memory augmented transformer network. In: SIGIR, pp. 2397–2406 (2020)

18. Xia, L., et al.: Knowledge-enhanced hierarchical graph transformer network for multi-behavior recommendation. In: AAAI, pp. 4486–4493 (2021)

19. Xia, L., Xu, Y., Huang, C., Dai, P., Bo, L.: Graph meta network for multi-behavior recommendation. In: SIGIR, pp. 757–766 (2021)

20. Xue, H., Dai, X., Zhang, J., Huang, S., Chen, J.: Deep matrix factorization models for recommender systems. In: IJCAI, pp. 3203–3209 (2017)

21. Yu, S., et al.: Leveraging tripartite interaction information from live stream e-commerce for improving product recommendation. In: KDD, pp. 3886–3894 (2021)

22. Zang, Y., Liu, Y.: GISDCN: a graph-based interpolation sequential recommender with deformable convolutional network. In: DASFAA, vol. 13246, pp. 289–297 (2022)

23. Zhang, W., Mao, J., Cao, Y., Xu, C.: Multiplex graph neural networks for multi-behavior recommendation. In: CIKM, pp. 2313–2316 (2020)

Eir-Ripp: Enriching Item Representation for Recommendation with Knowledge Graph

Kaiwen Li, Chunyang Ye$^{(\boxtimes)}$, and Jinghui Wang

Hainan University, Haikou 570203, HN, China
cyye@hainanu.edu.cn

Abstract. To improve the performance of recommendation models and enhance the interpretability of results, knowledge graph is often used to add rich semantic information to the recommended items. Existing methods either use knowledge graph as an auxiliary information to mine users' interests, or use knowledge graph to establish relationships between items via their hidden information. However, these methods usually ignore the interaction between users and items. As a result, the hidden relationship between users and items are not well explored in the item representation. To address this issue, we propose an enhancement model to learn item representation based on RippleNet (Eir-Ripp). By mining the users' historical behavior and user characteristics, users' preference and the correlation between users and items are extracted to complement the semantic information of items. We conduct extensive experiments to evaluate our proposal on three public data sets. Experimental results show that our model outperforms the baseline methods in terms of an up to 8.8% improvement in the recommendation.

Keywords: Knowledge graph · User interest · Item representation · Recommendation algorithm

1 Introduction

With the development of information technology, information overload is becoming more and more prominent. To address this issue, recommendation systems are proposed to mine useful information from a large amount of information to meet users' interests and preferences. In recent years, many recommendation algorithms are designed, amongst which the most popular traditional recommendation system is Amazon's coordinated filtering (CF) [1]. CF analyzes users' historical interaction and recommends items according to users' common preferences. However, the CF based recommendation algorithm suffers from the problems of cold start and sparsity of user item interaction, which compromises the applicability of personalized recommendation systems.

In order to solve the above problems, researchers integrate knowledge graph (KG) [2] into recommendation systems as auxiliary information. By enriching the semantic association information between entities, the problem of data sparsity can be alleviated, and the performance of recommendation systems is improved [3–5]. KG is a heterogeneous graph, in which nodes are represented as entities and edges are represented as corresponding relationships between entities. Mapping items and their attributes into KG establishes a relationship between items [6], which enhances the interpretability of the recommended results and has important research significance and value [4].

At present, KG based recommendation systems mainly apply KG in three ways: embedded method, path based method and unified method [6]. The embedding based method uses KG to enrich the representation of items or users, but ignores the association of information in the graph. The path based method explores the relationship between various entities in KG and provides users with the interpretation of the results, but it relies too much on the manually designed meta path. By combining the embedding based method and the path based method, the unified method uses the relationship between entities to mine the expression of users' interest or the expression of items' implicit information. It can explain the recommendation results, but ignores the interaction between users and items' implicit information.

To address the shortcomings of the above methods, we propose in this paper the Eir-Ripp model, which obtains users' historical behavior preference through multi information interaction, and then enhances the representation of items via graph convolution to the interaction between the users' historical behavior preference and the implicit information of items. Note that we regard the user history behavior as a collection of items that users like in the past. These collections and user characteristics form a user portrait to represent a specific user. We believe that the user history behavior and user characteristics may have similar characteristics in the low-level feature space and therefore propose a user interactive learning unit (Uil unit) to learn such similar characteristics. In particular, the Uil unit can extract user historical behavior preferences and user characteristics from multi information interaction. Then we design a graph convolution network to learn the correlation between users' historical behavior preferences and entities to enhance the KG semantic information of items. The enhanced item representation vector is used to discover user's potential hierarchical interest. Finally, we design a full connection network to integrate user interest, user characteristics of multi information interaction and user historical preference into a final user vector. Such user vector is helpful to recommendation systems because it expresses user information effectively, including, the information integrated by users in many aspects in the feature extraction process.

The main contributions of this paper are two-folded: 1) We propose a novel model to enrich item representation for recommendation systems with knowledge graph. By designing a user interactive learning unit to learn the high-order feature interaction between user historical behavior and user characteristics, an item enhancement method is proposed to enhance the KG semantic information of items via graph convolution. 2) We conduct extensive experiments to evaluate the Eir-Ripp model based on three public data sets. The results show that our Eir-Ripp model outperforms the baseline solutions. Specifically, compared with the best baseline solution, our average ACC in film, book and music recommendation is increased by 13.9%, 28.4% and 34.3% respectively, and the AUC is increased by 0.7%, 0.6% and 1.3% respectively.

The rest of this paper is organized as follows: Sect. 2 reviews the state-of-the-art research efforts on recommendation systems. Section 3 defines the research problem formally. Section 4 presents the Eir-Ripp model. Section 5 evaluates our proposal with experiments. Section 6 concludes the work and highlights some future research directions.

2   Related Work

In traditional recommendation algorithms, the most classical method is the CF [1] algorithm. The intuitive of CF is that users with similar behavior imply they have similar interests. CF calculates the score of items to measure the similarity between users. When the number of users and items are increasing, each user is associated with only a few items. The corresponding user-item scoring matrix becomes sparse and can not well represent the similarity of users. FM [7] is proposed to explore low-order feature interaction, and then decompose the interaction matrix to solve the problem of data sparsity, but it cannot extract features for highly sparse user behavior.

Recently, deep neural network (DNN) has also been applied to various recommended scenarios. FNN [8] introduces deep learning on the basis of FM to explore high-order combination and reduce feature engineering, but it focuses on high-order features and does not integrate low-order features into the model. In this regard, Wide & Deep [9] proposed by Google is a joint training framework for the integration of shallow model and deep model, which combines the memory ability of shallow model and the generalization ability of deep model. However, the interactive features of artificial feature engineering need to be used in the shallow layer, and some sparse features only have few cross feature samples. Designing paired feature interaction manually however is time-consuming. AFM [10] and DeepFM [11] are proposed to address this issue. AFM introduces attention mechanism based on FM model to automatically help FM distinguish the importance of interaction features, so that different interaction features have different effects on the prediction results. DeepFM avoids the work of artificial feature engineering in the Wide & Deep model, and replaces the LR in the wide part with FM with the ability to automatically learn cross features.

Although the introduction of deep learning can mine the potential information of users and items, it cannot explain the model results well. Knowledge graph can help to address this issue. Knowledge graph captures the relationship between two different entities and enriches the semantic association information between them. Knowledge graph embedding [12] (KGE) learns the special semantic information contained in the knowledge graph for vectorial representation. KGE is mainly divided into two categories: translation distance model and semantic matching model. The translation distance model calculates the distance between two entities according to the scoring function of distance, which represents the measurement of the rationality of facts. For example, the semantic representation of entities is obtained by models such as TransE [13], TransH

[14], TransR [15] and TransD [16]. Semantic matching model measures the credibility of facts according to the relationship between the potential semantics of matching entities and vector representation according to the scoring function of similarity, such as RESCAL [17], DistMult [18] and HoLE [19]. KGE is integrated into recommendation systems to represent entity information, including CKE [4] and DKN [20].

Wang et al. classify the training methods of knowledge graph on the recommendation systems into three categories: sequential training, joint training and alternating training. The representative methods of them include DKN, RippleNet [21] and KGCN [22], MKR [23]. RippleNet starts from the items that users are interested in, and then spreads to other items on the item knowledge graph. The diffused items belong to the user's potential preference by default, which is expressed by KGE and trained together with the recommendation model. However, the model does not explore the importance of relationship. KGCN captures local proximity structure aggregation through the neighbor information. In each entity, the weight of different neighbors is determined by specific users and relationships, but the interaction between user behavior and items is not considered in information aggregation. MKR divides KGE and recommendation into two models. By conducts high-order feature interaction for item embedding and item entity embedding through a cross compression unit, it trains the two models alternatively. However, the cross unit needs item and entity correspondence, and cannot contain entities other than the recommended target, which limits the diversification of KG information and reduces the interpretability.

3   Problem Definition

In recommendation scenarios, suppose we have a set of users $U = \{u_1, u_2, ..., u_m\}$ and a set of items $V = \{v_1, v_2, ..., v_n\}$. According to user implicit feedback, user set U and item set V are defined to form user item interaction matrix $Y \in R^{m \times n}$:

$$y_{uv} = \begin{cases} 1 & \text{if interaction } (u, v) \text{ is observed.} \\ 0 & \text{otherwise} \end{cases} \tag{1}$$

where $y_{uv} = 1$ indicates that there is implicit interaction between user u and item v, such as clicking, viewing or buying, otherwise, $y_{uv} = 0$. In addition to the interaction matrix Y, we also have a knowledge graph G with additional information, which is composed of $\{(h, r, t)|h \in E, r \in R, t \in E\}$ triples. E and R represent the entity set and relationship set in the knowledge graph respectively. For a given user item interaction matrix Y and knowledge graph G, the task of our model is to predict whether user u has potential interest in items v that he or she has not interacted with before. Specifically, our model mainly studies the prediction function $\hat{y_{uv}} = F(u, v|\theta, Y, G)$, where $\hat{y_{uv}}$ represents the probability of user u clicks on item v, F represents the function, and θ represents the parameters of the model.

4   Model

4.1   Overview

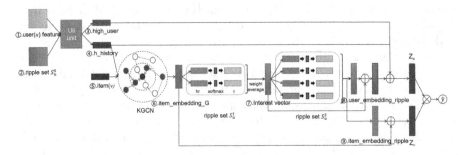

Fig. 1. Framework of Eir-Ripp.

The framework of Eir-Ripp is shown in Fig. 1. Eir-Ripp takes the user feature of a user u, the user's historical interest S_u^0 and an item v as inputs, and outputs the prediction probability of user u clicking on item v, where $u \in U$, $v \in V$, U represents all users and V represents all items. ① the user feature of user u interacts with ② S_u^0 through Uil unit to obtain ③ the user vector $high\_user$ of multi information interaction and ④ the user historical interest $h\_history$ of information. $h\_history$ and ⑤ item V are input to KGCN to output ⑥ the vector $item\_embedding\_G$ with enriched item representation. $item\_embedding\_G$ interacts with ripple set iteratively to obtain ⑦ the Interest vector of user u to item v, then accumulates them to represent ⑧ the user interest vector $user\_embedding\_ripple$. The combined $user\_embedding\_ripple$, $high\_user$ and $h\_history$ are input to the user's full connection layer Z_u to obtain the final user embedding. By combining $item\_embedding\_G$ and ⑨ RippleNet iterative $item\_embedding\_ripple$, the final item embedding is obtained in the full connection layer Z_v of the input item. Finally, we use two final embeddings (Z_u and Z_v) to calculate the predicted probability $\hat{y}$.

4.2   User Interactive Learning Unit

In order to obtain useful and user characteristics and historical interests, we propose the Uil unit in Eir-Ripp, as shown in Fig. 2. By inputting the feature $uf^{(u)}$ of user u and the historical interest S_u^0 of user u into the cross & compress unit (C unit), the feature of user u interacts with the historical interest at a low level. In this way, the additional information carried by both sides is learned, and the interactive user feature $uf_c^{(u)}$ and the interactive historical interest $S_{u_c}^0$ are obtained. Due to the imperfect information of users and items in the public data, the vectorized information contains some noise. Here, $S_{u_c}^0$ extracts the global information of positive features and the useful information of the global information of negative features through APReLU [24], so as to retain the user's features and historical information, and find the useful user information in the noise. Then, the model pays attention to useful information through Multi-Head Self-attention in the Transformer Layer [25], and uses forward propagation to find the implicit information of items that users like in history, so as to obtain the user's history vector $h\_history$. $uf_c^{(u)}$ is input into APReLU for information purification, and then TextCNN [26] is used for feature extraction to obtain $high\_user$. We did not use the Transformer Layer here because there is not much correlation between the user's features. Although more comprehensive user features are obtained by interacting with historical interests, the correlation between these features is relatively weak. Therefore, TextCNN is used for feature extraction to reduce the scale of the network to speed up network training.

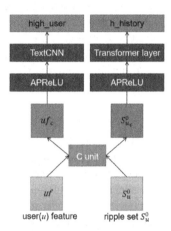

Fig. 2. User interactive learning unit (Uil unit).

Cross and Compress Unit. The cross compression unit is proposed in MKR model [23], which allows feature interaction between two similar information. Because both the user's characteristic matrix and the user's historical interest matrix express the user's interest information, the cross sharing between them can enable both to obtain additional information from each other, thereby enhancing their representation of user information. First, the original user feature matrix is obtained by splicing the user's single features. Then the user feature matrix is copied into a matrix $S_u^0 \in \mathbb{R}^{m \times d}$ of the same size as $uf \in \mathbb{R}^{m \times d}$. Finally, the cross matrix $C \in \mathbb{R}(d \times d)$ is constructed, where d is the dimension of S_u^0 and uf.

$$
C = uf^{\mathrm{T}} \cdot S_u^0 = \begin{bmatrix} uf^{(1)}S_u^{0(1)} & uf^{(1)}S_u^{0(2)} & & uf^{(1)}S_u^{0(d)} \\ uf^{(2)}S_u^{0(1)} & uf^{(2)}S_u^{0(2)} & & uf^{(2)}S_u^{0(d)} \\ \vdots & \vdots & \ddots & \vdots \\ uf^{(d)}S_u^{0(1)} & uf^{(d)}S_u^{0(2)} & & uf^{(d)}S_u^{0(d)} \end{bmatrix} \tag{2}
$$

uf and S_u^0 are crossed and input into the cross feature matrix C. C is then projected into the potential representation space of user feature and user historical interest. The interactive user feature uf_c and interactive user historical interest $S_{u_c}^0$ are output.

$$\begin{cases} uf_c = CW^{UU} + C^T W^{RU} + b^U = uf^T S_u^0 W^{UU} + \text{ uf } S_u^0 \text{ T} W^{RU} + b^U \\ S_{u_c}^0 = CW^{UR} + C^T W^{RR} + b^R = uf^T S_u^0 W^{UR} + \text{ uf } S_u^{0T} W^{RR} + b^R \end{cases} \quad (3)$$

where $W^{\cdots} \in \mathbb{R}^d$ and $b^\cdot \in \mathbb{R}^d$ are cross & compress unit weights and bias terms, respectively.

APReLU. APReLU [24] can automatically learn and use different linear transformations according to the input to extract positive information from the negative impact. The structure of APReLU uses ReLU and gap to map the input features to dimension 1 to obtain the global information of positive features. It then uses min (x, 0) and GAP to map the input feature to another dimension of 1 to obtain the global information of negative feature, which may contain some useful user information.

Transformer Layer. In NLP, Transformer [25] is designed to capture the dependence between words in a sentence through self attention mechanism [27]. In our model, we use the transformer encoder part only to capture the relevant characteristics of various items in the user's historical preferences. These characteristics are used to represent the correlation between items. Transformer encoder is composed of Multi-Head Self-attention mechanism, Add & Norm and feed-forward network.

Multi-Head Self-attention mechanism is composed of multiple Self-Attention, which divides the feature vector into multiple head expressions to form multiple subspaces. It then pays attention to their own information through Self-Attention, and finally integrates all aspects of information to make the model aware of various aspects of information. The formula of Self-Attention is defined as follows:

$$\text{Attention}(K, Q, V) = \text{softmax}\left(\frac{QK^T}{\sqrt{d}}\right) V \quad (4)$$

among them, Q, K and V represent query items, key items and value items in attention mechanism, respectively. In the calculation, by inputting the interactive historical interest $S_{u_c}^0$, we can obtain the weight matrix $W^Q, W^K, W^V \in \mathbb{R}^{(d \times d)}$ through linear projection. These three matrices are then used to calculate the important information of $S_{u_c}^0$ to obtain $S_{u_c}^0 W^Q, S_{u_c}^0 W^K, S_{u_c}^0 W^V$, which are then input into Multi-Head Self-attention mechanism for calculation:

$$\text{head}_i = \text{Attention}\left(S_{u_c}^0 W^Q, S_{u_c}^0 W^K, S_{u_c}^0 W^V\right) \quad (5)$$

$$S = MH\left(S_{u_c}^0\right) = \text{Concat}(\text{head}_1, \text{head}_2, \ldots, \text{head}_i) W^0 \quad (6)$$

After adding the input of the previous layer and the output of the previous layer through Add & Norm, the residual before and after attention mechanism is connected. Then the hidden layer is normalized to the standard positive distribution by LayerNormalization. Finally, the feed-forward network and Add & Norm are used for linear transformation.

$$X = AN \left(S_{u_c}^0 + S \right) \tag{7}$$

$$h\_history = \text{ReLU} \left(W_2 \cdot \text{ReLU} \left(W_1 \cdot X + b_1 \right) \right) + b_2 \tag{8}$$

TextCNN. TextCNN has a simple network structure, fast training speed, and can extract key information in sentences to capture the local correlation in many tasks. The information of a word needs a complete one-dimensional vector to express its meaning. Similarly, the user features are also composed of multiple one-dimensional vectors. Therefore, TextCNN can be used for feature extraction to obtain more comprehensive key information of user features.

4.3 Enhancement Module of Item Representation

In order to strengthen the interaction of implicit information between users and items, we design an item representation enhancement module. First, the historical interest $h\_history$ extracted by user u is obtained through Uil unit, and the predicted item v is convoluted with $h\_history$ in the local graph. In the local field, the weight of different neighbors depends on the historical interest of specific user u and the relationship between items [22], so that the implicit information between users and items affects each other.

$$\pi_r^{h\_history} = g(h\_history, r) \tag{9}$$

$\pi_r^{h\_history}$ indicates the user's preference for relationships, where $u, r \in \mathbb{R}^d$, $\pi_r^{h\_history} \in \mathbb{R}$. The g function here is used to calculate the preference degree of users with different relationships.

$$v_{N(v)}^u = \sum_{S_u^0 \in N(v)} \hat{\pi}_{r,S_u^0}^{h\_history} S_u^0 \tag{10}$$

$$\hat{\pi}_{r,S_u^0}^{h\_history} = \frac{\exp\left(\pi_{r,S_u^0}^{h\_history} \right)}{\sum_{S_u^0 \in N(v)} \exp\left(\pi_{r,S_u^0}^{h\_history} \right)} \tag{11}$$

$v_{N(v)}^{h\_history}$ represents the topological domain structure of user u to item entity V, $\hat{\pi}_{r,S_u^0}^{h\_history}$ represents the normalization of user relationship score, and $N(v)$ represents the entity set directly connected with item v in KG. The scale of the set is large. In order to reduce the computational cost caused by too many neighbors, K fixed neighbor nodes are randomly selected and the parameter hop_n is set to control the depth of the neighborhood.

$$S(v) = \{e | e \in N(v), |S(v)| = k\} \tag{12}$$

$v_{N(v)}^{h\_history}$ obtains the set of K nodes in the item v neighborhood of the entity represented by $v_{S(v)}^{h\_history}$ according to the adjacent office node set $S(v)$.

We add and aggregate entity v and its neighborhood $v_{S(v)}^u$ into a vector, and then perform nonlinear transformation:

$$item\_embedding\_G = \sigma \left(W^{Eir} \left(v + v_{S(v)}^{h\_history} \right) + b^{Eir} \right) \tag{13}$$

where $W^{Eir} \in \mathbb{R}^d$ is the weight of the full connection layer and $b^{Eir} \in \mathbb{R}^d$ is the offset term.

4.4   RippleNet

The items that user u has clicked and liked in the past are used as the user's historical interest S_u^0. S_u^0 is used as the seed of the user's historical interest concentration KG, and then spread to the periphery along the link to find items related to the user's interest, forming multiple ripple sets S_u^{hop}. Finally the direction of the user's interest is mined through the attention mechanism to strengthen the preference of expressing the user's interest.

$$\varepsilon_u^{hop} = \left\{ t \mid (h, r, t) \in G, h \in \varepsilon_u^{hop-1} \right\}, hop = 1, 2, \ldots, H \tag{14}$$

$$S_u^{hop} = \left\{ (h, r, t) \mid (h, r, t) \in G, h \in \varepsilon_u^{hop-1} \right\}, hop = 1, 2, \ldots, H \tag{15}$$

where G represents the KG of the entire item, ϵ_u^{hop} represents the entity set of user's hop hop, and S_u^{hop} represents the neighbor set of user's hop hop.

   $item\_embedding\_G$ is input to predict item v, the similarity p_i normalized to (h, r) in the first diffusion of ripple set is calculated.

$$\begin{aligned} p_i &= \text{softmax} \left(item\_embedding\_G^T h_i r_i \right) \\ &= \frac{\exp \left(item\_embedding\_G^T h_i r_i \right)}{\sum_{(h,r,t) \in S_u^1} \exp(item\_embedding\_G^T h_i r_i)} \end{aligned} \tag{16}$$

As shown in Fig. 1, after a weighted summation of t in the first layer ripple set according to similarity, the interest representation of user 1 hop is obtained (light green block):

$$o_u^1 = \sum_{(h_i, r_i, t_i) \in S_u^1} p_i t_i \tag{17}$$

whenever the interest representation of user 1 hop is obtained, the gap between the obtained vector and the originally strengthened item representation will gradually increase during the next hop calculation. In order to ensure the interaction of information between user and item, $item\_embedding\_G$ and o_u^{hop} are added for full connection to obtain the vector $item\_embedding\_ripple$ input in each hop in RippleNet:

$$item\_embedding\_ripple = W^{ripp} \left(item\_embedding\_G + o_u^{hop} \right) + b^{ripp} \tag{18}$$

where $W^{ripp} \in \mathbb{R}^d$ is the weight of the full connection layer and $b^{ripp} \in \mathbb{R}^d$ is the offset term. Finally, we repeat the above operation to obtain the RippleNet item vectors $item\_embedding\_ripple$ and $user\_embedding\_ripple$.

$$user\_embedding\_ripple = o_u^1 + o_u^2 + \ldots + o_u^H \tag{19}$$

4.5   Forecast

Based on the previous calculation, three user related vectors *high_user*, *h_history* and *user_embedding_ripple* and two item related vectors *item_embedding_G* and *item_embedding_ripple* are obtained.

The user related vector is plused with the item related vector. Then, one layer full connection Z_u and Z_v is designed to reduce the dimension to obtain the final user vector u' and the final item vector v'.

$$u' = W^{Z_u}(high\_user + h\_history + user\_embedding\_ripple) + b^{Z_u} \qquad (20)$$

$$v' = W^{Z_v}(item\_embedding\_G + item\_embedding\_ripple)) + b^{Z_v} \qquad (21)$$

where $W^{Z_u}, W^{Z_v} \in \mathbb{R}^d$ is the weight of the full connection layer and $b^{Z_u}, b^{Z_v} \in \mathbb{R}^d$ is the offset term. The final predicted value is multiplied by the user vector u' and the final item vector v', and then calculated by the sigmoid function σ:

$$\hat{y}_{u'v'} = \sigma(u'v') \qquad (22)$$

4.6   Loss Function

The Loss function of Eir-Ripp model includes three parts: the loss of recommended module, the loss of Eir module and the loss of RippleNet, as defined below:

$$L = L_{RS} + L_{Eir} + L_{Ripp} \qquad (23)$$

$$L_{RS} = \sum_{u \in U, v \in V} \mathcal{F}(\hat{y}_{u'v'}, y_{uv}) \qquad (24)$$

In the loss of recommended modules, u represents the user set, v represents the item set, and $\mathcal{F}$ represents the cross-entropy function.

$$L_{Eir} = \frac{\lambda_{l2}}{2}\left(||high\_user||_2^2 + ||v_{N(v)}^{h\_history}||_2^2\right) \qquad (25)$$

In the loss of Eir module, λ_{l2} represents the balance weight of L2 regular term, $high\_user \in \mathbb{R}^d$ is the user vector of multi information interaction extracted by Uil unit, and $v_{N(v)}^{h\_history}$ represents the set of entity v neighborhood nodes.

$$L_{Ripp} = \lambda_{kge}||MEAN(\sigma(hrt))||_2^2 + \lambda_{l2}\left(||h||_2^2 + ||r||_2^2 + ||t||_2^2\right) \qquad (26)$$

In the loss of Ripp module, λ_{kge} is the balanced weight of KGE, $MEAN$ function represents the average value, and hrt is the product of entities and relationships in triple (h, r, t) to represent KGE.

5   Evaluation

In this section, we evaluate Eir-Ripp's performance using three realistic recommended scenarios: movies, books and music.

5.1   Datasets

We used the following three data sets in the experiment: 1) MovieLens-1M is the most commonly used public data set in the research for recommendation systems. It mainly includes the movie data set, user data set and scoring data set of MovieLens website, in which the scoring data set has about 1 million explicit ratings (from 1 to 5). 2) Book-Crossing is a book data set collected by the book crossing community, including book data sets, user data sets and more than 1 million explicit book evaluation records (from 0 to 10). 3) Last.FM is the music data set provided by "last.FM" online music radio station, including music data set and user listening data set.

Table 1. Basic statistics settings for the three datasets.

Dataset	#users	#items	#interactions	#KG triples	#entities	#relations
MovieLens-1M	6,036	2,445	753,772	1,241,995	211,169	12
Book-Crossing	17,860	14,967	139,746	151,500	55,560	25
Last.FM	1,872	3,846	42,346	15,518	5,541	60

Table 2. Smaller KG datasets in MovieLens-1M and Book-Crossing.

Dataset	#KG triples	#entities	#relations
MovieLens-1M	20,195	4,928	7
Book-Crossing	19,793	12,454	10

MovieLens-1M, book-crossing and last.FM are explicit feedback data, which are converted into implicit feedback, where 1 indicates that the user's evaluation of the project is positive, and 0 indicates the unobserved set. For MovieLens-1M dataset, the positive score threshold is set to 4, whereas for Book-Crossing dataset and last.FM dataset, there is a certain sparsity, and the threshold is not set.

Our knowledge graph uses the data sets disclosed in RippleNet model and MKR model. These KG data sets are built for these three groups of real data sets using Microsoft satori, and a triplet with a confidence level greater than 0.9 is selected in each constructed KG. Due to MKR model factors, KG data used in MovieLens-1M and Book-crossing are different from Table 1, as shown in Table 2. We explain the reasons in Sect. 5.4.

5.2   Baseline

We compare Eir-Ripp model with the following baselines, which are divided into traditional recommendation model and recommendation model using knowledge map, as shown in Table 3. These models jointly use the same user item interaction

Table 3. Hyper-parameter settings of other models for the three datasets.

FM	$batch = \{2048, 128, 64\}$
FNN	$d = \{16, 32, 32\}$ $batch = \{256, 128, 64\}$ epoch $= 100$
Wide&Deep	$batch = \{128, 1024, 256\}$
AFM	$d = \{16, 32, 32\}$ $att\_d = 16$ $batch = 256$ epoch $= 100$
DeepFM	$batch = \{25600, 128, 1024\}$ epoch $= 100$
RippleNet	$d = \{16, 32, 32\}$ $hop = 2$ $\lambda_{kge} = 0.01$ $\lambda_{l2} = \{10^{-7}, 10^{-5}, 10^{-7}\}$ $batch = \{1024, 4096, 128\}$ epoch $= \{100, 200, 200\}$
KGCN	$d = \{64, 64, 32\}$ $hop = 2$ $\eta = \{0.02, 2 \times 10^{-4}, 10^{-4}\}$ $\lambda_{l2} = \{10^{-7}, 10^{-5}, 10^{-7}\}$ $batch = \{1024, 4096, 128\}$ epoch $= \{100, 200, 200\}$ $k = 8$
MKR	$L = \{1, 1, 2\}$ $H = \{1, 1, 1\}$ $d = \{8, 8, 4\}$ $\lambda_1 = \{0.5, 0.1, 0.1\}$ epoch $= \{3, 2, 2\}$

Table 4. Hyper-parameter settings of Eir-Ripp for the three datasets.

MovieLens-1M	$d = 8$ $hop = 2$ $\lambda_{kge} = 0.01$ $\eta = 0.02$ $S_u^{hop} = 32$ $batch = 2304$ $head = 1$ $t\_d = 16$ $hop_n = 2$ $k = 8$
Book-Crossing	$d = 32$ $hop = 2$ $\lambda_{kge} = 0.01$ $\lambda_{l2} = 10^{-5}$ $\eta = 0.02$ $S_u^{hop} = 16$ $batch = 256$ $head = 1$ $t\_d = 16$ $hop_n = 2$ $k = 8$
Last.FM	$d = 8$ $hop = 2$ $\lambda_{kge} = 0.01$ $\lambda_{l2} = 10^{-4}$ $\eta = 0.02$ $S_u^{hop} = 8$ $batch = 256$ $head = 1$ $t\_d = 8$ $hop_n = 2$ $k = 8$

data set, except that MKR uses a smaller KG data set in the recommended model of knowledge map. The traditional recommendation model parameters are not explained separately, that is, the dimension d is set to 32, the learning rate is 0.01, and the number of training cycles of all data sets is 50. L of MKR represents the number of network layers in the lower layer, H represents the number of network layers in the upper layer, and λ_1 represents the balance parameter.

5.3 Experiment Setup

Eir-Ripp is implemented based on the pytorch framework. In order to verify the effect of our model and ensure fairness, the above baselines are reproduced via pytorch, and the optimal hyper parameters are selected, as shown in Table 3. In Eir-Ripp, the hyper parameters we set are shown in Table 4, where η represents the learning rate, $batch$ represents the batch size, S_u^{hop} represents the number of diffusion nodes, $head$ represents the number of heads of multi head attention, and t_d represents the dimension of the attention layer in the Transformer layer, hop_n represents the item, and represents the number of enhanced diffusion. The setting of hyper parameters is determined by optimizing the AUC in the verification set. For each data set, training set, verification set and test set are divided according to 6:2:2. Each experiment is repeated five times. For the experimental results with a positive and negative error range of $[0.0001, 0.0005)$, we take the best experimental results in the five runs. If there is a result exceeding the positive and negative error range in a group of experiments, we conduct another five experiments, and take the average results of 10 experiments to represent the performance. We mainly use AUC [28] and accuracy (ACC) [29,30] as the evaluation model to evaluate the performance of CTR prediction.[1]

5.4 Results

In CTR prediction, the experimental results of all models in three different data sets are shown in Table 5. Under the pytorch framework, FM has achieved good results on three data sets, even surpassing some improved models based on FM. FNN adds DNN to FM for high-order combination. Although FNN has excessive influence on FM parameters, it enhances the ability of FM to a certain extent under this framework, which makes FNN have superior performance in the data sets of books and music. For the AFM model, although the attention network can distinguish which features are more important and pay attention to the full connection layer with greater weight, when the number of data features is small and sparse, it pays too much attention to the information brought by only the features and ignore the hidden information mined by more DNN, resulting in poor performance of AFM in book and music data sets. For Wide & Deep and DeepFM, although both models learn high and low-order features, the effect of high-order features is not very good on the data of books and music, and reduces the impact of low-order features on the prediction results, making the final prediction effect worse than FNN.

The results of RippleNet in MovieLens-1M are better than KGCN and MKR. It shows that RippleNet can accurately capture users' interests, especially in the case of intensive interaction between users and items. However, RippleNet performs worse than KGCN and MKR for non dense data sets (e.g., Book-crossing and last.FM). Although the KGCN is proposed for RippleNet's failure to pay attention to the importance of users to the relationship between items, the KG used by RippleNet is complex, and the number of recommended items accounts for only 1.8% of the number of KG entities, so it is impossible to effectively learn the whole complex KG information through less relationships. In Book-Crossing and last.FM, the proportion of recommended items in KG entities is 26.9% and 69.4% respectively, and the number of relationships is also large, which makes KGCN achieve better results on these two sets of data sets. MKR performed best in all baselines, even better than our model Eir-Ripp on Book-Crossing ACC. This shows that the cross unit of MKR can well interact with the entity features of the two tasks and transfer knowledge between tasks. However, the head entity in the KG triplet used by MKR in movielens-1m and Book-crossing can only be composed of recommended items, so the KG used is small, which limits the diversification of KG information and reduces the interpretability. As shown in Fig. 3(a) represents the common KG, (b) represents the KG used by MKR, and it cannot have a triplet with attribute nodes pointing to attribute nodes.

---

[1] The results of some baselines in this paper are different from those of the original paper because they are originally implemented using the tensorflow framework. We observe that the tensorflow and pytorch frameworks are different in the implementation of some network layers, the default parameters and random processes.

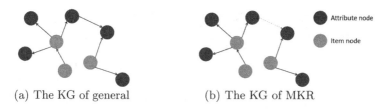

(a) The KG of general   (b) The KG of MKR

Fig. 3. The difference between the KG of general and the KG used by MKR.

Table 5. The results of AUC and Accuracy in CTR prediction.

Model	MovieLens-1M		Book-Crossing		Last.FM	
	AUC	ACC	AUC	ACC	AUC	ACC
FM	0.915	0.839	0.698	0.653	0.757	0.688
FNN	0.914	0.839	0.729	0.687	0.785	0.733
AFM	0.918	0.842	0.691	0.644	0.755	0.700
Wide&Deep	0.917	0.841	0.683	0.633	0.754	0.703
DeepFM	0.920	0.844	0.697	0.651	0.759	0.696
RippleNet	0.920	0.844	0.652	0.633	0.760	0.699
KGCN	0.892	0.817	0.669	0.616	0.796	0.721
MKR	0.914	0.839	0.734	**0.704**	0.795	0.738
Ours	**0.927**	**0.852**	**0.740**	0.681	**0.809**	**0.744**

Overall, our Eir-Ripp performed best on these three data sets. Specifically, compared with the best baseline solution, Eir-Ripp's average ACC in film, book and music recommendation is increased by 13.9%, 28.4% and 34.3% respectively, and AUC is increased by 0.7%, 0.6% and 1.3% respectively. In the same KG dataset, last.FM and Eir-Ripp are 1.3% and 0.6% higher than MKR in AUC and ACC, respectively. On the book crossing dataset, Eir-Ripp is 2.3% lower than the ACC of MKR. However, due to the small KG dataset used by MKR, Eir-Ripp is better than MKR in the recommended interpretability. On MovieLens-1M, both AUC and ACC Eir-Ripp are higher than MKR, and the KG data used is larger than MKR, which has better interpretability.

6   Conclusion

This paper proposes an enhanced Eir-Ripp for item representation model. Eir-Ripp is an end-to-end framework, which is composed of two parts: item representation enhancement module and interest module. The item representation enhancement module can obtain useful user high-order interaction features and user historical behavior preferences via our Uil unit. Then the graph convolution method is used to enhance the KG semantic information of items and strengthen the interaction of implicit information between users and items. The interest module uses RippleNet to discover users' potential interest. Finally, the user and item information output by the model are fused through a layer of nonlinear transformation. We conducted extensive experiments in three recommended scenarios. The results showed that Eir-Ripp was significantly better than the best baseline. In the future work, we plan to combine Eir-Ripp's item representation enhancement module into the MKR framework to overcome the limitation of KG dataset used by MKR.

Acknowledgment. This work was supported in part by the National Key Research and Development Program of China under Grant No. 2018YFB2100805, National Natural Science Foundation of China under the grant No. 61962017, 61562019, the Key Research and Development Program of Hainan Province under grant No. ZDYF2020008, Natural Science Foundation of Hainan Province under the grant No. 2019CXTD400, 2019RC088, and grants from State Key Laboratory of Marine Resource Utilization in South China Sea and Key Laboratory of Big Data and Smart Services of Hainan Province.

References

1. Linden, G., Smith, B., York, J.: Amazon. com recommendations: item-to-item collaborative filtering. IEEE Internet Comput. **7**(1), 76–80 (2003)
2. Wang, X., Wang, D., Xu, C., He, X., Cao, Y., Chua, T.S.: Explainable reasoning over knowledge graphs for recommendation. In: Proceedings of the AAAI Conference on Artificial Intelligence, vol. 33, pp. 5329–5336 (2019)
3. Gu, J., She, S., Fan, S., Zhang, S.: Recommendation based on convolution network of user's long-term and short-term interest and knowledge graph. Comput. Eng. Sci. **43**(03), 511–517 (2021)
4. Zhang, F., Yuan, N.J., Lian, D., Xie, X., Ma, W.Y.: Collaborative knowledge base embedding for recommender systems. In: Proceedings of the 22nd ACM SIGKDD International Conference on Knowledge Discovery and Data Mining, pp. 353–362 (2016)
5. Liu, Y., Li, B., Zang, Y., Li, A., Yin, H.: A knowledge-aware recommender with attention-enhanced dynamic convolutional network. In: Proceedings of the 30th ACM International Conference on Information & Knowledge Management, pp. 1079–1088 (2021)
6. Guo, Q., et al.: A survey on knowledge graph-based recommender systems. IEEE Trans. Knowl. Data Eng. 1 (2020)
7. Rendle, S.: Factorization machines. In: 2010 IEEE International Conference on Data Mining, pp. 995–1000. IEEE (2010)
8. Zhang, W., Du, T., Wang, J.: Deep learning over multi-field categorical data. In: Ferro, N., et al. (eds.) ECIR 2016. LNCS, vol. 9626, pp. 45–57. Springer, Cham (2016). https://doi.org/10.1007/978-3-319-30671-1_4
9. Cheng, H.T., et al.: Wide & deep learning for recommender systems. In: Proceedings of the 1st Workshop on Deep Learning for Recommender Systems, pp. 7–10 (2016)
10. Xiao, J., Ye, H., He, X., Zhang, H., Wu, F., Chua, T.S.: Attentional factorization machines: learning the weight of feature interactions via attention networks. In: Proceedings of the 26th International Joint Conference on Artificial Intelligence, pp. 3119–3125 (2017)

11. Guo, H., Tang, R., Ye, Y., Li, Z., He, X.: DeepFM: a factorization-machine based neural network for CTR prediction. In: Twenty-Sixth International Joint Conference on Artificial Intelligence, pp. 1725–1731 (2017)

12. Wang, H., et al.: GraphGAN: graph representation learning with generative adversarial nets. In: Proceedings of the AAAI Conference on Artificial Intelligence, vol. 32, pp. 2508–2515 (2018)

13. Bordes, A., Usunier, N., Garcia-Duran, A., Weston, J., Yakhnenko, O.: Translating embeddings for modeling multi-relational data. Adv. Neural. Inf. Process. Syst. **26**, 2787–2795 (2013)

14. Wang, Z., Zhang, J., Feng, J., Chen, Z.: Knowledge graph embedding by translating on hyperplanes. In: Proceedings of the AAAI Conference on Artificial Intelligence, vol. 28, pp. 1112–1119 (2014)

15. Lin, Y., Liu, Z., Sun, M., Liu, Y., Zhu, X.: Learning entity and relation embeddings for knowledge graph completion. In: Twenty-Ninth AAAI Conference on Artificial Intelligence, pp. 2181–2187 (2015)

16. Ji, G., He, S., Xu, L., Liu, K., Zhao, J.: Knowledge graph embedding via dynamic mapping matrix. In: Proceedings of the 53rd Annual Meeting of the Association for Computational Linguistics and the 7th International Joint Conference on Natural Language Processing (Volume 1: Long Papers), pp. 687–696 (2015)

17. Nickel, M., Tresp, V., Kriegel, H.P.: Factorizing YAGO: scalable machine learning for linked data. In: Proceedings of the 21st International Conference on World Wide Web, pp. 271–280 (2012)

18. Yang, B., Yih, W.T., He, X., Gao, J., Deng, L.: Embedding entities and relations for learning and inference in knowledge bases. arXiv preprint arXiv:1412.6575 (2014)

19. Nickel, M., Rosasco, L., Poggio, T.: Holographic embeddings of knowledge graphs. In: Proceedings of the AAAI Conference on Artificial Intelligence, vol. 30, pp. 1955–1961 (2016)

20. Wang, H., Zhang, F., Xie, X., Guo, M.: DKN: deep knowledge-aware network for news recommendation. In: Proceedings of the 2018 World Wide Web Conference, pp. 1835–1844 (2018)

21. Wang, H., et al.: RippleNet: propagating user preferences on the knowledge graph for recommender systems. In: Proceedings of the 27th ACM International Conference on Information and Knowledge Management, pp. 417–426 (2018)

22. Wang, H., Zhao, M., Xie, X., Li, W., Guo, M.: Knowledge graph convolutional networks for recommender systems. In: The World Wide Web Conference, pp. 3307–3313 (2019)

23. Wang, H., Zhang, F., Zhao, M., Li, W., Xie, X., Guo, M.: Multi-task feature learning for knowledge graph enhanced recommendation. In: The World Wide Web Conference, pp. 2000–2010 (2019)

24. Zhao, M., Zhong, S., Fu, X., Tang, B., Dong, S., Pecht, M.: Deep residual networks with adaptively parametric rectifier linear units for fault diagnosis. IEEE Trans. Industr. Electron. **68**(3), 2587–2597 (2020)

25. Vaswani, A., et al.: Attention is all you need. In: Proceedings of the 31st International Conference on Neural Information Processing Systems, pp. 6000–6010 (2017)

26. Zhang, Y., Wallace, B.: A sensitivity analysis of (and practitioners' guide to) convolutional neural networks for sentence classification. In: Proceedings of the Eighth International Joint Conference on Natural Language Processing (Volume 1: Long Papers), pp. 253–263 (2017)

27. Chen, Q., Zhao, H., Li, W., Huang, P., Ou, W.: Behavior sequence transformer for e-commerce recommendation in alibaba. In: Proceedings of the 1st International Workshop on Deep Learning Practice for High-Dimensional Sparse Data, pp. 1–4 (2019)
28. Li, Z.: Introduction and application of ROC curve under the background of bigdata. Sci. Educ. Guide **14**, 81–84 (2021)
29. Zhang, K., Su, H., Dou, Y.: A new multi-classification task accuracy evaluation method based on confusion matrix. Comput. Eng. Sci. **43**(11), 1910–1919 (2021)
30. Song, Y., Zhu, J., Zhao, L., Hu, A.: Centralized federated learning model based on model accuracy. J. Tsinghua Univ. (Nat. Sci. Ed.) **62**(5), 832–841 (2022)

User Multi-behavior Enhanced POI Recommendation with Efficient and Informative Negative Sampling

Hanzhe Li[1], Jingjing Gu[1(✉)], Haochao Ying[2(✉)], Xinjiang Lu[3], and Jingyuan Yang[4]

[1] Nanjing University of Aeronautics and Astronautics, Nanjing, China
{lihanzhe,gujingjing}@nuaa.edu.cn
[2] Zhejiang University, Hangzhou, China
haochaoying@zju.edu.cn
[3] Baidu Inc., Beijing, China
luxinjiang@baidu.com
[4] George Mason University, Fairfax, VA, USA
jyang53@gmu.edu

Abstract. Point-of-Interest (POI) recommendation plays a crucial role in the location-based social networks (LBSNs), while the extreme sparsity of user check-in data severely impedes the further improvement of POI recommendation. Existing works jointly analyse user check-in behaviors (i.e., positive samples) and POI distribution to tackle this issue. However, introducing user multi-modal behaviors (e.g., online map query behaviors), as a supplement of user preference, still has not been explored. Further, they also neglect to exploit why users don't visit the POIs (i.e., negative samples). To these ends, in this paper, we propose a novel approach, user multi-behavior enhanced POI recommendation with efficient and informative negative sampling, to promote recommendation performance. In particular, we first extract three types of relationships, i.e., POI-query, user-query and POI-POI, from map query and check-in data. After that, a novel approach is proposed to learn user and POI representations in each behavior through these heterogeneous relationships. Moreover, we design a negative sampling method based on geographic information to generate efficient and informative negative samples. Extensive experiments conducted on real-world datasets demonstrate the superiority of our approach compared to state-of-the-art recommenders in terms of different metrics.

Keywords: POI recommendation · User multi-behavior · Negative sampling

1 Introduction

With the rapid development of smart mobile device and communications technology (e.g., 4G/5G cellular data networks), Location-Based Social Networks (LBSNs) have been ubiquitous in our daily life, providing various services. Generally, users visit places, i.e. Point-of-Interests (POIs), and share their check-in records with experience to LBSN platforms through mobile devices. As a result, a large amount of user check-in data has been generated, which has become a basis to explore users' preferences and then recommend interesting but never visited POIs for users. This task, known as POI recommendation, has attracted serious attentions both in industry and academia.

Traditional methods of POI recommendation were based on the Collaborative Filtering (CF), and exploited the user check-in data to learn user preferences [2,6,26]. However, owing to the difficulty of collecting sufficient check-in logs and the limitation of the scope of user activities, the data sparsity significantly degrades the POI recommendation performance. To solve this problem, some works constructed ranking pairs for more sufficiently characterizing user preferences [5,22]. Meanwhile, [4,11,30] utilized context information, such as categories and geographical influences of POIs, to better understand users' visit behaviors. Recently, a few works proposed some graph-based methods for POI recommendation [19,25], which extracted relational graphs from various context and learned representations through graph embedding or GNN models for more precisely modeling.

In short, above works mined user check-in behaviors for recommending POIs based on historical check-in data. But the severely sparse check-in data still makes it difficult to fully mine user check-in behaviors. There are two challenges which are not well addressed:

(1) User Map Query Behaviors are Still not Effectively Studied. As far as we know, there are few works to study the effect of map query behaviors on user preferences in POI recommendation. Intuitively, users' map query records imply their potential travel intentions, especially the check-in preferences that lack historical regular travel records (such as office or home). Thus the map query data can be introduced as a supplement of user preference and increase the degree of personalization. Figure 1 shows an example of one user's daily life experiment at the weekend. She queried the shopping mall and preferred the POI: *Manhattan Mall*, which shows the consistency between user map query behaviors and check-in behaviors at LBSNs. Shortly after leaving Manhattan, she checked in a nearby restaurant (i.e., *Daniel*), which infers that she went to lunch after shopping. Although the user did not query the restaurant POI, as can be observed, the checked-in POI (e.g., *Daniel*) may locate inside or nearby the queried POI (e.g., *Manhattan Mall*), which indicates the potential relationships between user check-in and map query behaviors. Therefore, it's crucial to deeply explore the impact of user query behaviors on user preferences.

(2) Negative Samples are Still not Effectively Studied. Negative samples, the places that have not been visited by users, also imply users' personal preferences over different POIs, especially which are closer to the user but not selected. But most works generated negative samples randomly or based on the number of POI occurrences [5,27]. For instance, in Fig. 1, the *Soho Mall* in the query results, is closer to user than the *Manhattan Mall*, but it doesn't appear in users' check-ins, which means the user probably don't like *Soho Mall* and reflects her preference in another aspect. We argue that such negative samples

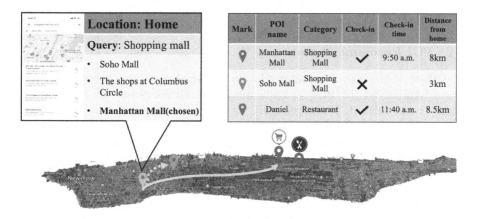

Fig. 1. An example of user daily life.

(e.g., *Soho Mall*) carry more user's disfavor which can accelerate model training process and improving the performance, namely informative negative samples.

To these ends, in this paper, we propose a novel approach to enhance POI recommendation by exploiting user map query and check-in behaviors with our negative sampling method, named as User Multi-behavior enhanced POI Recommendation with informative negative sampling (UMPRec). Specially, first we construct three graphs to capture different relationships, i.e., POI-query, user-query, POI-POI. Next, we propose a graph embedding model with a negative sampling method to learn the low dimensional representations of POIs and users in each behavior. Finally, we unify a pairwise ranking model, which is used to compute the score of each check-in record by user and POI representations, with the graph embedding model, to jointly learn parameters. Overall, the primary contributions are summarized as follows:

- We propose a novel approach UMPRec, which deeply studies user query behaviors and informative negative samples based on graph embedding and pairwise ranking models, to enhance POI recommendation. To the best of our knowledge, we are the first to utilize the context between user map query and check-in behaviors to address the data sparsity for the POI recommendation.
- We construct POI-query, user-query and POI-POI Graphs to extract three different types of relationships and learn multi-view user and POI representations for making a more precise recommendation.
- We design an efficient negative sampling methods based on datum points to generate more informative negative samples.
- We perform extensive experiments on real world dataset to evaluate our UMPRec model. The results demonstrate our model consistently outperforms state-of-the-art methods in terms of four different metrics.

The remainder of this paper is organized as follows. We first review the related work in Sect. 2. Section 3 introduces and elaborates our model in detail. Section 5 presents the experimental settings and results, followed by conclusions in Sect. 6.

2   Related Work

2.1   POI Recommendation

Compared to traditional recommendation scenarios, the data sparsity issue has seriously hindered the POI recommendation performance and there are many works try to solve it. First, generally, to deal with the user implicit feedback data, researchers utilized the one-class collaborative filtering approaches, such as weighted matrix factorization [6,26], the rank-based learning method [5,22]. Second, context information (e.g., spatio-temporal influence and sequential pattern) can be explored for profiling user complex preference to relieve the data sparsity problem. [11] regarded spatio-temporal pattern as a transition between user and POI. And [4] conducted an emotional semantic analysis of users' comments on POIs in LBSNs through CNNs. Moreover, some studies extracted relational graphs from multi-types of context information and learned representations through graph embedding or GNN models [1,14,19,25].

Recently, exploiting user mobility behaviors from heterogeneous data has provided a new perspective to address data sparsity. [17] explored the spatial influence from taxi and bus data for location prediction. [7] incorporated human trip activities, POI profiles, and demographic data to model POI demand of urban regions. However, to the best of our knowledge, there are few works to study POI recommendation from the user multi-behavior data at LBSNs.

Because user map query behavior strongly implied their daily needs and interests in urban life, more and more researchers have focused on studying user mobility and crowd patterns based on map query data. [18] demonstrated the correlation between POI visit and online map query, and integrated them for query-activity prediction task. [29] employed the query data to give early warning to the sudden large-scale human crowd events. Moreover, [24] jointly modeled dynamic context information, user's query keywords and clicked POIs, to solve the online POI query problem in dynamic scenarios. However, there are a few works to utilize map query behavior for POI recommendation.

2.2   Negative Sampling

Recently, negative sampling methods have been prevalent in recommendation for helping alleviate the data sparsity. Its main idea was to make full use of implicit negative samples at representation learning stage. [5] learned the factorization by ranking both visited and unvisited POIs correctly, as well as the geographic information. Similarly, [27] integrated the user sequential and temporal information into the POI embedding model, and considered geographical factors to sample unvisited POIs in the pairwise ranking model. However, these works selected negative samples randomly or just according to the frequency of occurrence. Different from previous works, we design a negative sampling method based on datum points which considers the geographic information and is more efficient.

3 Preliminaries

We start by providing the notations and POI recommendation problem definition in this section, and then present an overview of UMPRec.

Let U denotes a set of users and V denotes a set of POIs. m and n are the total numbers of users in U and POIs in V, respectively. For each user i, we denote the temporally-ordered check-in and the map query behaviors of i as $C_i = [c_1, c_2, ...]$ and $Q_i = [q_1, q_2, ...]$, where $c_j = (v_j^c, t_j^c)$ means user i checked in the POI $v_j^c \in V$ at timestamp t_j^c and $q_k = (v_k^q, t_k^q)$ means user i queried the POI $v_k^q \in V$ at timestamp t_k^q, respectively. Given the historical check-in C_i and map query records Q_i, the **goal** of POI recommendation is: given a user i, we compute the POI score vector $s_i = [p_1, p_2, ..., p]$, where each element represents the probability of visiting the corresponding POI in the future. Table 1 lists the notations in this paper.

Table 1. Description of notations

Notation	Description
U, V	The sets of users and POIs
m, n	The numbers of users and POIs
$c_j = (v_j^c, t_j^c)$	Check-in to POI v_j^c at timestamp t_j^c
$q_k = (v_k^q, t_k^q)$	Map query to POI v_k^q at timestamp t_k^q
$\boldsymbol{u}^c, \boldsymbol{v}^c$	User and POI vectors in check-in space
$\boldsymbol{u}^q, \boldsymbol{v}^q$	User and POI vectors in map query space
$G_{p-q} = (V, E_{p-q}, W_{p-q})$	POI-Query graph
$G_{u-q} = (V, E_{u-q}, W_{u-q})$	User-Query graph
$G_{p-p} = (V, E_{p-p}, W_{p-p})$	POI-POI graph

To exploit the user multi-behavior (i.e., check-in and map query) information for POI recommendation, we construct the POI-Query, User-Query and POI-POI graphs to encode user preferences, where V donates the set of vertices, E represents the set of edges for each graph and W means the weight of edges. Afterward, we can explore graph interactions to learn multi-view user and POI representations, where each view is corresponding to a particular behavior. Therefore, our model can discriminate possibly distinct patterns of the same users or POIs in the multi-behavior data.

The model skeleton overall is illustrated in Fig. 2. Our model consists of Four parts. First, we construct three context graphs based on the multi-behavior data. Then we generate informative negative samples through our negative sampling method, and pair with positive ones. After that, we employ a pairwise ranking loss (i.e., BPR loss) to learn user preference to POIs. Finally, we optimize our model under the supervision of the representation learning loss (i.e., Skip-gram [10]) and BPR loss in a joint training procedure.

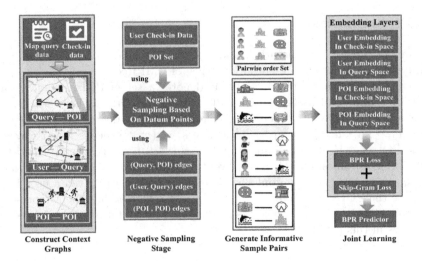

Fig. 2. An overview of UMPRec: The left part represents the construction process of the context graphs given the multi-behavior data. The two middle parts stand for the negative sampling stage and the generation of the sample pairs. Finally, the right part shows the joint learning stage.

4   Methodology

4.1   Context Graphs from User Multi-behavior

In consideration of both check-in and map query behaviors for POI recommendation task, we have the following intuitions.

a. Since map queries indicate users' implicit preference to POI visits, there is a high conversion rate from map queries to POI check-ins. Thus, for learning user preferences, we can mitigate the check-in data sparsity by studying User-Query correlations.

b. Different behaviors share the same POI preference in the same time slot. For instance, at dinner time, both the query and check-in POIs have a high probability to provide food service. Therefore, we should consider the POI-Query correlations.

c. Given that the importance of sequential patterns in human mobility [13,19], and the POIs visited in the same time slot may provide similar services, e.g., users usually check in bars after dinner, it's necessary to consider POI-POI correlations for modeling the user preference.

Inspired by above intuitions, we construct three context graphs with the multi-behavior data, each of which maintains a distinct type of relationships.

POI-Query Graph. In general, user multi-behavior reflects different aspects of user preferences. Thus, the assumption of universal relations between each query and check-in POI pair may introduce massive noise.

Regarding the temporal constraint of map query and check-in, we design a time-slicing scheme to capture the POI-Query correlations in the same time slot. On one hand, we differentiate the weekdays and weekends due to the completely different mobility patterns of working and rest [23]. On the other hand, for stimulating the human common regularity [28], we split a day into four sessions: from 3:00 a.m. to 10:59 a.m., 11:00 a.m. to 3:59 p.m., 4:00 p.m. to 9:59 p.m. and 10:00 p.m. to 2:59 a.m. Consequently, we have eight time sessions in total. For each user i, we extract all POI-Query pairs (v_j^c, v_k^q) such that POI v_k^q is queried in the same day or the same session as the check-in POI v_j^c. Afterward, we construct the POI-Query graph $G_{p-q} = (V, E_{p-q}, W_{p-q})$, where E_{p-q} is a set of POI-Query pairs (v_j^c, v_k^q), $W_{p-q}(v_j^c, v_k^q)$ is encoded with the number of the POI-query pair occurrences.

Finally, we can refer to POI-Query pairs to address the data sparsity issue of user check-in data. Given the POI-Query graph G_{p-q}, we augment the POI representations through the Skip-Gram model as the Eq. 1, which has been widely used for word embedding [10].

$$O_{p-q} = -\log p(v_k^q | v_j^c) = -\log \frac{\exp(v_j^c \cdot v_k^q)}{\sum_{v \in V} \exp(v_j^c \cdot v^q)}, \quad (1)$$

where v_j^c is the representation vector of POI v_j^c in the check-in space, v_k^q and v^q are the ones of POIs v_k^q and v in the map query space, respectively.

As the real number of POIs is large, it's impractical to compute the normalization term based on the entire set of POIs. To meet the efficiency requirement, we adopt negative sampling to approximate the intractable normalization, after which O_{p-q} is rewritten as:

$$O_{p-q} = -\log \sigma(v_j^c \cdot v_k^q) - \sum_{l=1}^{L} \mathbb{E}_{v \sim P_V}[\log \sigma(-v_j^c \cdot v^q)], \quad (2)$$

where $\sigma(x) = \frac{1}{1+e^{-x}}$ is the sigmoid function, L is the number of negative samples, and P_V is a distribution for drawing negative POIs according to geographical factors. We find that the Eq. 2 is aimed to jointly maximize $v_j^c \cdot v_k^q$ and minimize $-v_j^c \cdot v^q$, which means the probabilities of the observed edges will be greater than those of others. As a result, the check-in POIs can share the similar representations with the corresponding map query POIs.

User-Query Graph. Since the queried POIs imply users' future will of check-in, we introduce another graph to capture the correlations between users and map query POIs. Specifically, from the map query records Q_i of user i, we collect User-Query pairs (i, v_k^q) which means user i queried POI v_k^q. Then, we derive the User-Query graph $G_{u-q} = (U, V, E_{u-q}, W_{u-q})$, where E_{u-q} is a set of User-Query pairs (i, v_k^q), and $W_{u-q}(i, v_k^q)$ is encoded with the number of user query occurrences. However, the map query data only implies user intentions, most of which will not lead to a visit. To eliminate the query noise, we filter out edges from G_{u-q} with a lowest threshold.

Similar to the Eq. 2, we preserve User-Query relationships to set users and their query POIs close in the map query space by minimizing O_{u-q}:

$$O_{u-q} = -\log \sigma(\boldsymbol{u}_i^q \cdot \boldsymbol{v}_k^q) - \sum_{l=1}^{L} \mathbb{E}_{v \sim P_V}[\log \sigma(-\boldsymbol{u}_i^q \cdot \boldsymbol{v}^q)], \tag{3}$$

where $\boldsymbol{u}_i^q$ is the representation vector of user i in the map query space.

POI-POI Graph. Due to the temporal pattern of POI check-in, it is insufficient to just examine POI-POI relationships from the geographic and sequential perspectives [8,20,27]. Thus, we suggest to explore nearby POI relationships in a time-aware manner.

Concretely, we extract pairs of POIs within a certain geographical scope checked-in by same users: (a) in the same days, (b) in the same time sessions, since we assume these POI pairs can provide similar urban functions. Finally, we obtain the POI-POI graph $G_{p-p} = (V, E_{p-p}, W_{p-p})$ based on check-in behaviors. The weight of G_{p-p} denote the number of the POI pair co-occurrences. Next, we minimize O_{p-p} to regularize POI representations:

$$O_{p-p} = -\log \sigma(\boldsymbol{v}_j^c \cdot \boldsymbol{v}_k^c) - \sum_{l=1}^{L} \mathbb{E}_{v \sim P_V}[\log \sigma(-\boldsymbol{v}_j^c \cdot \boldsymbol{v}^c)], \tag{4}$$

where $\boldsymbol{v}^c$ means the representation vectors of POIs located in the check-in space.

4.2   Negative Sampling Based on Datum Points

In MF-based POI recommendation like BPR [12], the user-POI scores of visited pairs should be higher than those of others. However, an unlabeled user-POI pair may be caused by unawareness rather than dislike. Actually, users tend to know POIs close to them, but if a nearby POI doesn't appear in the user check-in history, it's likely to be disliked by the user. Accordingly, we can design informative negative samples to reduce the False Negative (FN) bias.

Specifically, we define the user trajectory center v_c for G_{u-q} based on user's historical check-in logs:

$$lon_{v_c} = \frac{lon_{s_1} + lon_{s_2} + ... + lon_{s_n}}{|S_i|}, lat_{v_c} = \frac{lat_{s_1} + lat_{s_2} + ... + lat_{s_n}}{|S_i|}, \tag{5}$$

where s_i means the trajectories of user i. Then, we define V_{u_i-neg} as the negative sample set of user i whose distance from the trajectory center is less than r.

While for G_{p-q} and G_{p-p}, the time complexity of distance calculation for each POI the user has not visited, is too high. We propose a two-stage sampling method which is efficient based on datum points.

In the first stage, we randomly select POIs in V with the proportion of τ. These POIs are denoted as datum points V^{base}. We treat each datum point as the center of a circle with the radius r. In each circle, POIs are recognized as geographical neighbors. Thereby, we obtain the datum points graph $G_{base} = (V, E_{nei})$. E_{nei} is defined as:

$$E_{nei} = \{(v_i, v_j)|v \in V^{base}, dis(v_i, v) < r, dis(v_j, v) < r\} \qquad (6)$$

where $dis(x, y)$ is the distance between two POIs.

In the second stage, for each pair (v_j, v_k) in G_{p-p} and G_{p-q}, if $v_j \in E_{nei}$, we generate negative samples from the adjacent set $V_{j-nei} = \{v|(v_j, v) \in E_{nei}\}$. Otherwise, we generate negative samples randomly. Finally, the Eq. 2~3 can be rewritten into:

$$O_{p-q} = -\log \sigma(\boldsymbol{v}_j^c \cdot \boldsymbol{v}_k^q) - \sum_{v_{neg}^q}^{V_{j-neg}} [\log \sigma(-\boldsymbol{v}_j^c \cdot \boldsymbol{v}_{neg}^q)], \qquad (7)$$

$$O_{u-q} = -\log \sigma(\boldsymbol{u}_i^q \cdot \boldsymbol{v}_k^q) - \sum_{v_{neg}^q}^{V_{u_i-neg}} [\log \sigma(-\boldsymbol{u}_i^q \cdot \boldsymbol{v}_{neg}^q)], \qquad (8)$$

where V_{u_i-neg} and V_{j-neg} are negative sample sets given the user trajectory center v_c and the adjacent set of POIs v_j^c, respectively. In addition, the Eq. 4 can be obtained in a similar way like the Eq. 3.

4.3   POI Recommendation with Multi-behavior

By incorporating map query behaviors into the latent matrix factorization framework for POI recommendation, the score that the user i will check in the POI v_j in the future is defined as:

$$x_{ij} = (w\boldsymbol{u}_i^c + (1 - w)\boldsymbol{u}_i^q) \cdot \boldsymbol{v}_j^c \qquad (9)$$

where w is the hyper-parameter for balancing the user preference from query and check-in behaviors. Compared with the form of $\boldsymbol{u}_i^c \cdot \boldsymbol{v}_j^c + \boldsymbol{u}_i^q \cdot \boldsymbol{v}_j^q$ which uses the POI representations from map query data, the Eq. 9 can empirically alleviate the noise of the query behaviors.

Due to the sparsity and the possible ambiguity of check-in data, we can hardly optimize the Eq. 9 directly. Thereby, regarding the goal of just a POI preference rank, we design a pairwise ranking objective following the BPR optimization criterion [12]. Specifically, we assume that users prefer the checked-in POIs than others, and define a ranking order $\succ_{u_i}$ over POIs v_j and v_k as follows:

$$v_j \succ_{u_i} v_k \leftrightarrow x_{ij} > x_{ik}, \qquad (10)$$

where v_j is a visited POI, and v_k is an unvisited POI generated by the negative sampling. (i, v_j) denotes user i has checked in the POI v_j, and we generate a set of pairwise preference orders $D = (u_i, v_j, v_k)$ and minimize the O_R.

$$O_R = - \sum_{(u_i,v_j,v_k)\in D} \log \sigma(x_{ij} - x_{ik}). \tag{11}$$

Finally, we combine the pairwise ranking loss with the three contracts of context graphs under a joint learning framework. The loss is computed as:

$$O = \arg\min_{\theta} O_R + \alpha O_{p-p} + \beta O_{p-q} + \gamma O_{u-q}, \tag{12}$$

where $\theta = \{\boldsymbol{U}^c, \boldsymbol{U}^q, \boldsymbol{V}^c, \boldsymbol{V}^q\}$ is the representation vectors of users and POIs in check-in and map query spaces, and α, β, γ are regularization parameters. The joint learning form can not only extract informative patterns from context graphs, but also avoid over-fitting through sharing user and POI presentations.

Once we have learnt the representations of users and POIs, we calculate the ranking score of POIs as the Eq. 9, and then choose the k POIs with the highest ranking scores for recommendation.

5   Experiments

In this section, we conduct extensive experiments to demonstrate the effectiveness of our UMPRec model. Using a large-scale real-world dataset, we evaluate the overall recommendation performance, the impacts of the multi-behavior data and the impacts of hyper-parameters.

5.1   Experimental Setting

Dataset. We utilized data from SHANGHAI, one of the largest cities in China, to conduct our experiments. The dataset contains POIs, user check-ins and user map queries records. Here, we just kept map queries and check-ins to POIs in the following categories: entertainment, shopping, hotel, restaurant, travel spot and fitness, while filtered out the rest because of the low user demands. Then we got total 474,969 check-ins and 443,304 queries with check-in sparsity is 99.93%, which calculated by $\frac{check-ins}{M \times N}\%$, the average number of user queries and check-ins are 15.4 and 16.5, respectively. We then used the check-ins in the last week for testing, those in the penultimate week for validation, and the rest of check-ins and queries for training. The data statistics are listed in Table 2.

Metrics. We adopted the Precision@k, Recall@k, MAP@k(Mean Average Precision) and NDCG@k(Normalized Discounted Cumulative Gain) metrics to test the performance. Essentially, Precision@k and Recall@k measure how correct the recommendation results are, while MAP@k and NDCG@k measure whether the ground-truth POIs are ranked at top places in the recommendation list.

Table 2. Statistics of dataset SHANGHAI

Description	Statistic
Time spanning	2018/07/01–2018/09/30
# of users	28786
# of POIs	23711
check-in sparsity	99.93%
avg. # of query POIs per user	15.4
avg. # of check-in POIs per user	16.5

Baselines. We compared our UMPRec model with the following baselines.

- **WRMF** [2], a point-wise latent factor model. It distinguished user observed and unobserved check-in data with different confidence values.
- **BPR** [12], a classic pairwise learning framework for implicit feedback data. Specifically, we employed matrix factorization as the internal predictor.
- **PACE** [20] combined user check-in behaviors and context information from users and POIs through a graph-based semi-supervised learning framework. Since the social networks were not available in our dataset, we only constructed the POI-POI spatial graph as the context.
- **SAE-NAD** [8] explicitly integrated the spatial information into an autoencoder framework and utilized the self-attention mechanism to generate user representation from historical check-in records.
- **CASER** [13], a CNN-based method which applied deep learning method on recommendation with convolutional networks.
- **NGCF** [15] adopted three GNN layers on the user-item graph to refine user and item representations via at most 3-hop neighbors' information.
- **DGCF** [16] developed disentangled GNN to capture independent user intentions, increasing the candidates in matching and guaranteeing the accuracy.
- **USG** [21] considered geographical influence by modeling the probability of a user visiting a new place given the user's historical check-ins.
- **MixGCF** [3] applied hop-mixing technique to generate more reliable negative samples from user-item graph structure and GNN aggregation process.
- **UltraGCN** [9] skipped message passing and directly approximated the limit of infinite message passing layers. Here we used it in POI recommendation.

Implementation. We implemented our model in PyTorch. First, to reduce noise, we remove all the edges with weight equal to 1 in the three constructed graphs. For the parameters setting, the number d of dimensions was fixed to 32 and set $\tau = 0.01$ for all methods. We used the Adam optimizer with $\beta_1 = 0.9$, $\beta_2 = 0.999$, $\epsilon = 10^{-6}$ and the learning rate was 0.001. To balance the data size from each task, we set $B = 512$ for graph G_{p-p} and $B = 128$ for others. Further, two types of regularization were employed during training: (a) an L_2 regularization with weight 5×10^{-6} on all parameters, (b) an early stopping if the

F_1 on validation set did not increase in successive 5 checks. For the regularization parameters, we fixed $\alpha = \gamma = 1$ and $\beta = 50$ by default for our UMPRec.

5.2   Performance Comparison

Table 3. Overall recommendation performance evaluation on Precision, Recall, MAP and NDCG (Pre@k means Precision@k and Rec@k means Recall@k)

Method	Pre@3	Pre@5	Pre@10	Rec@3	Rec@5	Rec@10
WRMF	0.0163	0.0141	0.0120	0.0203	0.0293	0.0485
BPR	0.0212	0.0176	0.0137	0.0275	0.0372	0.0561
CASER	0.0133	0.0111	0.0090	0.0137	0.0187	0.0301
PACE	0.0130	0.0110	0.0090	0.0175	0.0243	0.0389
NGCF	0.0213	0.0185	0.0147	0.0272	0.0389	0.0607
DGCF	0.0213	0.0174	0.0138	0.0280	0.0381	0.0587
SAE-NAD	0.0236	0.0202	0.0161	0.0303	0.0417	0.0651
USG	0.0227	0.0203	0.0161	0.0291	0.0424	0.0666
MixGCF	0.0249	0.0216	0.0170	0.0319	0.0454	0.0718
UltraGCN	0.0238	0.0203	0.0161	0.0314	0.0428	0.0669
UMPRec-r	0.0261	0.0218	0.0171	0.0329	0.0462	0.0702
UMPRec	**0.0265**	**0.0220**	**0.0175**	**0.0351**	**0.0468**	**0.0720**
Method	MAP@3	MAP@5	MAP@10	NDCG@3	NDCG@5	NDCG@10
WRMF	0.0160	0.0167	0.0188	0.0217	0.0248	0.0317
BPR	0.0219	0.0223	0.0243	0.0290	0.0321	0.0389
CASER	0.0114	0.0117	0.0124	0.0165	0.0178	0.0218
PACE	0.0127	0.0134	0.0152	0.0173	0.0198	0.0253
NGCF	0.0180	0.0209	0.0240	0.0284	0.0326	0.0404
DGCF	0.0198	0.0223	0.0252	0.0299	0.0331	0.0407
SAE-NAD	0.0242	0.0248	0.0273	0.0320	0.0359	0.0443
USG	0.0234	0.0246	0.0274	0.0311	0.0360	0.0447
MixGCF	0.0215	0.0249	0.0288	0.0334	0.0382	0.0478
UltraGCN	0.0256	0.0263	0.0290	0.0334	0.0374	0.0460
UMPRec-r	0.0273	0.0280	0.0306	0.0360	0.0397	0.0485
UMPRec	**0.0277**	**0.0283**	**0.0311**	**0.0365**	**0.0404**	**0.0495**

We first evaluate the overall recommendation performance. UMPRec-r is specifically a variants of UMPRec that generates negative samples by random sampling. The results under the four metrics with k = 3, 5, 10 are reported in Table 3. Note that, due to the diversity of POI categories and the influence of complicated context information to check-in behaviors, all methods for POI recommendation get low performance in general and thus researchers compare the relative improvement in this area, as we follow.

First, the PACE and CASER model do not work well, possibly because of the serious data sparsity and difficulty in extracting effective features. Note that, these two methods are quite similar in Precision@k, but differ greatly in other metrics. That's because the calculation of precision is affected by the number k of candidates. For each user, k is fixed, so a wrong prediction will make the two models, which perform poorly, have similar precision. Second, the pairwise ranking method is overall better than the point-wise model, comparing the point-wise WRMF with the pairwise BPR. Third, the results of USG and SAE-NAD are similar because they both mainly considered the geographical influence over POIs, and that's why they perform better than most of others. Moreover, NGCF, DGCF and UltraGCN are GNN-based and related to collaborative filtering. The first two methods have similar performance because the data sparsity is serious and they have the similar GNN structure. And UltraGCN performs better mostly because it respectively filters uninformative user-POI and POI-POI relationships, which avoids introducing too much noise. After that, MixGCF also performs well due to exploiting effective negative samples. Finally, our UMPRec model consistently performs the best under the four metrics, which also indicates the effectiveness of our negative sampling method by comparing with UMPRec-r.

5.3 Impacts of the Multi-behavior Data

A novel perspective of our UMPRec model is to exploit the multi-behavior data to address the sparsity issue. To evaluate the impacts of the multi-behavior data, i.e., check-in and map query records, we observe the results of check-in, map query, and negative sampling. But since the last one is already shown in Table 3, we only focus on the impact of the first two on the model. We specifically consider four variants of UMPRec: (1) **BPR** which does not consider the context graphs, (2) **WithCheck-in** which only incorporates POI-POI graph constructed by check-in behaviors, (3)**WithQuery** which considers POI-Query and User-Query Graphs, and (4) **ALL**, which is indeed our complete UMPRec model and combines the two behaviors. The results are reported in Fig. 3.

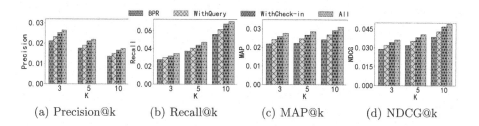

Fig. 3. Impacts of the multi-behavior data

In all our tests, exploiting the multi-behavior data consistently improves the performance compared with **BPR** under the four metrics. **WithCheck-in** performs better than **WithQuery**, which indicates the check-in behavior is more effective than the map query one, since check-ins indicate explicit preference of users to POIs while map queries are implicit feedbacks. Besides, because not all queries will turn into check-ins and therefore queries are more stochastic than check-ins. Moreover, the **ALL** model consistently outperforms **WithQuery** and **WithCheck-in** using either map query or check-in behaviors. This results show that these two behaviors are complementary. Recall that we extract three context graphs which are further combined in a joint learning framework. Finally, it is worth mentioning that, the (Precision@k, Recall@k, MAP@k, NDCG@k) of the complete UMPRec is on average (25.91%, 27.30%, 27.12%, 26.32%) higher than BPR, by utilizing the multi-behavior data.

In addition, we also empirically evaluate the impact of the balancing parameter w in the Eq. 11, and the negative samples L, which are shown in Fig. 4(a) and 4(b), respectively. Note that, we only report MAP@k because of same trend in four metrics and space limitation. As shown in Fig. 4(a), the extreme cases ($w = 0.0$ or $w = 1.0$) perform a sharp decline, implying that is not enough to only consider user check-in or query preference. Besides, as w increases, the performance gets better, and reaches the best at 0.8. That means though user query preference does contribute to the final performance, while user check-in preference plays a major role. From this perspective, we verify that combing user multi-behavior does enhance POI recommendation. And in Fig. 4(b), the model performs well when $5 \leq L \leq 20$, but the performance declines sharply when $L > 20$. This may because of the introduction of too much noise.

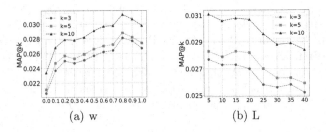

Fig. 4. Impact of hyper-parameter w, L

5.4 Impacts of Hyper-parameters

Finally, we evaluate the impacts of hyper-parameter α, β, γ of the Eq. 12, corresponding to POI-POI, POI-Query and User-Query graphs, respectively. In each time, we varied one of the three from 1 to 50, fixed the rest two to their default values and tested the performance of UMPRec with the four metrics when $k = 3$. The results are reported in Fig. 5.

First, the best result is achieved with a moderate $\alpha = 10$ and a moderate-to-large $\gamma \geq 10$. In the meanwhile, overall, the four metrics change sharply when varying β with respect to the other two, which means the performance of UMPRec is sensitive to the selection of β. All scores have a large increase when varying β from 1 to 50. The best performance under all the four metrics is when $\beta = 50$. As can be seen, the Recall and NDCG change the most. Note that, β regularizes the contribution of POI-Query in the overall objective. With β increasing, the model tends to characterize the relationship between check-in and query behaviors. Since the calculation of Recall is affected by the check-ins number of users, for users with few check-in records, correct prediction will greatly improve the metrics. With respect to NDCG, it measures the quality of ranking. Therefore, these results further verify that multi-behavior data can enhance the performance of POI recommendation.

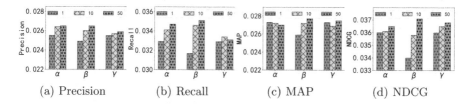

(a) Precision   (b) Recall   (c) MAP   (d) NDCG

Fig. 5. Impacts of hyper-parameters α, β, γ

6 Conclusion

In this paper, we proposed UMPRec, which utilized the user multi-behavior (i.e., map query and check-in) data to enhance the performance of POI recommendation with efficient and informative negative sampling. Specifically, we first extracted three context graphs from the multi-behavior data (i.e., map query and check-in behavior) to capture the distinct POI-Query, User-Query and POI-POI relationships. Then we designed a graph embedding method and an effective negative sampling method to learn multi-view user and POI representations better from these graphs. Besides, a pairwise ranking method was employed to estimate the probability that a user would check in a POI in the future. Finally, our proposed UMPRec model was optimized under the supervision of the representation learning and pairwise ranking in a joint training procedure. Our extensive experiments showed the superiority of our model, comparing with state-of-the-art methods in terms of four different metrics.

Acknowledgement. This work is supported in part by the National Natural Science Foundation of China under Grant (No. 62072235, No. 62106218).

References

1. Hang, M., Pytlarz, I., Neville, J.: Exploring student check-in behavior for improved point-of-interest prediction. In: SIGKDD, pp. 321–330 (2018)
2. Hu, Y., Koren, Y., Volinsky, C.: Collaborative filtering for implicit feedback datasets. In: ICDM, pp. 263–272 (2008)
3. Huang, T., et al.: MixGCF: an improved training method for graph neural network-based recommender systems. In: SIGKDD, pp. 665–674 (2021)
4. Jiao, X., Xiao, Y., Zheng, W., et al.: R2SIGTP: a novel real-time recommendation system with integration of geography and temporal preference for next point-of-interest. In: WWW, pp. 3560–3563 (2019)
5. Li, X., et al.: Rank-GeoFM: a ranking based geographical factorization method for point of interest recommendation. In: SIGIR, pp. 433–442 (2015)
6. Lian, D., Zhao, C., et al.: GeoMF: joint geographical modeling and matrix factorization for point-of-interest recommendation. In: SIGKDD, pp. 831–840 (2014)
7. Liu, Y., Liu, C., Lu, X., et al.: Point-of-interest demand modeling with human mobility patterns. In: SIGKDD, pp. 947–955 (2017)
8. Ma, C., et al.: Point-of-interest recommendation: exploiting self-attentive autoencoders with neighbor-aware influence. In: CIKM, pp. 697–706 (2018)
9. Mao, K., Zhu, J., Xiao, X., et al.: UltraGCN: ultra simplification of graph convolutional networks for recommendation. In: CIKM, pp. 1253–1262 (2021)
10. Mikolov, T., et al.: Distributed representations of words and phrases and their compositionality. In: Advances in Neural Information Processing Systems, vol. 26 (2013)
11. Qian, T., Liu, B., Nguyen, Q.V.H., et al.: Spatiotemporal representation learning for translation-based poi recommendation. TOIS $37(2)$, 1–24 (2019)
12. Rendle, S., Freudenthaler, C., Gantner, Z., et al.: BPR: Bayesian personalized ranking from implicit feedback. arXiv preprint arXiv:1205.2618 (2012)
13. Tang, J., Wang, K.: Personalized top-n sequential recommendation via convolutional sequence embedding. In: WSDM, pp. 565–573 (2018)
14. Wang, D., Wang, X., Xiang, Z., Yu, D., Deng, S., Xu, G.: Attentive sequential model based on graph neural network for next POI recommendation. World Wide Web $24(6)$, 2161–2184 (2021). https://doi.org/10.1007/s11280-021-00961-9
15. Wang, X., He, X., Wang, M., Feng, F., et al.: Neural graph collaborative filtering. In: SIGIR, pp. 165–174 (2019)
16. Wang, X., Jin, H., Zhang, A., et al.: Disentangled graph collaborative filtering. In: SIGIR, pp. 1001–1010 (2020)
17. Wang, Y., Yuan, N.J., Lian, D., et al.: Regularity and conformity: location prediction using heterogeneous mobility data. In: SIGKDD, pp. 1275–1284 (2015)
18. Wu, Z., Wu, H., Zhang, T.: Predict user in-world activity via integration of map query and mobility trace. In: UrbComp 2015 (1991)
19. Xie, M., Yin, H., Wang, H., et al.: Learning graph-based poi embedding for location-based recommendation. In: CIKM, pp. 15–24 (2016)
20. Yang, C., et al.: Bridging collaborative filtering and semi-supervised learning: a neural approach for POI recommendation. In: SIGKDD, pp. 1245–1254 (2017)
21. Ye, M., Yin, P., Lee, W.C., et al.: Exploiting geographical influence for collaborative point-of-interest recommendation. In: SIGIR, pp. 325–334 (2011)
22. Ying, H., Chen, L., Xiong, Y., et al.: PGRank: personalized geographical ranking for point-of-interest recommendation. In: WWW, pp. 137–138 (2016)

23. Yuan, Q., Cong, G., Ma, Z., et al.: Time-aware point-of-interest recommendation. In: SIGIR, pp. 363–372 (2013)
24. Yuan, Z., Liu, H., Liu, J., et al.: Incremental spatio-temporal graph learning for online query-poi matching. In: WWW, pp. 1586–1597 (2021)
25. Zhang, S., Cheng, H.: Exploiting context graph attention for poi recommendation in location-based social networks. In: DASFAA, pp. 83–99 (2018)
26. Zhao, P., Xu, X., Liu, Y., et al.: Exploiting hierarchical structures for poi recommendation. In: ICDM, pp. 655–664. IEEE (2017)
27. Zhao, S., Zhao, T., King, I., et al.: Geo-teaser: geo-temporal sequential embedding rank for point-of-interest recommendation. In: WWW, pp. 153–162 (2017)
28. Zhao, S., Zhao, T., Yang, H., et al.: STELLAR: spatial-temporal latent ranking for successive point-of-interest recommendation. In: AAAI (2016)
29. Zhou, J., Pei, H., Wu, H.: Early warning of human crowds based on query data from Baidu maps: analysis based on Shanghai stampede. In: Big Data Support of Urban Planning and Management. AGIS, pp. 19–41. Springer, Cham (2018). https://doi.org/10.1007/978-3-319-51929-6_2
30. Zhu, H., Li, W., Liu, W., Yin, J., Xu, J.: Top k optimal sequenced route query with poi preferences. Data Sci. Eng. 7(1), 3–15 (2022)

Neighborhood Constraints Based Bayesian Personalized Ranking for Explainable Recommendation

Tingxuan Zhang[1], Li Zhu[1(✉)], and Jie Wang[2]

[1] Xi'an Jiaotong University, Xi'an, China
zhuli@xjtu.edu.cn
[2] University of Glasgow, Glasgow, UK

Abstract. Model-based Collaborative Filtering (CF) methods such as Matrix Factorization (MF) have achieved promising ranking performance in recent years. However, these methods tend to be a black box that cant not provide any explanation for users. To obtain the trust of users and improve the transparency, recent research starts to focus on the explanation of recommendations. Explainable Bayesian Personalized Ranking (EBPR) leverages the relevant item to provide intuitive explanations for unexplainable model-based CF methods. It relies solely on feedback data and can be trained efficiently. However, EBPR ignores the neighborhood information of users and items in the latent space. In addition, the explainability of EBPR suffers from exposure bias. To address these issues, we propose a novel explainbale loss function and a corresponding Matrix Factorization-based model called Constrained Explainable Bayesian Personalized Ranking (CEBPR), which introduces neighborhood information in the latent space to enhance ranking performance and explainability. Furthermore, we analyze the impact of exposure bias on explainability and propose CEBPR+ to mitigate the bad effect. Finally, We conduct empirical experiments on three real-world datasets that demonstrate the advantages of our proposed methods in terms of accuracy and explainability.

Keywords: Recommender systems · Collaborative filtering · Explainability · Neighborhood information

1 Introduction

Recommender systems have played a vital role in people's daily life with the development of Internet technology. They provide an effective way to deal with the intractable problem of information overload. The earliest Collaborative Filtering (CF) methods provide recommendations to users according to the rating-based similarity between users or items, giving rise to neighborhood-based CF recommendation methods [9,14]. Subsequently, with the development of machine learning technology, model-based collaborative filtering methods [4,6,8] have

B. Li et al. (Eds.): APWeb-WAIM 2022, LNCS 13423, pp. 166–173, 2023.
https://doi.org/10.1007/978-3-031-25201-3_12

achieved more accurate recommendations than traditional neighborhood-based CF methods. However, model-based methods tend to be black boxes that only achieve accurate recommendations without any explanation. Unexplainable recommendations reduce the transparency and ultimately make users lose trust [1]. Therefore, recent researches begin to focus on the interpretability of recommender systems.

Explainable recommendations can be divided into different types depending on the data source [15]. Some recommender systems use content information or attention mechanisms to generate explanations [2,13,16]. But these methods require additional data sources or complex neural network structures. To address these issues, Damak et al. [3] proposed Explainable Bayesian Personalized Ranking (EBPR) provide neighborhood-based explanations for unexplainable model-based CF methods. EBPR relies solely on implicit feedback data and can be trained efficiently. However, EBPR ignores the neighborhood information of users and items in the latent space. Besides, exposure bias in EBPR can result in unobserved items that are highly explainable being sampled as negative, which in turn affect the overall explainability.

To address these issues, we propose a novel explainbale loss function and a corresponding Matrix Factorization-based model called Constrained Explainable Bayesian Personalized Ranking (CEBPR), which introduces neighborhood-information in the latent space to enhance ranking performance and explainability. In addition, we design a new constraint term for the negative sample with strong explainability and propose CEBPR+ to mitigate the adverse effects of exposure bias. Our contributions are summarized as follows:

- We propose two novel explainable pairwise loss functions based on Explainable Bayesian Personalized Ranking (EBPR) along with corresponding Matrix Factorization-based models. To the best of our knowledge, it is the first time to introduce neighborhood information in the latent space into explainable pairwise ranking methods.
- We analyze the impact of exposure bias on explainability. Based on this analysis, we design a novel constraint term for the negative sample with strong explainability and mitigate the adverse effect of exposure bias.
- We conduct empirical experiments on three real-world datasets to demonstrate the effectiveness of our proposed models in terms of ranking performance and explainability.

2 Preliminary

2.1 Explainability Matrix

In the field of Collaborative Filtering (CF), the earliest neighborhood-based method provides users with straightforward interpretation [12,14]. As shown in Fig. 1, a user-side explanation tells Bruce that similar users Lucy and Eric have bought this item, and a item-side explanation persuades Bruce to buy the mouse by showing the similar item keyboard he has bought recently. We adopt the item side explanation [7,11] since users may know the interacted items but know nothing about users with similar interests. The measuring way of explainability of item i for user u can be defined in Eq. (1).

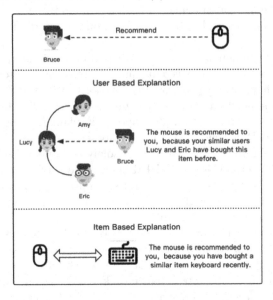

Fig. 1. Traditional neighborhood-based explanations. On the middle panel, the explanation is given by the relevant user. On the bottom panel, the explanation is given by the relevant item.

$$E_{ui} = P(Y_{uj} = 1 \mid j \in N_i^{\eta}). \tag{1}$$

Given a certain similarity measurement such as cosine similarity, N_i^{η} denotes a set of item i's η most similar items. These similar items are considered neighbors of the item i. Y_{uj} is a Bernoulli random variable and represents the positive or negative of an item. If the user u has interacted with item j, Y_{uj} is equal to 1, otherwise, it is equal to 0. This definition is formulated in Eq. (2).

$$Y_{uj} = \begin{cases} 1 & if \quad j \in I_u^+, \\ 0 & otherwise. \end{cases} \tag{2}$$

E_{ui} stands for the measure of explainability for user u to item i. Given the set of neighbors of item i, we consider the conditional probability which is obtained by the count of neighbors interacted by user u dividing the total count of neighbors as explainability of user u to item i. Thus, we can obtain an explainability matrix $E^{m \times n}$ for all users and items.

2.2 Explainable Bayesian Personalized Ranking

Rendle et al. [10] proposed Bayesian Personalized Ranking (BPR) can directly "optimized for ranking" and is widely used in various recommendation models. Although BPR appropriately captures and models the ranking-based preference, it can not provide any explaination. Recent years, Damak et al. proposed Explainable Bayesian Personalized Ranking (EBPR) [3], which introduced neighborhood-based explainability into BPR. EBPR captures explainable preference by adjusting the contribution of different items to the loss function. The formulation is given as follows:

$$L_{EBPR} = \frac{1}{|D|} \sum_{(u,i^+,i^-)\in D} -E_{ui^+}(1 - E_{ui^-})log\sigma(f_\Omega(u, i^+, i^-)). \quad (3)$$

where D represents the training data, which contains all triples of (u, i^+, i^-). I_u^+ represents the set of positive items of user u and I_u^- represent negative items of user u such that $I_u^- = I\backslash I_u^+$. σ is the Sigmoid function and f_Ω is a hypothesis parameter that quantifies the preference of user u for item i^+ relative to i^-. EBPR attempt to weigh the contribution of a triplet (u, i^+, i^-) by explainable coefficient $E_{ui^+}(1 - E_{ui^-})$. The higher explainability E_{ui^+} and the lower explainability E_{ui^-} should contribute more to the loss in the training process.

3   Proposed Methods

3.1   Constrained Explainable Bayesian Personalized Ranking (CEBPR)

Matrix Factorization (MF) [8] is a classic representation learning method. Given a rating matrix $R^{m\times n}$, MF factorizes $R^{m\times n}$ into two lower-rank matrices $P^{m\times k}$ and $Q^{n\times k}$ linearly. Here k represents the dimension of the latent space. We use $p_u \in P^{m\times k}$ and $q_i \in Q^{m\times k}$ denotes a user u and a item i respectively. Hence, the preference for user u to item i can be easily acquired by dot product $p_u \cdot q_i$. The recent Explainable Bayesian Personalized Ranking (EBPR) [3] integrates neighborhood-based explainability into Bayesian Personalized Ranking (BPR).

We argue that EBPR ignores inner correlation between the user representation and the item representation in the latent space. According to research [1], if item i is explainable to user u, then their representations in the latent space should be close to each other. In other words, $| p_u - q_i |$ should be close to zero as much as possible. Thus, idea is to give constraints to the user representation and positive item representation in latent space. Furthermore, the explanation is generated from neighbors of the positive item i, these neighbors should also be explainable to the target user. We design the function $W(i, j)$ to quantify the contributions of different neighbors. The definition of $W(i, j)$ is in Eq. (4):

$$W(i, j) = \frac{Sim(i, j)}{\sum_{j\in neighbors(i)} Sim(i, j)} * E_{uj}. \quad (4)$$

where item j is one of neighbors of the positive item i and $Sim(i, j)$ represents the cosine similarity between item i and item j. As mentioned in Sect. 2.1, we use E_{uj} represents the explainability for user u to the item j, thus we can obtain the Constrained Explainable Bayesian Personalized Ranking as follows:

$$L_{CEBPR} = L_{EBPR} + \sum_{j \in neighbors(i^+)} \frac{\alpha}{2} \|p_u - q_j\|^2 * W(i^+, j). \tag{5}$$

In Eq. (5), we take positive items and their neighbors into consideration. Besides, their contribution to the loss will be automatically adjusted according to their interpretability. A larger E_{uj} will make them contribute more to the loss and vice versa. In order to control the strength of constraints, we set a coefficients α as hyperparameter.

3.2   Adding Constraint for Explainable Negative Item (CEBPR+)

BPR is based on the Missing Completely At Random (MCAR) assumption [3]. BPR assume all non-interacted items as irrelevant, which introduces exposure bias into the recommender system. Some non-interacted items which may cater the user taste have no chance to be observed due to exposure bias. If these items were considered as user-dislike ones in negative sampling, the overall explainability of recommender system would deteriorate. Hence, we attempt to take this part of explainable negative items into consideration. To ensure a sufficient degree of explainability, we design a judgment function $J(u, i)$ to decide whether the negative samples are explainable or not. The definition of $J(u, i)$ is given in Eq. (6).

$$J(u, i) = \begin{cases} E_{ui} & if \quad E_{ui} \geq \theta, \\ 0 & if \quad E_{ui} < \theta. \end{cases} \tag{6}$$

The idea is to incorporate the latent space representations of explainable negative samples into the computation. Hence, we apply this additional constraint to CEBPR and obtain the formula of CEBPR+ in Eq. (7). For negative items, only a fraction of items that can achieve the requirement of threshold. Thus, the explainability of their neighbors is relatively small and can be ignored. We also introduce a coefficients β as hyperparameters to control the constraint strength of negative samples. These hyperparameters can make appropriate trade-offs between explainability and ranking performance.

$$L_{CEBPR+} = L_{EBPR} + \sum_{j \in neighbors(i^+)} \frac{\alpha}{2} \|p_u - q_j\|^2 * W(i^+, j) + \frac{\beta}{2} \|p_u - q_{i^-}\|^2 * J(u, i^-). \tag{7}$$

4   Experiments

4.1   Experiment Settings

We conduct our experiment on three publicly available datasets: The Movielens 100K [5] (ml-100k), the Movielens 1M [5] (ml-1m) datasets and the Last.FM 2K (lastfm-2k) [11]. We have intentionally transformed explicit data into implicit feedback data.

In our experiments, we compared our proposed CEBPR and CEBPR+ with the following methods: the classic Matrix Factorization (MF) [8], Explainable Matrix Factorization (EMF) [1], Bayesian Personalized Ranking (BPR) [10], Explainable Bayesian Personalized Ranking (EBPR) [3].

We adopt the widely-used Leave-One-Out (LOO) procedure [4,6]. For each user, we held out the latest interaction as the test item and leave the remaining data for training. Then, we rank the latest interacted item of the user with 100 randomly sampled non-interacted items. We fix the neighborhood size equal to 25 and the threshold θ equal to 0.1 to ensure the fair comparison [8].

We adopt the widely-used Hit Rate *(HR@K)* and Normalized Discounted Cumulative Gain *(NDCG@K)* to evaluate the ranking performance of proposed methods. In terms of explainability, we follow previous research [3], using Mean Explainability Precision *(MEP@K)* for evaluation. *MEP@K* calculate the proportion of explainable items in the Top K recommendation list. It is defined as follows:

$$MEP@K(TopK) = \frac{1}{|U|} \sum_{u=1}^{|U|} \frac{|\, i \in TopK(u) \cap E_{ui} > 0 \,|}{K}. \tag{8}$$

In Eq. (8), *TopK* denotes the top k items in the recommendation list. However, *MEP@K* only considers whether the item can be explained, regardless of how different their explainability are. Thus, the authors in [3] have further extended *MEP@K* to weight the items' different contributions to overall explainability. The definition is given in Eq. (9).

$$WMEP@K(TopK) = \frac{1}{|U|} \sum_{u=1}^{|U|} \frac{|\, i \in TopK(u) \cap E_{ui} > 0 \,|}{K} * E_{ui}. \tag{9}$$

The proposed evaluation metric was called Weighted Mean Explainability Precision *(WMEP@K)*. *WMEP@K* capture the degree of explainability difference by multiplying the explainability E_{ui} of each user-item pair.

4.2   Results and Discussion

Table 1 displays the evaluation results in terms of ranking performance. For ml-100k and ml-1m datasets, Both CEBPR and CEBPR+ achieve competitive ranking performance. Especially, CEBPR achieves the best HR and NDCG in these two datasets. Overall, pairwise methods outperform pointwise methods on ml-100k and ml-1m datasets in ranking performance evaluation, which demonstrates the powerful ability of pairwise ranking to deal with implicit feedback data.

Table 2 displays the evaluation results in terms of explainability. The proposed CEBPR+ consistently outperforms other methods on three datasets. Compared to EBPR, CEBPR+ make a significant improvement on the dataset ml-1m (2.17% MEP and 1.57% WMEP respectively). However, on the lastfm-2k

Table 1. Experiment results in terms of ranking performance on the three real-world datasets. We truncate the top 10 ($K = 10$) items from the final recommendation list. The best results are in bold and second to best results are underlined.

Dataset	ml-100k		ml-1m		lastfm-2k	
Model	HR	NDCG	HR	NDCG	HR	NDCG
MF	0.4984	0.2817	0.4910	0.2849	0.8853	0.6609
EMF	0.5175	0.2846	0.5090	0.2899	0.8506	0.6530
BPR	0.6936	0.4032	0.6841	0.4062	0.8954	0.6964
EBPR	0.6914	0.4034	0.6856	0.4043	0.8474	0.6339
CEBPR	0.7063	0.4127	0.6921	0.4135	0.8526	0.6542
CEBPR+	0.6965	0.4068	0.6843	0.4037	0.8453	0.6443

Table 2. Experiment results in terms of explainability on the three real-world datasets. We follow the same settings in Table 2.

Dataset	ml-100k		ml-1m		lastfm-2k	
Model	MEP	WMEP	MEP	WMEP	MEP	WMEP
MF	0.7490	0.2550	0.7437	0.2535	0.1963	0.0282
EMF	0.7943	0.2410	0.7809	0.2456	0.2035	0.0287
BPR	0.9049	0.3144	0.8804	0.2431	0.2595	0.0336
EBPR	0.9229	0.3397	0.9123	0.2624	0.3044	0.0367
CEBPR	0.9292	0.3458	0.9316	0.2767	0.3240	0.0381
CEBPR+	0.9373	0.3537	0.9340	0.2781	0.3470	0.0394

dataset, the performance of non-explainable models (MF, BPR) is better than any explainable model. This can be attributable to the extremely high sparsity of the lastfm-2k dataset (99.7%). However, the explainability improvement of our proposed methods in the lastfm-2k dataset is still significant. Compared with EBPR, CEBPR+ increase MEP from 30.44% to 34.70% and WMEP from 3.67% to 3.94% respectively.

5   Conclusion

In this paper, we first propose a novel explainable pairwise ranking called Constrained Explainable Bayesian Personalized Ranking. Furthermore, we analyze the impact of exposure bias on explainability, and proposed a corresponding improved method CEBPR+ to mitigate the effect. In the experiments, we compare our results with four baselines that are most related. According to the experiment results, our proposed methods achieved competitive ranking performance (HR and NDCG) and outperform other methods in terms of explainability (MEP and WMEP). In future, we will further discover the relationship between representation learning and neighborhood-based explainability.

Acknowledgement. This work was supported in part by the National Key Research and Development Program, China under Grant 2019YFB2102500.

References

1. Abdollahi, B., Nasraoui, O.: Using explainability for constrained matrix factorization. In: Proceedings of the Eleventh ACM Conference on Recommender Systems, pp. 79–83 (2017)
2. Cheng, Z., Chang, X., Zhu, L., Kanjirathinkal, R.C., Kankanhalli, M.: MMALFM: explainable recommendation by leveraging reviews and images. ACM Trans. Inf. Syst. (TOIS) **37**(2), 1–28 (2019)
3. Damak, K., Khenissi, S., Nasraoui, O.: Debiased explainable pairwise ranking from implicit feedback. In: Fifteenth ACM Conference on Recommender Systems, pp. 321–331 (2021)
4. Deng, Z.H., Huang, L., Wang, C.D., Lai, J.H., Philip, S.Y.: DeepCF: a unified framework of representation learning and matching function learning in recommender system. In: Proceedings of the AAAI Conference on Artificial Intelligence, vol. 33, pp. 61–68 (2019)
5. Harper, F.M., Konstan, J.A.: The MovieLens datasets: history and context. ACM Trans. Interact. Intell. Syst. (TIIS) **5**(4), 1–19 (2015)
6. He, X., Liao, L., Zhang, H., Nie, L., Hu, X., Chua, T.S.: Neural collaborative filtering. In: Proceedings of the 26th International Conference on World Wide Web, pp. 173–182 (2017)
7. Herlocker, J.L., Konstan, J.A., Riedl, J.: Explaining collaborative filtering recommendations. In: Proceedings of the 2000 ACM Conference on Computer Supported Cooperative Work, pp. 241–250 (2000)
8. Koren, Y., Bell, R., Volinsky, C.: Matrix factorization techniques for recommender systems. Computer **42**(8), 30–37 (2009)
9. Liu, Y., Li, B., Zang, Y., Li, A., Yin, H.: A knowledge-aware recommender with attention-enhanced dynamic convolutional network. In: Proceedings of the 30th ACM International Conference on Information & Knowledge Management, pp. 1079–1088 (2021)
10. Rendle, S., Freudenthaler, C., Gantner, Z., Schmidt-Thieme, L.: BPR: Bayesian personalized ranking from implicit feedback. arXiv preprint arXiv:1205.2618 (2012)
11. Resnick, P., Iacovou, N., Suchak, M., Bergstrom, P., Riedl, J.: GroupLens: an open architecture for collaborative filtering of netnews. In: Proceedings of the 1994 ACM Conference on Computer Supported Cooperative Work, pp. 175–186 (1994)
12. Sarwar, B., Karypis, G., Konstan, J., Riedl, J.: Item-based collaborative filtering recommendation algorithms. In: Proceedings of the 10th International Conference on World Wide Web, pp. 285–295 (2001)
13. Seo, S., Huang, J., Yang, H., Liu, Y.: Interpretable convolutional neural networks with dual local and global attention for review rating prediction. In: Proceedings of the Eleventh ACM Conference on Recommender Systems, pp. 297–305 (2017)
14. Su, X., Khoshgoftaar, T.M.: A survey of collaborative filtering techniques. Adv. Artif. Intell. **2009**, 421425 (2009)
15. Zhang, Y., Chen, X., et al.: Explainable recommendation: a survey and new perspectives. Found. Trends® Inf. Retrieval **14**(1), 1–101 (2020)
16. Zhang, Y., Lai, G., Zhang, M., Zhang, Y., Liu, Y., Ma, S.: Explicit factor models for explainable recommendation based on phrase-level sentiment analysis. In: Proceedings of the 37th International ACM SIGIR Conference on Research & Development in Information Retrieval, pp. 83–92 (2014)

Hierarchical Aggregation Based Knowledge Graph Embedding for Multi-task Recommendation

Yani Wang[1], Ji Zhang[2(✉)], Xiangmin Zhou[3], and Yang Zhang[4]

[1] Nanjing University of Aeronautics and Astronautics, Nanjing 211106, China
[2] The University of Southern Queensland, Toowoomba 4350, Australia
Ji.Zhang@usq.edu.au
[3] RMIT University, Melbourne 3001, Australia
[4] Zhejiang Lab, Hangzhou 311121, China

Abstract. Recently, knowledge graph has been used for alleviating the problems such as sparsity faced by the recommendation. Multi-task learning, which is an important emerged frontier research direction, helps complement the available information of different tasks and improves recommendation performance effectively. However, the existing multi-task methods ignore high-order information between entities. At the same time, the existing multi-hop neighbour aggregation methods suffer from the problem of over-smoothing. Also, the existing knowledge graph embedding methods in multi-task recommendation ignore the attribute triples in knowledge graph and recommendation tends to neglect the learning of user attributes. To mitigate these problems, we propose a multi-task recommendation model, called AHMKR. We use hierarchical aggregation and high-order propagation to alleviate the over-smoothing problem and obtain a better entity representation that integrates high-order information for multi-task recommendation. We leverage the text information of attribute triples, to improve the performance of knowledge graph in expanding the features of recommendation items. For users, we conduct fine-grained user learning based on the user attributes to capture user preferences in a more accurate matter. The experiments on the real-world datasets demonstrate the good performance of AHMKR.

Keywords: Recommender systems · Knowledge graph · Multi-task learning · Graph neural network

1 Introduction

Currently, many different kinds of side information are applied to alleviate the sparsity and cold start problems and help provide better recommendation, such as item category [2], social network, and knowledge graph. The use of knowledge graph in recommendation has attracted great research attention. Knowledge graph is a kind of structured knowledge base, which organizes scattered knowledge effactually. Its main goal is to describe various concepts existing in the real world and their relationships.

Supported by Natural Science Foundation of China (No. 62172372) and Zhejiang Provincial Natural Science Foundation (No. LZ21F030001).

The knowledge graph consists of entities, relations and attributes. Entities are objects in the actual world. Relations mean the relationships between objects or describe attributes of objects [5]. Knowledge graph is usually encoded as triples. There are two types of knowledge graph triples [7]. One is relation triples meaning triples which contain relationships between entities, e.g., ("Say It, Mo Yan", Author, Mo Yan). And the other one is attribute triples meaning triples which contain entities and its attributes, e.g., ("Say It, Mo Yan", Date of Publication, "July, 2007"). The existing recommendation models try many ways to improve the recommendation performance by combining knowledge graphs. Among these, the multi-task learning approach learns recommendation and KG-related tasks jointly and enables the knowledge graph to improve recommendation effectually. KG-related tasks can help recommendation get rid of local minimum, prevent recommendation from overfitting, and increase the generalization ability of recommendation. It is a relatively new and cutting-edge research direction.

However, the existing multi-task recommendation methods only model direct relations between entities, which ignore high-order information [6]. And the items associated with the entities are not enough to capture the complex and diversified preferences of users. At the same time, the existing multi-hop neighbour aggregation methods have the problem of over-smoothing. Obviously, the ignorance of relation is one of the causes [15]. Also, the existing knowledge graph embedding methods in multi-task recommendation neglect the attribute triples. Only structural information is unable to differentiate the meanings of relations and entities in different attribute triples [7]. It leads to unsatisfied recommendation effect affected by the sparsity and incompleteness of the knowledge graph. Moreover, the existing multi-task recommendation methods tend to ignore the fine-grained learning of user attributes. User attributes as part of user-side information are greatly vital. This results in the failure of fusing users' stable preferences contained in user attributes.

To mitigate these issues, we propose a multi-task recommendation model, AHMKR. Our model alternately trains recommendation and knowledge graph embedding task. Also, model shares features between items and entities. We take node-level and relation-level attention to aggregate and propagate neighbour information, so our model can alleviate over smoothing and obtain a better entity representation integrating high-order information. We process the text information of attribute triples, the performance of the knowledge graph is improved which is helpful for expanding the features of the recommendation items. In user-side, the user attributes are assigned different weights to obtain fine-grained user embedding and capture users' stable preferences. The experiments on the real-world datasets demonstrate the good performance of AHMKR.

2 Related Work

Our model can be counted as a kind of special embedding based recommendation. Embedding methods have gained many development. Most of the existing embedding based methods apply varieties of side information to improve the item representation. Meanwhile, the information also help learn user representation exactly. Some models learn user preferences straightly through bringing in users to construct user-item graph. Papers [1, 10, 12] apply multi-task learning approach to train recommendation and KG-related tasks which are the works closest to our method. The aggregation and propagation in our method utilizes the idea of Graph Neural Network (GNN). GNN is applied for recommendation increasingly. In recent research, the recommender systems based on GNN build models from two perspectives, one is social recommendation [3,4], another one is knowledge-graph-aware recommendation [9,11]. Our model can be counted as the latter.

3   Our Approach

In this part, We will introduce the proposed model AHMKR which is shown in the Fig. 1. The model contains four modules: user feature processing unit, recommendation, cross-compress unit and hierarchical aggregation based knowledge graph embedding.

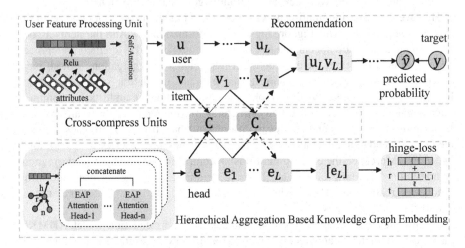

Fig. 1. Illustration of the AHMKR model.

3.1   User Feature Processing Unit

In the user feature processing unit, we process user attributes to get more accurate user feature vectors that can reflect users' stable preferences.

We apply multi-layer perceptron to extract features. Then we integrate the user attribute information and get the user feature vectors. However, we find the lack of expression about correlation and importance in attributes will make us left off information when mining user attribute features. So we take self-attention which is naturally suited to capture the internal relevance of data.

3.2   Recommendation

In the recommendation sub-module, we process the user vectors and item vectors to get the prediction result reflecting the user preferences for items. The input of recommendation is the feature vector **u** of user u which is the output result of the user feature processing unit and the feature vector **v** of item v which is the original feature of item.

For user u, we use L-layers multi-layer perceptron to process the feature. For item v, we acquire the feature through the processing of cross-compress units:

$$\mathbf{v}_L = E_{e \sim S(v)} \left[Cross^L [\mathbf{v}] \right] \tag{1}$$

Here, the associated entity set of item v is expressed by $S(v)$.

After processing the features of user u and item v, we predict the probability of user u participating in item v through applying the prediction function:

$$\hat{y}_{\mathrm{uv}} = \sigma f_{\mathrm{RS}} \left(\mathbf{u}_L, \mathbf{v}_L \right) \tag{2}$$

3.3   Hierarchical Aggregation Based Knowledge Graph Embedding

In this sub-module, we extract text information in attribute triples for better recommendation. As different meanings of relation and entity in different triples can't be distinguished by only using structural information, and it is easily affected by sparsity and incompleteness of knowledge graph. We utilize Long Short-Term Memory (LSTM) to encode attribute values of attribute triples into vectors.

To alleviate the over-smoothing problem and obtain a better entity representation, we conduct embedding aggregation propagation (EAP). Our embedding aggregation propagation is composed of node-level aggregation and relation-level aggregation.

In the node-level aggregation, we aggregate the neighbour nodes of head entities in specific relation. $\mathbf{h}_{\mathcal{N}(h_{r_i})}$ means the neighbour embedding of specific relation.

$$\mathbf{h}_{\mathcal{N}(h_{r_i})} = \sum_{t \in \mathcal{N}(h_{r_i})} \pi^{r_i}(h, r, t) \mathbf{t} \tag{3}$$

We use softmax function on the significance of entity t to h in specific relation r_i:

$$\pi^{r_i}(h, r, t) = \frac{\exp \left(LeakyRelu \left((\mathbf{Wh})^T \mathbf{W}(\mathbf{r} + \mathbf{t}) \right) \right)}{\sum_{t' \in \mathcal{N}(h_{r_i})} \exp \left(LeakyRelu \left((\mathbf{Wh})^T \mathbf{W}(\mathbf{r} + \mathbf{t}') \right) \right)} \tag{4}$$

We aggregate the embedding of the corresponding head entity h with the neighbour embedding of specific relation, and the result is expressed as $\mathbf{h}'_{\mathcal{N}(\mathbf{h}_{r_i})}$.

$$\mathbf{h}'_{\mathcal{N}(h_{r_i})} = f \left(\mathbf{h}, \mathbf{h}_{\mathcal{N}(h_{r_i})} \right) \tag{5}$$

After node-level aggregation, We treat the enhanced entity embedding that incorporates neighbour nodes under particular relation as virtual entity, so that each type of relation is contained merely once in the allocation to alleviate the over-smoothing problem. And according to the relation between the entity and the virtual entity, we assign different attention weights to carry out relation-level aggregation.

We acquire different weights between entity and virtual nodes and normalize α_{r_i} to get normalization coefficient between the entity h and the neighbour of specific relation.

$$\alpha_{r_i} = \frac{\exp\left((\mathbf{h})^T \operatorname{Diag}(\mathbf{r})\mathbf{h}'_{\mathcal{N}(h_{r_i})}\right)}{\sum_{t=1}^{T} \exp\left((\mathbf{h})^T \operatorname{Diag}(\mathbf{r})\mathbf{h}'_{\mathcal{N}(h_{r_t})}\right)} \tag{6}$$

We get the final head entity embedding $\mathbf{H}$ after the relation-level aggregation.

$$\mathbf{H} = \sum_i \alpha_{r_i} \mathbf{h}'_{\mathcal{N}(h_{r_i})} \tag{7}$$

To obtain the high-order connectivity information in the knowledge graph, we utilize embedding propagation to gather the deeper information of the neighbours so that we can capture the complex and diverse preferences of users.

$$\mathbf{H}^{(l)} = f\left(\mathbf{H}^{(l-1)}, \mathbf{h}'^{(l-1)}_{\mathcal{N}_h}\right) \tag{8}$$

After the aggregation and propagation, we use the cross-compress unit to share features of items and entities:

$$\mathbf{e}_L = E_{v \sim \mathcal{S}(h)}\left[Cross^L[\mathbf{e}]\right] \tag{9}$$

Here, the associated item set of entity h is expressed by $\mathcal{S}(h)$ and $\mathbf{e}$ means the feature of entity h.

3.4 Cross-compress Unit

The cross-compress unit shares the latent features of items and entities. It contains two operations, cross operation and compress operation.

In the cross operation, unit constructs a cross feature matrix. This is a fusion of item features and entity features. After the cross operation, unit conducts compress operation which projects the cross feature matrix into vectors of items and entities in their latent representation spaces to get the next layer feature vectors.

3.5 Optimization

We introduce the following loss function to train our model.

$$\mathcal{L} = \sum_{u \in U, v \in V} \mathcal{J}\left(\hat{y}_{uv}, y_{uv}\right) + \lambda_1 \sum_{(h,r,t) \in T} \sum_{(h',r,t') \in T'} [\gamma + d(h + r, t) - d\left(h' + r, t'\right)]_+$$
$$+ \lambda_2 \|W\|_2^2 \tag{10}$$

Here, the first term is the loss of recommendation, the second term is the loss of knowledge graph embedding, $\lambda_2 \|W\|_2^2$ is the regularization term for preventing overfitting.

4   Experiments

4.1   Datasets

In our experiments, we use MovieLens dataset and Book-Crossing dataset. MovieLens-1M has 6036 users, 3152 items and 796132 interactions. The KG for it has 215332 triples. Book-Crossing contains 19676 users, 19875 items and 163724 interactions in the Book-Crossing community. The KG for it has 69754 triples.

4.2   Experiments Setup

The ratio of training, validation and test set in AHMKR is 6:2:2. We process each experiment three times to obtain the average result. We set $\lambda_2 = 10^{-7}$, and f_{RS} is the inner product. Aggregating two-order information is beneficial enough to get satisfied performance and keep relative low computation, so we set $l = 2$. We evaluate AHMKR in two scenarios. One is the CTR prediction. Another one is the top-K recommendation.

Table 1. Results of AUC and ACC in CTR prediction scenario.

Model	Movielens-1M		Book-crossing	
	AUC	ACC	AUC	ACC
MKR	0.917 (−1.93%)	0.843 (−1.86%)	0.734 (−2.00%)	0.704 (−2.36%)
Wide&Deep	0.898 (−3.96%)	0.820 (−4.54%)	0.712 (−4.94%)	0.624 (−13.45%)
DKN	0.655 (−29.95%)	0.589 (−31.43%)	0.622 (−16.96%)	0.598 (−17.06%)
CKE	0.801 (−14.33%)	0.742 (−13.62%)	0.671 (−10.41%)	0.633 (−12.21%)
PER	0.710 (−24.06%)	0.664 (−22.70%)	0.623 (−16.82%)	0.588 (−18.45%)
AHMKR	**0.935**	**0.859**	**0.749**	**0.721**
AHMKR-U	0.927 (−0.86%)	0.852 (−0.81%)	0.741 (−1.06%)	0.714 (−0.97%)
AHMKR-E(a)	0.925 (−1.07%)	0.851 (−0.93%)	0.739 (−1.34%)	0.711 (−1.39%)
AHMKR-E(h)	0.929 (−0.64%)	0.854 (−0.58%)	0.744 (−0.67%)	0.716 (−0.69%)

4.3   Performance

Overall Comparison. Table 1 is the AUC and ACC of various models in CTR prediction scenario. Figure 2 is the result of Recall@K and Precision@K in top-K scenario. We can observe that PER [13] gets unsatisfactory results as rational design of meta-paths is difficult. DKN [8] has poor performance as it is better proper for recommendation about long text. CKE [14] performs better than DKN as it utilizes the structured content in the knowledge graph. Wide&Deep merely connects the attributes, so the effect is mediocre. MKR shares features between user-item interaction and knowledge graph so that MKR [10] performs well. Our method outperforms all baselines, because we take hierarchical aggregation propagation to obtain high-order information and process the text information of attribute triples to expand the features of the recommendation items. Also, we extract user attributes to capture users' more accurate preferences.

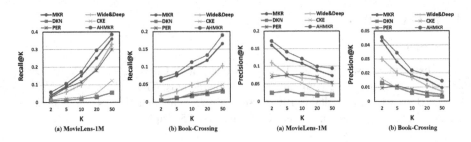

Fig. 2. The results of Recall@K and Precision@K in top-K recommendation scenario.

Ablation Study. Our model is a framework composed of sub-modules. For different modules, we conduct ablation study to assess whether they are beneficial to the goal. AHMKR-U is a variant that doesn't process attribute triples or conduct hierarchical aggregation propagation. AHMKR-E(a) is a variant that doesn't process the user attributes or conduct hierarchical aggregation propagation. AHMKR-E(h) is another variant which doesn't process the user attributes or process attribute triples. Table 1 shows the performance of the variants. We can find that the user feature extraction which captures stable preferences is helpful for better recommendation according to the performance of AHMKR-U. AHMKR-E(a) proves the importance of the text information in attribute triples which expands the features of recommendation items. The variant AHMKR-E(h) has a relatively good performance as hierarchical aggregation propagation obtains better entity representation for multi-task recommendation. Above all, the satisfying performance is achieved when we combine all the sub-modules.

Table 2. Results of AUC in movie CTR prediction with different ratios of training set r.

Model	20%	40%	60%	80%	100%
MKR	0.874	0.882	0.897	0.908	0.917
Wide & Deep	0.802	0.815	0.840	0.876	0.898
DKN	0.582	0.601	0.620	0.638	0.655
CKE	0.692	0.716	0.754	0.775	0.801
PER	0.607	0.638	0.662	0.688	0.710
AHMKR	**0.887**	**0.895**	**0.911**	**0.921**	**0.935**

Results in Sparse Scenarios. We verify the performance of models in sparse scenarios by adjusting the proportion of the training set. We test the proportions of the training set as 100%, 80%, 60%, 40%, 20% respectively. Table 2 is the performance in sparse scenarios. It can be found that with the raising of the sparsity, the overall performance of all models declines. However, our approach outperforms other baselines consistently. More importantly, our method still performs well when the training set is 20%, indicating that AHMKR can sustain satisfying performance even when the data is badly sparse.

5   Conclusions

In this work, we propose a multi-task recommendation model. We take hierarchical aggregation propagation to alleviate over smoothing and obtain entity representation integrating high-order information. Also, we extract the text information of attribute triples which is helpful for expanding the features of the items. In user-side, we process fine-grained attribute embedding to capture users' stable preferences. Experiments on the real-world datasets demonstrate that AHMKR has accomplished better performance than other models.

References

1. Cao, Y., Wang, X., He, X., Hu, Z., Chua, T.S.: Unifying knowledge graph learning and recommendation: towards a better understanding of user preferences. In: The World Wide Web Conference (2019)
2. Du, J., Zheng, L., He, J., Rong, J., Wang, H., Zhang, Y.: An interactive network for end-to-end review helpfulness modeling. Data Sci. Eng. **5**, 261–279 (2020)
3. Fan, W., et al.: Graph neural networks for social recommendation. In: The World Wide Web Conference (2019)
4. Fan, W., et al.: A graph neural network framework for social recommendations. IEEE Trans. Knowl. Data Eng. **34**, 2033–2047 (2020)
5. Ji, S., Pan, S., Cambria, E., Marttinen, P., Yu, P.S.: A survey on knowledge graphs: representation, acquisition and applications. IEEE Trans. Neural Netw. Learn. Syst. **33**, 494–514 (2020)
6. Liu, W., Cai, H., Cheng, X., Xie, S., Yu, Y., Zhang, H.: Learning high-order structural and attribute information by knowledge graph attention networks for enhancing knowledge graph embedding. ArXiv abs/1910.03891 (2019)
7. Sun, Z., Hu, W., Li, C.: Cross-lingual entity alignment via joint attribute-preserving embedding. In: SEMWEB (2017)
8. Wang, H., Zhang, F., Xie, X., Guo, M.: DKN: deep knowledge-aware network for news recommendation. In: Proceedings of the 2018 World Wide Web Conference (2018)
9. Wang, H., et al.: Knowledge graph convolutional networks for recommender systems with label smoothness regularization. ArXiv abs/1905.04413 (2019)
10. Wang, H., Zhang, F., Zhao, M., Li, W., Xie, X., Guo, M.: Multi-task feature learning for knowledge graph enhanced recommendation. In: The World Wide Web Conference (2019)
11. Wang, X., He, X., Cao, Y., Liu, M., Chua, T.S.: KGAT: knowledge graph attention network for recommendation. In: Proceedings of the 25th ACM SIGKDD International Conference on Knowledge Discovery & Data Mining (2019)
12. Xin, X., He, X., Zhang, Y., Zhang, Y., Jose, J.M.: Relational collaborative filtering: Modeling multiple item relations for recommendation. In: Proceedings of the 42nd International ACM SIGIR Conference on Research and Development in Information Retrieval (2019)
13. Yu, X., et al.: Personalized entity recommendation: a heterogeneous information network approach. In: Proceedings of the 7th ACM International Conference on Web Search and Data Mining (2014)
14. Zhang, F., Yuan, N.J., Lian, D., Xie, X., Ma, W.Y.: Collaborative knowledge base embedding for recommender systems. In: Proceedings of the 22nd ACM SIGKDD International Conference on Knowledge Discovery and Data Mining (2016)
15. Zhao, X., Jia, Y., Li, A., Jiang, R., Chen, K., Wang, Y.: Target relational attention-oriented knowledge graph reasoning. Neurocomputing **461**, 577–586 (2021)

Mixed-Order Heterogeneous Graph Pre-training for Cold-Start Recommendation

Wenzheng Sui[1,2], Xiaoxia Jiang[2], Weiyi Ge[2], and Wei Hu[1(✉)]

[1] State Key Laboratory for Novel Software Technology, Nanjing University,
Nanjing, China
`whu@nju.edu.cn`
[2] Science and Technology on Information Systems Engineering Laboratory,
Nanjing, China

Abstract. The cold-start problem is a fundamental challenge in recommendation. Heterogeneous information networks (HINs) provide rich side information in addition to sparse user-item interactions, which can be used to alleviate the cold-start problem. However, most existing models based on graph neural networks (GNNs) only consider the user-item interactions as supervision signals, making them unable to effectively exploit the side information. In this paper, we propose a novel pre-training model, named MHGP, for cold-start recommendation in a self-supervised manner. The key idea is to leverage the mixed-order information in a HIN. We first use GNNs with a hierarchical attention mechanism to encode the first-order and high-order structures of a user-item HIN. Then, we pre-train the embeddings of users and items by contrasting the two structure views and maximizing the agreement of positive samples in each view. Afterwards, the embeddings are fine-tuned together with the recommendation model. Experiments show that our model can consistently improve the performance of cold-start recommendation and outperform other state-of-the-art pre-training models.

Keywords: Recommender systems · Graph pre-training · Contrastive learning

1 Introduction

Recommender systems, which are used to search in a large amount of information and provide personalized recommendation services, play an indispensable role in web services nowadays. Among the recommendation models, collaborative filtering (CF) is one of the most widely-used algorithms, which leverages users' historical interactions to obtain their preferences. However, CF suffers from the cold-start problem as some users may have few interactions.

With the rapid development of web services, various kinds of side information have become available for recommender systems, which form the so-called

This work is supported by Science and Technology on Information Systems Engineering Laboratory (No. 05202006).

B. Li et al. (Eds.): APWeb-WAIM 2022, LNCS 13423, pp. 182–190, 2023.
https://doi.org/10.1007/978-3-031-25201-3_14

heterogeneous information networks (HINs) [1] and can be used to alleviate the cold-start problem. Recently, some efforts try to use metapaths [8] or graph neural networks (GNNs) [1,8] to learn the embeddings of users and items on HINs, since both metapaths and GNNs are capable of capturing high-order semantic relations. As the cold-start users and items may have much more high-order neighbors, aggregating these neighbors can help learn better embeddings of the cold-start users and items.

However, most existing models exploit the rich side information in a supervised manner [1,8] where the supervision signals are still user-item interactions. As the cold-start users and items have few interactions, they are not fully trained during the training process. Thus, the side information is not fully exploited, especially for the cold-start users and items with rich side information. Besides, the user-item interactions merely describe the direct interaction relations between users and items, while various side information describes many other first-order and high-order relations, which reflect different aspects of users and items. Therefore, the interactions can help learn the direct interaction relation better, but introduce some noises when guiding the process of learning other relations.

To tackle the above problems, a feasible solution is to design a pre-training task which is specifically designed for assisting the aggregation of rich side information. However, most of the existing pre-training models are not for the HIN-based recommendation scenario [6,8], where the first-order neighbors of a user can directly reflect one part of this user's preference, and the high-order neighbors of a user implies another part of his/her preference. They describe this user's preference from two perspectives and together form a more complete one. Therefore, a key challenge is how to jointly consider the two kinds of neighbors in the pre-training task.

In this paper, we propose a novel pre-training model named MHGP to exploit the rich side information in a HIN for cold-start recommendation. We first encode users and items in both first-order and high-order structure views with GNNs and three different attention mechanisms. Then, we collect users and items which are connected with each other by multiple metapaths as the positive samples and leverage contrastive learning to make the embeddings of the first-order structure view of positive samples similar; meanwhile align their embeddings in the high-order structure view. Once the pre-training process converges, the pre-trained embeddings will be fine-tuned with the recommendation model.

We conduct comparative experiments on three real-world datasets. The results demonstrate that our pre-trained model can improve the performance of recommendation models in the cold-start recommendation scenario and outperform several state-of-the-art pre-training GNNs models.

2   Related Work

2.1   Pre-training GNNs

Recently, pre-training GNNs has attracted plenty of attentions which aims to improve the performance of GNNs. The pre-training task can be performed with

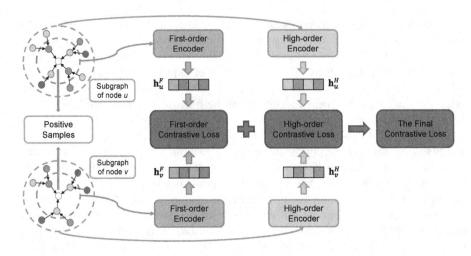

Fig. 1. The MHGP framework.

contrastive learning, such as DGI [7], DMGI [5] and GraphCL [10]. There are also some works performing the pre-training task in other ways, such as L2P-GNN [4] in a meta-learning way.

A few works aim to improve the recommendation task. The work in [2] simulates the cold-start scenario and takes the embedding reconstruction as the pre-training task. SGL [9] performs graph data augmentation for contrastive learning which can be implemented in a pre-training manner. In overall, they cannot fully exploit various types of nodes and relations for pre-training on a HIN to enhance the recommendation task.

2.2 Cold-Start Recommendation

In recent years, the studies on the cold-start problem mainly focus on two directions. One is how to leverage side information to learn better embeddings of users and items, such as DisenHAN [8] and HGRec [6]. The other direction is how to exploit the underlying patterns in the interactions. Most studies adopt GNNs to mine the high-order collaborative information behind the user-item bipartite graph, such as LightGCN [3]. However, these models exploit the high-order information in a supervised manner. For the cold-start users and items, their embeddings are rarely trained as they have very few interactions.

3   The Proposed MHGP Model

In this section, we introduce our mixed-order heterogeneous graph pre-training (MHGP) model to pre-train the embeddings of users and items. The overall architecture is illustrated in Fig. 1.

3.1 First-order Structure View Encoding

As our purpose is to pre-train the embeddings of users and items for recommen-dation, we do not consider the embeddings of other node types. Therefore, in the HIN-based recommendation scenario, a user's first-order neighbors can be users or items, while an item's first-order neighbors can simply be users.

Item's First-Order Structure Encoding. The importance of different users who have interacted with the same item may be different. Therefore, we apply a node-level attention mechanism to encode the first-order structures of items:

$$\mathbf{h}_i^F = \sum_{u \in \mathcal{N}_i} \alpha_{i,u} \mathbf{h}_u, \tag{1}$$

$$\alpha_{i,u} = \frac{\exp(LeakyReLU(\mathbf{c}_n[\mathbf{h}_i \,\|\, \mathbf{h}_u]))}{\sum_{u \in \mathcal{N}_i} \exp(LeakyReLU(\mathbf{c}_n[\mathbf{h}_i \,\|\, \mathbf{h}_u]))}, \tag{2}$$

where $\mathbf{h}_u \in \mathbb{R}^d$ is the embedding of user u. $\mathcal{N}_i$ is the first-order neighbor set of item i. $\mathbf{c}_n \in \mathbb{R}^{2d \times 1}$ is the attention vector, and $\|$ denotes the concatenation operation.

User's First-Order Structure Encoding. For a user, the two types of first-order neighbors contribute differently to a user's preference. Therefore, we design a hierarchical attention mechanism consisting of node-level attention and type-level attention to fully capture the influence from users' first-order neighbors:

$$\mathbf{h}_u^F = \beta_1 \sum_{i \in \mathcal{N}_u^I} \alpha_{u,i} \mathbf{h}_i + \beta_2 \sum_{v \in \mathcal{N}_u^U} \alpha_{u,v} \mathbf{h}_v, \tag{3}$$

where $\mathbf{h}_i, \mathbf{h}_v \in \mathbb{R}^d$ are the embeddings of item i and user v, respectively. $\mathcal{N}_u^I$ and $\mathcal{N}_u^U$ denote user u's first-order neighbor sets of items and users, respectively. $\alpha_{u,i}$ and $\alpha_{u,v}$ are the node-level attention values of which the calculations are similar to $\alpha_{i,u}$. β_1 and β_2 are the type-level attention values:

$$\beta_i = \frac{\exp(\mathbf{w}_i)}{\sum_{j \in \{1,2\}} \exp(\mathbf{w}_j)}, \tag{4}$$

$$\mathbf{w}_i = \frac{1}{|\mathcal{U}|} \sum_{u \in \mathcal{U}} \mathbf{c}_t \tanh(\mathbf{W}^F \mathbf{h}_u^i + \mathbf{b}^F), \tag{5}$$

where $\mathbf{c}_t \in \mathbb{R}^d$ is the attention vector, $\mathbf{W}^F \in \mathbb{R}^{d \times d}$ and $\mathbf{b}^F \in \mathbb{R}^{d \times 1}$ are the learnable parameters.

3.2 High-order Structure View Encoding

In a HIN, we can obtain the high-order neighbors by exploiting the rich metapath-based neighbors [1]. As each metapath carries a specific semantic relation, different kinds of metapath-based neighbors imply different preference characteristics.

Metapath-Based Neighbor Generation. In a HIN, there are several adja-cency matrices including user-item interaction matrix $\mathbf{Y}$ to describe the whole graph, with each describing one kind of first-order relation. Thus, we can obtain the metapath-based neighbors by matrix multiplication of these adjacency matrices, such as $\mathbf{YY}^T$ for metapath "user-item-user". Afterwards, we set all nonzero values to 1 to form the final adjacency matrix.

Metapath-Based Neighbor Aggregation. Assuming that there are M metapaths $\{\Phi_1, \Phi_1, ..., \Phi_M\}$ and their corresponding matrices are obtained. For each metapath Φ_m, we use a GCN to aggregate the corresponding neighbors to obtain $\mathbf{h}_u^{\Phi_m}$. Then, we apply a semantic-level attention mechanism to fuse the embeddings of all kinds of metapaths starting from user u:

$$\mathbf{h}_u^H = \sum_{m=1}^{M} \beta_{\Phi_m} \mathbf{h}_u^{\Phi_m}, \tag{6}$$

where β_{Φ_m} is the semantic-level attention values of which the calculation is similar to β_1 and β_2. The embeddings of all kinds of metapaths of items are calculated in the same way, denoted by $\mathbf{h}_i^H$.

3.3   Pre-training with Contrastive Learning

There are always some users sharing similar preferences in the recommendation scenarios. Therefore, the embeddings of the first-order and high-order structure views of these users should be similar, respectively. This also applies to items. We treat these users and items as positive samples and leverage contrastive learning to force the two kinds of embeddings of positive nodes to be consistent.

We first count how many kinds of metapaths are connected between each pair of nodes i and j, and the result is denoted by $connectivity(i, j)$. For each node i, we select all the nodes where $connectivity(i, j) > 0$ and sort them in descending order to form $\mathcal{S}_i$. As $\mathcal{S}_i$ can be very large and the nodes with lower $connectivity(i, j)$ values may introduce some noises, we set a threshold $T_{\mathcal{S}}$. If $|\mathcal{S}_i| > T_{\mathcal{S}}$, we select the top-$T_{\mathcal{S}}$ nodes as the positive nodes of node i.

After obtaining the embeddings of the first-order and high-order structure views, we feed them into a feed-forward neural network to project them into the same semantic space. Then, the final loss is calculated as follow:

$$\mathcal{L} = \lambda \mathcal{L}_u + (1 - \lambda)\mathcal{L}_i, \tag{7}$$

where $\mathcal{L}_u$ and $\mathcal{L}_i$ denote the losses from user side and item side, respectively. λ is a learnable parameter to adaptively balance the importance of the two sides. The calculation of $\mathcal{L}_u$ is given as follows:

$$\mathcal{L}_u = \lambda_u \mathcal{L}_u^F + (1 - \lambda_u)\mathcal{L}_u^H, \tag{8}$$

$$\mathcal{L}_u^F = \frac{1}{|\mathcal{U}|} \sum_{u \in \mathcal{U}} - \log \frac{\sum_{v \in \mathcal{S}_u} \exp(sim(\mathbf{h}_u^F, \mathbf{h}_v^F)/\tau)}{\sum_{w \in \mathcal{U}} \exp(sim(\mathbf{h}_u^F, \mathbf{h}_w^F)/\tau)}, \tag{9}$$

$$\mathcal{L}_u^H = \frac{1}{|\mathcal{U}|} \sum_{u \in \mathcal{U}} - \log \frac{\sum_{v \in \mathcal{S}_u} \exp(sim(\mathbf{h}_u^H, \mathbf{h}_v^H)/\tau)}{\sum_{w \in \mathcal{U}} \exp(sim(\mathbf{h}_u^H, \mathbf{h}_w^H)/\tau)}, \tag{10}$$

Table 1. Statistics of the datasets.

Datasets	Nodes	Edges (Sparsity)	Metapaths
Last.FM	User (U): 1,892	U-I: 92,834 (0.2722%)	UIU, UUU,
	Item (I): 18,022	U-U: 25,434	UITIU, IUI,
	Tag (T): 9,749	I-T: 186,479	ITI, IUUI
Ciao	User (U): 2,375	U-I: 36,065 (0.0900%)	UIU, UUU,
	Item (I): 16,861	U-U: 57,544	UICIU, IUI,
	Category (C): 6	I-C: 36,065	ICI, IUUI
Douban Movie	User (U): 13,367	U-M: 286,743 (0.1692%)	UMU, UMTMU,
	Movie (M): 12,677	U-U: 4,085	UMDMU,
	Actor (A): 6,311	M-A: 33,587	UMAMU,
	Director (D): 2,449	M-D: 11,276	MAM, MUM
	Type (T): 38	M-T: 27,668	MDM, MTM

where $sim(\cdot)$ denotes the cosine similarity. τ denotes the temperature hyperparameter. λ_u is a learnable parameter to adaptively balance the importance of the two kinds of embeddings of users. The calculation of $\mathcal{L}_i$ is similar to $\mathcal{L}_u$.

3.4 Fine-Tuning with Recommendation Models

Many existing GNN-based recommendation models initialize the embeddings of users and items randomly, which may lead to local optima during training and further affect the performance of recommendation. To alleviate this problem, we use the pre-trained embeddings to initialize the recommendation model. The embeddings are further fine-tuned with the recommendation model under the supervision of interactions.

4 Experiments and Results

4.1 Experiment Settings

We conduct the experiments on three real-world datasets: Last.FM, Ciao and Douban Movie. All the datasets contain few interactions and rich side information. The statistics of the datasets are summarized in Table 1.

We choose LightGCN [3] as the base recommendation model and choose three pre-training models DGI [7], DMGI [5] and SGL [9] for comparison.

For each dataset, we randomly choose $x\%$ interactions as the training set, $\frac{1-x\%}{2}$ as the validation set, and $\frac{1-x\%}{2}$ as the testing set. To simulate various cold-start environments, we set x to 20 and 40, respectively.

All the pre-training models are trained from scratch. The early stopping patience is 20 epochs. We tune the learning rate in $\{0.01, 0.001, 0.0001\}$. For our MHGP, the GNN layer is set to 1 for both encoders. For LightGCN, the number of GNN layer is 2 and the embedding size is fixed to 64. We fine-tune other hyperparameters according to the original papers.

Table 2. Performance of top-20 recommendation with LightGCN as the base model.

Datasets	Metrics	LightGCN	+DGI	+DMGI	+SGL	+MHGP
Last.FM (20%)	Precision	0.1144	0.0829	0.1143	0.1177	**0.1347**
	Recall	0.1177	0.0853	0.1176	0.1211	**0.1387**
	NDCG	0.1320	0.0940	0.1330	0.1369	**0.1577**
Last.FM (40%)	Precision	0.1055	0.1034	0.1037	0.1062	**0.1103**
	Recall	0.1423	0.1398	0.1398	0.1433	**0.1486**
	NDCG	0.1389	0.1364	0.1369	0.1395	**0.1462**
Ciao (20%)	Precision	0.0081	0.0105	0.0121	0.0076	**0.0135**
	Recall	0.0227	0.0317	0.0336	0.0212	**0.0423**
	NDCG	0.0177	0.0246	0.0254	0.0164	**0.0315**
Ciao (40%)	Precision	0.0103	0.0108	0.0124	0.0106	**0.0127**
	Recall	0.0375	0.0407	0.0440	0.0390	**0.0470**
	NDCG	0.0264	0.0283	0.0310	0.0273	**0.0322**
Douban Movie (20%)	Precision	0.0672	0.0658	0.0650	0.0665	**0.0684**
	Recall	0.1510	0.1351	0.1262	0.1466	**0.1664**
	NDCG	0.1230	0.1141	0.1100	0.1195	**0.1301**
Douban Movie (40%)	Precision	0.0553	0.0544	0.0544	0.0542	**0.0558**
	Recall	0.1603	0.1469	0.1465	0.1587	**0.1704**
	NDCG	0.1158	0.1092	0.1089	0.1129	**0.1201**

Table 3. Ablation study results with 20% interactions as the training set. P for Precision@20, R for Recall@20 and N for NDCG@20.

Models	Last.FM			Ciao			Douban Movie		
	P	R	N	P	R	N	P	R	N
MHGP	**0.1347**	**0.1387**	**0.1577**	**0.0135**	**0.0423**	**0.0315**	**0.0684**	**0.1664**	**0.1301**
MHGP$_f$	0.1344	0.1381	0.1569	0.0134	0.0419	0.0308	0.0667	0.1468	0.1205
MHGP$_h$	0.1256	0.1292	0.1461	0.0130	0.0421	0.0310	0.0671	0.1505	0.1218

4.2 Overall Performance Comparison

The overall performance is shown in Table 2. We can see that our pre-training model MHGP can consistently improve the performance of LightGCN, which demonstrates the effectiveness of MHGP for cold-start recommendation. In addition, the relative improvement increases as the training data decreases. This indicates that when the user-item interactions are sparse, our model can learn better embeddings of users and items by reasonably exploiting the rich side information. Besides, in most cases, our proposed model outperforms other state-of-the-art pre-training models. This indicates that MHGP is more suitable for the

HIN-based recommendation task. By contrasting the first-order and high-order structures of the positive samples, MHGP can effectively capture the inherent structure information in a HIN and further benefit the recommendation task.

4.3 Ablation Study

We design two variants $MHGP_f$ and $MHGP_h$ to perform ablation study. $MHGP_f$ only considers the first-order neighbors while $MHGP_h$ only considers the high-order neighbors. We compare them with MHGP and the results are given in Table 3. We can see that MHGP always achieves the best performance, indicating the necessity of jointly considering the two kinds of neighbors. Furthermore, all of them can improve the performance of LightGCN, which demonstrates the effectiveness of aggregating each kind of neighbors in pre-training. We also observe that $MHGP_h$ performs better than $MHGP_f$ on the Ciao and Douban Movie datasets. However, on the Last.FM datasets, the performance of $MHGP_f$ is better. This is reasonable since the interactions are sparser while the side information is richer on Ciao and Douban Movie than on Last.FM.

5 Conclusion and Future Work

In this paper, we introduce a novel pre-training model MHGP to exploit the rich information in a HIN for enhancing cold-start recommendation. MHGP uses contrastive learning to force the embeddings of first-order and high-order structures of positive nodes to be similar. Thus, it can learn the better embeddings of users and items. Experiments show that MHGP outperforms other state-of-the-art pre-training GNN models. In future work, we will explore whether MHGP can benefit other recommendation scenarios such as sequential recommendation.

References

1. Fu, X., Zhang, J., Meng, Z., King, I.: MAGNN: metapath aggregated graph neural network for heterogeneous graph embedding. In: WWW, pp. 2331–2341 (2020)
2. Hao, B., Zhang, J., Yin, H., Li, C., Chen, H.: Pre-training graph neural networks for cold-start users and items representation. In: WSDM, pp. 265–273 (2021)
3. He, X., Deng, K., Wang, X., Li, Y., Zhang, Y., Wang, M.: LightGCN: simplifying and powering graph convolution network for recommendation. In: SIGIR, pp. 639–648 (2020)
4. Lu, Y., Jiang, X., Fang, Y., Shi, C.: Learning to pre-train graph neural networks. In: AAAI, pp. 4276–4284 (2021)
5. Park, C., Kim, D., Han, J., Yu, H.: Unsupervised attributed multiplex network embedding. In: AAAI, pp. 5371–5378 (2020)
6. Shi, J., Ji, H., Shi, C., Wang, X., Zhang, Z., Zhou, J.: Heterogeneous graph neural network for recommendation. CoRR arXiv:2009.00799 (2020)
7. Velickovic, P., Fedus, W., Hamilton, W.L., Liò, P., Bengio, Y., Hjelm, R.D.: Deep graph infomax. In: ICLR (2019)

8. Wang, Y., Tang, S., Lei, Y., Song, W., Wang, S., Zhang, M.: DisenHAN: disentangled heterogeneous graph attention network for recommendation. In: CIKM, pp. 1605–1614 (2020)
9. Wu, J., et al.: Self-supervised graph learning for recommendation. In: SIGIR, pp. 726–735 (2021)
10. You, Y., Chen, T., Sui, Y., Chen, T., Wang, Z., Shen, Y.: Graph contrastive learning with augmentations. In: NeurIPS, pp. 5812–5823 (2020)

MISRec: Multi-Intention Sequential Recommendation

Rui Chen[1], Dongxue Chen[1], Riwei Lai[1], Hongtao Song[1(✉)], and Yichen Wang[2]

[1] Harbin Engineering University, Harbin, Heilongjiang, China
{ruichen,cdx,lai,songhongtao}@hrbeu.edu.cn
[2] Hunan University, Changsha, Hunan, China
yichenwang.hnu@gmail.com

Abstract. Learning latent user intentions from historical interaction sequences plays a critical role in sequential recommendation. A few recent works have started to recognize that in practice user interaction sequences exhibit multiple user intentions. However, they still suffer from two major limitations: (1) negligence of the dynamic evolution of individual intentions; (2) improper aggregation of multiple intentions. In this paper we propose a novel **Multi-Intention Sequential Recommender** (MISRec) to address these limitations. We first design a multi-intention extraction module to learn multiple intentions from user interaction sequences. Next, we propose a multi-intention evolution module, which consists of an intention-aware remapping layer and an intention-aware evolution layer. The intention-aware remapping layer incorporates position and temporal information to generate multiple intention-aware sequences, and the intention-aware evolution layer is used to learn the dynamic evolution of each intention-aware sequence. Finally, we produce next-item recommendations by identifying the most relevant intention via a multi-intention aggregation module. Extensive experimental results demonstrate that MISRec consistently outperforms a large number of state-of-the-art competitors on three public benchmark datasets.

Keywords: Recommender system · Sequential recommendation · Intention modeling

1 Introduction

In the Internet era, recommender systems have found their way into various business applications, such as e-commerce, online advertising, and social media. Recently, sequential recommendation has emerged as the mainstream approach for next-item recommendation. Learning a user's latent intentions from his/her temporally ordered interactions lies in the core of sequential recommendation.

Supported by the National Key R&D Program of China under Grant No. 2020YFB1710200.

In real-world scenarios, a user normally exhibits multiple intentions in his/her historical interactions. To this end, some very recent studies [4, 7, 10] have started to explore a user's multiple latent intentions in different ways.

While these studies have confirmed that modeling a user's multiple intentions is a rewarding research direction, we argue that they still suffer from two major limitations. First, they largely neglect the dynamic evolution of individual intentions. While previous studies emphasize the extraction of multiple intentions from user interaction sequences, they overlook the benefits of modeling the dynamic evolution of each individual intention, which is essential for next-item recommendation. Second, modeling a user's intention to interact with an item as a weighted sum of multiple intentions is counter-intuitive. While a user exhibits multiple intentions in his/her historical interaction sequence, the interaction with a particular item is usually driven by a single intention.

In this paper, we propose a novel **M**ulti-**I**ntention **S**equential **Rec**ommender (MISRec) to address these two limitations. We first design a multi-intention extraction module to extract multiple intentions from user interaction sequences. Next, we propose a multi-intention evolution module, consisting of an intention-aware remapping layer and an intention-aware evolution layer. The intention-aware remapping layer incorporates position information and recommendation time intervals to generate multiple intention-aware sequences, where each sequence corresponds to a learned intention. The intention-aware evolution layer is used to learn the dynamic evolution of each intention-aware sequence. Finally, we produce next-item recommendations by explicitly projecting a candidate item into multiple intention subspaces and determining its relevance to each intention. Empowered by Gumbel-softmax, we devise a multi-intention aggregation module to adaptively determine whether each intention is relevant to the target item or not. We perform a comprehensive experimental study on three public benchmark datasets and demonstrate that MISRec consistently outperforms representative state-of-the-art competitors.

2   Related Work

Sequential recommendation has been an emerging paradigm for next-item recommendation. GRU4Rec [1] is the first to employ gated recurrent units (GRUs) to extract sequential patterns from user interaction sequences. Caser [8] considers convolutional neural networks (CNNs) as the backbone network to learn sequential patterns as local features of recent items. NARM [6] uses an attention mechanism to capture more flexible sequential patterns from user interaction sequences. SASRec [2] proposes to leverage self-attention to adaptively consider interacted items. All the above works assume that a user has only a monolithic intention and thus a single embedding representation, which does not reflect the reality well, leaving much room for further improvement. As such, some recent works have started to explore how to better model users using multiple intentions. MCPRN [10] designs a dynamic purpose routing network to capture different user intentions. SINE [7] activates sparse user intentions from a given concept pool and then aggregates the intentions for next-item recommendations.

3   Proposed Method

3.1   Problem Setting

Let $\mathcal{U} = \{u_1, u_2, \cdots, u_{|\mathcal{U}|}\}$ and $\mathcal{I} = \{i_1, i_2, \cdots, i_{|\mathcal{I}|}\}$ be the set of all users and the set of all items, respectively. Given a sequence of user u's historically interacted items $S^u = (s_1^u, s_2^u, \cdots, s_l^u)$ with $s_i^u \in \mathcal{I}$ and the corresponding time sequence $T^u = (t_1^u, t_2^u, \cdots, t_l^u)$ with $t_1^u \leq t_2^u \leq \cdots \leq t_l^u$, the goal of sequential recommendation is to predict the next item with which user u is most likely to interact next. In addition, the recommendation time t is important for recommendation. We transform the interaction time sequence T^u into a new time interval sequence $Tiv^u = (tiv_1^u, tiv_2^u, \cdots, tiv_l^u)$, where $tiv_i^u = \min(t - t_i^u, \tau)$ with τ being a hyperparameter controlling the maximum time interval.

3.2   Embedding Layer

Following previous works, we first transform the user u's interaction sequence $(s_1^u, s_2^u, \cdots, s_l^u)$ into a fixed-length sequence $(s_1^u, s_2^u, \cdots, s_n^u)$, where n denotes the maximum length that our model handles. In the embedding layer, we create an item embedding matrix $\mathbf{E}_i \in \mathbb{R}^{|\mathcal{I}| \times d}$ based on the one-hot encodings of item IDs, where d is the dimension of embedding vectors. Then we retrieve the interaction sequence embedding matrix $\mathbf{E}_{S^u} = \left[e_{s_1^u}, e_{s_2^u}, \cdots, e_{s_n^u}\right] \in \mathbb{R}^{n \times d}$, where $e_{s_i^u}$ is the embedding of item s_i^u in $\mathbf{E}_i$. We also establish two embedding matrices, $\mathbf{E}_P = \left[e_{p_1}, e_{p_2}, \cdots, e_{p_n}\right] \in \mathbb{R}^{n \times d}$ for absolute positions and $\mathbf{E}_{Tiv^u} = \left[e_{tiv_1^u}, e_{tiv_2^u}, \cdots, e_{tiv_n^u}\right] \in \mathbb{R}^{n \times d}$ for relative time intervals.

3.3   Multi-Intention Extraction Module

To capture multiple intentions behind a user's historical interaction sequence, we propose a multi-intention extraction module based on multi-head attention mechanism. Specifically, we map the embedding matrix of a user's interaction sequence $\mathbf{E}_{S^u}$ into different latent subspaces using multiple heads, where each head represents an intention of a user. Let γ be the number of heads and thus the number of intentions. We generate the kth intention m_k^u via

$$m_k^u = head_k^u \mathbf{W}_t, \tag{1}$$

$$head_k^u = Attention(\mathbf{E}_{S^u} \mathbf{W}_k^Q, \mathbf{E}_{S^u} \mathbf{W}_k^K, \mathbf{E}_{S^u} \mathbf{W}_k^V), \tag{2}$$

where $head_k^u \in \mathbb{R}^{1 \times \frac{d}{\gamma}}$ is the output of kth head through a multi-head attention layer. Note that, to match the dimension of item embeddings, a transformation matrix $\mathbf{W}_t \in \mathbb{R}^{\frac{d}{\gamma} \times d}$ is proposed to transform $head_k^u$ from $\mathbb{R}^{1 \times \frac{d}{\gamma}}$ to $\mathbb{R}^{1 \times d}$. $Attention(\cdot)$ is an attention function, and $\mathbf{W}_k^Q$, $\mathbf{W}_k^K$, and $\mathbf{W}_k^V \in \mathbb{R}^{d \times \frac{d}{\gamma}}$ are the trainable transformation matrices of the kth head's query, key and value, respectively. Inspired by previous works [9], we adopt scaled dot-product as the attention function:

$$Attention(\mathbf{Q}, \mathbf{K}, \mathbf{V}) = softmax(\frac{\mathbf{Q}\mathbf{K}^\top}{\sqrt{d}})\mathbf{V}. \tag{3}$$

After the multi-intention extraction module, we obtain a user u's γ intentions, denoted by $(m_1^u, m_2^u, \cdots, m_\gamma^u)$.

3.4   Multi-Intention Evolution Module

With the extracted multiple intentions from the multi-intention extraction module, we next capture the dynamic evolution of each intention via a multi-intention evolution module, which consists of an intention-aware remapping layer and an intention-aware evolution layer.

Intention-Aware Remapping Layer. Simply capturing the sequential patterns on the learned intentions lacks guarantees to model the dynamic evolution of user intentions precisely [5]. Therefore, we first design an intention-aware remapping layer to explicitly inject sequentiality and temporal information into intention-aware interaction sequences. In particular, we devise an extended scaled dot-product attention mechanism, where the learned intentions play the role of query vectors, and the key and value of the scaled dot-product attention are the interaction sequence injected with positional and temporal information:

$$(\textbf{Key} : \textbf{Value}) : (\mathbf{E}_{S^u}\mathbf{W}_S^K + \mathbf{E}_P\mathbf{W}_P^K + \mathbf{E}_{Tiv^u}\mathbf{W}_T^K : \mathbf{E}_{S^u}\mathbf{W}_S^V + \mathbf{E}_P\mathbf{W}_P^V + \mathbf{E}_{Tiv^u}\mathbf{W}_T^V),$$
$$(4)$$

where $\mathbf{E}_{S^u}$, $\mathbf{E}_P$, $\mathbf{E}_{Tiv^u} \in \mathbb{R}^{n \times d}$ are the embedding matrices of the interaction sequence, position sequence and time interval sequence, respectively. $\mathbf{W}^K$ and $\mathbf{W}^V \in \mathbb{R}^{d \times d}$ are the trainable matrices for keys and values, where the subscripts S, P and T indicate the matrices for the interaction sequence, position sequence and time interval sequence, respectively. Then we compute a new intention-aware interaction sequence $\mathbf{S}_k^u = (s_{k1}^u, s_{k2}^u, \cdots, s_{kn}^u)$ via

$$\mathbf{S}_k^u = softmax\left(\frac{(m_k^u \mathbf{W}_{S_k}^Q)\textbf{Key}^\top}{\sqrt{d}}\right)\textbf{Value},$$
$$(5)$$

where $\mathbf{W}_{S_k}^Q \in \mathbb{R}^{d \times d}$ is the trainable matrix for intention m_k^u.

Intention-Aware Evolution Layer. To capture the dynamic evolution of each intention, we employ gated recurrent units (GRUs) to model the dependencies between interacted items under each individual intention. Specifically, the input to the GRU for the kth intention is the kth intention-aware interaction sequence $\mathbf{S}_k^u$. We utilize the last hidden state h_k^u of the GRU to represent the user u under the kth intention. We further adopt a point-wise feedforward network (FFN) to endow the model with non-linearity and consider interactions between different latent dimensions:

$$m_k^u = h_k^u + Dropout(FFN(LayerNorm(h_k^u))), \tag{6}$$

$$LayerNorm(x) = \alpha \odot \frac{x - \mu}{\sqrt{\sigma^2 + \epsilon}} + \beta, \tag{7}$$

$$FFN(x) = ReLU(x\mathbf{W}_1 + b_1)\mathbf{W}_2 + b_2, \tag{8}$$

where $\mathbf{W}_1, \mathbf{W}_2 \in \mathbb{R}^{d \times d}$ are learnable matrices, and b_1, b_2 are d-dimensional bias vectors. μ and σ are the mean and variance of x, α and β are the learned scaling factor and bias term, respectively. We apply layer normalization to the input h_k^u before feeding it into the FFN, apply dropout to the FFN's output, and add the input h_k^u to the final output.

3.5 Multi-Intention Aggregation Module

Intuitively, a user's interaction with an item is usually driven by a single intention. Directly combining the multiple intention representations as the final intention representation is counter-intuitive and cannot maximize the benefits of extracting multiple intentions. In addressing this issue, we adopt the Gumbel-softmax to adaptively determine whether an intention is relevant to the candidate item or not. Specifically, we first explicitly project the candidate item's embedding e_{n+1} into different intention subspaces (see Eq. 9), and then calculate the relevance between each intention representation and the candidate item's embedding in each intention subspace via the inner product operation (see Eq. 10). Distinct from the previous methods using softmax to aggregate the multiple intention representations, we adopt the Gumbel-softmax to identify the most relevant intention (see Eqs. 11 and 12). Finally, we obtain the final representation m^u of user u at the finer granularity of intentions.

$$e_{n+1}^k = e_{n+1} \mathbf{W^k}, \tag{9}$$

$$r_{n+1}^k = e_{n+1}^k m_k^{u\top}, \tag{10}$$

$$a_k = \frac{\exp((\log(r_{n+1}^k) + g_i)/\tau)}{\sum_{j=1}^{\gamma} \exp((\log(r_{n+1}^j) + g_j)/\tau)}, \tag{11}$$

$$m^u = \sum_{k=1}^{\gamma} a_k * m_k^u. \tag{12}$$

3.6 Model Training

After we get the final representation m_u of user u, prediction scores are calculated as the inner product of the final user representation m_u and the candidate item's embedding e_i:

$$r_{u,i} = e_i m_u^\top. \tag{13}$$

We use the pairwise Bayesian personalized ranking (BPR) loss to optimize the model parameters. Specifically, it encourages the predicted scores of a user's historical items to be higher than those of unobserved items:

$$\mathcal{L}_{BPR} = \sum_{(u,i,j) \in \mathcal{O}} -\ln \sigma(r_{u,i} - r_{u,j}) + \lambda \|\Theta\|_2^2, \tag{14}$$

where $\mathcal{O} = \{(u,i,j)|(u,i) \in \mathcal{O}^+, (u,j) \in \mathcal{O}^-\}$ denotes the training dataset consisting of the observed interactions $\mathcal{O}^+$ and sampled unobserved items $\mathcal{O}^-$, $\sigma(\cdot)$ is the sigmoid activation function, Θ is the set of embedding matrices, and λ is the L_2 regularization parameter.

4 Experiments

4.1 Experimental Setup

Datasets and Evaluation Metrics. We evaluate our framework on three public benchmark datasets that are widely used in the literature. **Amazon-Review** datasets[1] contain product reviews from the online shopping platform Amazon, and we use two representative datasets, Grocery and Gourmet Food (referred to as **Grocery** and **Beauty**). **MovieLens**[2] datasets contain a collection of movie ratings from the website MovieLens. and we use MovieLens-1M (referred to as **ML1M**) in our experiments. Following previous works [2,5], we filter out cold-start users and items with fewer than 5 interactions and sort the interactions of each user by timestamps. Similarly, we use the most recent item for testing, the second most recent item for validation, and the remaining items for training. We evaluate our framework by two widely-adopted ranking metrics, Hit Ratio@N (**HR@N**) and Normalized Discounted Cumulative Gain@N (**NDCG@N**).

Baselines. To demonstrate the effectiveness of our solution, we compare it with a wide range of representative sequential recommenders, including four single-intention-aware methods (**GRU4Rec** [1], **NARM** [6], **Caser** [8], and **TiSASRec** [5]) and a multi-intention-aware method, **SINE** [7].

Implementation Details. Identical to the settings of previous methods, the embedding size is fixed to 64. We optimize our method with Adam [3] and set the learning rate of Grocery, Beauty, and ML1M to 10^{-4}, 10^{-3} and 10^{-4}, respectively, and the mini-batch size to 256 for all three datasets. The maximum length of interaction sequences of Grocery, Beauty, and ML1M is set to 10, 20, and 50, respectively. The maximum time interval is set to 512 sec for all three datasets. The temperature parameter τ in the Gumbel-softmax is set to 0.1. To address overfitting, we use L_2 regularization with the regularization coefficients of 10^{-5} for Grocery and ML1M and 10^{-4} for Beauty.

4.2 Main Results

Overall Comparison. We report the overall comparison in Table 1, where the best results are boldfaced and the second-best and third-best results are underlined. We can draw a few interesting observations: (1) TiSASRec achieves the best performance among single-intention-aware methods, indicating the effectiveness of the self-attention mechanism and temporal information in capturing sequential patterns. However, without considering multiple user intentions, these methods cannot identify a user's true intention accurately, leading to suboptimal recommendations. (2) As a multi-intention-aware method, SINE performs generally better than most single-intention-aware methods, which shows

---

[1] http://jmcauley.ucsd.edu/data/amazon/links.html.
[2] https://grouplens.org/datasets/movielens/1m/.

Table 1. Performance of different models on the three datasets. All the numbers in the table are percentages with % omitted.

	Grocery				Beauty				ML1M			
	Metric@10		Metric@20		Metric@10		Metric@20		Metric@10		Metric@20	
	HR	NDCG	HR	NDCG	HR	NDCG	HR	NDCG	HR	NDCG	HR	NDCG
GRU4Rec	4.79	2.41	7.89	3.19	3.98	2.09	6.38	2.69	14.17	6.90	23.06	9.13
NARM	6.21	3.21	9.72	4.09	7.28	4.18	10.23	4.92	15.23	7.10	25.66	9.71
Caser	5.65	2.85	9.02	3.69	5.92	3.15	8.91	3.90	15.62	7.48	27.72	11.00
TiSASRec	7.32	3.20	11.00	4.06	8.25	4.23	11.31	5.00	22.37	10.82	33.54	13.64
SINE	6.27	2.96	9.89	3.73	5.84	2.57	8.73	3.30	16.64	7.18	27.91	10.00
MISRec	7.63	3.37	11.37	4.26	8.83	4.72	12.42	5.45	22.81	11.62	34.37	14.45
Improv.	4.23	4.98	3.36	4.16	7.03	11.58	9.81	9.00	1.97	7.39	2.47	5.94

Table 2. Performance of different variants of MISRec. The results of HR@20 and NDCG@20 are omitted due to the space limitation.

	Grocery		Beauty		ML1M	
	HR@10	NDCG@10	HR@10	NDCG@10	HR@10	NDCG@10
w/o PE	7.47	3.30	8.29	4.35	22.50	10.74
w/o TIE	7.46	3.31	8.38	4.49	22.48	10.80
w/o GS	7.59	3.34	8.78	4.68	22.68	11.50
MISRec	**7.63**	**3.37**	**8.83**	**4.72**	**22.81**	**11.62**

that explicitly exploring multiple user intentions is a rewarding direction. However, the performance of SINE is still worse than TiSASRec. We deem that it is caused by the negligence of the dynamic evolution of individual intentions and the improper aggregation of multiple intentions. (3) By addressing the two issues mentioned above, MISRec maximizes the benefits of extracting multiple intentions and consistently yields the best performance on all datasets, which well justifies our motivation.

Ablation Study. To investigate the contributions of different components on the final performance, we conduct an ablation study to compare the performance of different variants of our MISRec model on the three datasets. The variants include: (1) **w/o PE** removes positional embeddings in the multi-intention evolution module. (2) **w/o TIE** removes time interval embeddings in the multi-intention evolution module. (3) **w/o GS** replaces the Gumbel-softmax with the softmax in the multi-intention aggregation module. Table 2 shows the performance of all variants and the full MISRec model on the three datasets. By comparing the performance of different variants, we can derive that both positional embeddings and time interval embeddings lead to performance improvement, which demonstrates the significance of explicitly modeling the dynamic evolution of different intentions. Furthermore, identifying the most relevant intention rather than aggregating multiple intentions consistently improves the performance by a significant margin, which justifies our motivation.

5   Conclusion

In this paper, we proposed a novel Multi-Intention Sequential Recommender (MISRec) to address the limitations of existing works that leverage users' multiple intentions for better next-item recommendations. We made two major contributions. First, we designed a multi-intention evolution module that effectively models the evolution of each individual intention. Second, we proposed to explicitly identify the most relevant intention rather than aggregate multiple intentions to maximize the benefits of extracting multiple intentions. A comprehensive experimental study on three public benchmark datasets demonstrates the superiority of the MISRec model over a large number of state-of-the-art competitors.

References

1. Hidasi, B., Karatzoglou, A., Baltrunas, L., Tikk, D.: Session-based recommendations with recurrent neural networks. In: Proceedings of the 4th International Conference on Learning Representations (ICLR) (2016)
2. Kang, W., McAuley, J.J.: Self-attentive sequential recommendation. In: Proceedings of the 18th IEEE International Conference on Data Mining (ICDM), pp. 197–206 (2018)
3. Kingma, D.P., Ba, J.: Adam: a method for stochastic optimization. In: Proceedings of the 3rd International Conference on Learning Representations (ICLR) (2015)
4. Li, C., et al.: Multi-interest network with dynamic routing for recommendation at Tmall. In: Proceedings of the 28th International Conference on Information and Knowledge Management (CIKM), pp. 2615–2623 (2019)
5. Li, J., Wang, Y., McAuley, J.J.: Time interval aware self-attention for sequential recommendation. In: Proceedings of the 13th International Conference on Web Search And Data Mining (WSDM), pp. 322–330 (2020)
6. Li, J., Ren, P., Chen, Z., Ren, Z., Lian, T., Ma, J.: Neural attentive session-based recommendation. In: Proceedings of the 26th International Conference on Information and Knowledge Management (CIKM), pp. 1419–1428 (2017)
7. Tan, Q., et al.: Sparse-interest network for sequential recommendation. In: Proceedings of the 14th International Conference on Web Search And Data Mining (WSDM), pp. 598–606 (2021)
8. Tang, J., Wang, K.: Personalized top-n sequential recommendation via convolutional sequence embedding. In: Proceedings of the 11th International Conference on Web Search And Data Mining (WSDM), pp. 565–573 (2018)
9. Vaswani, A., et al.: Attention is all you need. In: Proceedings of the 31st International Conference on Neural Information Processing Systems (NIPS), pp. 5998–6008 (2017)
10. Wang, S., Hu, L., Wang, Y., Sheng, Q.Z., Orgun, M.A., Cao, L.: Modeling multi-purpose sessions for next-item recommendations via mixture-channel purpose routing networks. In: Proceedings of the 28th International Joint Conference on Artificial Intelligence (IJCAI), pp. 3771–3777 (2019)

MARS: A Multi-task Ranking Model for Recommending Micro-videos

Jiageng Song[1,2], Beihong Jin[1,2(✉)], Yisong Yu[1,2], Beibei Li[1,2],
Xinzhou Dong[1,2], Wei Zhuo[3], and Shuo Zhou[3]

[1] State Key Laboratory of Computer Science, Institute of Software,
Chinese Academy of Sciences, Beijing, China
Beihong@iscas.ac.cn
[2] University of Chinese Academy of Sciences, Beijing, China
[3] MX Media Co., Ltd., Singapore, Singapore

Abstract. Micro-videos have become very popular recently. While using a micro-video app, the user experiences are strongly affected by the ranking of micro-videos. Moreover, the micro-video recommendation is often required to satisfy multiple business indicators. The existing models mainly utilize multi-modal features whose acquisition cost is too high for start-up companies. In the paper, we propose a multi-task ranking model MARS for recommending micro-videos. MARS aims at two tasks: finishing playing prediction and playback time prediction. For providing high accuracy in performing these two tasks, MARS adopts the multi-expert structure and mines historical statistical information besides interactions between users and micro-videos. Results of offline experiments and online A/B tests show that MARS can achieve good performances on two tasks. Further, MARS has been deployed in a real-world production environment, serving thousands of users.

Keywords: Recommender systems · Micro-video ranking · Multi-task learning

1 Introduction

As watching micro-videos via apps on mobile phones becomes prevalent, the numbers of users and micro-videos continue to grow rapidly. An abundant supply of micro-videos extends the breadth of mind and depth of vision, and also brings the information overload issue. Good recommendations can help users find their favorite micro-videos, allowing them to enjoy entertainment. Further, while the time of users using the app increases, the high user engagement and then the high retention rate will be reached. Therefore, recommending micro-videos for users becomes a hot topic both in academia and industry.

In the immersive micro-video apps such as TikTok and MX TakaTak, a user watches the micro-videos according to the order of micro-videos that the app proactively sends. If the user has an interest in the micro-videos with high-ranking, then the user's attention will be caught quickly and the user might

B. Li et al. (Eds.): APWeb-WAIM 2022, LNCS 13423, pp. 199–214, 2023.
https://doi.org/10.1007/978-3-031-25201-3_16

keep on watching micro-videos. Therefore, the ranking of recalled micro-videos is very important, which has been reflected in practices of industry, that is, the micro-video recommendation has a specialized phase targeted for ranking. In the paper, we focus on the ranking of recalled micro-videos.

So far, some models have been proposed for micro-video ranking. Unfortunately, they are all for a single task, i.e., optimizing a single metric given a group of positive samples. However, in essence, the micro-video recommendation is a multi-task problem. At least, achieving a high playing finishing rate and long average playback time can be treated as two tasks that the micro-video recommendation should concern about. The reason is obvious. For a user watching a micro-video, either a high playing finishing rate or long average playback time can represent the degree of users accepting and liking the micro-video. These two metrics form the basis of the retention rate, where the retention rate, the most valuable business indicator, cannot be directly optimized by a ranking model. We believe that using a single-task model to solve a multi-task problem will not reach satisfactory performance.

Recently, multiple multi-task models including MMoE [10], PLE [14] and SNR [9] are proposed to model multiple tasks in parallel. Some models have been applied in industry including the e-commercial scenario [12] and the video streaming scenario [17]. Existing models adopt multi-modal information (e.g., video cover or image features) or manually-labeled information (e.g., item category) as input. However, while building a multi-task model for micro-video ranking, we have to face three challenges. Firstly, we cannot force producers of micro-videos to provide more details (e.g., category, hashtag) about published micro-videos besides micro-videos themselves. Secondly, the acquisition of multi-modal features from micro-videos requires a large amount of calculation. Considering the time cost and hardware cost, it is infeasible for a start-up company to obtain these multi-modal features. As a result, only interactions between users and micro-videos including implicit interactions (such as watching) of large quantities and explicit interactions (such as like and favorite) of small quantities can be utilized. Thirdly, given two tasks, i.e., realizing a high playing finishing rate and long average playback time, we find that the finishing rate and playback time have different implications. Guided by the finishing rate, the ranking model will be inclined to recommend micro-videos with short duration. Otherwise, if optimizing the playback time, the ranking model will tend to recommend micro-videos with long duration. Conflicting recommendation results show that it is formidably difficult to fulfill two tasks simultaneously in a ranking model.

In the paper, we propose MARS, a multi-task micro-video ranking model, to predict the finishing rate and playback time simultaneously. MARS trims historical interaction information to augment training signals and then models multiple tasks by adjusting the original MMoE. To the best of our knowledge, this is the first work on optimizing both the finishing rate and playback time for recommending micro-videos.

Our contributions are summarized as follows.

- We leverage user historical interactions to extract different information corresponding to multiple tasks to serve the multiple prediction tasks.
- We adopt the structure of MMoE and differentiate the experts by different historical information extracted, forcing them to learn different aspects of the knowledge related to different tasks.
- We conduct offline experiments on two micro-video datasets and online A/B tests on an actual production environment. The experimental results show our model is reasonable and effective. Furthermore, our model has been deployed on MX TakaTak, one of micro-video apps for Indian users.

The rest of the paper is organized as follows. Section 2 introduces the related work. Section 3 describes the MARS model in detail. Section 4 gives the experimental evaluation. Finally, the paper is concluded in Sect. 5.

2   Related Work

Our work is mainly related to the research under two topics: micro-video recommendation and multi-task learning.

Micro-video Recommendation: With the rapid development of micro-videos applications, recommending models for micro-videos [3,6,7] have appeared one after another. Some models utilize the multi-modal information of micro-videos, including image, audio and text to recommend micro-videos. For example, Jiang et al. [4] take multi-scale time effects, user interest group modeling and false positive interactions into consideration, take micro-video multi-modal embeddings as input and propose a Multi-scale Time-aware user Interest modeling Network (MTIN). Wei et al. [15] design a Multi-Modal Graph Convolution Network (MMGCN) which can yield modal-specific representations of users and micro-videos. Liu et al. [8] propose the User-Video Co-Attention Network (UVCAN) which can learn multi-modal information from both users and micro-videos using an attention mechanism. Huang et al. [5] propose a hierarchical model depending on the multi-modal features to describe user interests in viewing micro-videos. Liu and Chen [7] propose a new variant of Transformer to model user sequential behaviors for next micro-video recommendation. However, these models are all for a single task, i.e., optimizing a single metric given a group of positive samples, while the micro-video recommendation is a multi-task problem.

Multi-task Learning: Multi-task learning exploits the information in the training signals of related tasks, thus improving generalization performance [1]. For example, Share Bottom (SB) model [1] is a classical model which shares hidden bottom layers across all task layers. Misra et al. [13] propose a cross-stitch module that fuses the information from different single-task models. Ma et al. [10] propose a multi-task model MMoE, which splits the shared low-level layers into sub-networks, which are called experts, and uses different gating networks for different tasks to utilize different sub-networks. Ma et al. [9] propose a novel framework called Sub-Network Routing (SNR) to achieve more flexible parameter sharing while maintaining the computational advantage of the SB model. In recent years, the requirement of performing multiple recommendation tasks at the same time pushes forward with the multi-task models [2,18]. Tang et al. [14] propose a Progressive Layered Extraction (PLE) model, which designs a novel sharing structure to address the seesaw phenomenon and negative transfer in multi-task learning. Some methods from the industry are also presented to perform multi-task learning from business processes. For example, Ma et al. [12] propose ESMM, which models multiple prediction tasks, including predicting Click-Through Rate (CTR), ConVersion Rate (CVR) and post-view Click-Through&ConVersion Rate (CTCVR) and employs a feature representation transfer learning strategy, to eliminate the sample selection bias and data sparsity problems in the e-commerce scenario. Zhao et al. [17] develop a large-scale multi-objective ranking system for recommending videos on YouTube based on the Wide&Deep model architecture and the MMoE. Currently, there are also some researchers trying to model the micro-video recommendation from the perspective of multi-task modeling. For example, Ma et al. [11] propose a coarse-to-fine multi-task micro-video recommendation model to jointly optimize click-through rate and playtime prediction.

3 Methodology

3.1 Definition

In this section, we give a formal description of the data we employ, formulating our goal of multi-task micro-video recommendations.

Definition 1. (Interactions) We denote the user set as $\mathcal{U}$ and the micro-video set as $\mathcal{V}$. An interaction between a user $u_i \in \mathcal{U}$ and a micro-video $v_j \in \mathcal{V}$ at moment t can be represented as $(u_i, v_j, t, r_{ij}, c_{ij})$, in which r_{ij} indicates whether u_i finishes playing v_j. In detail, if u_i does, $r_{ij} = 1$, otherwise, $r_{ij} = 0$. $c_{ij} \in [0,1]$ denotes the normalized playback time.

Definition 2. (Historical Interaction Sequences) For a user $u_i \in \mathcal{U}$, we collect all the micro-videos he/she has interacted with, and after sorting the historical micro-videos by interaction timestamp in ascending order, we obtain the historical interaction sequence $s_i = \left[v_{i1}, v_{i2}, \ldots, v_{i|s_i|}\right]$.

Definition 3. (User Features) User features include age, gender, location, etc. Generally, the features can be divided into continuous features and discrete features. We get the feature values by designing a set of continuous feature mapping functions and a set of discrete feature mapping functions. Specifically, the value of the k-th continuous feature of user u_i is defined as $f_k^u(u_i)$, where $k \in [1, n_u]$ and n_u is the number of continuous user features. The value of the k-th discrete feature of user u_j is $g_k^u(u_i)$, where $k \in [1, m_u]$, and m_u is the number of discrete features.

Definition 4. (Micro-video Features) The features of a micro-video include tags, duration, etc. For each micro-video $v_j \in \mathcal{V}$, we construct the mapping functions for n_v continuous micro-video features and m_v discrete micro-video features, denoted as $f_k^v(v_j)$ and $g_k^v(v_j)$, respectively.

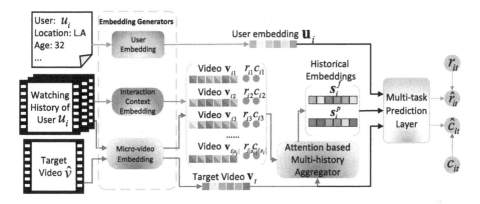

Fig. 1. The overall structure of the MARS.

Definition 5. (Interaction Context Features) For each interaction, we collect interaction context features, such as the percentage of playback time to duration, whether the user likes or downloads the micro-video, etc. Suppose the numbers of continuous and discrete features of the interaction context are n_c and m_c, respectively, the k-th continuous and discrete interaction feature between user u_i and micro-video v_j are defined as $f_k^c(u_i, v_j)$ and $g_k^c(u_i, v_j)$, respectively. Note that the indicator of finishing playing r_{ij} and the one of playback time c_{ij} are a discrete feature and a continuous feature of the interaction context, respectively.

Our goal is to build a multi-task model that can simultaneously predict the finishing playing probability and playback time, given a user and a micro-video. These two predictions are straightforwardly called finishing playing prediction and playback time prediction tasks.

The input and output of the model are as follows. The input includes historical interaction sequence s_i of user $u_i \in \mathcal{U}$, the user features $(\cup_{k=1}^{n_u} f_k^u(u_i)) \cup (\cup_{k=1}^{m_u} g_k^u(u_i))$, target micro-video v_t and its features $(\cup_{k=1}^{n_v} f_k^v(v_t)) \cup (\cup_{k=1}^{m_v} g_k^v(v_t))$, features of each interacted micro-video $(\cup_{k=1}^{n_v} f_k^v(v_{ij})) \cup (\cup_{k=1}^{m_v} g_k^v(v_{ij}))$ and the features of interaction context of each interacted micro-video $(\cup_{k=1}^{n_c} f_k^c(u_i, v_{ij})) \cup (\cup_{k=1}^{m_c} g_k^c(u_i, v_{ij}))$, in which $j \in [1, |s_i|]$. The output includes the predicted finishing playing probability $\hat{r}_{it}$ and the predicted playback time $\hat{c}_{it}$ when user u_i watches the target micro-video v_t.

3.2 Model

Overviews: The overall structure of the model is shown in Fig. 1, which consists of three main modules, i.e., the embedding generators, the attention-based multi-history aggregator, and the multi-task prediction layer. First, we obtain the user embedding by inputting user features into the user embedding generator, and micro-video embeddings by inputting the features and IDs of each interacted micro-video and target micro-video into the micro-video embedding generator. In particular, being different from the target micro-video, the embedding of each interacted micro-video is appended with the corresponding interaction context feature embedding. Then, we input the embeddings of interacted micro-videos and the target micro-video into the attention-based multi-history aggregator to generate multiple historical embeddings corresponding to multiple tasks, i.e., the finishing playing-based historical embedding and the playback time-based historical embedding. Then, we input the user embedding, the target micro-video embedding and the multiple historical embeddings to the multi-task prediction layer to obtain the predictions of both the finishing playing probability and the playback time. The score of the target micro-video is calculated by weighting the prediction values of the multiple tasks. Finally, the candidate micro-videos are ranked according to their scores as the recommendation results.

Embedding Generators: Let the embedding matrix of the k-th discrete feature of the user be $\mathbf{E}_k^u \in \mathbb{R}^{h_k^u \times d}$, where h_k^u is the number of categories of the k-th discrete user feature. The function $LP(\mathbf{E}, x)$ is defined to look up the x-th row of the two-dimensional matrix $\mathbf{E}$. Then the k-th discrete feature embedding of user u_i is $LP(\mathbf{E}_k^u, g_k^u(u_i))$. For the k-th continuous user features, we apply the maximum normalization to obtain $f_k^{'u}(u_i) = f_k^u(u_i) / \max(\{f_k^u(u), u \in \mathcal{U}\})$. The generator splices the embeddings of the discrete and normalized continuous user features to yield the user embedding, as shown in Eq. (1).

$$\mathbf{e}_i^u = \left[LP(\mathbf{E}_1^u, g_1^u(u_i)), \ldots, LP(\mathbf{E}_{m_u}^u, g_{m_u}^u(u_i)), f_1^{'u}(u_i), \ldots, f_{n_u}^{'u}(u_i) \right] \quad (1)$$

Let the ID embedding matrix of micro-videos be $\mathbf{E}_{id}^v \in \mathbb{R}^{|\mathcal{V}| \times d}$, and the embedding matrix of the k-th discrete micro-video feature be $\mathbf{E}_k^v \in \mathbb{R}^{h_k^v \times d}$, where h_k^v is the number of categories of the k-th discrete micro-video feature. Then for a micro-video $v_j \in \mathcal{V}$, its ID embedding vector is denoted as $LP(\mathbf{E}_{id}^v, v_j)$, the embedding vector of its k-th discrete micro-video feature is denoted as $LP(\mathbf{E}_k^v, g_k^v(v_j))$, and the k-th continuous micro-video feature after maximum normalization is denoted as $f_k^{'v}(v_j)$. We splice the ID embedding and feature embedding of the micro-video v_j as its embedding $\mathbf{e}_j^v$, as shown in Eq. (2).

$$\begin{aligned} \mathbf{e}_j^v = \big[& LP(\mathbf{E}_{id}^v, v_j), LP(\mathbf{E}_1^v, g_1^v(v_j)), \\ & \ldots, LP(\mathbf{E}_{m_v}^v, g_{m_v}^v(v_j)), f_1^{'v}(v_j), \ldots, f_{n_v}^{'v}(v_j) \big] \end{aligned} \quad (2)$$

Following the above approach, the embedding of each interacted micro-video of user u_i is obtained in turn. Thus we obtain $\mathbf{S}_i = \left[\mathbf{e}_{i1}^v, \mathbf{e}_{i2}^v, \ldots, \mathbf{e}_{i|s_i|}^v \right]$. The embedding of the target micro-video can also be derived, denoted as $\mathbf{e}_t^v$.

Similarly, let $\mathbf{E}_k^c \in \mathbb{R}^{h_k^c \times d}$ denote the embedding matrix of the k-th discrete interaction feature. Splicing the embeddings of each discrete and continuous interaction feature of the interaction between user u_i and micro-video v_{ij}, we obtain the context feature embedding of this interaction, denoted as $\mathbf{e}_{ij}^c$.

For each interacted micro-video $v_{ij} \in s_i$, we construct its embedding as $cat(\mathbf{e}_{ij}^v, \mathbf{e}_{ij}^c)$, where $cat(\cdot)$ is a vector concatenation function.

Next, we get the embedding $\mathbf{u}_i$ of user u_i and map the embeddings of interacted micro-videos and the target micro-videos into the same hidden space by Eqs. (3).

$$\mathbf{u}_i = \mathbf{W}_1^\top \mathbf{e}_i^u + \mathbf{b}_1$$
$$\mathbf{v}_{ij} = \mathbf{W}_2^\top \left(cat(\mathbf{e}_{ij}^v, \mathbf{e}_{ij}^c)\right) + \mathbf{b}_2 \tag{3}$$
$$\mathbf{v}_t = \mathbf{W}_3^\top \mathbf{e}_t^v + \mathbf{b}_3$$

where $\mathbf{W}_1 \in \mathbb{R}^{d^u \times d^u}, \mathbf{W}_2 \in \mathbb{R}^{(d^v + d^c) \times d^v}, \mathbf{W}_3 \in \mathbb{R}^{d^v \times d^v}, \mathbf{b}_1 \in \mathbb{R}^{d^u}, \mathbf{b}_2, \mathbf{b}_3 \in \mathbb{R}^{d^v}$ are trainable parameters, where d^v is the number of dimensions of mirco-video embeddings and d^u is the number of dimensions of user embedding.

Attention-based Multi-history Aggregator: We generate historical embeddings for a user by aggregating the embeddings of interacted micro-videos. Micro-videos in the historical interactions, which are related to the target micro-video should play a larger role in the aggregation. Therefore, we design an attention network, which calculates the weight of each interacted micro-video successively based on the correlation between the one and the target micro-video. Then we do weighted summation of the embeddings of historical interacted micro-videos to form the historical embedding.

We first take the feature interactions of each interacted micro-video and the target micro-video as the input for the attention network. Specifically, the feature interactions between the j-th micro-video $v_{ij} \in s_i$ interacted by user u_i and the target micro-video v_t is calculated by Eq. (4).

$$\text{Inter}\,(\mathbf{v}_{ij}, \mathbf{v}_t) = cat\,(\mathbf{v}_{ij}, \mathbf{v}_t, \mathbf{v}_{ij} - \mathbf{v}_t, \mathbf{v}_{ij} \odot \mathbf{v}_t) \tag{4}$$

We design a novel attention score calculation method which can fuse the information related to the two tasks from the historical interaction sequence, and then generate two different historical embeddings based on finishing playing and playback time.

The historical embedding based on finishing playing only aggregates the historical interacted micro-videos that have been finished by the user. We calculate the attention score a_{ij}^f between historical interacted micro-video of user u_i, which is denoted as $v_{ij} \in s_i$, and the target micro-video v_t via Eq. (5).

$$a_{ij}^f = I\,(r_{ij})\,ReLU\,\left(\mathbf{W}_f^T \text{Inter}\,(\mathbf{v}_{ij}, \mathbf{v}_t) + b_f\right) \tag{5}$$

where $\mathbf{W}_f^T \in \mathbb{R}^{4d^v \times 1}$, b_f is a learning parameter. $I(\cdot)$ is an indicator function, as shown in the Eq. (6), in which an unfinished interacted historical micro-video v_{ij} whose r_{ij} is equal to 0, has an attention score of negative infinity.

$$I(x) = \begin{cases} -\infty & \text{if } x = 0 \\ x & \text{otherwise} \end{cases} \tag{6}$$

The historical embedding based on finishing playing, denoted as s_i^f, is calculated by Eq. (7). First we input the attention score into the softmax function to calculate the weight of each historical interacted micro-video by Eq. (8) and then do the weighted aggregation of the embeddings of the interacted micro-videos. Unfinished micro-videos will be ignored since the softmax function will return 0 if using the negative infinity as the input.

$$w_{ij}^f = \frac{\exp\left(a_{ij}^f\right)}{\sum_{v_{i*} \in s_i} \exp\left(a_{i*}^f\right)} \tag{7}$$

$$\mathbf{s}_i^f = \sum_{v_{ij} \in s_i} w_{ij}^f \mathbf{v}_{ij} \tag{8}$$

The longer the playback time, the stronger the user preferences. So we use the playback time to scale the attention score. The attention score based on the watching time between $v_{ij} \in s_i$ and v_t is calculated as Eq. (9), where $\mathbf{W}_p \in \mathbb{R}^{4d \times 1}$, b_p is a learning parameter.

$$a_{ij}^p = I\left(c_{ij}\right) ReLU\left(\mathbf{W}_p^T \text{Inter}\left(\mathbf{v}_{ij}, \mathbf{v}_t\right) + b_p\right) \tag{9}$$

The historical embedding $\mathbf{s}_i^p$ based on watching time is calculated similarly as the one based on finishing playing.

Therefore, we obtain two historical embeddings of the same list of historical interactions, whereas they have different meanings, i.e., the historical embedding based on finishing playing $\mathbf{s}_i^f$ and the historical embedding $\mathbf{s}_i^p$ based on playback time.

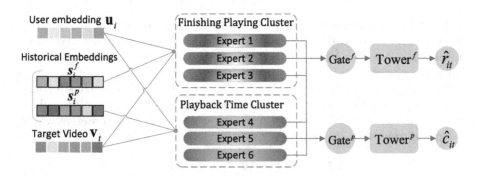

Fig. 2. The structure of the multi-task prediction layer

Multi-task Prediction Layer: As shown in the Fig. 2, the multi-task prediction layer mainly includes three components, i.e., multi-expert clusters, expert gate networks and task towers.

We can extract information of different aspects from the same data source by designing multiple experts [10]. We refer to the set of experts with the same input as the multi-expert cluster. In this paper, we construct two multi-expert clusters based on two tasks, namely, the finishing playing expert cluster and the playback time expert cluster. We denote the number of experts in the two clusters as e_f and e_p, respectively. The input of the finishing playing cluster is $cat([\mathbf{s}_i^f, \mathbf{v}_t, \mathbf{s}_i^f \odot \mathbf{v}_t, \mathbf{u}_i])$. The playback time expert cluster replaces the historical embedding of finishing playing with the one of playback time as a part of the input, that is, its input is $cat([\mathbf{s}_i^f, \mathbf{v}_t, \mathbf{s}_i^f \odot \mathbf{v}_t, \mathbf{u}_i])$. The two kinds of input data increase the difference between the knowledge of the two clusters. The k-th finishing playing expert and the k-th playback time expert can be represented as functions $\text{Expert}_k^f(\cdot)$ and $\text{Expert}_k^p(\cdot)$, respectively. In this paper, the two kinds of experts are multi-layer perceptrons (MLPs), sharing the same structure with different parameters, which transforms the input to d-dimensional hidden vectors. The output of each expert within the expert cluster forms the output of the cluster. For user u_i, the output of the two expert clusters, $\mathbf{C}_i^f \in \mathbb{R}^{e_f \times d}$ and $\mathbf{C}_i^p \in \mathbb{R}^{e_p \times d}$ is calculated via Eq. (10), where $x \in \{f, p\}$.

$$\mathbf{C}_i^x = \left[\text{Expert}_k^x \left(cat \left(\left[\mathbf{s}_i^f, \mathbf{v}_t, \mathbf{s}_i^f \odot \mathbf{v}_t, \mathbf{u}_i \right] \right) \right) | 1 \le k \le e_x \right] \tag{10}$$

In addition, we concatenate the two output matrices $\mathbf{C}_i^f$ and $\mathbf{C}_i^p$ by row and get the expert output matrix $\mathbf{C}_i \in \mathbb{R}^{(e_f + e_p) \times d}$.

Each task needs to extract information from $e_f + e_p$ experts and aggregates it for prediction. The role of the gates in Fig. 2 is to calculate the weight of each expert. So the finishing playing prediction task and the playback time prediction task each has a gate. The input for both is same, which is $cat(\mathbf{s}_i^f, \mathbf{s}_i^p, \mathbf{s}_i^f \odot \mathbf{v}_t, \mathbf{s}_i^p \odot \mathbf{v}_t, \mathbf{v}_t, \mathbf{u}_i)$. The two gates share the same structure, both of which are an MLP with softmax as the last layer, denoted as $\text{Gate}^f(\cdot)$ and $\text{Gate}^p(\cdot)$. For the two tasks of predictions of u_i watching v_t, the weight vectors of experts, $\mathbf{g}_i^f$ and $\mathbf{g}_i^p$ are calculated by Eq. (11).

$$\mathbf{g}_i^x = \text{Gate}^x \left(cat \left(\left[\mathbf{s}_i^f, \mathbf{s}_i^p, \mathbf{s}_i^f \odot \mathbf{v}_t, \mathbf{s}_i^p \odot \mathbf{v}_t, \mathbf{v}_t, \mathbf{u}_i \right] \right) \right) \tag{11}$$

where $x \in \{f, p\}$, $\mathbf{g}_i^f, \mathbf{g}_i^p \in \mathbb{R}^{e_f + e_p}$ and the k-th dimension of the two vectors represents the weight of the k-th expert for each task. For each prediction task, we rely on the weight of each expert calculated by the Gate to have a weight summation of the output of each expert in $\mathbf{C}_i$.

Next, we construct a corresponding task tower networks for each task, abbreviated as a tower. We denote the towers for the two tasks of finishing playing and the playback time as functions $\text{Tower}^f(\cdot)$ and $\text{Tower}^p(\cdot)$, respectively, both with the MLP structure and a single variable output. Among them, the tower for the finishing playing contains a sigmoid activation function to convert the predicted score into the finishing probability. The prediction of finishing playing and playback time is calculated by Eq. (12).

$$\begin{aligned} \widehat{r}_{it} &= \text{Tower}^f(\text{sum}_{row}(\mathbf{g}_i^f \odot \mathbf{C}_i)) \\ \widehat{c}_{it} &= \text{Tower}^p(\text{sum}_{row}(\mathbf{g}_i^p \odot \mathbf{C}_i)) \end{aligned} \tag{12}$$

Finally, we comprehensively consider the finishing playing probability and playback time of the target micro-video as its final score, as shown in Eq. (13), where λ_* is a hyperparameter.

$$\widehat{y}_{it} = \lambda_1 \widehat{r}_{it} + \lambda_2 \widehat{c}_{it} \tag{13}$$

Training Loss: For the finishing playing prediction, we model it as a binary classification task, optimized by cross-entropy loss, as shown in Eq. (14).

$$\mathcal{L}^f = \sum_{u_i \in \mathcal{U}} \sum_{v_t \in \mathrm{pos}_i \cup \mathrm{neg}_i} -(r_{it} \log \widehat{r}_{it} + (1 - r_{it}) \log(1 - \widehat{r}_{it})) \tag{14}$$

where pos_i and neg_i are the sets of positive and negative samples sampled by user $u_i \in \mathcal{U}$, respectively.

As for the playback time prediction, we model it as a regression task, optimized by MSE loss as shown in Eq. (15).

$$\mathcal{L}^p = \sum_{u_i \in \mathcal{U}} \sum_{v_t \in \mathrm{pos}_i \cup \mathrm{neg}_i} (\widehat{c}_{it} - c_{it})^2 \tag{15}$$

The total loss of the model is shown in Eq. (16).

$$\mathcal{L} = \omega_1 \mathcal{L}^f + \omega_2 \mathcal{L}^p + \omega_3 \mathcal{L}_{reg} \tag{16}$$

In Eq. (16), ω_* is a hyperparameter and $\mathcal{L}_{reg}$ is the regularization term loss for all parameters of the model.

4 Experimental Evaluation

In this section, we conduct extensive offline experiments and online A/B tests to answer the following four research questions:

RQ1: Does MARS outperform existing multi-task models?

RQ2: What is the influence of various components in the MARS architecture?

RQ3: How does the number of experts impact on the performance?

RQ4: How does MARS perform in the live production environment, i.e., when serving MX TakaTak?

4.1 Experimental Setup

Datasets: We adopt the following datasets for our offline experiments.

- WeChat-Channels: It is a dataset released by WeChat Big Data Challenge 2021. It contains two-week interactions from 20,000 anonymous users on WeChat Channels, which is a popular micro-video platform in China.
- MX-TakaTak: It is collected from one of the largest micro-video platforms in India, MX TakaTak. We collect real historical interaction records of TakaTak from Sept. 15, 2021 to Sept. 24, 2021, then randomly sample 23,632 users and form the Takatak dataset using their interactions.

We filter the data satisfying the following constraints from the original datasets.

Table 1. Statistics of datasets

Dataset	#Users	#Items	#Actions	Avg. action per user
WeChat-Channels	20000	70253	6.93M	346.58
MX-TakaTak	23632	64513	9.59M	405.62

- the users whose interactions are less than five,
- the micro-videos whose times of user watching is less than five,
- the interactions whose playback time is too long,
- the micro-videos whose duration is too long or too short.

The statistics of the two cleaned datasets are listed in Table 1. As shown in Table 1, the MX-TakaTak dataset contains more actions than WeChat-Channels.

For the WeChat-Channels dataset, we choose the video ID, video duration, author ID, background music ID, ID of publisher of background music, tags, user ID as features. For the MX-TakaTak dataset, we choose the video ID, video duration, publisher ID, IDF tag, user ID as features, where the tag whose IDF (Inverse Document Frequency) is the highest among all the tags of the micro-video is chosen as the IDF tag of this micro-video.

To avoid the data penetration during the model training and evaluation [16], we strictly maintain the timing relationship among the three data sets. In detail, we construct the test set using interactions of the last day, the validation set using interactions of the second-to-last day, and the training set using other interactions.

Metrics: Currently, we have two prediction tasks. One is finishing playing prediction, that is, predicting whether the micro-video finishes playing. Another is playback time prediction, predicting the playback time of the micro-video. If the playback time is longer than the running time of a micro-video, then we mark the micro-video with *finishing playing*. Further, considering that the playback time follows the long tail distribution, we take a logarithm operation of playback time x, i.e., $\log(x+1)$, and then normalize it to a range of $[0, 1]$ by math.log(301). These information is used to calculate the performance metrics.

For offline evaluation, we employ different metrics for two tasks. We adopt AUC (Area Under the Curve), GAUC (Group AUC) and Logloss as metrics of the task of finishing playing prediction, and adopt MSE (Mean Square Error), NRMSE (Normalized Root Mean Square Error) and NMAE (Normalized Mean Absolute Error) to evaluate the performance of task of playback time prediction.

For online evaluation, we calculate the average percentage of user finishing playing and adopt it to evaluate the task of finishing playing prediction, and calculate the average playback time of all users to evaluate the task of predicting playback time.

Implementation Details: We implement MARS with PyTorch. We initialize the parameters with the uniform distribution $U(-1/\sqrt{d}, 1/\sqrt{d})$. We set the number of experts to 6. We set L2 penalty to 10^{-5} and the loss weight of two tasks to 1.0 and 1.0, respectively. We optimize the model through Adam. The learning rate is set to 0.001 and will decay by 0.1 after every 10000 steps. Batch size is set to 256.

4.2   Performance Comparison

We choose two multi-task models, i.e., MMoE and the share-bottom model, as competitors. The experimental results are listed in Table 2.

Table 2. Recommendation performance. The number in a bold type is the best performance in each column.

Datasets	Model	GAUC	AUC	Logloss	MSE	NRMSE	NMAE
WeChat	MMoE	0.64746	0.70601	0.62178	0.05395	0.46879	0.66021
-Channels	Share-bottom	0.64715	0.70642	0.62205	0.05448	0.47108	0.66043
	MARS	**0.64963**	**0.74479**	**0.59149**	**0.04933**	**0.44827**	**0.64431**
	Improv.	0.34%	5.43%	4.87%	8.56%	4.38%	2.41%
MX	MMoE	0.68491	0.68269	0.64524	0.02468	0.35902	0.55925
-TakaTak	Share-bottom	0.68515	0.68616	0.63974	0.02503	0.3616	0.5608
	MARS	**0.6873**	**0.75581**	**0.57706**	**0.02025**	**0.32518**	**0.52989**
	Improv.	0.31%	10.15%	9.80%	17.95%	9.43%	5.25%

For both datasets, our model outperforms the other multi-task models in all the metrics. For example, on the WeChat-Channels dataset, the AUC of our model is 5.49% higher than the AUC of MMoE, and NRMSE of our model is 4.38% lower than the NRMSE of MMoE. On the MX-TakaTak dataset, when compared with MMoE, our model increases in AUC by 10.71% and decreases in NRMSE by 9.43%.

4.3   Ablation Study

We conduct the ablation study to observe the contributions of different components in MARS. Therefore, we build a series of variants of MARS.

We remove task related information of attention-based multi-history aggregator from the original MARS and obtain the first variant, denoted by MARS-A. The second variant is the one which removes the structure of MMoE from MARS, denoted by MARS-B. We compare our model with these two variants and the experimental results are listed in Table 3. From Table 3, we find that the performance of these two variants gets worse as expected, no matter whether the metric is AUC, GAUC, MSE or NRMSE.

Table 3. Ablation study on the MX-TakaTak dataset

Model	GAUC	AUC	MSE	NRMSE
MARS	**0.6873**	**0.75581**	**0.02025**	**0.32518**
MARS-A	0.68268	0.75505	0.02032	0.32574
MARS-B	0.68702	0.75529	0.02026	0.32531

Then, we observe the performance of MARS, if let MARS perform only a single task. Thus, we remove L^p from the loss function and form the variant MARS-C to observe its performance of performing the task of finishing playing prediction. Also, we obtain the variant MARS-D by removing L^f from the loss function and observe its performance on predicting playback time.

Experimental results on the MX-TakaTak dataset are listed in Table 4. From the results, we find although MARS and two variant have same network structure, MARS outperforms MARS-C on the task of finishing playing prediction, and MARS and MARS-D have the similar performance on predicting the playback time.

Table 4. Ablation study of single task on the MX-TakaTak dataset

Model	GAUC	AUC	Logloss	MSE	NRMSE	NMAE
MARS	**0.6873**	**0.75581**	**0.57706**	**0.02025**	**0.32518**	0.52989
MARS-C	0.68476	0.75434	0.57778	–	–	–
MARS-D	–	–	–	0.02025	0.32522	**0.52888**

4.4 Impact of Number of Experts

The number of experts is the most important hyper-parameter in MARS. We change the number of experts in MARS and conduct the experiments on the WeChat dataset to observe the impact of different number of experts on performance. The experimental results are listed in Table 5. From Table 5, we find that while the number of experts is set to 6, the model behaves best, taking all the metrics together.

Table 5. Impact of number of experts on the WeChat-Channels dataset.

#Experts	GAUC	AUC	Logloss	MSE	NRMSE	NMAE
4	0.65014	0.74452	0.59224	0.04937	0.44847	0.64434
6	0.64963	0.74479	0.59149	0.04933	0.44827	0.64431
8	0.64986	0.74436	0.59130	0.04941	0.44866	0.64457
10	0.64993	0.74317	0.59388	0.04962	0.44958	0.64354

4.5 Deployment and Online A/B Tests

Our model has been deployed on the MX-TakaTak platform, working together with other models.

While migrating our model from offline to online, we add some extra features of micro-videos and users, including continuous features from 7-day statistical data and discrete features. Some added discrete features are listed as follows.

- the state where the micro-video is uploaded,
- the state where the user stays,
- the type of user mobile phone.

Some added continuous features are listed as follows.

- the average CTR on feed page within 7 days,
- the average CTR on the micro-video of feed page,
- average valid playing rate within 7 days,
- average playing finishing rate within 7 days.

We apply the uniform discretization transform to the continuous values, and then treat the transformed results as discrete features.

The deployed version of MARS uses more interactions between users and micro-videos, including liking, favorite, sharing, downloading, giving a comment, reading a comment, entering the homepage of publisher, loop times, the hour when the video is watched. We also apply the uniform discretization transform to these values, and then build the corresponding embedding matrices which will be a part of features. We collect interaction records and use Spark to generate the training set. After training, we perform the recommending for the users logged on the previous day. The serving process is depended on SageMaker provided by Amazon.

Table 6. Online A/B tests on MX TakaTak

Model	Metrics	Jan. 1	Jan. 2	Jan. 3	Jan. 4	Jan. 5
MARS	Average playback time (s)	**943.904**	**1001.19**	**972.359**	**998.581**	**996.011**
	Playing finishing rate	**0.3585**	**0.3745**	**0.3778**	**0.3844**	**0.3854**
Baseline	Average playback time (s)	937.456	996.183	971.888	993.023	988.684
	Playing finishing rate	0.3515	0.3712	0.3729	0.379	0.3804
Improv.	Average playback time	0.69%	0.50%	0.05%	0.56%	0.74%
	Playing finishing rate	1.99%	0.89%	1.31%	1.42%	1.31%

We conduct online A/B tests from Jan. 1, 2022 to Jan. 5, 2022. Experimental results in Table 6 show that our model greatly outperforms the baseline model in terms of average playback time and playing finishing rate and achieves good recommendation results.

5   Conclusion

In this paper, we examine the micro-video ranking and propose a multi-task ranking model MARS for recommending micro-videos. MARS takes full advantage of user interactions and historical statistics, and guides different experts to learn different knowledge for different tasks under the structure of MMoE. We conduct offline experiments on two real-world datasets. The results show that MARS improves the accuracy of two prediction tasks. We also give a number of key points of deploying the model to the production environment and conduct online A/B tests. The results illustrate that MARS is feasible and effective.

Acknowledgements. This work was supported by the National Natural Science Foundation of China under Grant No. 62072450 and the 2021 joint project with MX Media.

References

1. Caruana, R.: Multitask learning. Mach. Learn. **28**(1), 41–75 (1997)
2. Chen, X., Gu, X., Fu, L.: Boosting share routing for multi-task learning. In: Companion Proceedings of the Web Conference 2021, pp. 372–379 (2021)
3. Chen, X., Liu, D., Zha, Z.J., Zhou, W., Li, Y.: Temporal hierarchical attention at category- and item-level for micro-video click-through prediction. In: 2018 ACM Multimedia Conference (2018)
4. Jiang, H., Wang, W., Wei, Y., Gao, Z., Nie, L.: What aspect do you like: multi-scale time-aware user interest modeling for micro-video recommendation. In: MM 2020: The 28th ACM International Conference on Multimedia (2020)
5. Lei, H., Luo, B.: Personalized micro-video recommendation via hierarchical user interest modeling. In: Pacific Rim Conference on Multimedia (2017)
6. Li, Y., Liu, M., Yin, J., Cui, C., Xu, X.S., Nie, L.: Routing micro-videos via a temporal graph-guided recommendation system. In: Proceedings of the 27th ACM International Conference on Multimedia (2019)
7. Liu, S., Chen, Z.: Sequential behavior modeling for next micro-video recommendation with collaborative transformer. In: 2019 IEEE International Conference on Multimedia and Expo (ICME) (2019)
8. Liu, S., Chen, Z., Liu, H., Hu, X.: User-video co-attention network for personalized micro-video recommendation. In: The World Wide Web Conference (2019)
9. Ma, J., Zhao, Z., Chen, J., Li, A., Hong, L., Chi, E.H.: SNR: sub-network routing for flexible parameter sharing in multi-task learning. In: Proceedings of the AAAI Conference on Artificial Intelligence, vol. 33, pp. 216–223 (2019)
10. Ma, J., Zhao, Z., Yi, X., Chen, J., Hong, L., Chi, E.H.: Modeling task relationships in multi-task learning with multi-gate mixture-of-experts. In: Proceedings of the 24th ACM SIGKDD International Conference on Knowledge Discovery & Data Mining, pp. 1930–1939 (2018)
11. Ma, S., Zha, Z., Wu, F.: Knowing user better: jointly predicting click-through and playtime for micro-video. In: 2019 IEEE International Conference on Multimedia and Expo (ICME), pp. 472–477. IEEE (2019)
12. Ma, X., et al.: Entire space multi-task model: An effective approach for estimating post-click conversion rate. In: The 41st International ACM SIGIR Conference on Research & Development in Information Retrieval, pp. 1137–1140 (2018)
13. Misra, I., Shrivastava, A., Gupta, A., Hebert, M.: Cross-stitch networks for multi-task learning. In: Proceedings of the IEEE Conference on Computer Vision and Pattern Recognition, pp. 3994–4003 (2016)
14. Tang, H., Liu, J., Zhao, M., Gong, X.: Progressive Layered Extraction (PLE): a novel multi-task learning (MTL) model for personalized recommendations. In: Fourteenth ACM Conference on Recommender Systems, pp. 269–278 (2020)
15. Wei, Y., Wang, X., Nie, L., He, X., Hong, R., Chua, T.S.: MMGCN: multi-modal graph convolution network for personalized recommendation of micro-video. In: the 27th ACM International Conference (2019)
16. Wu, S., Tang, Y., Zhu, Y., Wang, L., Xie, X., Tan, T.: Session-based recommendation with graph neural networks. In: Proceedings of the AAAI Conference on Artificial Intelligence, vol. 33, pp. 346–353 (2019)

17. Zhao, Z., et al.: Recommending what video to watch next: a multitask ranking system. In: Proceedings of the 13th ACM Conference on Recommender Systems, pp. 43–51 (2019)
18. Zhou, T., et al.: JUMP: a joint predictor for user click and dwell time. In: Proceedings of the 27th International Joint Conference on Artificial Intelligence. AAAI Press, pp. 3704–3710 (2018)

Security, Privacy, and Trust
and Blockchain Data Management
and Applications

How to Share Medical Data Belonging to Multiple Owners in a Secure Manner

Changsheng Zhao, Wei Song$^{(\boxtimes)}$, and Zhiyong Peng

School of Computer Science, Wuhan University, Wuhan, China
{2016302580172,songwei,peng}@whu.edu.cn

Abstract. Over the past few years, we have witnessed that the big data analysis technologies based on machine learning and deep learning have brought great change to medical research. The medical data is a special kind of sensitive data, which is contributed by multiple users. For example, the prescriptions are generated by doctors but also contain the patients' symptoms and history of diseases. However, the existing data sharing solutions focus on protecting the privacy of the single data owner while ignoring the privacy belonging to multiple owners. In this paper, we first identify this problem and propose a novel secure medical data sharing scheme based on the Chinese remainder theorem to protect the privacy of multiple owners in the medical data. To better support the scenario of medical data sharing, we implement the proposed scheme over a medical blockchain system to build a medical data sharing prototype, and dynamically manage two kinds of subscribers based on logical key hierarchy. The performance analysis validates the security and efficiency of the proposed method for the data sharing scenario of medical data from multiple owners.

Keywords: Multiple owners · Medical data sharing · Chinese remainder theorem · Blockchain

1 Introduction

Medical data has great value in use or research such as recommendations for medical advertisements, social network and research of specific diseases. As an attractive paradigm for digital information processing, in healthcare systems, as the owner of the data, users are able to decide whether and with whom they share data according to their own needs. For example, insurance companies evaluate whether users are eligible for insurance by using their health information.

We assume that there is lots of medical data generated from the processes of Doctor Bob serving his patient Alice. In this case, both of them want to share the data with others. However, medical data in healthcare systems is complicated and varied. In order to achieve flexible and secure medical data sharing, it should be categorized into several different types by the owner of data: 1) The first category is Alice's personal data including her medical history, medication and allergies, and original health data collected by various sensor devices. We call the patient's health data used for diagnostic purpose M_1. 2) The second category is Bob's personal data which is generally not shared with others and will not be discussed in this paper. 3) The last category is diagnosis or prescription generated by Bob examining Alice's health data. We call it M_2. Apparently, M_2 is contributed by Bob, moreover it contains Alice's privacy, so the owners of M_2 are not only Alice but also Bob. However, all the existing studies on medical data sharing have ignored this point. In these solutions, the patient is allowed to individually share diagnosis with others without the consent of the doctor. Motivated by this, the main contribution of this work is to design a sharing scheme for medical data belonging to multiple owners in a secure manner.

There is no doubt that having patients and doctors as co-owners of data poses tremendous challenges to the flexible and secure data sharing. Diagnosis and sharing are totally different functions for the users. Thus, simply using symmetric cryptography to encrypt M_1 and M_2 causes the two functions to interfere with each other, especially in the case of updating the secret key. In our work, we design the master key encryption (MKE) algorithm based on the Chinese remainder theorem (CRT) [1] to generate the master key pair and two slave key pairs for Alice and Bob respectively. One of their slave key pairs is the same, so that they can access M_2 and M_1 expediently, while the other slave key pair is private for flexible sharing data with the special subscribers.

To achieve scalable and secure medical data sharing, only designing an encryption scheme is not enough. Traditional healthcare systems store medical data in the cloud or the local data center, which suffers from scalability and security issues. The major drawbacks of centralized infrastructure have motivated us to use blockchain and decentralized storage technology. Meanwhile, by verifying the signatures [2] of all data owners recorded on the blockchain, it can effectively prevent any owner sharing data without the permission of other owners.

In addition, it is noteworthy that medical data continuously generated by the fixed doctor and patient will be appreciated by a great many data demanders. Failure to manage these data subscribers will lead to secret keys abuse and dangerous authorization. For two kinds of data subscribers who subscribe to all data and subscribe only to M_1, it is a great challenge to authorize them to access the medical data without revealing the privacy. In this paper, we put forward an innovative tree structure based on logical key hierarchy and the corresponding management scheme. We summarize our contributions as follows.

- To our knowledge, we are the first to consider how to share data that belongs to multiple owners. We design a novel encryption scheme based on CRT to achieve medical data sharing with this concern.
- We propose a blockchain-based medical data sharing system by which the patient can share M_1 alone or jointly share M_1 and M_2 with the permission of the doctor. And we provide a concrete tree structure and the key management scheme that enables the patient and the doctor to jointly manage the two types of data subscribers dynamically and efficiently.
- We show the performance of our scheme through implementation and demonstrate that it is feasible to apply the scheme to a real healthcare applications.

2   Related Work

In healthcare systems, attribute based-encryption (ABE) [3–6] is often used to achieve flexible access control over encrypted data. For example, Li *et al.* [4] proposed a sharing scheme by using ABE technology to encrypt users' EHR data. However, traditional ciphertext-policy ABE leads to the problem of revealing access policies. Zhang *et al.* [7] proposed an effective solution to conceal the sensitive attribute values in the access policy to solve this problem. However, ABE-based schemes are only suitable for one master and multiple users. And the scheme based on preset thresholds is not universal. In 2020, Ma *et al.* [8] proposed an innovative access control model. In this model, in order to optimize the communication and computation cost of users, complex computation is outsourced to public cloud servers to make it compatible with Android devices. These technologies have driven the development of digital healthcare systems to a certain extent, but as mentioned earlier, these schemes can't be used for multiple owners and there are various problems with centralized storage which need to be solved with decentralized storage technologies.

In recent years, some efforts are devoted to proposing architectures of healthcare systems based on blockchain because of its good prospects for privacy protection. Lvan *et al.* [9] proposed a secure storage scheme of medical data based on blockchain, which can prevent medical data from being tampered with. Zhang *et al.* [10] proposed a blockchain-based healthcare system, which realized stable operation of the system by using the alliance chain and Byzantine consensus mechanism. To protect users' health data and achieve flexible data sharing, Xu *et al.* [11] proposed a privacy protection scheme for health data, where the records of users' health data and diagnosis are stored in different chains. Meanwhile, it uses the distributed file system IPFS to avoid the problem of a single point of failure. In 2021, Guo *et al.* [12] proposed a secure sharing model, which realizes the sharing between users and third parities by using proxy re-encryption algorithm. While the framework of the above works gives us some inspiration, none of them took into account the ownership of doctors in data sharing. Similarly, there is only one kind of subscriber in their schemes and we need to design a tree structure to manage two kinds of subscribers dynamically.

3   Preliminaries

3.1   MKE Scheme Based on the Chinese Remainder Theorem

The MKE scheme is a RSA-based public key crypto system leveraging the CRT, which allows multiple decryption keys to decrypt the same message encrypted by the master key [13]. The main idea of CRT-based MKE is to generate the master key with n public-private key pairs. Let e_i, d_i be the ith public and private key pair with p_i, q_i being the ith prime number pair. Especially, $e_i d_i \equiv 1$ mod $\phi(p_i q_i)$, where $\phi(x)$ is Euler's totient function. We need to generate the master key pair e_M, d_M and the following congruence equations should be satisfied.

$$e_M \equiv e_i (mod\phi(p_i q_i)) \qquad d_M \equiv d_i (mod\phi(p_i q_i)) \qquad (1)$$

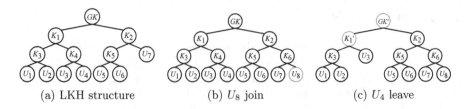

Fig. 1. Logical key hierarchy.

Let $k_i = \phi(p_i q_i)$. For the following congruence equations, we can calculate e_M by using CRT.

$$\begin{cases} e_M \equiv e_1 \quad (mod\ k_1) \\ e_M \equiv e_2 \quad (mod\ k_2) \\ \quad\quad \vdots \\ e_M \equiv e_n \quad (mod\ k_n) \end{cases} \tag{2}$$

3.2 Logical Key Hierarchy

Logical Key Hierarchy (LKH) [14] was proposed by D.Wallne *et al.* to manage group members, which is applicable to publication and subscription of data.

LKH Structure. LKH is usually a balanced binary tree and represents users as leaf nodes in the tree structure. The group key GK is located at the root node of the tree. Each internal node is associated with a key encryption key(KEK) which is used to update group keys efficiently. The owner needs to maintain the complete tree. Each group member has a unique key shared with the owner and also needs to maintain keys throughout the branch path respectively. For example, there are 7 users in a tree as shown in Fig. 1(a). User 6 has its unique secret key U_6, group key GK and KEKs: K_2, K_5.

Update. When a new user joins the group, the owner first establishes a new shared key with the user. In order to add a new node to a non-complete branch of the tree or a new layer, the data owner needs to transmit GK and KEKs on the path from the root to the new user securely. When a user leaves the group, the owner needs to update GK and the KEKs related to the leaving user as shown in Fig. 1(c). Finally, the owner multicasts the changed keys encrypted by KEK or U_i to the relevant users.

4   System Model

4.1   System Overview

Considering there are the patient Alice, the doctor Bob and two types of data subscribers in the instance, the entities in our system are described as follows.

Lightweight Users. Lightweight users are such a set of lightweight nodes that only store the block headers, and they can only generate and publish transactions.

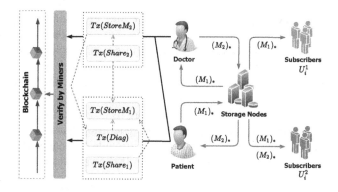

Fig. 2. System model of blockchain-based healthcare system

They usually want access to medical service, or subscribe to other users' data, or both. In one instance, lightweight users can be divided into the patient Alice and many subscribers U_i.

- *Patient:* The patient Alice can aggregate and encrypt data M_1 from IoT devices and store them in a storage node. Then, Alice authorizes M_1 to the doctor Bob. At last, Alice is allowed to share M_1 or jointly share M_1 and M_2 with the consent of Bob.
- *Subscribers:* There are two kinds of data subscribers, U_i^1 and U_i^2. They are allowed to use medical data for insurance evaluation, research, and so on.

Storage Nodes. The space owners of storage nodes are a set of peers that have strong storage capacity and want to provide storage service for economic reasons. They provide download service only for data owners and authorized users recorded on the blockchain.

Doctor. The doctor Bob can provide continuous diagnosis M_2 based on the patient's health data M_1. When Bob uploads M_2 and publishes a transaction $Tx(StoreM_2)$, it is necessary to indicate the source data M_1 of the diagnosis. The owner of M_1 and the doctor Bob collectively serve as the owner of M_2.

Blockchain. All storage and authorization transactions are recorded on the blockchain and can't be altered due to the tamper-proof feature of blockchain. The uniqueness of the hash value also allows it to be used as the storage address.

Miners. There are also some user nodes with powerful computing and storage resources, called miners. They can store complete blockchain and generate, publish, verify transactions and provide query service to lightweight users.

As shown in Fig. 2, here we briefly show data flows in our scheme. The patient Alice encrypts the health data M_1 and sends them to a storage node. At the same time, She publishes a transaction $Tx(StoreM_1)$ which contains the hash of encrypted M_1. After being verified by miners, the transaction is recorded on the blockchain forever. Then the authorized doctor Bob can download the health data from the storage node, decrypt it and give online diagnosis M_2. After he sends the encrypted diagnosis to the storage node and generates a transaction $Tx(StoreM_2)$, Alice can get the address of encrypted M_2 from the verified transaction on the blockchain and download the encrypted diagnosis. At last, Alice and Bob can share these medical data with others and generate sharing transaction $Tx(Share_1)$ or $Tx(Share_2)$ with their signatures.

With the storage transaction, data owners are recorded on the blockchain and they can't be tampered with. Thus, during the generation phase of authorization transaction, the transaction publisher needs to provide the signatures of all real data owners recorded in the source transaction. If the validation is successful, it reveals that the transaction has been approved by all data owners. And storage nodes are able to refuse to provide downloads to unauthorized users who are not recorded on the blockchain. By effectively combining these technologies of blockchain, multi-signatures verification and transactions, we further prohibit one of the owners sharing data without the consent of all data owners.

4.2   Transactions

The templates for several different transactions recorded on the blockchain are described in detail below. These transactions contain the timestamp ts, hash of the encrypted medical data $H((M)_*)$, the signature of the data owner σ. $DatasetId$ is the identifier of the dataset to which the data belongs, which is customized by users and convenient to accomplish batch authorization of data.

- $StoreM_1$: The patient Alice stores M_1 in a storage node.
  $Tx(StoreM_1) = \{ID_{Alice}, H((M_1)_*), DatasetId, ts, \sigma_{Alice}\}$
- $Diag$: Alice authorizes M_1 to the doctor Bob. $H(Tx(StoreM_1))$ indicates the store transaction of M_1.
  $Tx(Diag) = \{H(Tx(StoreM_1)), ID_{Bob}, ts, \sigma_{Alice}\}$
- $StoreM_2$: Bob stores M_2 in a storage node. $H(Tx(Diag))$ indicates the corresponding $Diag$ transaction.
  $Tx(StoreM_2) = \{H(Tx(Diag)), H((M_2)_*), DatasetId, ts, \sigma_{Alice}, \sigma_{Bob}\}$
- $Share_1$: Alice shares M_1 with other users. This transaction contains $\{ID_{U_i^1}\}$, an identifier list of authorized subscribers U_i^1.
  $Tx(Share_1) = \{DatasetId, \{ID_{U_i^1}\}, ts, \sigma_{Alice}\}$
- $Share_2$: Alice and Bob share M_1 and M_2 with other users. This transaction contains $\{ID_{U_i^2}\}$, an identifier list of authorized subscribers U_i^2.
  $Tx(Share_2) = \{DatasetId, \{ID_{U_i^2}\}, ts, \sigma_{Alice}, \sigma_{Bob}\}$

4.3   Signature and Verification

All transactions published on blockchain need to be verified by miners. The steps for signature and verification are as follows.

1) $KeyGen$: Every registered user has a secret key s and a public key $v = g^s$.
2) $Sign$: All data owners need to sign the main message m of the transaction with their secret key. The signature on the transaction is $\sigma \leftarrow g^{(s+H(m))^{1/2}}$. For example, in transaction $Tx(StoreM_1)$, $m = \{ID, H((M_1)_*), DatasetId, ts\}$.

3) $Verify$: Firstly, the verifier looks for all owners of the transaction on the blockchain. If they do not match the owners of these signatures, the verification fails. Then the verifier verify multiple signatures as follows.

$$\prod_{i=1}^{n} e(\sigma_i, \sigma_i) = e(\prod_{i=1}^{n} v_i \cdot g^{n \cdot H(m)}, g) \tag{3}$$

where n is the number of signatures.

Specially, if Alice is the only owner of the transaction, such as $Tx(StoreM_1)$, the verification structure is as follows.

$$e(\sigma_{Alice}, \sigma_{Alice}) = e(v_{Alice} \cdot g^{H(m)}, g) \tag{4}$$

Moreover, if the data owners are Alice and Bob like $Tx(StoreM_2)$, the verification structure is as follows.

$$e(\sigma_{Alice}, \sigma_{Alice}) \cdot e(\sigma_{Bob}, \sigma_{Bob}) = e(v_{Alice} \cdot v_{Bob} \cdot g^{2H(m)}, g) \tag{5}$$

5   Our Proposed Scheme

The proposed scheme is composed of initialization, diagnosis and sharing. The owners of medical data Alice and Bob generate secret keys with MKE scheme based on the CRT for encryption, and group keys for management and sharing with subscribers. The detailed notations and descriptions are listed in Table 1.

5.1   Initialization

System Initialization. Generate the secret random number t and the public parameter g^t for the system.

User Registration. The system generates a unique ID for each registered user as an identifier. Then, we generate a random secret s for signing and the public key g^s. At last, for this user, our system generates another secret random number k and secret key pair (r, g^r) to establish a trusted channel for the first communication between users, where $r = s \oplus k$.

5.2   Diagnosis

Store M_1. After aggregating the original health data M_1, Alice needs to generate the secret key to encrypt M_1. Firstly, Alice generates two slave key pairs (e_1, d_1) and (e_2, d_2), and two random numbers k_1 and k_2. Specially, $e_1 = e_2$ (mod $gcd(k_1, k_2)$). We can set up the following remainder equations and obtain the unique solution e_{M_1} based on the Chinese remainder theorem.

$$\begin{cases} e_{M_1} \equiv e_1 \pmod{k_1} \\ e_{M_1} \equiv e_2 \pmod{k_2} \end{cases} \tag{6}$$

Table 1. Summary of notations used in our construction.

M_1	Original health data generated by the patient
M_2	Diagnosis generated by the doctor
$Share_1$	Patient shares M_1 with subscribers
$Share_2$	Patient and doctor share M_1 and M_2 with subscribers
U_i^1	Subscriber who describes to M_1
U_i^2	Subscriber who describes to M_1 and M_2
Tx	Transaction
σ	Signature on transaction Tx
(s, v)	Key pair used for signing
(r, g^r)	Key pair used to generate traffic encripion key
KEY	Temporary traffic encryption key
e_{M_1}	The first master key used to encrypt M_1
e_{M_2}	The second master key used to encrypt M_2
(e_i, d_i)	Slave key pair used to generate master key
GK	Group key
GK_1	Group key for all U_i^1
GK_2	Group key for all U_i^2
KEK	Key encryption key
KEY_U_i	Shared secret key between user i and data owners
$H(\cdot)$	Hash function
$(\cdot)_*$	Encryption function using encryption key $*$

Then, Alice encrypts M_1 with e_{M_1}, and uploads encrypted M_1 to a storage node. The one who has d_1 or d_2 can decrypt and read M_1. Meanwhile, she publishes a storage transaction $Tx(StoreM_1)$ with her signature and waits for verification to pass.

Diag. Alice chooses the doctor Bob for treatment. At first, they generate a temporary communication key KEY based on the key agreement protocol. By the way, in order to avoid man-in-the-middle attacks [15], it's better to provide their own signature when exchanging the communication public key g^r .

$$KEY = (g^t, g^{r_{Alice}})^{r_{Bob}} = (g^t, g^{r_{Bob}})^{r_{Alice}} \tag{7}$$

Then, Alice sends slave key pair (e_1, d_1) and k_1 encrypted with KEY to Bob. At last, she generates and publishes a diagnosis transaction $Tx(Diag)$. Once verified, the transaction is recorded on the blockchain.

StoreM$_2$. The doctor Bob requests a download from the corresponding storage node based on the hash value of encrypted M_1 in the transaction $Tx(Diag)$. The Storage Node queries whether Bob is an authorized user of M_1 on the blockchain to decide whether to provide downloads. When the download is completed, he decrypts and reads M_1 with d_1, and generates a prescription M_2.

Similarly, before uploading M_2, Bob needs to generate the other master key e_{M_2}. In addition to the slave key pair (e_1, d_1) sent by Alice, Bob also needs to generate a slave key pair (e_3, d_3) and k_3. We can establish the following remainder equations and calculate e_{M_2}.

$$\begin{cases} e_{M_2} \equiv e_1 \pmod{k_1} \\ e_{M_2} \equiv e_3 \pmod{k_3} \end{cases} \tag{8}$$

Then, the doctor encrypts M_2 with e_{M_2}, and uploads it to a storage node. The one who has d_1 or d_3 is able to decrypt and read M_2. At last, M_2 is collectively held by Bob and Alice, so the storage transaction $Tx(StoreM_2)$ published by Bob must be signed by both of them. Otherwise, the verification will fail.

Finally, as one of the owners of M_2, the patient requests a download from the storage node based on the hash value of encrypted M_2. Since Alice holds the slave key d_1, she can directly decrypt and check M_2.

5.3   Sharing

Corresponding to data types, data sharing can also be defined as two types, 1)$Share_1$: Alice shares M_1 with data subscribers of U_i^1, 2)$Share_2$: Alice and Bob share M_1 and M_2 with data subscribers of U_i^2. It makes no sense to share M_2 alone because M_2 is the prescription for M_1. Alice and Bob generate a special LKH tree to manage all data subscribers. In this section, we first introduce the tree structure and then explain how it dynamically handles membership changes.

Structure. In our structure, the group key GK located at the root node of the tree. The left and right subtrees with GK as the root node correspond to two types of subscribers respectively. At the same time, GK_1 and GK_2 are located at the root of the left and right subtrees, respectively. The secret key KEY_U_i shared by U_i and owners is located at the leaf node, which is unique and confidential to other subscribers, where $KEY\_U_i^v = H(GK_v \| g^{ru_i})$, GK_v is the group to which user U_i^v belongs. All secret keys in the tree are symmetric keys and the generation of them is very efficient. In addition, there is no relationship between the doctor and subscribers U_i^1 who only subscribe to M_1, so Alice holds the whole tree and Bob only holds the right subtree.

As shown in Fig. 3(a), user U_{1-3}^1 only subscribes to M_1 and U_{1-3}^2 subscribes to M_1 and M_2. All subscribers establish a shared secret key KEY_U_i with data owners. Users also need to store KEKs on the path from their node to their group root GK_1 or GK_2 for updating purposes.

User U_i^1 Join. After Alice adds U_i^1 to the left subtree of the tree, she and U_i^1 exchange their public key g^r, and establish a temporary traffic encryption key KEY based on the Diffie-Hellman key agreement protocol. Then, Alice uses KEY to encrypt d_2, GK, GK_1 and all KEKs on the path from the leaf node U_i^1 to its group root GK_1, and sends them to U_i^1. They can compute their shared secret key KEY_U_i. In the end, she publishes $Tx(Share_1)$. U_i^1 can download encrypted M_2, decrypt and read it with d_2.

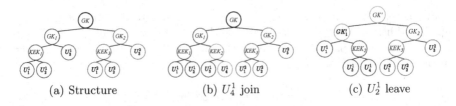

Fig. 3. Example of structure update for user U_i^1 join/leave events.

For example, as shown in Fig. 3(b), when U_4^1 subscribes to M_1, Alice and U_4^1 establish the temporary traffic encryption key: $KEY = (g^t, g^{rv})^{r_{Alice}} = (g^t, g^{rv})^{r_{Alice}}$. Then she sends $(d_2, GK, GK_1, KEK_2)_{KEY}$ to U_4^1. The shared secret key $KEY\_U_4^1 = H(GK_1 || g^{r_{U_4^1}})$. Finally, she sends $(KEK_2)_{KEY\_U_3^1}$ to U_3^1.

User U_i^1 Leave. After removing U_i^1 from the tree, to ensure afterward security, Alice needs to update d_2 and e_{M_1}, GK and GK_1, and $KEKs$ on the path from the leaf node U_i^1 to the group root GK_1. For remaining users U_j^1, $j \neq i$, she then sends the newly generated keys encrypted with KEK or $KEY\_U_j^1$ to them. And for users U_i^2, she broadcasts GK' and d_2' via encrypted message $(d_2', GK')_{GK_2}$. Similarly, Alice needs to publish a transaction $Tx(Share_1)$, in which the id list includes all remaining subscribers. After that, the storage node is able to identify all authorized users through this transaction.

For example, as shown in Fig. 3(c), when U_2^1 leaves, Alice generates new slave key d_2' and master key e_{M_1}', and changes GK and GK_1. Then she needs to multicast the new keys to the existing users in the tree. Specifically:

- unicast $(d_2', GK', GK_1')_{KEY\_U_1^1}$ to U_1^1
- multicast $(d_2', GK', GK_1')_{KEK_2}$ to U_{3-4}^1
- multicast $(d_2', GK')_{GK_2}$ to all U_i^2

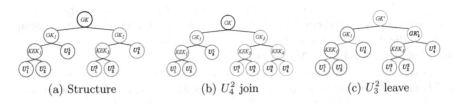

Fig. 4. Example of structure update for user U_i^2 join/leave events.

User U_i^2 Join. First of all, data owners add U_i^2 to the right subtree of the tree, such as U_4^2 shown in Fig. 4(b). After setting up a temporary communication key KEY with U_i^2, Alice and Bob do the following steps:

- Alice unicasts $(d_2, GK, GK_2, KEK_4)_{KEY}$ to U_4^2

- Alice unicasts $(KEK_4)_{KEY\_U_3^2}$ to U_3^2
- Bob unicasts $(d_3)_{KEY\_U_4^2}$ to U_4^2

At last, they publish a transaction $Tx(Share_2)$ with their signatures. Once verified, U_i^2 is able to download encrypted M_1 and M_2 and decrypt them with d_2 and d_3 respectively. If the verification fails, it indicates that Alice or Bob does not admit this transaction, and the download request from U_i^2 will be rejected.

User U_i^2 Leave. After removing U_i^2 from the tree, Alice needs to update d_2, e_{M_1}, GK, GK_2, and related KEKs. Meanwhile, Bob changes d_3 and e_{M_2}. Take U_3^2 for example, Alice and Bob perform the following steps:

- Alice multicasts $(d_2', GK')_{GK_1}$ to all U_i^1
- Alice multicasts $(d_2', GK', GK_2')_{KEK_3}$ to U_{1-2}^2
- Alice unicasts $(d_2', GK', GK_2')_{KEY\_U_4^2}$ to U_4^2
- Bob multicasts $(d_3')_{GK_2'}$ to all U_j^2, j≠i

At last, they publish $Tx(Share_2)$ that contains their signatures, in which the id list includes all remaining subscribers U_j^2.

Secret Key Update. To ensure the security and ownership of the data, data owners are allowed to update the slave key. If Alice updates d_2 used to decrypt $(M_1)_{e_{M_1}}$, then she multicasts $(d_2')_{GK}$ to all subscribers. On the other hand, if Bob updates d_3 used to decrypt $(M_2)_{e_{M_2}}$, then he multicasts $(d_3')_{GK_2}$ to all U_i^2.

6   Performance Analysis

In this section, we evaluate our scheme by comparing it with some state-of-the-art schemes. Our experiments were implemented in Java language under IntelliJ IDEA compiler. Our project runs on a desktop, which is 64-bit Win 10 operating system with 8-GB RAM, the processor is Intel Core i7-7700 CPU @ 2.80 GHz.

We use the medical data set of diabetes in Ruijin Hospital as our experimental data set. This data set comes from the artificial intelligence competition held by Ruijin Hospital in 2018. In our experiments, we annotated and extracted entities in the data. Treat symptoms as patient's health data M_1. Treat disease name, drug name and instruction in drug use as diagnosis M_2.

6.1   Security Features

We establish our evaluation metrics to measure the functionality and security features of our scheme according to various schemes [8,11,12,16,17], which are all used to manage medical data. From Table 2, all schemes provide access control and privacy preserving to users and most of them allow users to share data with third parties. But only our scheme achieves medical data sharing from multiple owners. In a word, our scheme provides more functionality and security features than other existing schemes.

Table 2. Comparisons of security features.

Security feature	[8]	[11]	[12]	[16]	[17]	Proposed
Access control	✓	✓	✓	✓	✓	✓
Privacy Preserving	✓	✓	✓	✓	✓	✓
Distributed storage	✗	✓	✗	✗	✓	✓
Data cannot be tampered with	✗	✓	✓	✗	✓	✓
Ensure the authenticity of the data	✗	✓	✓	✗	✓	✓
Security of communication	✓	✗	✓	✓	✓	✓
Diagnosis	N/A	✓	N/A	✓	✓	✓
Data sharing	✓	N/A	✓	✓	N/A	✓
Data sharing from multiple owners	✗	✗	✗	✗	✗	✓
Management of data subscribers	✗	✗	✗	✓	✗	✓

N/A means not considered.
✓ means the scheme supports the functionality.
✗ means the scheme does not support the functionality.

6.2 Master Key Encryption

RSA-based MKE schemes can be better applied to our system than asymmetric key encryption/decryption. For the plaintext of medical data set with an average length of about 5,000 bytes, we tested the average time taken for encryption and decryption of our scheme and the following schemes. Park et al.'s scheme [13] and Kung et al.'s scheme [18] use traditional MKE in which the number of slave keys will be changed when users join or leave. As shown in Fig. 5(a), the computation costs of them are similar and within the acceptable range of users.

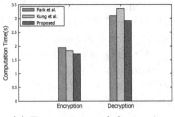
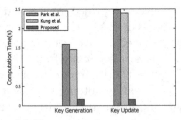

(a) Encryption and decryption   (b) Key generation and update

Fig. 5. Computation overhead comparison of master key encryption.

In addition, one of the weakness of MKE is the huge cost when the number of slave keys needs to be changed. However, in our scheme, there are fixed two slave keys whether users join or leave. Thus, the operation of key update has a similar time cost to the operation of key generation, both of them have lower computation overhead than traditional MKE-based schemes, see Fig. 5(b).

6.3   Blockchain-Based System

First, referring to the design of Bitcoin [19], we designed the structures of block and the key parameters' length in the block header is shown in Table 3. In addition, it also shows the key parameters' length in transactions stored in the block body. Thus, the size of $Tx(StoreM_1)$, $Tx(Diag)$ and $Tx(StoreM_2)$ are 164, 100 and 164 bytes. When sharing data with a subscriber, the size of $Tx(Share_1)$ and $Tx(Share_2)$ are 100 and 132 bytes. After considering the structure of merkle tree and so on, we can conclude that a block of 1M bytes can contain about 7800 transactions. We know that a complete diagnosis process consists of three transactions. Assuming a block is generated every minute, the throughput can reach 60 complete diagnoses per second.

Table 3. The key parameters in the block header and transactions.

Block header	Parameters	PreHash	MerkleRoot	Index	Timestamp	Nonce
	Length(Bytes)	32	32	32	4	4
Transactions	Parameters	ID	hash	σ	ts	$DatasetId$
	Length(Bytes)	32	32	32	4	32

Then we generated our own private blockchain, creating a unique merkle hash root and a genesis block on a testnet. Because of the real-time needs of the medical system, we measured the average time of several important processes, as shown in Table 4. T_D represents the time from when a user generates $Tx(Diag)$ to when the doctor reads the health data. T_R indicates the time from when the doctor generates $Tx(StoreM_2)$ to when the user reads the diagnosis. T_{S_1} and T_{S_2} represents the time from when data owners generate the sharing transaction $Tx(StoreM_1)$ or $Tx(StoreM_2)$ to when the subscriber reads the medical data. The major time-consuming operations include verifying signatures, packaging transactions, downloading and decrypting data. Obviously, our system can totally meet users' real-time needs for all functions.

Table 4. The time cost (min) of several important processes.

Process	T_D	T_R	T_{S_1}	T_{S_2}
Time	0.82	0.73	1.39	1.45

6.4   Overhead of Sharing

There are two kinds of subscribers and different sharing and management methods for them in our scheme. Thus, the overhead of sharing with n_1 subscribers of U_i^1 and n_2 subscribers of U_i^2 needs to be described separately. We assume that $n = n_1 = n_2$, the results of the experiment are shown in Fig. 6.

In sharing, the main computation overhead comes from t_{Enc}, t_{Dec}, $t_{Rekeying}$ and t_{KEY}, where t_{Enc} and t_{Dec} are the computation time for an AES encryption and decryption for a 32-byte block of data which take 0.133 ms and 0.137 ms respectively, $t_{Rekeying}$ is the computation time for updating the master key which takes 0.157 ms, and t_{KEY} is the computation time for computing the temporary traffic encryption key which takes 13.47 ms.

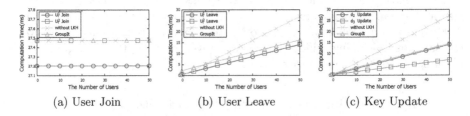

(a) User Join   (b) User Leave   (c) Key Update

Fig. 6. Computation overhead comparision of sharing.

Then we compare the update overhead of our scheme with traditional schemes which are not based on LKH and a related group key management scheme (GroupIt) [18] which is based on traditional LKH. When a user of U_i^2 joins, the patient needs to establish a temporary traffic encryption key with the doctor, then they need to do three symmetric encryption operations in total to send messages to users. The computational overhead can be represented as $2t_{KEY}+3t_{Enc}+3t_{Dec}$, which is the same as the traditional scheme and GroupIt scheme.

When a user leaves or a data owner updates the secret key, the owner needs to update secret key and encrypt it by the shared key with every user in traditional schemes. Thus, the computational overhead can be represented as $t_{Rekeying}+(n_1+n_2)\times t_{Enc}+(n_1+n_2)\times t_{Dec}$, where the cost of encryption is proportional to the number of users. In addition, our scheme further reduces the number of encryption times compared with the traditional LKH scheme.

In summary, our proposed scheme expends less computation cost than those of other existing schemes.

7   Conclusion

In this paper, we propose a blockchain-based healthcare system scheme with distributed storage. We put forward a MKE-based encryption scheme for medical data and a LKH-based management scheme for subscribers, which achieves flexible data sharing from multiple owners. Performance evaluation shows that our scheme is efficient and feasible in practice. There are two owners of medical data in our work, but it's easy to extend our scheme to more owners. In fact, it could be further applied to other fields besides medical data with multiple owners. In the future, we plan to expand our sharing and management schemes to improve the system's efficiency and to obtain empirical data for further studies.

References

1. Vinoth, R., Deborah, L.J.: Secure multi-factor authenticated key agreement scheme for industrial IoT. IEEE Internet Things J. **8**, 3801–3811 (2020)
2. Aitzhan, N.Z., Svetinovic, D.: Security and privacy in decentralized energy trading through multi-signatures, blockchain and anonymous messaging streams. IEEE Trans. Depend. Sec. Comput. **15**(5), 840–852 (2018)
3. Padhya M, Jinwala D.C.: R-OO-KASE: revocable online/offline key aggregate searchable encryption. Data Sci. Eng. **5**, 391–418 (2020)
4. Li, M., Yu, S., Zheng, Y., Ren, K., Lou, W.: Scalable and secure sharing of personal health records in cloud computing using attribute-based encryption. IEEE Trans. Parallel Distrib. Syst. **24**, 131–143 (2013)
5. Phuong, T., Rui, N., Xin, C., Wu, H.: Puncturable attribute-based encryption for secure data delivery in Internet of Things. In: IEEE INFOCOM 2018 - IEEE Conference on Computer Communications (2018)
6. Song, X., Li, B., Dai, T., et al.: A trust management-based route planning scheme in LBS network. In: ADMA (2022)
7. Zhang, Y., Zheng, D., Deng, R.H.: Security and privacy in smart health: efficient policy-hiding attribute-based access control. IEEE Internet Things J. **5**, 2130–2145 (2018)
8. Hui, M., Rui, Z., Guomin, Y., Zishuai, S., Kai, H., Yuting, X.: Efficient fine-grained data sharing mechanism for electronic medical record systems with mobile devices. IEEE Trans. Depend. Sec. Comput. **17**, 1026–1038 (2018)
9. Ivan, D.: Moving toward a blockchain-based method for the secure storage of patient records (2016). http://www.healthit.gov/sites/default/files/9-16-drew-ivan_20160804_blockchain_for_healthcare_final.pdf
10. Zhang, C., Li, Q., Chen, Z., Li, Z., Zhang, Z.: Medical chain: alliance medical blockchain system. Acta Automatica Sinica **45**(8), 1495–1510 (2019)
11. Xu, J., Xue, K., Li, S., Tian, H., Yu, N.: HealthChain: a blockchain-based privacy preserving scheme for large-scale health data. IEEE Internet Things J. **6**(5), 8770–8781 (2019)
12. Guo, H., Zhang, Z., Xu, J., An, N., Lan, X.: Accountable proxy re-encryption for secure data sharing. IEEE Trans. Depend. Sec. Comput. **18**(1), 145–159 (2021)
13. Park, M.H., Park, Y.H., Jeong, H.Y., Seo, S.W.: Key management for multiple multicast groups in wireless networks. IEEE Trans. Mob. Comput. **12**(9), 1712–1723 (2013)
14. Agee, R., Wallner, D., Harder, E.: Key management for multicast: issues and architectures. RFC (1998)
15. Vinoth, R., Deborah, L.J., Vijayakumar, P., Kumar, N.: Secure multi-factor authenticated key agreement scheme for industrial IoT. IEEE Internet Things J. **8**, 3801–3811 (2020)
16. Yang, Y., Zheng, X., Liu, X., Zhong, S., Chang, V.: Cross-domain dynamic anonymous authenticated group key management with symptom-matching for e-health social system. Future Gener. Comput. Syst. **84**, S0167739X1730554X (2017)
17. Zhang, A., Lin, X.: Towards secure and privacy-preserving data sharing in e-health systems via consortium blockchain. J. Med. Syst. **42**(8), 140 (2018)
18. Kung, Y.H., Hsiao, H.C.: GroupIt: lightweight group key management for dynamic IoT environments. IEEE Internet Things J. **5**, 5155–5165 (2019)
19. Nakamoto, S.: Bitcoin: a peer-to-peer electronic cash system (2008)

LAP-BFT: Lightweight Asynchronous Provable Byzantine Fault-Tolerant Consensus Mechanism for UAV Network Trusted Systems

Lingjun Kong, Bing Chen$^{(\boxtimes)}$, and Feng Hu

The College of Computer Science and Technology, Nanjing University of Aeronautics and Astronautics, Nanjing 211106, China
{Lingjun,cb_china,huf}@nuaa.edu.cn

Abstract. An UAV network performing missions in complex environments is an Ad hoc network of a set of lightweight nodes that is not only exposed to external physical interference and network attacks, but also to the problem of dynamically generated error nodes within the network. The UAV network is essentially an asynchronous Byzantine distributed system. Completing the mission relies on the trustworthiness of the participating UAVs. The timely identification and isolation of errant nodes is necessary to ensure the overall performance of the UAV network during the mission. Assessing the latest status of UAVs and reaching consensus across the network is the key to solving the trust problem. It is a major challenge to break through the limitation of insufficient resources of UAV networks to achieve efficient consensus and accurate evaluation of UAV trusted status. The approach proposed in this paper is lightweight and asynchronous provable Byzantine fault-tolerant consensus algorithm that achieves global trusted state evaluation by obtaining an asynchronous generic subset by consensus on the local status data of nodes. It effectively reduces the communication and computational overhead. Through QualNet UAV network simulation experiments, comparing existing asynchronous consensus algorithms with better practicality, the recommended lightweight asynchronous provable consensus algorithm has better performance in terms of consensus latency and energy consumption rate.

Keywords: UAV networks · Byzantine distributed systems · Node status detection · Lightweight asynchronous provable consensus

1 Introduction

UAV networks are a special kind of mobile Ad hoc networks that are important for tasks such as search and rescue, exploration and military purposes due to their quick networking and ease of deployment. However, the complex mission environment exposes the UAV network not only to network partitioning caused by physical interference, but also to the risk of malicious network attacks by external nodes. At the same time, the mission process can turn legitimate UAV nodes into faulty or selfish nodes due to external interference and energy consumption; the open nature of wireless networks also makes UAV nodes more vulnerable to network attacks (e.g., link layer attacks) and compromise them to become Byzantine nodes. With no central support, dynamically generating error nodes, and arbitrary behavior of Byzantine nodes, the UAV network in the mission process is essentially an asynchronous Byzantine system. Messages passed between nodes may be dropped, delayed, or even tampered with. It is therefore necessary to build highly trusted distributed Byzantine fault-tolerant systems for UAV networks, so that they can guarantee robust, highly trusted UAV networks with constrained resources in unfriendly mission environments. Real-time sensing of node status changes, accurately identifying untrustworthy nodes and isolating them from the mission network in time to efficiently reach a consensus on the latest status of all participating nodes is the key. The asynchronous nature and resource constraints of UAV networks in Byzantine environments are a major challenge to achieve lightweight and efficient consensus. The consensus protocol for UAV networks explored in this paper is a BFT protocol that not only needs to resist malicious attacks from external nodes, but also faces interference from erroneous nodes that are dynamically generated internally.

Byzantine Fault Tolerance protocol (BFT) achieves deterministic consensus through message interaction between all nodes in an authentication environment where there is no more than $1/3$ of the system nodes, but with a communication complexity of $O(n^3)$. PBFT [4] is a simple and efficient Byzantine consensus scheme proposed by Castro and Liskov, and it is the first state machine that can operate correctly in an asynchronous, Byzantine error-ridden scenario. However, PBFT is a leader-based Byzantine protocol, which can only guarantee system activity in a weakly synchronous network environment, while in a fully asynchronous network environment was demonstrated by [9] that PBFT is vulnerable to a class of delayed attacks against the leader, rendering the system inoperable.

More research on asynchronous consensus is unfolding in different applications [6,7,11,12] to address the respective problems, but there are still problems of inefficient protocols, high communication complexity (up to $O(n^2)$ or even $O(n^3)$), and high computational overhead, making the performance of these protocols drop dramatically when the system scales up, and thus difficult to enter the practical usable stage. Miller et al. constructed an efficient Byzantine protocol, HoneyBadger BFT(HB-BFT) [9], in a fully asynchronous environment, based on the MVBA framework which uses the efficient RBC protocol from the literature [1] for the broadcast of proposed values and innovatively combines the idea of "apportionment" with an asynchronous common subset (ACS) protocol [2,10] to reduce transmission costs [1,3,9]. However, HB-BFT has the following drawback: while the expected number of "rounds" of each ABA protocol has a constant number of "rounds", the expected number of rounds to run n concurrent ABA sessions can be huge. More than that, these ABA instances are not executed in a fully concurrent manner. When n becomes large and the network is unstable, this leads to a difficult determination of the ACS runtime. The implementation of [8] confirms that the

ABA protocol has a greater practical impact on system performance. The nodes of the UAV network are too lightweight to support the parallel computing of multiple ABA instances. [8] employing preferential agents, provably reliable broadcast protocols, etc. effectively reduce the number of ABA instances and speed up the efficiency of asynchronous Byzantine consensus. Dumbo-BFT [5] enhances the improvement of the communication model based on [8] by proposing an optimized multi-valued verified Byzantine asynchronous consensus algorithm. Asynchronous consensus algorithms in [5,8,9] based on the optimization and improvement of the ACS protocol are practical in Byzantine asynchronous environments, but the scenarios in which they are applied do not consider the limitation of computing power and bandwidth, the network topology are relatively stable, and the consensus data are third-party customer transactions, independent of the nodes involved in the consensus. In contrast, the nodes of the UAV network have limited computing power, the topology changes dynamically, and the consensus data is a time-sensitive assessment of the nodes' trustworthiness status. They all use randomly selected sets of transactions, threshold signatures, encryption, and decryption through RS code coding techniques, which will reduce the throughput of the UAV network and increase the overall computational workload. Inspired by these asynchronous consensus algorithms, the UAV network in the authentication environment implements MVBA with external proving by means of smart contracts provided by the blockchain system to complete the generation of ACS, avoiding threshold encryption and RS-Code operations. The dispersed packets are split according to the size of the delegate agents and distributed in order of the list of agent nodes to enable lightweight reliable broadcasting (LD-RBC) within the delegate agents. Smart contracts validate the legitimacy of nodes' trusted status datasets (PMVBA).

Our contributions can be summarized as follow:

– Introduce a node trusted state detector, which sets the node's trusted status triple, including local reputation assessment of neighboring nodes, the node's own residual energy and the number of neighboring nodes. The routing protocol adds a detection function where neighboring nodes monitor each other's forwarding behavior, assess the local reputation of neighboring nodes in real time and collect their own running status. The latest reputation status data of the current node is used as the consensus object, from which a common subset is extracted as the consensus result to create a new block of the node's global reputation status. As the basis for the subsequent trusted running of the UAV network, the new block is synchronized to all mission nodes to update the blockchain and isolate untrustworthy nodes from the mission network in time to maintain the trustworthiness of the UAV network during the mission process.

– Adopt dynamic selection of multiple delegate agents to complete consensus, balance network resource consumption and improve consensus fault tolerance. Based on the latest nodes' trustworthy status information, the system periodically updates the set of delegate agent nodes for each consensus round by comparing the reputation value, remaining energy and the number of neighboring nodes. Dynamically electing different delegated authority members to undertake each round of consensus operations disperses the computational consumption and extends the effective working time of the mission system. This improves the overall resource usage efficiency of the UAV network and effectively reduces the probability of erroneous nodes becoming delegated agents. More importantly, the asynchronous consensus algorithm remains active when more than one-third of the total number of erroneous nodes exist in the network.

- Propose a lightweight asynchronous provable Byzantine fault-tolerant consensus algorithm (LAP-BFT) that adapts to the resource constraint of UAV networks. The algorithm sorts the trusted status data records in order of node registration, and disperses the local trusted status data set based on the size of the delegated agents. The algorithm achieves a lightweight and reliable broadcast of the dispersed data among the delegated agents, and efficiently completes a multi-valued Byzantine consensus protocol by means of external proving in the form of a smart contract with a computational complexity of O(1), and finally constructs an asynchronous common subset of locally trusted status data. LAP-BFT not only reduces the bandwidth requirements of individual nodes, but also avoids the high computational power consumption caused by threshold key operations, while ensuring the activity and consistency of consensus. This effectively reduces communication complexity, improves throughput and reduces consensus latency.

The organization of the content charpters of this thesis: The paper is outlined as follows. The system model and a specific description of the recommended scheme (LAP-BFT) is in Sect. 2. The proof and analysis of the nature of the system is presented in Sect. 3. Section 4 observes the consensus latency, throughput and energy consumption rate. Section 5 is conclusion.

2   System Model and Recommended Solution

2.1   System Model

The mission-oriented UAV network consists of multiple (N) lightweight UAVs, denoted as $\{U^i\}_{i\in[N]}$, [N] is the set of integers $\{1, 2, ..., N\}$,and can be considered as a P2P overlay network based on a mobile Ad hoc network. The trusted system of the UAV network is a permission blockchain system. The identity vector commitment and the registered node base information of the UAV network are stored in the Genesis block and synchronised to all The identity vector commitment and other information are stored in the genesis block. The security environment is set up with an elliptical public key cryptosystem, where the registration server allocates public and private keys for UAVs, and generates a unique identity for UAVs by hashing the IP and public key of UAVs and mapping them to a point of the elliptical curve (finite exchange group G). The system constructs UAV identity vector commitment and provides a witness of existence for all registered UAVs. A set of UAV nodes are selected as delegate agents responsible for the first round of consensus at the start of the mission. The UAV network runs in an unfriendly Byzantine problematic environment and in most cases exhibits asynchrony. Its trustworthiness depends on the current trusted status of the UAV nodes, including the global reputation assessment, the residual energy and the number of neighboring nodes. As the mission progresses, the UAV trusted status keeps changing. A trusted system for the UAV network blockchain based on a lightweight asynchronous provable Byzantine fault-tolerant consensus mechanism that achieves consensus on the local status of nodes and counts the global trusted status through a dynamically elected group of delegated agent nodes to ensure high trustworthiness of the UAV network during the mission.

2.2   Threat Model

In an unfriendly mission environment, external malicious nodes can not only use their powerful performance to perform replay attacks and DOS attacks, but also exploit the openness of the wireless network to implement intrusions, such as link layer attacks. As a special kind of mobile Ad hoc network, the UAV network is also characterized by multi-hop forwarding and dynamic routing, so dynamically generated Byzantine nodes during the mission cannot prevent the delivery of messages between trusted nodes, but can discard messages sent by the correct node. Sending inconsistent messages to different nodes or deliberately delaying the delivery of messages, as well as tampering with the content of forwarded messages, changing the content of the nodes' own forwarded messages, changing the status of the nodes themselves, etc. Byzantine nodes can even collude with each other to increase the trustworthiness of each other's status or maliciously sing the praises of the correct node. The behavior of selfish nodes does not lead to malicious attacks, but can consume network resources. Too many incorrect nodes in the system will not only seriously affect the overall performance of the UAV network, but will also cause the system to crash due to the inability of the consensus algorithm to terminate.

2.3   Blockchain Structure

The registration of the UAV's identity, the deployment of the blockchain program and the synchronization of the genesis block are UAV in a secure environment. The genesis block contains basic information about the registered UAVs, a vector commitment of the UAVs' identities, and a set of delegated agents for the first round of authorization to perform consensus operations. The smart contracts required by the system are also deployed in the genesis block, including a local reputation evaluation smart contract, a node global state statistics smart contract, and an authentication smart contract based on vector commitments of identities. The block structure and blockchain form is shown in Fig. 1.

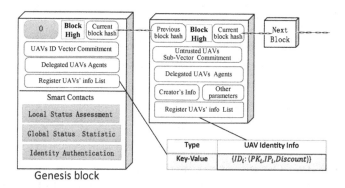

Fig. 1. UAV network blockchain structure

2.4   Lightweight Asynchronous Provable Byzantine Fault-Tolerant Consensus Mechanism

The lightweight asynchronous provable Byzantine fault-tolerant consensus mechanism (LPA-BFT) recommended in this paper operates as a blockchain in a UAV network trusted system, called a node status detector. All nodes monitor the data forwarding behavior of their neighboring nodes in real time to assess local trusted state values; delegate agent nodes to collect local state data for global reputation statistics and Byzantine fault-tolerant consensus, including dispersed reliable transmission, provable multi-value Byzantine consensus, global trusted state assessment, and synchronization of the blockchain after building new blocks. A node state detector continuously updates the global reputation assessment, providing the basis for the trusted running of the UAV network. After a brief explanation of the UAV network trust system setup, this sections describes the details of the UAV network trusted blockchain system consensus algorithm in turn. The LAP-BFT consensus process is shown in Fig. 2.

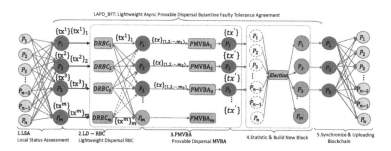

Fig. 2. A round of consensus process.

1) Collection and evaluation of local trusted status values of nodes:
All mission nodes implement local reputation assessment by monitoring the forwarding behavior of their neighboring nodes and collecting their latest operational status, mainly the remaining energy of the node and the number of

Table 1. Classification of data forwarding behavior and penalty rules

Data forwarding behavior	Reputation discount
Normal forwarding	0
Delayed forwarding	−1
Sending data, not forwarding data	−2
Forwarding of incorrect data	−3

neighboring nodes. The assessment method is a reputation discount for bad behavior. Initially, all nodes are trusted with an initial reputation of 10.0, which is recorded in the genesis block. The final global reputation assessment is determined to be an untrustworthy node when it is below equal to 0. UAV neighbor nodes are classified into four types of forwarding behavior, normal forwarding, delayed forwarding, sending data but not forwarding data and forwarding incorrect data, for which a reputation discount penalty is applied, and the discount scoring rules are shown in Table 1. for the non-responsive behavior of faulty nodes, and the lying behavior of colluding nodes, local evaluation is difficult to screen, but can be done by delegating agents in a round of local state data consensus, after global reputation status statistics to detect problems. Suppose ID_x is a neighbour node of ID_i, $CurScore_i^x$ denotes the reputational loss score for the current round, and $Discount_{i-x}^k$ is the discount valuation of ID_i for the k-th forwarding behavior of ID_x. k indicates the number of times the forwarding behavior of the neighbor node is detected in the current round. Due to the nature of wireless networks, there are receiving and sending conflicts, and there is a possibility of receiving the forwarded data not returned by the neighboring nodes and generating misjudgment of the neighboring nodes' discarding the forwarded data behavior. Treating the local reputation assessment in the cycle as an average value can weaken the impact of misjudgment on the assessment, and at the same time only doing reputation discount penalty according to the nature of the behavior, which also effectively avoids the high score assessment of malicious nodes colluding with each other. the reputation value in the local trusted state of neighboring nodes at the end of a round:

$$CurScore_i^x = \frac{\sum_k^{nCount} Discount_{i-x}^k}{nCount} \tag{1}$$

where nCount is the number of times ID_i detected ID_x forwarding behavior in the current round of detection. The local reputation evaluation of all neighboring nodes is denoted $\{CurScore_i^x\}_{x\in[Nei]}$ neighboring nodes set. The node obtains the current energy value $Energy_i$, constructs a local reputation state record for the current round of ID_i:

$$LSA_i^t = \{ID_i||Energy_i||Neighbors_i||curBlockHigh||\{CurScore_i^x\}_{x\in[Nei]}Sign_{SK_i}\}, \tag{2}$$

where W_i is existential witness, denoted as ID_i exists in the identity vector commitment.t, and the authenticity of the data source is verified by the identity authentication smart contract in the genesis block; $Sign_{SK_i}$ is the signature of ID_i on the local state data used to ensure the integrity of the local trusted data; curBlockhigh corresponds to the height of the blockchain is used to prevent replay attacks.

2) Global Trusted State Assessment

The current global trusted status assessment of all nodes is done by a group of delegated agents authorized by the system. During mission execution, honest UAV nodes send the local state data collected in this round to all delegated agents in a multicast manner in each round of the blockchain. The delegated agents receive the same node's local status data consistently, but there is no guarantee that all agents will receive exactly the same record of the nodes' local status data. The complex mission environment, interference from errant nodes in the asynchronous UAV network, and the dynamic nature of the UAV network topology may result in different delegated agents collecting different nodes and different numbers of data records. To obtain local status data that is consistent across all nodes, consensus must be reached among the delegate agents. Due to the energy consumption of the delegated agent nodes and the risk of becoming the errant node, after each round of consensus, the nodes with the best status are selected from among the trusted UAV nodes to reconstitute the delegated agents. The delegate agents are responsible for collecting local trust status data from all nodes and generating consensus results between honest delegate agents, i.e. an asynchronous common subset of the local trust status dataset, and finally calculating the global trust status values for all nodes to create new blocks and update the blockchain. The purpose of this design is to avoid single point risks, to balance the consumption of network resources, to minimize the probability of erroneous nodes becoming delegate agents and, more importantly, to enable honest delegate agent nodes to have deterministic and consistent outputs for all collected local trusted state data. The process of asynchronous consensus consists of three major steps, lightweight dispersed reliable transmission, provable multi-valued Byzantine consensus and new block creation and blockchain updates. The Lightweight Dispersed Reliable Broadcast subprotocol (LD-RBC): The purpose of the Reliable Broadcast subprotocol (RBC) is to reliably transmit the proposed values proposed by each node to other nodes in the system. The proposed LD-RBC scheme builds on the traditional RBC protocol of Bracha [5], dropping RS code scheme of [5,8,9], with identity vector commitment authentication and data integrity verification instead of threshold encryption. Instead of randomly selecting encrypted transactions from a pool of transactions, the collected trusted state records are segmented according to the order of the delegated agent list and the length of the list to form dispersed packets of consistent length, reducing computation and improving transmission efficiency. First, verify the smart contract and hash function for legitimate validation, including identity authentication and data hash validation, based on the identity commitments of all registered UAVs in the node's local creation block, with a validation cost of $O(1)$. Secondly, check if there are duplicate data, as only the data collected in the current round is stored locally, it is easy to determine if it is the required data set for the agent based on the block height in the submitted data, and discard it if it is duplicate data to avoid replay attacks. If the block height in the received packet does not match the delegated agent block height and is higher than the agent block height, simply discard the process; if it is lower than the agent block height, perform blockchain synchronization on the sending node. After confirming that the received status data is valid and legitimate, it is saved to the local database. First, the size of the dispersal packet is calculated based on the number of agent nodes, and second, the local trusted status dispersal packet is obtained by the position ordinal number of ID_d in the delegated agents list, finally, the agent node ID_d sends its own dispersed packet to the other agent nodes according to RBC protocol.

Provable Multi-Value Byzantine Consistent Sub-Agreement (PMVBA): After each agent node receives a dispersed packet from another agent, it verifies the identity of the agent node sending the packet and validates each record in the dispersed packet as legitimate through a smart contract in the Genesis block. First, the sender's identity is verified, and then the received data records are verified with records from the recipient's local corresponding region. If there are local state data records that are not identical, the sender is deemed to have tampered with the data and the data submitted by this agent node is rejected, otherwise the broadcast continues after adding its own signature. When the number of additional signatures is, for example, greater than two-thirds of the total number of delegated agents, $|M| * 2/3$, it indicates that a deterministic consensus result is obtained.

Assessing the global trusted state and updating the blockchain, At the end of a consensus round, for all records in the asynchronous common subset, the reputation discount expectation and variance of a node is calculated by counting all local reputation discount assessments obtained by that node. Combined with the existing reputation values in the blockchain, the current global reputation of the node is calculated and global discounts are enforced for lying and faulty nodes, resulting in a trusted status record for each node:

$$GStatus = \{GReputation_x^{curhigh}||Energy_x||Neighbors_x\}. \tag{3}$$

Table 2. The rules for global reputation discounts

Reasons for reputation discount	Global discount score
Excessive local discount deviation variance ($Discount_\sigma^x > 5\sigma^x$)	−0.5
Faulty node suyspected (no record) $Discount_{faulty}^x$	−1

The statistics start with $Discount_\sigma^x$, $Discount_{faulty}^x$ initialised to 0. The global reputation assessment is as in Eq. (1–2).

$$GDiscount^x = \left(\sum_{i=0}^{n} CurScore_i^x\right)/n \tag{4}$$

$$\sigma^x = \sqrt{\sum_{i=0}^{n} (CurScore_i^x - GDiscount^x)^2 / n} \qquad (5)$$

where $i, x \in [N]$, Then the total reputation loss and global reputation assessment of UAV node ID_x after this round of consensus are (3) and (4), respectively.

$$GDiscount^x = GDiscount^x + Discount_\sigma^x + Discount_{faulty}^x \qquad (6)$$

$$GReputation_x^{curhigh} = GReputation_x^{curhigh-1} + GDiscount^x \qquad (7)$$

where curhigh is the height of the current blockchain. The statistics form the set of global trusted status triples for all registered nodes: $\{GStatus\}_{x \in [N]}$, shaped as (5):

$$GStatus = \{GReputation_x^{curhigh} || Energy_x || Neighbors_x\} \qquad (8)$$

Based on the global status dataset, a new set of delegated agents is selected. The witness W_x of nodes with $GReputation_x^{curhigh}$ less than or equal to 0 is aggregated to the identity subvector commitment of the untrustworthy node. Based on the statistical results, new blocks were created to provide the basis for the next round of UAV network trustworthiness. Statistical analysis of consensus results, asynchronous common subsets of local trusted status data, construction of new blocks, and synchronization of blockchains to provide a trusted foundation for the continued operation of the UAV network.

3   Proof of System Properties

The ultimate design aim of this solution is to achieve atomic broadcast of the latest state assessment of UAV nodes in an asynchronous UAV network, and to establish consistent network-wide block data of the global section trusted state to support continuous trusted running of the UAV network. LAP_BFT must satisfy the properties Agreement, Total Order, External-Validity, which represent security, and Termination, which represents activity.

3.1   Proof of Security

Theorem 1. The system satisfies the following Agreement property: At the end of a round of consensus protocol, if there exists an honest node that outputs a locally trusted set of state records $\{LStatusPacket_d\}_{|R|}$, then all honest nodes output $\{LStatusPacket_d\}_{|R|}$.

Proof: M delegated agents arrange the collected local trusted status data records (LSAs) into the dataset $(LSA)_N$, in the order of the nodes' identity vector commitments in the Genesis block. Disperse them into M subsets of records of size $B = N/M$. The delegated agent nodes take the corresponding B records in the order of their position in the current delegated agent list, e.g., if the position of the agent node ID_d requires Index, then $LStatusPacket_d = \{LSA\}_{index+B}$. After RBC communication, there are R sets of $LStatusPacket_d$ data records in R honest nodes, which are combined into a consistent data set $\{LStatusPacket_d\}_{|R|}$ in the order of Index size and proven.

Theorem 2. The system satisfies the following Total order property: If an honest node outputs a sequence of messages $\{v_0, v_1, ..., v_j\}$, another honest node outputs a sequence of messages

$$\{v_0', v_1', ..., v_j'\}, then v_0 = v_0', v_1 = v_1', ..., v_j = v_j'.$$

Proof: The local trusted state dataset collected by the delegated agent group is ordered based on the order in which the identity vector commitments were generated at the time of UAV registration. Honest nodes transmit dispersed packets via a reliable broadcast protocol, which are then verified by a provable multi-value agreement protocol and finally concatenated according to the order of the delegated agent nodes' positions in their lists, so that the order of the data in all honest nodes is consistent and proven.

Theorem 3. Corresponding to the resilience of censorship in atomic broadcasting, ACS requires verifiability of consensus results, and this scheme provides external validation. The system satisfies the following External-Validity property, such that if an honest node outputs a value v, then $Ext\_Verify(v) = true$, where Ext_Verify is external verification.

Proof: The UAV network runs in an authenticated environment with an authentication function stored in the Genesis block as a blockchain smart contract. This includes authentication of the data sender; node data integrity verification, i.e., the node performs a $Hash(curScore_i^x)$ process on the reputation assessment values of its local data against neighboring nodes, and during the consensus process, the agent nodes receiving the dispersed packets compare the corresponding hash values to determine whether the sender has tampered with them. The legitimacy of the data sent by the honest node is confirmed by the data legitimacy verification smart contract provided by the blockchain, which is proven.

3.2   Proof of Activity

Theorem 4. The system satisfies the following Termination property. Let f be the number of error nodes in the delegated agent nodes. If $f + 1$ activates the PMVBA protocol and all messages between honest nodes (trusted delegated agents) arrive, then the honest nodes output $\{LStatusPacket_d\}_{|R|}$, where $|R|$ is the number of honest nodes.

Proof: The PMVBA protocol is executed in all delegated agent nodes, and if at least $f + 1$ delegated agents are trusted honest nodes, then there are honest nodes that receive $|R|$ copies of the signed and confirmed scattered packets, where $|R| = f + 1$, and the final honest node outputs $\{LStatusPacket_d\}_{|R|}$. If $f < (N+1)/3$, the prime protocol terminates execution.

4   Simulation Experiments

In this scheme, the final asynchronous generic subset of ACS is generated by executing DRBC and PMVBA sub-protocols. Let the length of the data to be consented be |m| and the simulation experiments focus on the message complexity, communication complexity and computational complexity in the related algorithms during the consensus process. Comparing the latency and computational overhead of the asynchronous consensus algorithms in [5,8] in one round of consensus.

4.1   The Base Latency

A round of consensus latency is the average time interval from the first node starting the protocol to the $n - th$ node getting the result. The transactions are the nearest local status data of the nodes, and thus the more nodes the larger the transaction volume. The experimental design sets different network sizes and configures maximum number of error nodes allowed. The experimental results are shown in Fig. 3. It is obvious that the asynchronous consensus latency of HBBFT is much higher than other schemes, mainly because the ABA sub-protocol in the ACS protocol has multiple instances in each node, which not only consumes a large number of operations, but also increases the consensus latency. In contrast, [8] algorithm uses an agent approach to reduce the number of ABA instances, but because the consensus result of the final generated ACS is achieved by all participating nodes randomly selecting a set of transactions using the RS Code technique, and the threshold encryption and decryption operations are added to the consensus algorithm. LAP-BFT only runs on the selected agents and provides smart contracts through the blockchain for external proof to achieve the final ACS consensus result, which takes less time to run.

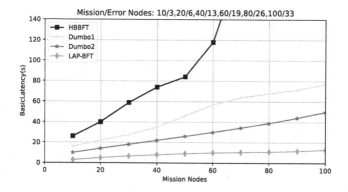

Fig. 3. Comparison of consensus base delays

4.2 Latency Under Different Error Nodes

Each round will generate a different number of error nodes, simulating the dynamic generation of error nodes. The UAV network in each round of consensus error nodes gradually increased, comparing the latency of the four consensus algorithms, as shown In Fig. 4, it can be seen that the consensus algorithms of HBFBT and Dumbo-HBT gradually increase the delay with the increase of error nodes, and the slope of the increase is also larger, when the error nodes exceed one-third of the total number of participating nodes, the terminability of the consensus cannot be satisfied and the infinite delay is delayed. LAP-BFT uses each round to dynamically select the optimal group of nodes based on reputation as the delegate agent for asynchronous consensus, which maximizes the possibility of Error nodes, especially Byzantine nodes, are excluded from the list of authorized delegate agents, and the probability of generating more than one-third of error nodes in the delegate agents is small. Therefore, even if there are erroneous nodes that exceed Byzantine fault tolerance, LAP-BFT asynchronous consensus can satisfy the terminable condition with little change in latency.

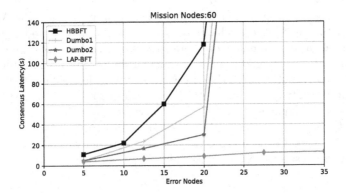

Fig. 4. Comparison of delay times with different error nodes

4.3 The Rate of Energy Consumption

The UAV network is energy supply is limited, and extending the runtime of the UAV network is also an important manifestation of lightweighting. The verification environment of a UAV network of 50 registered UAVs generates 14 erroneous nodes at some stage, 5 faulty nodes, which do not process forwarded data, 5 Byzantine nodes, which randomly tamper with forwarded data and 4 selfish nodes, which send data, but do not forward it. The UAVs move on a given path, without considering obstacle avoidance. The energy consumption corresponds to the running time and the experiments compare the consumption rates of different asynchronous consensus algorithms for a given (time) energy. This is shown in Fig. 5.

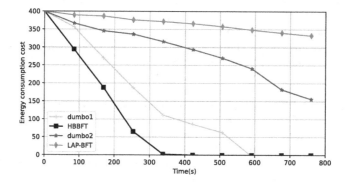

Fig. 5. Comparison of energy consumption rates

Dumbo1 uses fixed proxies and its computational complexity is related to the number of proxies, but because the consensus algorithm runs all the time, the proxy nodes are consumed quickly and the consensus process cannot continue when the proxy consumption ends. Dumbo2 uses more energy than LAP-BFT because the threshold key is still computed. This is because LAP-BFT uses only external verification to prove the consensus result, and more importantly, dynamically selects groups of delegated agents to share the consensus computation, extending the running time of the entire network.

5   Conclusion

The goal of this solution is to create an asynchronous fault-tolerant system to maintain the trustworthiness of the UAV network during mission execution. Through mutual monitoring between drone nodes during data transmission, nodes assess the behavior of their respective neighboring nodes and collect the latest current local trusted state. A lightweight asynchronous provable consensus is used to reach a network-wide agreement on the nodes' trusted state, providing a trusted environment for the next round of drone network operations. The transmission efficiency and computational overhead of three practical asynchronous consensus algorithms operating in a UAV network environment are compared. Smart contracts are used for many aspects of the scheme, such as authentication, proof of consensus results, etc. How to dynamically execute smart contracts in an asynchronous environment according to practical needs to be explored in depth.

References

1. Alhaddad, N., Duan, S., Varia, M., Zhang, H.: Practical and improved Byzantine reliable broadcast and asynchronous verifiable information dispersal from hash functions. IACR Cryptolology ePrint Archive, p. 171 (2022). https://eprint.iacr.org/2022/171

2. Ben-Or, M., Kelmer, B., Rabin, T.: Asynchronous secure computations with optimal resilience (extended abstract). In: Anderson, J.H., Peleg, D., Borowsky, E. (eds.) Proceedings of the Thirteenth Annual ACM Symposium on Principles of Distributed Computing, Los Angeles, California, USA, August 14–17, 1994. pp. 183–192. ACM (1994). https://doi.org/10.1145/197917.198088

3. Cachin, C., Tessaro, S.: Asynchronous verifiable information dispersal. In: Fraigniaud, P. (ed.) DISC 2005. LNCS, vol. 3724, pp. 503–504. Springer, Heidelberg (2005). https://doi.org/10.1007/11561927_42

4. Castro, M., Liskov, B.: Practical byzantine fault tolerance. In: Seltzer, M.I., Leach, P.J. (eds.) Proceedings of the Third USENIX Symposium on Operating Systems Design and Implementation (OSDI), New Orleans, Louisiana, USA, 22–25 February 1999, pp. 173–186. USENIX Association (1999), https://dl.acm.org/citation.cfm?id=296824

5. Guo, B., Lu, Z., Tang, Q., Xu, J., Zhang, Z.: Dumbo: faster asynchronous BFT protocols. IACR Cryptology ePrint Archive, p. 841 (2020). https://eprint.iacr.org/2020/841

6. Kashyap, R., Arora, K., Sharma, M., Aazam, A.: Security-aware GA based practical Byzantine fault tolerance for permissioned blockchain. In: 4th International Conference on Control, Robotics and Cybernetics, CRC 2019, Tokyo, Japan, 27–30 September 2019, pp. 162–168. IEEE (2019). https://doi.org/10.1109/CRC.2019.00041

7. Li, B., Liang, R., Zhu, D.: Blockchain-based trust management model for location privacy preserving in VANET. IEEE Trans. Intell. Transp. Syst. 22(6), 3765–3775 (2021)

8. Lu, Y., Lu, Z., Tang, Q., Wang, G.: Dumbo-MVBA: optimal multi-valued validated asynchronous Byzantine agreement, revisited. IACR Cryptology ePrint Archive, p. 842 (2020). https://eprint.iacr.org/2020/842

9. Miller, A., Xia, Y., Croman, K., Shi, E., Song, D.: The honey badger of BFT protocols. IACR Cryptology ePrint Archive, p. 199 (2016). https://eprint.iacr.org/2016/199

10. Rabin, M.O.: Efficient dispersal of information for security, load balancing, and fault tolerance. J. ACM 36(2), 335–348 (1989). https://doi.org/10.1145/62044.62050

11. Xu, X., Zhu, D., Yang, X., Wang, S., Qi, L., Dou, W.: Concurrent practical Byzantine fault tolerance for integration of blockchain and supply chain. ACM Trans. Internet Tech. 21(1), 7:1–7:17 (2021). https://doi.org/10.1145/3395331

12. Zhang, Z., Zhu, D., Fan, W.: QPBFT: practical byzantine fault tolerance consensus algorithm based on quantified-role. In: Wang, G., Ko, R.K.L., Bhuiyan, M.Z.A., Pan, Y. (eds.) 19th IEEE International Conference on Trust, Security and Privacy in Computing and Communications, TrustCom 2020, Guangzhou, China, 29 December 2020–1 January 2021, pp. 991–997. IEEE (2020). https://doi.org/10.1109/TrustCom50675.2020.00132

A Secure Order-Preserving Encryption Scheme Based on Encrypted Index

Haobin Chen, Ji Liang, and Xiaolin Qin[✉]

Nanjing University of Aeronautics and Astronautics, Nanjing, China
{chen604133140,qinxcs}@nuaa.edu.cn

Abstract. With the development of cloud computing, data owners usually choose to store data in the cloud. But cloud servers may not be fully trusted. Due to privacy concerns, data owners need to encrypt the data before storing it in the cloud. There is still a lack of solutions that satisfy both safety and efficiency. To perform secure queries on encrypted data, we propose a scheme. The scheme includes an encrypted index based on B^+ tree, which can hide the order information of sensitive data. It also includes a secure search algorithm based on the encrypted index, which is implemented through the interaction protocol between the client and the server. We analyze the scheme's security and prove that the scheme can protect the privacy of data, queries, results, and access patterns. At last, we perform experiments on five datasets to verify that the scheme is effective.

Keywords: Encrypted database · Encrypted index · Secure query · Cloud database · Privacy protection

1 Introduction

With the rise of cloud computing, driven by lower cost, higher reliability, better performance, and faster deployment, many individuals and organizations store data on the cloud such as Microsoft Azure, Amazon EC2, and Google AppEngine. However, the data owner loses absolute control over the data in this way. The cloud service provider can access the original data directly without supervision, which will lead to the leakage of privacy. On the one hand, the data owners need to take advantage of storage capabilities provided by the cloud. On the other hand, the leakage leads to data owners not being able to fully trust the cloud.

To protect the privacy of data, a straightforward approach is to encrypt the databases before storing them on the cloud [7]. But it is difficult to query encrypted data. A simple solution is to return all the ciphertext to the client. The client decrypts the data and performs queries on plaintexts. However, the storage overhead, computation overhead, and communication overhead required by this method are unacceptable.

B. Li et al. (Eds.): APWeb-WAIM 2022, LNCS 13423, pp. 247–261, 2023.
https://doi.org/10.1007/978-3-031-25201-3_19

Many researchers proposed schemes for this problem. These schemes can be divided into three categories. The first is to operate directly on the encrypted data without decrypting it. Many existing schemes focus on this kind, including homomorphic encryption [6,23,26] and order-preserving encryption [2,3,8,16,22,24,27,31]. The second scheme is the oblivious random access memory (ORAM) [21,30]. The third scheme filters unnecessary data using the encrypted index [7,13–15,17,29]. However, these works have weak privacy protection, such as leaking the order information and access patterns of the data.

To achieve secure queries, we propose a scheme based on an encrypted index that can protect the privacy of data, queries, results, and access patterns. First, we design an encrypted index based on the B^+ tree, by introducing an obfuscation mechanism to hide the order information in the index. We further propose a secure search algorithm to protect the data access pattern. The main contributions of our scheme are summarized as follows.

- We design an encrypted index OHBPTree (order hidden B^+ tree) based on the B^+ tree, which can hide the order information of sensitive data.
- To protect the security, we propose a secure search algorithm based on our encrypted index, which is executed through client-server interaction. The algorithm does not reveal privacy to the cloud.
- We analyze the security of the scheme and perform experiments on real datasets to evaluate the efficiency of our scheme.

The rest of the paper is organized as follows. We give an overview of related work in Sect. 2. In Sect. 3, we present our scheme, including the encrypted index and the secure search algorithm. In Sect. 4, we prove that our scheme is secure. We discuss the performance of the proposed scheme based on experiments in Sect. 5. We conclude the paper in Sect. 6.

2   Related Work

2.1   Encrypted Database

The CSAIL lab at MIT [25] proposed the first encrypted database CryptDB and the Onion model. CryptDB improves the efficiency by pre-designing Onion models for different operations (e.g. query, sorting). The encrypted database system named MONOMI was proposed at [28]. MONOMI has introduced four new technologies to speed up the querying of encrypted data. In addition, many encrypted database systems had been proposed, such as KafaDB [32], Crypt-JDBC [4]. Some database vendors added similar features to their products, such as Always Encrypted [1].

2.2   Operation on Encrypted Data

Rivest et al. [26] proposed the concept of homomorphic encryption. It can support directly calculating encrypted data without decryption. ElGamal [6] proposed a multiplicative homomorphic encryption algorithm based on Differ-Huffman key exchange. This algorithm is based on the intractability of discrete logarithms in finite fields. Pailler [23] proposed an additively homomorphic encryption scheme based on composite factorial modulus. However, partially homomorphic encryption algorithms only support a limited class of computations. Although recent research on fully homomorphic encryption shows it is possible to perform all types of computations over encrypted data [9], the performance overhead is prohibitively high, about 10^9 times that of AES [10].

Agrawal et al. [22] first proposed the concept of order-preserving encryption and designed the bucket partition scheme. They divided the data into different buckets, flattened the buckets into a uniform distribution, then mapped to a set of uniformly distributed ciphertexts. Bolydyreva et al. [2] proposed the first provably secure order-preserving encryption scheme BCLO. This scheme is based on hypergeometric distribution sampling. Popa et al. [24] proposed the mutable order-preserving encryption scheme mOPE. This scheme maintained a binary tree in the server through the interaction between the client and the server to generate ciphertext for the data. Liu et al. [18,19] proposed a scheme to achieve order-preserving encryption by combining linear functions and noise. Although the ciphertext obtained by order-preserving encryption can be directly used for comparison, it will leak order information and be vulnerable to inference attacks [12,20].

2.3   Encrypted Index

This type of scheme filters unnecessary data using the encrypted index. Hacigumus et al. [13] proposed a bucket index-based scheme, which divides the index data into different buckets. When querying, first find the bucket that meets the conditions, return all the ciphertext data in the bucket, and the client decrypts data and filters. Hore et al. [14] proposed an optimized division method. However, this scheme requires the data owner to store indices locally, rather than in the cloud. Kerschbaum et al. [17] proposed a scheme based on an array, which retains data order information through the array. But their solution leaks access patterns. A range coverage technique was proposed in [29]. It can convert range search to keyword search by constructing a full binary tree over attribute's values bottom-up and selecting a set of nodes whose subtrees cover the given range entirely. However, the scheme of indexing data inversely by range has a higher demand for storage capacity, while leaks access patterns.

3   Secure Query Scheme

3.1   Scheme Overview

Adversary Model. There mainly exist two types of adversaries: semi-honest and malicious [5]. In the semi-honest model, each participant explicitly and correctly implements the secure protocol specification, but the cloud server intends to obtain the additional information of intermediate results and uses them to analyze the messages. In the malicious model, the adversary can violate the protocol specification arbitrarily, but in practice, it is inefficient to be employed. In this paper, we adopt the semi-honest adversary model because it is a widely used model. We assume that the cloud server is not fully trusted and the client is trusted.

Privacy Specification. We propose a secure query scheme over encrypted data, which attempts to protect the privacy of data, query, and result. Specifically, the scheme should achieve the following privacy specification.

- **Data Privacy.** The cloud servers just have encrypted data and do not know the exact data.
- **Query Privacy.** The cloud servers know nothing about the plaintext information of the query.
- **Result Privacy.** The cloud servers do not know the plaintext information of the result.
- **Access Patterns Privacy.** Access patterns refer to the data corresponding to the query. For the cloud server, it knows nothing about which exact data matches the query.
- **Index Privacy.** The index does not reveal information related to plaintext to the cloud server.

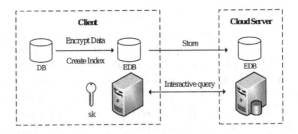

Fig. 1. The model of scheme. The client creates the OHBPTree index and encrypts the data. The cloud server is responsible for storage. Queries are implemented by the interaction between the client and the server.

The scheme model is shown in Fig. 1, which consists of two parts, the cloud server CS and the client C. The client C is responsible for creating the encrypted index and encrypting the data, which is then storing it on CS. When querying, C needs to participate in the whole process. C receives the ciphertext returned by CS, then decrypts and compares it. The cloud server CS is responsible for storing and maintaining encrypted data and indexes. CS interacts with C when querying, and completes the search on the index under the guidance of C.

3.2   Order-Hiding Encrypted B$^+$ Tree

For queries, the order information of the data can speed up the search. Some index structures store order information through the relationship of nodes. Replacing the data in the node with ciphertext can make use of the order information retained in the index structure to achieve efficient queries. But this also leaks order information. In this paper, we design an encrypted index based on the B$^+$ tree and hide the order information in the index from the adversary by an obfuscation mechanism.

Definition 1. *Move the first r elements of an array to the end of the array, which is called the rotation of the array. After the rotation, the array obtained is the rotated array, and the position where the elements move is called the rotation position.*

The scheme includes an improved encrypted index base on the B$^+$ tree, we call it OHBPTree. The OHBPTree has the following properties.

- In OHBPTree, each non-leaf node contains m indexed values and m pointers to subordinate nodes, each indexed value is the maximum value among the indexed values in its subordinate node. Leaf nodes contain indexed values and records, the data is encrypted by the encryption algorithm that satisfies semantic security and stored in the node in the form of ciphertext, the cloud cannot obtain any information related to the plaintext from the ciphertext [11].
- To preserve order information in the index, a different random offset r is generated for each node of the index. The indexed value array in the node is rotated by r to change the position of the indexed values, thereby hiding the order information of the data in the node. Because r is a key-like existence, r is not stored to protect it.
- The leaf nodes of the index form a circular chain, and the attacker cannot distinguish the head and tail of the leaf nodes, thus hiding the order information between the nodes.

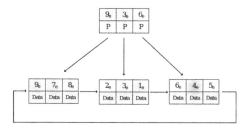

Fig. 2. An example of OHBPTree

For example, considering the plaintext space $\{1, 2, ..., 9\}$, the index constructed on the plaintext space is shown in Fig. 2. In each node of the index, the indexed values are stored as ciphertext. Also, the indexed values are no longer in the order from smallest to largest because they are rotated by random offsets. Besides, the rotation position of each node is different.

In OHBPTree, the position of the indexed values within each node is obfuscated by a random offset r. The arrangement of the indexed values is different from the original order, and the data are encrypted in the index, so the adversary cannot obtain the order information of the data. Meanwhile, the search on the index becomes complicated because the order of the indexed values changes. To realize the query in the index after changing the order, we design a new search algorithm.

Theorem 1. *In OHBPTree, there are n^2/m kinds of possible sequences, where n is the size of encrypted data and m is the node order.*

Proof. Assuming that the ciphertext set is E, $|E| = n$, and the order of the nodes in the index is m, then the number of leaf nodes in the index is n/m. Since the data in each leaf node passes through random numbers r rotates, so there are m ways of data arrangement in the leaf node. Leaf nodes form a circular linked list, so there are n/m arrangements of leaf nodes. Then the arrangement of data obtained by the attacker from the index is $n/m * m * n/m = n^2/m$.

The process of creating an index is described as Algorithm 1. In lines 1 and 2, a basic plaintext B$^+$ tree is first constructed based on the dataset V. For each node in T, each index value v_i in the node is encrypted by the encryption algorithm $Enc()$ to obtain the encrypted data e_i and stored using e_i instead, as shown in lines 4 to 6. A random offset r is generated for each node, and the index value array and pointer array are rotated according to r, hiding the order information in the index, as shown in lines 7 and 8. C indexes T by Algorithm 1, and sends T to CS for storage.

Algorithm 1. OHBPTree Create Algorithm

Input: Set of plaintext V, Secure key sk
Output: OHBPTree T
 1: **for** each value v_i in V **do**
 2:     Insert v_i into T;
 3: **for** each node n in T **do**
 4:     **for** each value v_i in n **do**
 5:         $e_i \leftarrow Enc(v_i, sk)$;
 6:         Replace v_i with e_i;
 7:     Generate a random num r;
 8:     Change order of N by r;
 9: **return** T;

3.3 Query Algorithm Based on OHBPTree

To protect the query privacy, C needs to encrypt query conditions. The query is completed through an interactive protocol between C and CS. C performs the necessary decryption during the search.

Logarithmic time complexity is an important property of queries on indexes. When using the index to query, the binary search method is often used to speed up the positioning in the node, thereby improving the efficiency of the entire query. According to the property of OHBPTree, the data sequence in the node is obfuscated by the offset r, so it is impossible to locate the position of the indexed values to be queried according to the traditional binary search. Here, we propose a novel binary search-based algorithm that enables logarithmic time-complexity queries on OHBPTree indexes. The algorithm $Search$ is shown in Algorithm 2.

Algorithm 2. OHBPTree Search Algorithm

Input: OHBPTree T, Secure key sk, Query value k
Output: Node n
1:   $n \leftarrow T$ in CS;
2: **while** n is not NULL **do**
3:     $low \leftarrow 0$; $high \leftarrow n.num - 1$;
4:     $kl \leftarrow Dec(low, n.key[low], sk)$ in C;
5:     **while** $high - low > 1$ **do**
6:       $mid \leftarrow low + (high - low)/2$;
7:       $km \leftarrow Dec(mid, n.key[mid], sk)$ in C;
8:       **if** $kl <= k <= km \parallel k <= km < kl \parallel km < kl <= k$ **then**
9:         $high \leftarrow mid$;
10:      **else** $low \leftarrow mid + 1$; $kl \leftarrow km$;
11:     **if** n is Leaf **then**
12:       break;
13:     $n \leftarrow n.child[high]$ in CS;
14: **return** n;

The search algorithm starts searching from the index root node. In each node, the corresponding location is found by binary search as shown in lines 3 to 10. In line 3, set low to 0 and $high$ to $n.num - 1$, where $n.num$ is the number of index values in node n. kl is the value obtained after decrypting the encrypted data at position low in the node. In line 4, C decrypts the index value of position low by the decryption algorithm $Dec()$ to obtain kl. km is the value obtained after decrypting the encrypted data at position mid in the node. In each loop, C decrypts the index value on position mid to km and compares the magnitudes of kl, km, and the query value k. Then C decides the direction of interval contraction based on the comparison result, as shown in lines 6 to 10. When the lookup in the leaf node is finished, the query result is returned. The comparison results in the binary search can be divided into the following cases.

$kl \leq k \leq km$. This situation means that the rotation position is not between kl and km, and k is in the interval between kl and km, so the search interval is reduced forward.

$k \leq km < kl$. This case indicates that the rotation is between kl and km, but k is between the rotation position and km, forward reduction.

$km < kl \leq k$. This case indicates that the rotation is between positions kl and km, but k is between kl and the rotation position, forward reduction.

Other Situations. Backward reduction.

The algorithm implements the query by binary search, and OHBPTree is a tree structure. So the time complexity of the algorithm is O(logN) and N is the size of the data set.

During the search process, the decryption and comparison of the ciphertext are completed by C. In this process, C only tells CS the reduced direction of the search interval. Due to the reduction direction does not uniquely correspond to the comparison result, CS cannot infer the partial order relationship between the data according to the result returned by C, nor can it infer the rotation position. So the search process does not reveal order information.

Algorithm 3. OHBPTree Range Query Algorithm

Input: OHBPTree T, Secure key sk, Query range $[a, b]$
Output: Query result R
1: $low \leftarrow Search(T, a, sk)$;
2: $high \leftarrow Search(T, b, sk)$;
3: **for** each node n between low and $high$ **do**
4:     CS inserts all data of n in R;
5: Decrypt and Filter data in C;
6: **return** R;

The range query algorithm is shown in Algorithm 3. To protect the security of the access patterns, we design a two-stage range query algorithm. The lower bound low and the upper bound $high$ of the query range are found in the index by the algorithm $Search()$ described in Algorithm 2, as shown in lines 1 and 2. For all nodes n between low and $high$, the index values in n are added to the result set R, as shown in lines 3 and 4. Finally, C decrypts and filters the data, as shown in line 5. Algorithm 3 queries the upper and lower bounds of the range by calling Algorithm 2 twice, so the time complexity of Algorithm 3 is log(N) and N is the size of the data set.

4   Security Analysis

4.1   Security Proof

The security of our scheme is analyzed as follows. Data and query conditions are encrypted by an encryption system that satisfies semantic security before being submitted to CS. Therefore, CS cannot gain information related to plaintext from ciphertext [11]. The necessary decryption in the query process and the decryption of the query result are performed by C, and this process will not reveal the information related to the plaintext to CS. Therefore, the solution can protect the privacy of data, queries, and results.

In our range query scheme, CS will return to C all data in leaf nodes involved in the query range. CS only knows which leaf is involved in the query scope. Since the leaves contain records that do not meet the query conditions, CS cannot judge whether the two queries are the same through the returned results. For example, when querying [2, 9] and [1, 7] in the index of Fig. 2, the results returned by CS are all the ciphertexts in the three leaves. From the view of CS, the results of the two queries are the same, and the CS gets obfuscated query results. Therefore, the scheme can protect the privacy of the access mode.

In the index, the data is encrypted by the encryption algorithm that satisfies semantic security, and sensitive information will not be revealed. The order preserved in the index is obfuscated by random number r, and the data sequence gained by the adversary from the static index file consists of n^2/m kinds. The adversary cannot uniquely determine the sequential information of the ciphertext. In the search process, as described in Sect. 3.3, CS cannot analyze the partial order relationship between kl and km according to the shrinking direction of the search interval. So the index is secure.

To illustrate that our scheme satisfies the security of Indistinguishability under chosen-plaintext attack (IND-CPA). We design the challenger game as follows. Challenger D has the secure key sk, the encryption algorithm and the index creation algorithm. The adversary A selects two plaintext sets $V_0 = \{v_0^1, v_0^2, ..., v_0^n\}$ and $V_1 = \{v_1^1, v_1^2, ..., v_1^n\}$ of the same length and sends them to D. D chooses b $\in \{0, 1\}$, encrypts V_b and builds an index, and then returns the ciphertext set E and index T to A. A selects the ciphertext $e_i \in E$ and $v_j \in V_b$. If there is $Dec(e_i, sk) = v_j$, then the adversary A wins the game.

Assuming that the size of the V_0 and V_1 is n, the node order in the index is m, and the number of times v_j appears in V_0 and V_1 is s, then the adversary can at most use the probability $Pr[Dec(e_i, sk) = v_j] = s/2n + \epsilon$ if the adversary wins the game with a negligible advantage ϵ, the method is IND-CPA security.

Theorem 2. *The query scheme based on the OHBPTree index can satisfy IND-CPA security.*

Proof. In the scheme, the data is encrypted using the IND-CPA security encryption algorithm. Defining the attacker's advantage from the ciphertext as ϵ_1, that ϵ_1 can be neglected. In our scheme, the order information of the ciphertext is hidden in the index. If A can know the correct order of the ciphertext, then A can correctly guess the mapping of the plaintext. Therefore, the probability that A gets the correct order of ciphertexts can be regarded as the probability that A wins. We prove that there are n^2/m ciphertext sequences that A can parse from T. Then the probability that A gets the order of the ciphertext is m/n^2, which is the probability p that A wins the game. Obviously $m/n^2 < s/2n$. So the advantage ϵ_2 of A wins the game by analyzing the index can be neglected. In summary, the scheme meets the IND-CPA security.

4.2 Preventing Inference Attack

Although our scheme uses the index to save the order information of the data, by introducing the obfuscation mechanism, the order information is hidden from the attacker, and the attacker cannot carry out inference attacks.

Definition 2 *(δ-dense). We call an encrypted column δ-dense if it contains the encryptions of at least a δ fraction of its message space. If $\delta = 1$, we simply say that the column is dense.*

Definition 3 *(Sorting-Attack). If there is a dense column, the adversary sorts the dense column and message space and outputs a function that maps each ciphertext e to the element of the message space with the same rank.*

Taking the month as an example, the adversary can arrange the ciphertext of the month into $e_1, e_2, ..., e_{12}$ according to its order information, while the plaintext sequence is $1, 2, ..., 12$. The adversary can map the plaintexts with the same sequence to achieve the purpose of the attack.

In order-preserving encryption and traditional indexing, the ciphertext or the relationship between nodes can reveal the order information of the ciphertext. And the order information within the index is changed by random numbers in OHBPTree. The adversary cannot uniquely get the ciphertext order and cannot recover the ciphertext by sorting-attack.

Definition 4 *(Cumulative-Attack). The adversary knows the order information of the ciphertext and calculates the cumulative distribution function $F_E(e_i) = P(E < e_i)$ of each ciphertext e_i by combining the order information and frequency. Adversaries can improve their ability to match plaintext to ciphertext using the cumulative distribution function of the data and the cumulative distribution function of the auxiliary data set.*

For example, if a given ciphertext is greater than 90% of the ciphertexts in the encrypted dataset, then we should match it to a plaintext that also is greater than about 90% of the auxiliary data z.

In the previous section, we demonstrate that the attacker cannot uniquely determine the correct order information from the index. Based on the wrong order information, the cumulative distribution ψ' calculated by the attacker for each ciphertext is different from the actual cumulative distribution ψ for that ciphertext. Combining ψ' with the cumulative distribution function θ of the auxiliary data, the correct plain-ciphertext mapping relationship cannot be inferred. So the cumulative attack will not pose a threat to the scheme.

5   Experiments

5.1   Experiment Setup

To evaluate the performance, we select five datasets and selected an attribute in each dataset to build the index. The datasets are described in Table 1. The edit-ky and edit-hi are edit datasets from Wikipedia, and each record represents the user's edit record on the page. We build the index on the user attribute, the former has 53,813 users and 106,433 records, and the latter has 203,193 users and 357,938 records. Wiki-OC is an occupational dataset derived from Wikipedia, and each record represents the correspondence between people and occupations. We build indexes on occupation attributes. The dataset has a total of 101,730 values and 250,945 records. Wiki-LO is a bidirectional network of entities and their locations in Wikipedia. We build the index on the entity attribute and the dataset has a total of 172,091 entities and 293,697 records. Emai-Eu is an email dataset from European research institutions. We build the index on the sender attribute, and the dataset has a total of 265,214 values and 420,045 records.

Table 1. Description of dataset

Data set	Num of keys	Dataset size
edit-ky	53,813	106,433
Wiki-OC	101,730	250,945
Wiki-LO	172,091	293,697
edit-hi	203,193	357,938
Email-Eu	265,214	420,025

We experiment with the run time of index creation, query one value, and the range query. For creation, the run-time includes the time to build the index and the encryption time. For query one value, the run-time includes the communication overhead and decryption time. For range query, the run-time includes communication overhead and decryption time. We repeat each experiment 5 times. In the query experiment, we randomly select 5 data or ranges each time and finally calculate the average time of the query. We compare the running time on OHBPTree with the B^+ tree and the scheme ESEDS in [17]. The ESEDS is an array-indexed scheme that leaks access patterns.

All experiments are conducted on a machine with a 6-core Intel i7 CPU at 2.6 GHz and 12 GB of RAM.

5.2   Evaluation

Figure 3 shows the time to create the index and the size of the space occupied by the index. As shown in Fig. 3(a), without encryption, ESEDS builds the

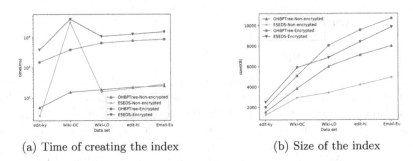

(a) Time of creating the index   (b) Size of the index

Fig. 3. Create the index over different datasets

index in a time that gradually exceeds OHBPTree as the dataset size increases. OHBPTree takes 13% less time to build the index than ESEDS over the Email-Eu dataset. With encryption, ESEDS consumes more time to build the index than OHBPTree. OHBPTree is 46% to 52% more efficient than ESEDS. When building an index, ESEDS takes a lot of time if the attribute values in the dataset are not in order. Because ESEDS is array-based, it takes a lot of time to sort. OHBPTree, on the other hand, is not affected. For example, on Wiki-OC, ESEDS-non-encrypted takes much longer to build the index than OHBPTree-non-encrypted. As shown in Fig. 3(b), whether in the encrypted or unencrypted cases, the size of OHBPTree is slightly larger than ESEDS. With encryption, OHBPTree's space overhead is 8% to 17% more than ESEDS.

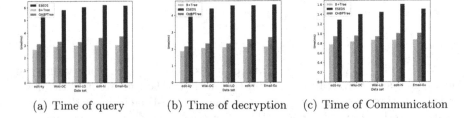

(a) Time of query   (b) Time of decryption   (c) Time of Communication

Fig. 4. Query over different datasets

Consider a query SELECT * FROM *table* WHERE *val* = *v*. Figure 4 shows the run-time of query value v. As shown in Fig. 4(a) and 4(b), compared with the ESEDS scheme, the efficiency on OHBPTree is about 40% to 42% higher. The main source of this gap is the overhead of decryption. The decryption time required by OHBPTree is 42% to 46% less than that of ESEDS. The communication overhead is shown in Fig. 4(c). OHBPTree's communication overhead costs 26% to 37% less than ESEDS. Compared with the B$^+$ tree, OHBPTree takes 17% to 22% more time in querying. Among them, the decryption time of OHBPTree is 15% to 22% more than that of B$^+$ tree and the communication overhead is 15% to 21% more. This shows that the obfuscation operation has an acceptable impact on the query.

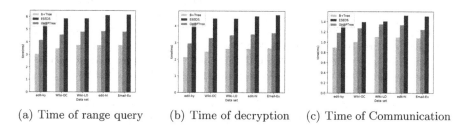

(a) Time of range query   (b) Time of decryption   (c) Time of Communication

Fig. 5. Range query over different datasets

Consider a range query SELECT * FROM *table* WHERE *val* BETWEEN *a* AND *b*. As shown in Fig. 5(a), the range query efficiency of OHBPTree is 22% to 24% higher than that of ESEDS and 29% to 36% lower than that of B^+ tree. The difference in efficiency mainly comes from the decryption overhead. From Fig. 5(b), it can be seen that OHBPTree takes 23% to 28% less decryption time than ESEDS and 33% to 38% more than B^+ tree during range query. The communication overhead is shown in Fig. 5(c). The communication overhead of OHBPTree is 11% to 17% less than ESEDS and 14% to 31% more than B^+ trees. So the effect of obfuscation on range queries is acceptable. The above experimental results show that our scheme performs well in index building and querying.

6 Conclusion

In this paper, we propose a privacy-preserving query scheme. First, we design an index structure that can hide the order information. The index is based on a B^+ tree and introduces an obfuscation mechanism to change the order of indexed values with random offsets. A secure query algorithm with logarithmic time complexity is proposed based on this index. In addition, we analyze the security of the scheme, which protects the data, queries, results, and access patterns. The final experimental results show that the scheme has efficient query performance. We will work on stronger security and better query efficiency in our future work.

Acknowledgement. The work is supported by the National Natural Science Foundation of China under Grant 61972198.

References

1. Antonopoulos, P., et al.: Azure SQL database always encrypted. In: Proceedings of ACM SIGMOD International Conference on Management of Data, pp. 1511–1525 (2020)

2. Boldyreva, A., Chenette, N., Lee, Y., O'Neill, A.: Order-preserving symmetric encryption. In: Joux, A. (ed.) EUROCRYPT 2009. LNCS, vol. 5479, pp. 224–241. Springer, Heidelberg (2009). https://doi.org/10.1007/978-3-642-01001-9_13

3. Boldyreva, A., Chenette, N., O'Neill, A.: Order-preserving encryption revisited: improved security analysis and alternative solutions. In: Rogaway, P. (ed.) CRYPTO 2011. LNCS, vol. 6841, pp. 578–595. Springer, Heidelberg (2011). https://doi.org/10.1007/978-3-642-22792-9_33

4. Chen, H., Tian, X., Yuan, P.: Crypt-JDBC model: optimization of onion encryption algorithm. J. Front. Comput. Sci. Technol. 11(8), 1246–1257 (2017)

5. Cui, N., Yang, X., Wang, L., Wang, B., Li, J.: Secure range query over encrypted data in outsourced environments. In: Pei, J., Manolopoulos, Y., Sadiq, S., Li, J. (eds.) DASFAA 2018. LNCS, vol. 10828, pp. 112–129. Springer, Cham (2018). https://doi.org/10.1007/978-3-319-91458-9_7

6. ElGamal, T.: A public key cryptosystem and a signature scheme based on discrete logarithms. IEEE Trans. Inf. Theory 31(4), 469–472 (1985)

7. Elmehdwi, Y., Samanthula, B.K., Jiang, W.: Secure k-nearest neighbor query over encrypted data in outsourced environments. In: IEEE International Conference on Data Engineering, pp. 664–675. IEEE (2014)

8. Ge, Y., Cao, J., Wang, H., Chen, Z., Zhang, Y.: Set-based adaptive distributed differential evolution for anonymity-driven database fragmentation. Data Sci. Eng. 6, 380–391 (2021). https://doi.org/10.1007/s41019-021-00170-4

9. Gentry, C.: Fully homomorphic encryption using ideal lattices. In: Proceedings of ACM Symposium on Theory of Computing (STOC), pp. 169–178 (2009)

10. Gentry, C., Halevi, S., Smart, N.P.: Homomorphic evaluation of the AES circuit. In: Safavi-Naini, R., Canetti, R. (eds.) CRYPTO 2012. LNCS, vol. 7417, pp. 850–867. Springer, Heidelberg (2012). https://doi.org/10.1007/978-3-642-32009-5_49

11. Goldreich, O.: Foundations of Cryptography, vol. 2. Cambridge University Press, Cambridge (2004)

12. Grubbs, P., Sekniqi, K., Bindschaedler, V., Naveed, M., Ristenpart, T.: Leakage-abuse attacks against order-revealing encryption. In: IEEE Symposium on Security and Privacy, pp. 655–672. IEEE (2017)

13. Hacigümüş, H., Iyer, B., Li, C., Mehrotra, S.: Executing SQL over encrypted data in the database-service-provider model. In: Proceedings of the 2002 ACM SIGMOD International Conference on Management of Data, pp. 216–227 (2002)

14. Hore, B., Mehrotra, S., Tsudik, G.: A privacy-preserving index for range queries. In: Proceedings of the Thirtieth International Conference on Very Large Data, pp. 720–731 (2004)

15. Hu, H., Xu, J., Ren, C., Choi, B.: Processing private queries over untrusted data cloud through privacy homomorphism. In: IEEE International Conference on Data Engineering, pp. 601–612. IEEE (2011)

16. Kerschbaum, F., Schröpfer, A.: Optimal average-complexity ideal-security order-preserving encryption. In: Proceedings of the 2014 ACM SIGSAC Conference on Computer and Communications Security, pp. 275–286 (2014)

17. Kerschbaum, F., Tueno, A.: An efficiently searchable encrypted data structure for range queries. In: Sako, K., Schneider, S., Ryan, P.Y.A. (eds.) ESORICS 2019. LNCS, vol. 11736, pp. 344–364. Springer, Cham (2019). https://doi.org/10.1007/978-3-030-29962-0_17

18. Liu, D., Wang, S.: Programmable order-preserving secure index for encrypted database query. In: IEEE International Conference on Cloud Computing, pp. 502–509. IEEE (2012)

19. Liu, D., Wang, S.: Nonlinear order preserving index for encrypted database query in service cloud environments. Concurr. Comput. Pract. Exp. **25**(13), 1967–1984 (2013)
20. Naveed, M., Kamara, S., Wright, C.V.: Inference attacks on property-preserving encrypted databases. In: Proceedings of the 22nd ACM SIGSAC Conference on Computer and Communications Security, pp. 644–655 (2015)
21. Ostrovsky, R.: Efficient computation on oblivious RAMs. In: Proceedings of ACM Symposium on Theory of Computing, pp. 514–523 (1990)
22. Padhya, M., Jinwala, D.C.: R-OO-KASE: revocable online/offline key aggregate searchable encryption. Data Sci. Eng. **5**, 391–418 (2020). https://doi.org/10.1007/s41019-020-00136-y
23. Paillier, P.: Public-key cryptosystems based on composite degree residuosity classes. In: Stern, J. (ed.) EUROCRYPT 1999. LNCS, vol. 1592, pp. 223–238. Springer, Heidelberg (1999). https://doi.org/10.1007/3-540-48910-X_16
24. Popa, R.A., Li, F.H., Zeldovich, N.: An ideal-security protocol for order-preserving encoding. In: IEEE Symposium on Security and Privacy, pp. 463–477. IEEE (2013)
25. Popa, R.A., Redfield, C.M., Zeldovich, N., Balakrishnan, H.: CryptDB: protecting confidentiality with encrypted query processing. In: ACM Symposium on Operating Systems Principles, pp. 85–100 (2011)
26. Rivest, R.L., Shamir, A., Adleman, L.: A method for obtaining digital signatures and public-key cryptosystems. Commun. ACM **21**(2), 120–126 (1978)
27. Roche, D.S., Apon, D., Choi, S.G., Yerukhimovich, A.: POPE: partial order preserving encoding. In: Proceedings of the 2016 ACM SIGSAC Conference on Computer and Communications Security, pp. 1131–1142 (2016)
28. Tu, S.L., Kaashoek, M.F., Madden, S.R., Zeldovich, N.: Processing analytical queries over encrypted data. Proc. VLDB Endow. **6**(5), 289–300 (2013)
29. Shi, E., Bethencourt, J., Chan, T.H., Song, D., Perrig, A.: Multi-dimensional range query over encrypted data. In: IEEE Symposium on Security and Privacy, pp. 350–364. IEEE (2007)
30. Stefanov, E., et al.: Path ORAM: an extremely simple oblivious ram protocol. J. ACM (JACM) **65**(4), 1–26 (2018)
31. Tueno, A., Kerschbaum, F.: Efficient secure computation of order-preserving encryption. In: Proceedings of the 15th ACM Asia Conference on Computer and Communications Security, pp. 193–207 (2020)
32. Zheguang, Z., Seny, K., Tarik, M., Stan, Z.: Encrypted databases: from theory to systems. In: Conference on Innovative Data Systems Research (2021)

A Deep Reinforcement Learning-Based Approach for Android GUI Testing

Yuemeng Gao[1], Chuanqi Tao[1,2,3,4(✉)], Hongjing Guo[1], and Jerry Gao[5]

[1] College of Computer Science and Technology, Nanjing University of Aeronautics and Astronautics, Nanjing, China
{gaoyuemeng,taochuanqi,guohongjing}@nuaa.edu.cn
[2] Ministry Key Laboratory for Safety-Critical Software Development and Verification, Nanjing University of Aeronautics and Astronautics, Nanjing, China
[3] Collaborative Innovation Center of Novel Software Technology and Industrialization, Nanjing University, Nanjing, China
[4] State Key Laboratory for Novel Software Technology, Nanjing University, Nanjing, China
[5] Computer Engineering Department, San Jose State University, San Jose, USA
Jerry.gao@mail.sjsu.edu

Abstract. The mobile application market is booming, and Android applications occupy a vast market share. However, the applications may contain many errors. The task in the testing phase is to find these errors as soon as possible. It is urgent to test the application rapidly and effectively. Otherwise, it may affect user experience and cause substantial economic losses. Mobile applications iterate continuously to consummate performance and functional requirements, which leads to the increased complexity of applications and explosive growth of state combinations. Reinforcement learning aims to learn strategies to achieve specific goals by maximizing rewards. This paper applies it to Android GUI testing. In this paper, we propose ATAC. It is black-box based and adopts Advantage Actor-Critic (A2C) algorithm, which contains an actor (policy) and a critic (value function) to generate test cases automatically through deep reinforcement learning. To verify the validity of the proposed approach, we conducted our experiment on seventeen open-source applications from Github. Compared with ARES and Monkey, ATAC shows higher code coverage and detects more errors.

Keywords: Reinforement learning · Android GUI testing · Advantage actor critic

1 Introduction

The scale of Android applications keeps growing. Android applications are developed and iterated continuously to meet the demand of users and conform to the trend of the times. Many applications may contain failures, or new ones may be brought in during the iteration. The failures in applications may lead to poor

B. Li et al. (Eds.): APWeb-WAIM 2022, LNCS 13423, pp. 262–276, 2023.
https://doi.org/10.1007/978-3-031-25201-3_20

user experience and even substantial economic losses. We should consider the performance, reliability, and security of applications and portability for Android devices in the process of testing applications.

Graphical User Interface (GUI) is a part of the application that is visible to the user. GUI may contain many widgets (menus, buttons, text boxes, images, etc.). GUI testing mainly performs functional testing for applications under test (AUT). It examines the app's behavior by interacting (e.g., click, long-click, scroll, and input strings) with the GUI. An application consists of activities, and activity is combined with fragments that contain many widgets. As a result, exploring all states of an application is complex, and it can be called explosion combinations of states. The cost of testing Android GUI is high because of limited human resources and time pressure. We are faced with challenges when we try to exploit with machine tools instead of human resources. The challenges include explosion combinations of states and limitations of exploration. Some application behavior can only be revealed upon providing specific inputs. It is of great significance to test the mobile app GUI effectively and efficiently.

GUI testing has been attracted massive attention from researchers. Some researchers propose random testing, which generates random events on the GUI. One of the most famous random test tools is Monkey [16] which is provided by Google for stability and stress testing. It generates random user events such as key presses and random inputs. However, random tests like Monkey may generate large amounts of invalid and redundant events. It is ineffective for them to explore more states and detect failures. Model-based strategies [2,5,13,18,19,28] build precise or abstract GUI models by static or dynamic methods to generate test cases. Nevertheless, the strategies are influenced by two problems. One of the problems is inherent state explosion. Another one is that the effectiveness of generated test cases depends on the built model's integrity and the representation of the application's state.

Reinforcement learning (RL) contains the agent learning strategies to maximize returns or achieve specific goals by interacting with the environment. It is extensively used in Android GUI testing. Unlike supervised learning, which requires labeled datasets, RL is able to learn automatically by interacting with the environment. Though RL is applied to GUI testing, most of them are implemented by using Q-learning. Q-learning uses a table to record expected values of actions that are called action values in a specific state. The Q table occupies a lot of memory. Some researchers consider replacing tabular methods with Deep Neural Networks (DNN). The agent utilizes DNN to learn the action-value function automatically through past experiences. Driven by a reward function, the agent guides us to explore the AUT, unlike random testing which explores the AUT without purpose. The optimal action to perform can be predicted by the agent even if the state has never been visited before. It can effectively solve the state explosion problem. RL allows us to test GUI effectively and efficiently. DeepGUIT [8] adopts Deep Q Network (DQN) to represent the value function, and ARES [26] applies TD3, DDPG, and SAC to fit value functions. However, these methods utilize a replay buffer that records the pairs of states and rewards. The replay buffer also occupies a lot of memory, even if less than Q-table. We propose ATAC (Automatic

Testing based on A2C), a novel approach to test Android GUI. In this paper, we propose ATAC (**A**utomatic **T**esting based on **A2C**), a novel approach based on deep reinforcement learning to testing Android GUI.

ATAC applies the A2C algorithm to Android GUI testing to avoid using a relay buffer. A2C algorithm that introduces the idea of parallelism constructs multiple threads, and the thread interacts with the environment respectively. We evaluate our approach in seventeen apps in the environment of Android 10. Our approach achieves higher coverage and detects more failures than state-of-the-art test generation tool Monkey and RL-based tool ARES. Besides, ATAC does not need a replay buffer, and it can save memory space and consume fewer resources. The neural network it constructs is much smaller than DeepGUIT and ARES. It can work even if the agent does not equip with GPU.

The remainder of the paper is organized as follows: Sect. 2 surveys related work in Android GUI testing. Section 3 introduces the background of Reinforcement Learning. In Sect. 4, our proposed approach is described in detail. Section 5 shows the evaluation of our approach and discusses the results. Section 6 concludes our work, and Sect. 7 presents the threat of validity.

2   Background

Reinforcement Learning (RL) is a branch of Machine Learning which tries to maximize the reward it obtains in a complex and uncertain environment. It consists of the agent and the environment. RL trains the agent by constantly interacting with the environment. The purpose of the agent is to receive as much reward from the environment as possible through trial and error.

Markov decision process is one of the most basic theoretical models of reinforcement learning, and most problems can be regarded as or transformed into the Markov decision process. A Markov decision process is represented by a 4-tuple $<S, A, P, R>$.

S: Set of all possible states

A: Set of possible actions that the agent can perform in all states.

$P(S \times A \times S \rightarrow [0, 1])$: State transition function represents the probability of taking an action to transfer to some state.

$R(S \times A \times S \rightarrow R)$: Reward function represents the reward received from the environment after taking the action which changes the current state.

At a time step t, the agent observes the current state $s_t \in S$, and then selects and performs the action from action space A. Then a reward is obtained, the environment moves to a new state, and then the agent continues to repeat the above process until the terminal state or timeout and restarts. The process above can be represented by a trajectory: $s_0, a_0, r_1, s_1, a_1, r_2, s_3....$ a_t means the action performed and r_t denotes the reward when performing a_t in the state s_t. The accumulated return Rt from the time step t with discount factor $\gamma \in (0, 1]$.

The action-value function $Q^\pi(s, a)$ is used to estimate the expected return when we take an action in a state when following the current policy. The state-value function $V^\pi(s)$ is to evaluate the expected return in the current state

s when following the π. The action-value function or the value function can be eliminated by substitution. Then, the Bellman optimal equation can be derived:

$$\begin{cases} V^*(s) = \max_{a \in A}[r(s,a) + \gamma \sum s'[p(s'|s,a)V^*(s')], & s \in S, \\ Q^*(s,a) = r(s,a) + \sum_{s',r} p(s',r|s,a) \max_{a'} Q^*(s',a'), & s \in S, a \in A. \end{cases} \quad (1)$$

We use the symbol $*$ to indicate the estimated value. $r(s,a)$ indicates the reward when performing a in the state s. We can find an optimal policy through Bellman's expectation and optimal function in theory. However, the Bellman equation is difficult to obtain. Therefore, we need other ways to estimate the optimal value function, such as policy iteration.

Most of work [1,20,25,31] based on RL adopts Q-learning as agent. Q-learning uses a table to record expected values of actions that are called action values in a specific state. Recently, some researchers replace tabular methods with Deep Neural Networks (DNN). DeepGUIT represents the value function by DQN, and ARES adopts TD3, DDPG, and SAC to fit value functions.

A standard policy-based method [26] updates θ by $\nabla_\theta log\pi(a_t|s_t;\theta)$, it is an unbiased estimate of $\nabla_\theta E[R_t]$. The parameter θ of policy is updated by $\nabla_\theta(R_t - b_t(s_t))log\pi(a_t|s_t;\theta)$. The advantage $A(a_t,s_t)$ can be calculated by $Q(s_t,a_t) - V(s_t)$. The method [9,29] introduced above can be viewed as an Actor-Critic architecture with policy π as an actor and the baseline $V(s_t)$ as a critic.

Advantage Actor-Critic (A2C) [23] is a typical Actor-Critic method, which is a synchronous, deterministic variant of Asynchronous Advantage Actor-Critic (A3C). It uses multiple workers to avoid the use of a replay buffer. Figure 1 shows the mechanism of A2C.

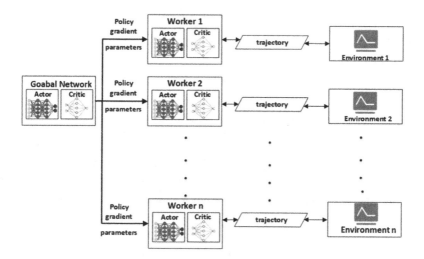

Fig. 1. Mechanism of A2C algorithm

3　Approach

The process of testing Android applications can be considered the Markov Decision Process. This section will introduce ATAC and the details to model the testing process by reinforcement learning. ATAC adopts Advantage Actor-Critic (A2C) for test generation. ATAC extracts information from AUT installed in Android devices or emulators by Appium, which is an automated testing tool for native or hybrid mobile applications. Android devices and Appium communicate with each other through Android Debug Bridge. The GUI states composed of GUI elements are in the form of XML files. The file includes the package name, executable events (clickable, long-clickable, scrollable), bounds, and resources-id. ATAC defines states, rewards, and actions through the information above.

At a time step t, the environment gets the current GUI of the Android device through Appium, extracts the state from it, and passes the observation to the agent. Then, the agent (Actor-network follows policy π) predicts an action that will likely archive more accumulated rewards. Once the action is performed, the agent will obtain a reward as feedback from the environment. The agent will adjust the actor-network and critic network according to the rewards. The workflow of the ATAC is shown in Fig. 2.

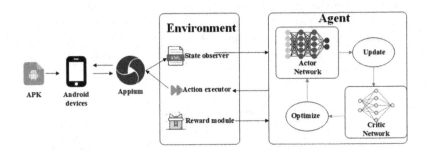

Fig. 2. Workflow of ATAC

3.1　Representation of States and Actions

As mentioned above, the GUI of the application under test is composed of GUI elements. ATAC only considers the executable GUI elements to represent the states so that they can be passed to and processed by the neural network. ATAC denotes the abstract state by all executable widgets from the GUI of the application:

$$s_t = (w_1, w_2, w_3, ..., w_i, ...w_n) \tag{2}$$

w_i represents the widget which is executable, it is a 3-dimensions binary vector, and three dimensions represent whether the widget i is clickable, long-clickable, or scrollable. If the widget i is clickable, long-clickable, or scrollable, the corresponding dimension will be marked as 1, or we will denote the dimension as 0. For example, if the widget i is clickable, long-clickable, but not scrollable, we denote w_i as $[1, 1, 0]$. At a time step t, the agent can observe the state s_t, composed of the n widgets, where n is the total number of executable widgets.

Both system-level actions and typical actions are considered in our approach. The system-level actions include switching the internet connection state, screen rotation, and return. There is no return button in many applications explicitly, and the applications may keep in a stalemate state, which prevents them from exploring more space. ATAC adopts a similar action representation as ARES [26]. A 3-dimensions vector denotes each action, and the first dimension specifies which widget or system-level action will be operated. The second dimension works when the widget expects an input since the dimension indicates the index of a string in a predefined dictionary. The third dimension acts as a compliment. When a widget is clickable and long-clickable, the third dimension decides which actions to perform. When a widget is scrollable, the third dimension specifies the scrolling direction.

3.2   Reward Function

The design of the reward function is a vertical procedure. The agent adjusts the strategy according to the feedback (reward) received after performing a specific action. The reward tells the agent which actions are encouraged and discouraged. Rewards mainly depend on the change of states and whether they can detect errors. ATAC defines two forms of reward function according to the particularity of return. There are two aspects to consider when the agent chooses to return. First, we hope it returns in time when the application enters a deadlock state. Second, we do not expect frequent returns, or it may affect our application exploration.

The definition of reward function is as follows:

a)   When the agent performs actions:

$$r_t = \begin{cases} R_1, & crash \\ -R_4, & s_{t+1} \notin AUT \\ -R_3, & stopTime < N \\ R_3, & stopTime \geq N \end{cases} \qquad (3)$$

b)   When the agent performs other actions:

$$r_t = \begin{cases} R_1, & S_{t+1} \notin visitedState \quad or \quad crash \\ R_2, & S_{t+1} \in visitedState \quad and \quad a_t \notin visitedWidgets \\ -2^{repeat_{time}}, & S_{t+1} \in visitedState \quad and \quad a_t \in visitedWidgets \\ -R_4, & S_{t+1} \notin AUT \end{cases} \qquad (4)$$

The numbers in the function satisfies the condition that $R_1 > R_2 > R_4 > R_3$. $R_1 = 1000$. $R_2 = 500$, $R_4 = 100$ and $R_3 = 50$. When the application crashes, either the agent explores new states or operates a widget that has not been performed before, we encourage the behaviors. The *stop_time* indicates how long it has been in the current state. If the application is stuck in the same states for a long time and the agent performs the return, a positive reward is given. If the agent frequently returns when the application is in a new state, we give it a negative reward. The *repeat_time* indicates the times of performing the same action continuously.

3.3   Advantage Actor-Critic (A2C) Based Testing

Advantage Actor-Critic (A2C) is a synchronous variant of Asynchronous Advantage Actor-Critic (A3C). They both maintain a policy network and a value function network. They choose the same advantage function $A(a_t, s_t) = (r_{t+1} + \gamma V(s_t) - V(s_t))$ and employ multithreading to perform gradient descent. The difference between them is the time to update our global policy θ_π and value function θ_v. A2C is synchronous, and A3C is asynchronous.

Advantage Actor-Critic constructs multiple threads, and the thread interacts with the environment, respectively. In each iteration, the global network waits for threads to finish their episodes and update the global network by the gradients uploaded by threads. Afterward, the global network sends the latest network parameters to all the threads simultaneously. The detailed algorithm for each thread is shown in Algorithm 1.

Algorithm 1. Advantage Actor-Critic Algorithm for each worker(thread)

Input: global actor network θ_π, global critic network θ_v
Output: accumulated actor network gradient $d\theta_\pi$, critic network gradient $d\theta_v$
 1: **repeat**
 2:     Reset gradients:$d\theta_\pi \leftarrow 0$, $d\theta_v \leftarrow 0$
 3:     Observe the GUI state S_t
 4:     **repeat**
 5:         $a_t \leftarrow \pi\left(a_t | s_t, \theta_\pi'\right)$
 6:         Perform an action a_t and Receive a reward r_t
 7:         Observe a new GUI state s_{t+1}
 8:     **until** s_t is terminal or timeout
 9:     **for** each step i of episode **do**
10:         $R \leftarrow r_i + \gamma R$
11:         $d\theta_\pi \leftarrow d\theta_\pi + \nabla_{\theta_\pi'} \log\pi(a_i | s_i; \theta_\pi')(R - V(s_i; \theta_\pi'))$
12:         $d\theta_v \leftarrow d\theta_v + \partial(R - V(s_i; \theta_\pi'))^2 / \partial\theta_\pi'$
13:     **end for**
14: **until** timeout
15: **return** $d\theta_\pi, d\theta_v$

---

4   Empirical Study

In this section, we mainly focus on evaluating the proposed approach. ATAC is compared with the state-of-art tools, Monkey, and ARES in code coverage (instruction coverage, branch coverage, line coverage, and method coverage) and fault detection (numbers of failures). Monkey is a random tests generation tool, and ARES is a tool based on deep reinforcement learning. Code coverage and fault detection are used as evaluation metrics in researches [8,25,26,31,32]. Therefore, our empirical study is designed to answer the following research questions:

RQ1: Does ATAC achieve higher instruction coverage, branch coverage, line coverage, and method coverage than state-of-art testing tools?
RQ2: Can ATAC reveal more failures than state-of-art testing tools?

4.1 Applications Under Test

Though the approach is black-box, to compare with other approaches, we have to access source codes to collect coverage information by Jacoco [12]. ATAC selects and evaluates our approach on seventeen open-source F-Droid applications from Github [15]. Other researchers have also chosen these applications with the same goal, and ATAC has excluded the out-of-date applications that cannot run. Some of the applications are still being revised and iterated. Detailed information of the application under test is shown in Table 1. The AUT can be found at https://github.com/RL-ATAC/AUT-of-ATAC. We have uploaded the apks and the source code of the applications, which will help calculate coverage of applications.

Table 1. Target applications for evaluation

Applications	Instructions	Braches	Lines	Methods	Classes
QuickSettings	470	28	115	17	4
MunchLife	954	63	214	34	12
Silent-ping-sms	1,732	137	324	39	7
BatteryDog	2,945	164	600	72	18
AnyCut	1,527	97	379	72	18
Afwall	63,182	5,397	14,004	2,155	337
Jamendo	18,382	1,103	4,467	909	205
Drawablenotepad	2,498	132	563	124	22
AnyMemo	50,311	2,732	11,251	2,476	547
AmazeFileManager	85,537	7,418	20,471	3,018	488
zooborns	3,543	259	780	132	24
AntennaPod	59,826	4,909	14,830	2,587	385
vanilla	48,376	4,794	10,718	1,477	211
BudgetWatch	5,248	319	1,194	168	51
sanity	28,002	2,972	4,949	1,233	263
timber	54,178	4,209	11,966	2,217	427
Drawablenotepad	2,498	132	563	124	22

4.2 Evaluation Setup

We experiment on a 64-bit computer with Ubuntu 18.04.5 LTS, processor Intel core i7-9700 CPU 3.00 GHz GPU Nvidia GeForce RTX 2080Ti and memory RAM 62.6 GB. Since the emulator is not stable, the applications in the experiment are run on real Android devices with operation system Android 10, 6 GB memory.

The framework of ATAC is based on ARES. It realizes the A2C algorithm mainly by OpenAI Gym [6]. The communication between applications and agent is through Appium [24] and Android Debug Bridge [10]. The structures of the Actor and Critic networks are the same, and ATAC adopts two 2-layers neural networks. There are 64 neurons in each layer. ATAC chooses relu as the activation function in both neural networks. The number of actions that can be performed changes from time to time in each state, so Gaussian distribution is adopted as output, representing the probability distribution of actions in the current state. Other parameters ATAC uses are shown in Table 2.

Table 2. Specific parameter settings

learning_rate	0.0007	use_rms_prop	True
n_steps	5	max_grad_norm	0.5
gamma	0.99	total_timesteps	5000
ent_coef	0	n_eval_episodes	5
vf_coef	0.5	rms_prop_eps	1e-05

The length of an episode is set as 250. The agent explores and tests the application within 5000-time steps, limiting the time to one hour. The same settings are used in ARES. The throttle setting was 200 in Monkey, and the procedure also lasted one hour. The tool was configured to ignore any crash, system timeout, and security exceptions until the timeout was reached.

4.3 Experiment Results

In order to answer RQ1 and avoid the randomness of the results as much as possible, we repeat all the experiments ten times and use the average of ten executions to represent the final result, including the comparative experiment. Table 3 shows the average instruction coverage of AUT. Table 3 shows the average instruction coverage of AUT. Inst represents instruction coverage, bran represents branch coverage, line represents line coverage, and meth represents method coverage.

ATAC gets higher instruction, branch, line, and method coverage in thirteen of seventeen apps. It can be noticed that ATAC has higher average instruction coverage (45.7%) than Monkey (11%) and ARES (40.4%). The Average branch average (32.4%) of ATAC is also higher than Monkey (6.5%) and ARES (26.3%).

Table 3. Coverage of AUT(%)

Apps under test	ATAC				ARES				Monkey			
	inst	bran	line	meth	inst	bran	line	meth	inst	bran	line	meth
QuickSettings	**88.3**	**70.3**	**87.8**	82.0	83.1	55.7	83.1	82.0	72.9	45.0	72.6	**86.8**
MunchLife	**83.7**	**59.8**	**84.9**	**88.0**	80.1	51.2	80.2	87.7	59.0	33.8	53.5	66.4
Silent-ping-sms	**42.0**	**23.0**	**46.0**	**59.0**	38.4	19.1	41.4	53.6	30.8	11.5	30.3	42.4
BatteryDog	**71.1**	**55.1**	**70.1**	**74.1**	70.1	53.0	68.4	72.8	64.6	49.7	60.4	65.5
AnyCut	**68.7**	**51.8**	**69.8**	**75.0**	68.0	49.0	69.0	75.0	39.4	22.2	39.8	50.1
Afwall	**18.9**	**14.4**	**19.5**	**23.4**	7.2	4.8	7.2	10.4	6.0	4.0	7.0	10.0
Jamendo	**23.0**	**14.0**	**20.0**	**23.0**	16.7	9.0	16.3	22.3	10.9	6.5	11.6	14.9
microMathematics	41.0	28.7	40.2	48.2	**44.9**	**32.3**	**44.2**	**55.2**	24.4	15.0	24.3	31.6
AnyMemo	44.7	**31.2**	**45.3**	**49.7**	24.6	16.3	25.7	28.9	17.7	11.0	19.3	23.3
AmazeFileManager	**29.6**	**21.1**	**28.5**	**35.7**	26.1	18.3	24.9	32.7	15.4	8.9	14.6	20.6
zooborns	**17.8**	**7.7**	**19.8**	**24.5**	17.4	7.4	18.4	23.9	11.5	4.8	12.5	17.1
AntennaPod	**19.3**	**9.4**	**18.4**	**22.2**	16.7	7.9	16.3	18.9	10.4	5.6	9.2	9.8
vanilla	51.9	39.9	52.0	58.3	**55.5**	**42.9**	**56.1**	**63.3**	19.8	12.7	21.3	28.5
BudgetWatch	44.0	29.7	45.4	60.0	**45.1**	**30.6**	**47.1**	**60.7**	19.8	9.1	20.9	25.9
sanity	**20.1**	**8.8**	**19.0**	**23.1**	17.6	7.0	16.7	21.4	6.8	3.0	7.5	10.9
timber	26.9	15.0	28.2	32.5	**31.3**	**17.2**	**31.3**	**35.7**	11.0	6.5	11.9	14.9
Drawablenotepad	**86.3**	**72.1**	**84.4**	**83.1**	44.1	25.2	43	47.6	57.4	30.5	56.3	60.7
AVERAGE	45.7	32.5	45.8	50.7	40.4	26.3	40.5	46.6	28.1	16.5	27.8	34.1

At the same time, ATAC does better in average line coverage and method coverage. ATAC performs best in 13 of 17 apps while ARES performs best in 4 of 17 apps. By applying A2C algorithm, ATAC achieves higher the instruction coverage, branch coverage, line coverage, and method coverage than state-of-art tools Monkey and ARES.

To answer RQ2, ATAC explores ten times on each app. In each iteration, ATAC may find the same failure many times. We only record it once because it has been repeated many times. We counted the different failures detected in each iteration and calculated the average value of ten iterations. We use the average value to evaluate the ability of methods to detect failures. ATAC detects 47 failures, and ARES detects 34 failures. AUT is widely used in Android GUI testing, and many researchers and papers point out the failures in AUT. Many bugs that researchers detect have been fixed, and the source code of most apps is still being modified. In seven of seventeen apps, ATAC, Monkey, and ARES detect many failures. ATAC reveals more failures than ARES in BatteryDog, Anymemo, AmazeFileManager, zooborns, and Drawablenotepad. The failures include RuntimeException, NullPointerException, and NumberFormatException. ATAC can reveal more failures than state-of-art testing tools.

To explore the specific reasons, the A2C algorithm encourages the execution of events that lead to new or partially explored states. The control that does not repeat the same action in the same state context provides access to most of an AUT's functionality. ATAC achieved the highest code coverage in 13/17 apps and the highest average of failures in apps 5/7. The experiment shows that RL-based methods are better than the random method and ATAC works better than ARES (Fig. 3 and Table 4).

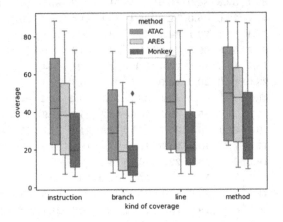

Fig. 3. Code coverage achieved by Monkey, ARES and ATAC

Table 4. Failures detected in AUT

Applications	ATAC	ARES	Monkey
QuickSettings	0	0	0
MunchLife	0	0	0
Silent-ping-sms	0	0	0
BatteryDog	1.0	0.1	0
AnyCut	0	0	0
Afwall	0	0	0
Jamendo	0	0	0
Drawablenotepad	0	0	0
AnyMemo	0.5	0.1	0
AmazeFileManager	0.1	0	0
zooborns	1.1	0.9	0
AntennaPod	0	0	0
vanilla	0	0	0
BudgetWatch	0	0	0
sanity	0.1	0.3	0
timber	1.2	0.8	0.2
Drawablenotepad	1.2	0.8	0
Total	47	34	2

5   Threats to Validity

Internal Threats. The threat to internal validity is the non-deterministic characteristic of reinforcement learning. The code coverage may be different in different executable iterations. To reduce the threat, ATAC executes ten times on each app with the same method and take the average as the final result.

External Threats. The threat to external validity is the applications used for evaluation. Though there are a large number of apps, ATAC only tests seventeen apps. We choose apps with different categories and different sizes to reduce the threat.

6   Related Work

Random Exploration Strategy. Dynodroid [21] focuses on the widgets and the input of the widgets. It uses a novel randomized algorithm to select a widget. Monkey is a widely-used tool provided by Google for stability and stress testing. It sends pseudo-random user events (such as key, gesture, touch screen, and other inputs) to the system. However, it may generate many invalid test generations, which are meaningless for test applications. Some researchers [17,27] introduce fuzz testing to test applications, and they generate fuzz intent input. Intent Fuzzer [17] combines static analysis and random test generation. Droidfuzzer [34] analyzes the Intent-filter tag in the AndroidManifest.xml file to targets activities.

Model-Based Exploration Strategy. Model-based methods [3,30], which test applications through building precise or abstract GUI models, are also widely used. They use dynamic or static strategies to model the behaviors and generate test cases. Beak et al. [5] proposes a set of multi-level GUI Comparison Criteria (GUICC) that provides the selection of multiple abstraction levels for GUI model generation. Stoat [28] combines dynamic analysis and static analysis to generate test cases. Espada et al. [13] models the behaviors of users by composing specially-designed state machines, which can be used by a model checking tool to generate all possible user interaction. Amalfitano et al. [2] explores the app's GUI intending to exercise the application structure. The accuracy and completeness are challenged for model-based or structured methods. Meanwhile, inherent state explosion is also a problem.

System Exploration Strategy. To alleviate the path explosion problem, Anand et al. [4] proposes an approach is based on concolic testing. Gao et al. [14] builds a dynamic symbolic execution engine for Android applications without manual modeling of the execution environment. Evolutionary approaches [11,22] are proposed in system testing of Android applications. Implementing a systematic strategy is able to explore behavior that would be hard to reach with random techniques. However, these tools are considerably less scalable than random techniques.

Machine Learning Based Strategy. Machine learning techniques including supervising learning, active learning, and reinforcement learning are widely used by researchers. SwiftHand [7] applies machine learning to learn a model of the app during testing and uses the model to generate user inputs that visit unexplored states of the app. Then, it uses the app's execution on the generated inputs to refine the model. More and more researchers [1,9,20,23,29,33] try to adopt reinforce learning into GUI test generation. Most of them [2,31] utilize Q-learning as an agent which maintains a Q-table to record Q value. Q-testing [25] uses Q-learning and divides different states at the granularity of functional scenarios to efficiently explore different functionalities. The Q-table needs huge memories if the states and actions space is large. [8,32] replace Q-learning with Deep Q Network, which utilizes Q network to predict actions in certain states. ARES [26] is a Deep RL approach for black-box testing of Android applications, and it employs DDPG, SAC, and TD3 algorithms as the agents. However, it needs a huge replay buffer to save the experience.

7   Conclusion

The paper proposed an approach ATAC based on reinforcement learning to generate test cases to test Android GUI automatically. Our approach applies a deep neural network as an agent, and ATAC uses the A2C algorithm to train our agent. Our approach is evaluated in seventeen apps and performs better than Monkey and ARES. It has higher instruction, branch, line, and method coverage. The performance of detecting failures is also better. We will compare ATAC with more tools and explore more apps in future work. In our work, state change is defined by the transitions of activities. In the next stage, we consider using functional scenarios transitions instead. We also consider transplanting our work to IOS apps.

References

1. Adamo, D., Khan, M.K., Koppula, S., Bryce, R.C.: Reinforcement learning for Android GUI testing. In: Proceedings of the 9th ACM SIGSOFT International Workshop on Automating TEST Case Design, Selection, and Evaluation, A-TEST@SIGSOFT FSE 2018, Lake Buena Vista, FL, USA, 05 November 2018, pp. 2–8. ACM (2018)
2. Amalfitano, D., Fasolino, A.R., Tramontana, P., Carmine, S.D., Memon, A.M.: Using GUI ripping for automated testing of Android applications. In: IEEE/ACM International Conference on Automated Software Engineering, ASE 2012, Essen, Germany, 3–7 September 2012, pp. 258–261. ACM (2012)
3. Amalfitano, D., Fasolino, A.R., Tramontana, P., Ta, B.D., Memon, A.M.: MobiGU-ITAR: automated model-based testing of mobile apps. IEEE Softw. **32**(5), 53–59 (2015)
4. Anand, S., Naik, M., Harrold, M.J., Yang, H.: Automated concolic testing of smartphone apps. In: 20th ACM SIGSOFT Symposium on the Foundations of Software Engineering (FSE-2012), SIGSOFT/FSE 2012, Cary, NC, USA, 11–16 November 2012, p. 59. ACM (2012)

5. Baek, Y.M., Bae, D.: Automated model-based Android GUI testing using multi-level GUI comparison criteria. In: Proceedings of the 31st IEEE/ACM International Conference on Automated Software Engineering, ASE 2016, Singapore, 3–7 September 2016, pp. 238–249. ACM (2016)
6. Brockman, G., et al.: OpenAI gym. CoRR abs/1606.01540 (2016)
7. Choi, W., Necula, G.C., Sen, K.: Guided GUI testing of Android apps with minimal restart and approximate learning. In: Proceedings of the 2013 ACM SIGPLAN International Conference on Object Oriented Programming Systems Languages & Applications, OOPSLA 2013, Part of SPLASH 2013, Indianapolis, IN, USA, 26–31 October 2013, pp. 623–640. ACM (2013)
8. Collins, E., Dias-Neto, A.C., Vincenzi, A., Maldonado, J.C.: Deep reinforcement learning based Android application GUI testing. In: SBES 2021: 35th Brazilian Symposium on Software Engineering, Joinville, Santa Catarina, Brazil, 27 September–1 October 2021, pp. 186–194. ACM (2021)
9. Degris, T., Pilarski, P.M., Sutton, R.S.: Model-free reinforcement learning with continuous action in practice. In: American Control Conference, ACC 2012, Montreal, QC, Canada, 27–29 June 2012, pp. 2177–2182. IEEE (2012)
10. Android Developers: Android debug bridge (ADB). https://developer.android.com/studio/command-line/adb
11. Dong, Z., Böhme, M., Cojocaru, L., Roychoudhury, A.: Time-travel testing of Android apps. In: ICSE 2020: 42nd International Conference on Software Engineering, Seoul, South Korea, 27 June–19 July 2020, pp. 481–492. ACM (2020)
12. EclEmma: Jacoco Java code coverage library. https://www.eclemma.org/jacoco/index.html
13. Espada, A.R., Gallardo, M., Salmerón, A., Merino, P.: Using model checking to generate test cases for Android applications. In: Proceedings Tenth Workshop on Model Based Testing, MBT 2015, London, UK, 18 April 2015, vol. 180, pp. 7–21. EPTCS (2015)
14. Gao, X., Tan, S.H., Dong, Z., Roychoudhury, A.: Android testing via synthetic symbolic execution. In: Proceedings of the 33rd ACM/IEEE International Conference on Automated Software Engineering, ASE 2018, Montpellier, France, 3–7 September 2018, pp. 419–429. ACM (2018)
15. GitHub. https://github.com
16. Google: UI/application exerciser monkey. https://developer.android.com/studio/test/monkey
17. NCC Group: Intent fuzzer. https://www.nccgroup.trust/us/our-research/intent-fuzzer/
18. Gu, T., et al.: AimDroid: activity-insulated multi-level automated testing for Android applications. In: 2017 IEEE International Conference on Software Maintenance and Evolution, ICSME 2017, Shanghai, China, 17–22 September 2017, pp. 103–114. IEEE Computer Society (2017)
19. Gu, T., et al.: Practical GUI testing of Android applications via model abstraction and refinement. In: Proceedings of the 41st International Conference on Software Engineering, ICSE 2019, Montreal, QC, Canada, 25–31 May 2019, pp. 269–280. IEEE/ACM (2019)
20. Li, X., et al.: RLINK: deep reinforcement learning for user identity linkage. World Wide Web **24**(1), 85–103 (2021). https://doi.org/10.1007/s11280-020-00833-8

21. Machiry, A., Tahiliani, R., Naik, M.: Dynodroid: an input generation system for Android apps. In: Joint Meeting of the European Software Engineering Conference and the ACM SIGSOFT Symposium on the Foundations of Software Engineering, ESEC/FSE 2013, Saint Petersburg, Russian Federation, 18–26 August 2013, pp. 224–234. ACM (2013)

22. Mahmood, R., Mirzaei, N., Malek, S.: EvoDroid: segmented evolutionary testing of Android apps. In: Proceedings of the 22nd ACM SIGSOFT International Symposium on Foundations of Software Engineering (FSE-22), Hong Kong, China, 16–22 November 2014, pp. 599–609. ACM (2014)

23. Mnih, V., et al.: Asynchronous methods for deep reinforcement learning. In: Proceedings of the 33nd International Conference on Machine Learning, ICML 2016, New York City, NY, USA, 19–24 June 2016. JMLR.org (2016)

24. Appium: Mobile app automation made awesome (n.d.). http://appium.io/

25. Pan, M., Huang, A., Wang, G., Zhang, T., Li, X.: Reinforcement learning based curiosity-driven testing of Android applications. In: ISSTA 2020: 29th ACM SIGSOFT International Symposium on Software Testing and Analysis, Virtual Event, USA, 18–22 July 2020, pp. 153–164. ACM (2020)

26. Romdhana, A., Merlo, A.: Keynote: ARES: a deep reinforcement learning tool for black-box testing of Android apps. In: 19th IEEE International Conference on Pervasive Computing and Communications Workshops and other Affiliated Events, PerCom Workshops 2021, Kassel, Germany, 22–26 March 2021, p. 173. IEEE (2021)

27. Sasnauskas, R., Regehr, J.: Intent fuzzer: crafting intents of death. In: Proceedings of the 2014 Joint International Workshop on Dynamic Analysis (WODA) and Software and System Performance Testing, Debugging, and Analytics (PERTEA), WODA+PERTEA 2014, San Jose, CA, USA, 22 July 2014, pp. 1–5. ACM (2014)

28. Su, T., et al.: Guided, stochastic model-based GUI testing of Android apps. In: Proceedings of the 2017 11th Joint Meeting on Foundations of Software Engineering, ESEC/FSE 2017, Paderborn, Germany, 4–8 September 2017, pp. 245–256. ACM (2017)

29. Sutton, R.S., Barto, A.G.: Reinforcement Learning - An Introduction. Adaptive Computation and Machine Learning. MIT Press, Cambridge (1998)

30. Takala, T., Katara, M., Harty, J.: Experiences of system-level model-based GUI testing of an Android application. In: Fourth IEEE International Conference on Software Testing, Verification and Validation, ICST 2011, Berlin, Germany, 21–25 March 2011, pp. 377–386. IEEE Computer Society (2011)

31. Vuong, T.A.T., Takada, S.: A reinforcement learning based approach to automated testing of Android applications. In: Proceedings of the 9th ACM SIGSOFT International Workshop on Automating TEST Case Design, Selection, and Evaluation, A-TEST@SIGSOFT FSE 2018, Lake Buena Vista, FL, USA, 05 November 2018, pp. 31–37. ACM (2018)

32. Vuong, T.A.T., Takada, S.: Semantic analysis for deep Q-network in Android GUI testing. In: The 31st International Conference on Software Engineering and Knowledge Engineering, SEKE 2019, Hotel Tivoli, Lisbon, Portugal, 10–12 July 2019, pp. 123–170. KSI Research Inc. and Knowledge Systems Institute Graduate School (2019)

33. Williams, R.J.: Simple statistical gradient-following algorithms for connectionist reinforcement learning. Mach. Learn. 8, 229–256 (1992). https://doi.org/10.1007/BF00992696

34. Ye, H., Cheng, S., Zhang, L., Jiang, F.: DroidFuzzer: fuzzing the Android apps with intent-filter tag. In: The 11th International Conference on Advances in Mobile Computing & Multimedia, MoMM 2013, Vienna, Austria, 2–4 December 2013, p. 68. ACM (2013)

RoFL: A Robust Federated Learning Scheme Against Malicious Attacks

Ming Wei[1], Xiaofan Liu[1], and Wei Ren[1,2,3(✉)]

[1] School of Computer Science, China University of Geosciences, Wuhan, China
{mingwei,lxfpolarbear,weirencs}@cug.edu.cn
[2] Henan Key Laboratory of Network Cryptography Technology, Zhengzhou, China
[3] Hubei Key Laboratory of Intelligent Geo-Information Processing, Wuhan, China

Abstract. Privacy protection is increasingly important in contemporary machine learning-based applications. While federated learning can provide privacy protection to some extent, it assumes that clients (and their updates) are trusted. However, we also need to consider the potential of malicious or compromised clients. In this paper, we propose a robust federated learning (RoFL) scheme, designed to detect multiple attacks and block malicious updates from being passed to the central model. To validate our scheme, we train a CNN classification model based on the MNIST dataset. We then conduct experiments focusing on the impacts of model parameters (e.g., malicious amplification factors, fractions of training clients, fractions of malicious clients, and data distribution characteristics (i.e., IID or Non-IID)) on the proposed (RoFL) scheme. The findings demonstrate that the proposed (RoFL) scheme can effectively protect federated learning models from malicious attacks.

Keywords: Federated learning · Privacy protection · Malicious detection · Edge computing

1 Introduction

As the application of machine learning becomes more pervasive in our society, there is a demand for such applications to support properties such as privacy protection. Hence, there has been a trend of designing machine learning models to also consider privacy protection, interpretability, ethics, etc. However, in the context of privacy protection, the latter conflicts with accuracy since we can obtain better models once we can access a wide variety of data (including sensitive data from different users) for training. Federated learning is an efficient method to help with privacy protection, which enables each user to train the model with their data locally and transmit the training parameter updates (instead of transmitting raw data) to a centralized server. It can also be deployed in edge-cloud based environments, which results in improved efficiency, reduced latency and computational cost, etc. [1–4]. Despite these advantages, there are several limitations associated with federated learning-based approaches such as

B. Li et al. (Eds.): APWeb-WAIM 2022, LNCS 13423, pp. 277–291, 2023.
https://doi.org/10.1007/978-3-031-25201-3_21

the potential for data leakage during the process of gradients transmission and the risk of malicious/compromised servers sharing incorrect parameter updates. While we can use encryption algorithms to minimize the risk of data leakage, this comes at the cost of significant communication cost increase.

In this paper, we design a robust federated learning scheme that allows one to identify malicious clients without incurring significant communication costs, while also ensuring the security of federated learning. The contributions of this paper are as follows:

1. The possible malicious attacks in the training process of the federated learning model are discussed and summarized into three adversary models.
2. A robust federated learning (**RoFL**) scheme which adds the detection of malicious attacks is proposed. In this way, the training accuracy and speed of the central server are guaranteed.

The rest of this paper will be organized as follows: Sect. 2 introduces the related works. Section 3 develops the **RoFL** algorithm and three possible adversary models. Section 4 proposes a robust scheme against malicious attacks and gives comprehensive security analysis about the given scheme. Section 5 presents the training procedures and experiments. Section 6 concludes the paper.

2   Related Works

The concept of federated learning was first proposed by McMahan et al. [5]. They developed a deep learning model, FederatedAveraging (FedAvg), which is communication-efficient for decentralized data named federated learning. There have been many research about the application of federated learning. Yang et al. [6] presented a very detailed framework including horizontal federated learning, vertical federated learning and federated transfer learning. They divided federated learning into different types, and described the suitable scenarios, advantages and disadvantages of each type. Bonawitz et al. [7] proposed a system that is applicable on mobile phones. They observed that bias or some other reasons may lead to the inaccuracy of model training. But they did not use additional methods to improve such situation. Mandal et al. [8] presented PrivFL system which emphasized security and privacy performance [9] of federated learning. However, the computation time of such system still needs to be reduced by implementing specific interfaces.

Besides the application fields and scenarios, many researchers were concerned about the security of federated learning. Their concerns may be mainly divided into two aspects, the security in the process of data transmission, and whether the server is malicious. Bonawitz et al. [10] improved safety performance in federated learning through the aggregation of high-dimensional data. Their protocol ensured security in the process of delivering data between servers. However, this kind of protocol increased the costs during data transmission. Furthermore, Bhowmick et al. [11] introduced the differential privacy algorithm which can protect privacy leakage caused by minor changes to improve the data safety level.

Agarwal et al. [12] used secure multiparty computation (SMC) to encrypt the transmitted data. Afterwards, Runhua et al. [13] and Truex et al. [14] combined both differential privacy and SMC to build a hybrid algorithm to protect the safety of gradients when transmitted between servers.

Different from those encryption algorithms and data protection algorithms adopted in the transmission period of federated learning, Li et al. [15] took into account the situations where malicious data exists. They compared the situation in which the server may be malicious with the situation in which no additional privacy preservation is added. The result turned out that the mechanism in which detection about whether the server is malicious is added performed better and reminded us to pay attention to the possible malicious data problem in federated learning. Mouthukuri et al. [16] do a survey on security and privacy of federated learning which concludes providing an overview of realizing federated learning, and identifying and examining security vulnerabilities and threats in federated learning environments. Besides, they also summarize the future scope for security and privacy of federated learning. To protect users' privacy, Mothukuri et al. [17] also propose a federated-learning-based anomaly detection. Different from training on the dataset from the server, the detection scheme can identify the intrusion on various Internet of things devices, and keep the integrity of by sharing the weight. However, there are few research about detection of malicious clients which may affect both the results and the speed in training processes of the federated learning model.

3   Problem Formulation

This section will focus on the problem formulation of robust models under the federated learning framework. First, we will introduce the general federated learning framework under deep learning settings. Then, we select several adversary models to represent what our model may encounter while transmitting privacy information.

3.1   Federated Learning Without Detection of Malicious Attacks

Privacy is one of the major concerns in applications of federated learning. To protect the localized data for each client, it shares the model updates instead of exchanging the raw data. Here we consider the local updating scheme and give the brief description of one round of the process: Step 1, the server distribute its global model to each client; Step 2, each client computes gradients from a minibatch of local data points and applies the gradients to update the local model iteratively; Step 3, the server performs a global aggregation after certain steps of local updates. Local updating schemes can reduce communication by training models locally.

Here we mathematically formulate the federated learning solution with the local update scheme. The goal of federated learning is to learn the parameters $\mathbf{W}$ of a neural network[1]. The training data, however, due to privacy concerns or expensive cost of sending raw data, are distributed over a large number K of clients and cannot be sent to the server. Here we assume K is a fixed number and each client $k \in [1, K]$ is stored with a fixed local dataset. Then we want to iteratively generalized the server model to fit the training data without explicitly delivering the raw data. Specifically, at the beginning of each round $t \geq 0$, the server distributes the current global parameters $\mathbf{W}_t$ to a fraction $C \in (0, 1]$ of clients, where the number of selected clients $m = \lceil CK \rceil$ should be a positive integer with rounding up for CK. We denote these random selected m clients using a subset $\mathcal{S}_t$ among K clients. Then each client in $\mathcal{S}_t$ independently performs E local epochs of training to fit their own local data. Once we have the well-trained local parameters $\mathbf{W}_t^1, \mathbf{W}_t^2, \cdots, \mathbf{W}_t^m$, they are sent back to the server and the server applies these updates to its global model by averaging them: $\mathbf{W}_{t+1} = \frac{1}{m} \sum_{i=1}^{m} \mathbf{W}_t^i$. This ends a round of training, and then the whole process iterates. The above method is known as **FedAvg**. Note that in literature [18], the updating scheme has another form: $\mathbf{W}_{t+1} = \mathbf{W}_t + \frac{1}{m} \sum_{i=1}^{m} (\mathbf{W}_t^i - \mathbf{W}_t)$, which is mathematically identical. For simplicity, here we use the first form in this paper.

3.2   Adversary Model

In the previous part, we have introduced the conventional local update scheme. However, this **FedAvg** scheme is based on the assumption that no malicious clients exists. In the following, we will introduce several adversary models that our model may potentially encounter during transmitting information between clients and the central server. And we will show how malicious clients would affect the updating performance of the global model.

Adversary Model 1: Origin Attack. In most cases, a certain portion of clients is initially and permanently malicious. Here we name this adversary as origin adversary, which attacks from the original client/data. For all m clients, we denote the portion as γ, thus γm clients are malicious and the other $(1-\gamma)m$ clients are normal. However, these malicious clients would randomly corrupt the data per transmission with a probability β every time.

Adversary Model 2: Transmission Attack. There is another situation that all of the clients are always trustworthy. However, malicious attackers may corrupt the data during transmission. Such that there would have a random fraction $\tilde{C} \in (0, 1]$ of the m selected clients whose transmitted data are corrupted, so that

---

[1] In this paper, we consider the federated learning for deep neural networks.

there would be a random $\tilde{m} = \lceil \tilde{C}m \rceil$ clients passing their corrupted gradients to the server[2].

Adversary Model 3: Cheater Attack. Some attackers may claim to be other clients involved in the training process and submit corrupted data to the server for multiple times. Here we name them as cheater attackers. For instance, the server may have initially collected a certain amount of data from a fixed number of clients m. However, there are more data from more than m clients, which confuses the server to verify the authenticity of these data.

4 Proposed Scheme

In this section, we introduce our robust federated learning method to cope with the above adversary models, which is with one more verification step before updating the global model. The server would use a small set of test datasets to verify the local parameters $\mathbf{W}_t^i$, $\forall i \in [1, m]$, such as to obtain its corresponding loss $\mathcal{L}(\mathbf{W}_t^i)$. The server would have a tolerance on the loss value, therefore parameters of any client that satisfies $\mathcal{L}(\mathbf{W}_t^i) \leq \xi \mathcal{L}(\mathbf{W}_t)$ can be adopted to update the server model, where ξ is the rejection factor determining the upper bound of the acceptable loss. Note that we may take similar procedure to apply to other metrics such as accuracy.

Case 1: Counter. In Origin Attack, suppose m clients take part in the procedure of federated learning while γm of them interfere the process by sending the wrong data to the server every time. We call these γm clients as "original attackers" $\{\mathcal{A}_1, \mathcal{A}_2, \cdots, \mathcal{A}_{\gamma m}\}$. Suppose attackers process the data by multiplying a factor α, and each attacker would select the different factor by herself/himself. Besides, different value of α will have the different influence on the results. However, with the standard range of the model data, server can find these abnormal data.

Based on the above steps, we may roughly determine which data are assumed corrupted, thus pointing to the potential malicious clients. We set a counter N for each client and increase the counting once the data transmitted from this client is assumed corrupted. As soon as the counter exceeds a certain threshold N_0 ($N_0 < m$), we will determine the particular client as a malicious client. And the server won't pass the updated parameters to all of these malicious clients nor collect their submitted data in the future process. Besides, all these clients will be marked as "attackers" and the information will be broadcasted among all m clients. Finally, all attackers will be removed from our learning system. Overall, the whole process can be summarized in Algorithm 1.

---

[2] Note that the malicious clients are randomly selected from the m clients for each round, which means the gradients of one client may be corrupted in the current round but be normal in the next round.

Algorithm 1: RoFL1. We denote C as the fraction of selected clients for each round; K as the total number of clients; E as the number of local epochs; ξ as the rejection factor; N^i as the number of times client i is considered malicious; N_0 as the threshold to determine the particular client as a malicious client;

Server Updates:
Initialize $\mathbf{W}_0$;
$m \leftarrow \lceil C \cdot K \rceil$;
$\{N^i\}_{i \in \mathcal{S}_t} \leftarrow 0$;
for $t = 1$ **to** T **do**
  $\mathcal{S}_t \leftarrow$ select a random set of m clients;
  for $i \in \mathcal{S}_t$ **do**
    $\mathbf{W}_t^i \leftarrow$ ClientUpdate$(i, \mathbf{W}_t)$;
    $F_t^i \leftarrow$ ClientCheck$(\mathbf{W}_t^i, \mathbf{W}_t, \xi, N^i, N_0)$;
  $N \leftarrow \sum_{i \in \mathcal{S}_t} F_t^i$; // N is the number of valid clients
  $\mathbf{W}_{t+1} \leftarrow \frac{1}{N} \sum_{i \in \mathcal{S}_t} F_t^i \mathbf{W}_t^i$;
return $\mathbf{W}_T$;

ClientUpdate$(i, \mathbf{W})$: // Local updates for client i
for $epoch = 1$ **to** E **do**
  $\mathbf{W} \leftarrow$ Gradient Decent;
return $\mathbf{W}$ to server;

ClientCheck$(\mathbf{W}^i, \mathbf{W}, \xi, N^i, N_0)$: // Server checks client i
Compute $\mathcal{L}(\mathbf{W}^i)$ and $\mathcal{L}(\mathbf{W})$;
if $\mathcal{L}(\mathbf{W}^i) \leq \xi \mathcal{L}(\mathbf{W})$ *and* $N^i \leq N_0$ **then**
  $F = 1$;
else
  $F = 0$; $N^i += 1$;
return F;

Case 2: Validator. In Transmission Attack, we denote an adversary $\mathcal{U}$ who wants to attack and control some honest clients to send malicious parameters to the server. Different from Origin Attack, $\mathcal{U}$ can randomly select different portion of all clients to send malicious data at one time, that is, one client can be selected several times in succession or not be selected for a long time and $\mathcal{U}$ will select different number of clients to control them and send the wrong data. The selected portion β belongs to $[0, 1]$. Thus, it is complicated and meaningless to find out all controlled clients, and it is more important to improve the robustness of the federated learning model.

In this case, given that the data are contaminated during transmission processes, we build up a validator to verify the trustworthiness of information for each round of update. The validator count the number of valid clients according to the above step. Then we set ψ as the tolerance ratio for corruption, which we usually take $\psi = 0.5$. Generally, we consider that less than half of the available

---

Algorithm 2: RoFL2. We denote C as the fraction of selected clients for each round; K as the total number of clients; E as the number of local epochs; ξ as the rejection factor; ψ as the tolerance ratio for contamination;

Server Updates:
Initialize $\mathbf{W}_0$;
$m \leftarrow \lceil C \cdot K \rceil$;
for $t = 1$ **to** T **do**
   $\mathcal{S}_t \leftarrow$ select a random set of m clients;
   for $i \in \mathcal{S}_t$ **do**
      $\mathbf{W}_t^i \leftarrow$ ClientUpdate(i, $\mathbf{W}_t$);
      $F_t^i \leftarrow$ ClientCheck($\mathbf{W}_t^i, \mathbf{W}_t, \xi$);
   $N \leftarrow \sum_{i \in \mathcal{S}_t} F_t^i$; // N is the number of valid clients
   if $N \leq m(1 - \psi)$ **then**
      | continue;
   else
      $\mathbf{W}_{t+1} \leftarrow \frac{1}{N} \sum F_t^i \mathbf{W}_t^i$;
return $\mathbf{W}_T$;

ClientUpdate($i, \mathbf{W}$): // Local updates for client i
for $epoch = 1$ **to** E **do**
   $\mathbf{W} \leftarrow$ Gradient Decent;
return $\mathbf{W}$ to server;

ClientCheck($\mathbf{W}^i, \mathbf{W}, \xi$): // Server checks client i
Compute $\mathcal{L}(\mathbf{W}^i)$ and $\mathcal{L}(\mathbf{W})$;
if $\mathcal{L}(\mathbf{W}^i) \leq \xi\mathcal{L}(\mathbf{W})$ **then**
   | $F = 1$;
else
   $F = 0$;
return F;

---

data is not credible. If once corrupted clients are over half of all clients after detection, the server will restart this round learning and send the model back again. However, in fact, it is hard for the adversary $\mathcal{U}$ to attack and control half of the all clients once since it needs strong computing power. Therefore, we set the tolerance ratio as 0.5.

Once the validator detects the portion of valid clients is smaller than 50%, we assume that the updating information are deeply corrupted, and we directly discard the updating information for this round. Otherwise, we will update the gradients using the transmitted information from those valid clients. Overall, the whole process can be summarized in Algorithm 2.

Case 3: Authenticator. Here we denote an adversary who wants to pretend as a selected client and send the wrong data to the server as $\mathcal{P}$. In this setting,

---

Algorithm 3: RoFL3. Here C, K, and E are the same meaning as defined in the previous algorithms;

Server Updates:
Initialize $\mathbf{W}_0$; $m \leftarrow \lceil C \cdot K \rceil$;
Assign secret key $\{sk_i^{(c)}\}_{i=1}^m$ to m clients via a secret channel;
Store a copy of $\{sk_i^{(c)}\}_{i=1}^m$ as $\{sk_i^{(s)}\}_{i=1}^m$ in server;
Make agreements with clients on default rules for updating secret keys;
for $t = 1$ **to** T **do**
$\quad$ $\mathcal{S}_t \leftarrow$ select a random set of m clients;
$\quad$ **for** $i \in \mathcal{S}_t$ **do**
$\quad\quad$ $\mathbf{W}_t^i \leftarrow$ ClientUpdate(i, $\mathbf{W}_t$);
$\quad\quad$ $hash_i^{(c)} \leftarrow$ HashUpdate($sk_i^{(c)}$);
$\quad\quad$ $F_t^i \leftarrow$ ClientCheck($hash_i^{(c)}$);
$\quad$ $N \leftarrow \sum_{i \in \mathcal{S}_t} F_t^i$; // N is the number of valid clients
$\quad$ $\mathbf{W}_{t+1} \leftarrow \frac{1}{N} \sum F_t^i \mathbf{W}_t^i$;
return $\mathbf{W}_T$;

ClientUpdate(i, $\mathbf{W}$): // Local updates for client i
for $epoch = 1$ **to** E **do**
$\quad$ $\mathbf{W} \leftarrow$ Gradient Decent;
return $\mathbf{W}$ to server;

HashUpdate(sk_i): // Hash updates for client i
Update sk_i using the default rule; Generate $hash_i$ using the updated sk_i;
return $hash_i$;

ClientCheck($hash_i^{(c)}$): // Server checks client i
$hash_i^{(s)} \leftarrow$ HashUpdate($sk_i^{(s)}$);
if $hash_i^{(s)} \equiv hash_i^{(c)}$ **then**
$\quad$ return 1;
else
$\quad$ return 0;

---

other honest clients that were chosen in the training process may potentially be framed by $\mathcal{P}$. We call $\mathcal{P}$ as a cheater.

Once there are cheater attackers, we cannot simply filter the corrupted data. In this model, we incorporate the hash value into identity authentication steps. In the beginning, the server assign secret keys to each client via absolutely safe channels and store a copy of these keys. Then the server make agreements with each client on certain rules of how to update secret keys after exchanging parameters of each round. In the updating scheme, clients generate hash values using the updated secret keys and send their gradients along with the hash values to the server. Overall, the whole process can be summarized in Algorithm 3.

According to the processes of malicious attacks, the cheater $\mathcal{P}$ wants to send malicious parameters to the client in the same way as other honest clients. In this

Table 1. Parameters of CNN architecture.

Layer	Kernel	Feature map	Padding
Input	/	$28 * 28 * 1$	/
Conv1 + MaxPooling1	$5 * 5 * 16$	$12 * 12 * 16$	Valid
Conv2 + MaxPooling2	$5 * 5 * 32$	$4 * 4 * 32$	Valid
Flatten+FullyConnect1&2+Softmax	/	$512{\rightarrow}128{\rightarrow}10$	/

*Activation functions: 'ReLu'.

case, $\mathcal{P}$ should send parameters with the hash value of the framed client's secret key to the server. Thus, if $\mathcal{P}$ wants to forge as other clients which are selected in the training process, $\mathcal{P}$ should obtain their secret keys which are difficult to achieve since secret keys are kept secretly from others. Another way for $\mathcal{P}$ to take part int the training process is to find the collision of the hash function. Here, we use SHA-256 as cryptographic hash function in Algorithm 3 which is considered very difficult to find collisions. Therefore, it is nearly impossible for $\mathcal{P}$ to send malicious data to the server as impostors.

5   Experiments and Analysis

In this section, we conduct experiments on MNIST digit recognition dataset [19] for image classification task using a convolutional neural network (CNN) model with architecture listed in Table 1. In the following, the specific experimental settings and results are detailed.

5.1   Experiments Settings

The experiments are conducted on MNIST dataset. This dataset has a total of $70,000$ small square 28×28 grayscale images of handwritten single digits between 0 to 9, with $60,000$ for training set and $10,000$ for test set. The image classification task is to classify a given image of a handwritten digit into one of 10 classes representing integer values from 0 to 9, inclusively. Following the setting of [5], we consider the heterogeneity of the clients, i.e., the clients may collect data in a highly non-independent and identically distributed (Non-IID) manner.

As for the parameter settings for training, we fixed the number of clients as $K = 100$, the local training epochs on each clients as 5, the local training batch size as 32, and the test batch size as 128. The training is starting with 0.01 learning rate using the SGD momentum optimizer training for 20 epochs. All the experiments are conducted with a fixed random seed 100. To compare the performances of malicious detection models, we mainly focus on four groups listed in Table 2. In the following part, we conduct experiments for each setting to evaluate the effects of different parameters.

Table 2. Experiments settings.

\		Detect malicious clients	
		True (RoFL)	False (FedAvg)
Malicious clients	True	Exp. A	Exp. B
	False	Exp. C	Exp. D

5.2   Case 1, Experiments About RoFL with the Counter

In this part, the malicious clients are initially and permanently existed. To simulate the malicious clients, we apply a modification of the local weights using: $\mathbf{W}_t^i = \alpha \mathbf{W}_t^i \times (1 + \mathrm{rand}(0,1))$, where α is the malicious amplification factor. In addition, we set $N_0 = 3$ as the tolerance threshold for potential malicious clients.

The Parameter α. Here we investigate the effect of the malicious amplification factor α. To fix the other parameters, we set moderate values that $\gamma = 0.5$ and $\beta = 0.8$ using IID data. As for α, they are selected from the set $\{0.5, 1.0, 2.0\}$. The results are shown in Fig. 1, with Fig. 1a the loss and Fig. 1b the accuracy. These results demonstrate the effectiveness of the detection scheme: The models with malicious clients without detection (the *Exp. B* with $\alpha = 1.0$ and 2.0) have bad performances (i.e., higher loss and lower accuracy) than those with detection scheme. Specially, we should note that the *Exp. B* with $\alpha = 0.5$ has similar performance comparing to *Exp. A*, *Exp. C*, and *Exp. D*. This is due to the malicious process: given $\alpha = 0.5$, the weights are randomly reduced to 50%–100% of its original value, this is similar to the effect of the weight decay regularization, thus this setting can be with a comparable performance.

In general, with the malicious amplification factor $\alpha \geq 1.0$, the models without malicious detection scheme would have awful performances, while the ones with malicious detection perform well. In the following experiments, we will fix $\alpha = 1.0$ as a moderate value.

The Parameter γ. Given the malicious scheme, we investigate the effect of the malicious portion γ. Thus, we fix $\beta = 0.8$ and set $\gamma = 0.2, 0.5, 0.8$. The results are shown in Fig. 2. In general, the results prove the effectiveness of the malicious detection module. Those without malicious detection module show much worse results than the other experiments with malicious detection module. However, we may find that the *Exp. B* with $\gamma = 0.2$ also has a good performance, which is reasonable. Recall that the parameter γ denotes the malicious portion; we can expect the experiments with a small amount of malicious clients to have good results since the malicious clients may not get much chance to be selected for training thanks to the randomization selection scheme. In addition, we also evaluate the detection rate for each experiments in *Exp. A*: when $\gamma = 0.2$, the detection rate of malicious clients is 100%; when $\gamma = 0.5$, the detection rate is

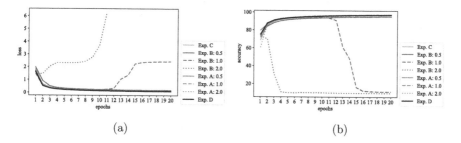

Fig. 1. Experiments with different malicious amplification factors α.

58%; and when $\gamma = 0.8$, the detection rate is 48.75%. Given that we only train the model for 20 epochs and we have set $N_0 = 3$, once we have a high portion of malicious clients, it would be difficult to detect all the malicious clients within limited training epochs. However, the detection module proved its detection ability in *Exp. A* with $\gamma = 0.2$, in which all the malicious clients are detected.

The Parameter β. Here we investigate the effect of the contaminating probability β. Thus, we fix $\gamma = 0.5$ and β is selected from the set $\{0.6, 0.8, 1.0\}$. The results are shown in Fig. 3. The results show that the malicious detection module is effective: those without the detection module have awful results (*Exp. B*) while the ones with the detection module perform excellent (*Exp. A*). We also evaluate the detection rate for each experiments in *Exp. A*: when $\beta = 0.6$, the detection rate of malicious clients is 52%; when $\gamma = 0.8$, the detection rate is 58%; and when $\gamma = 1.0$, the detection rate is 72%.

5.3 Case 2, Experiments About RoFL with the Validator

To simulate the malicious clients, we apply a modification of the local weights using: $\mathbf{W}_t^i = \alpha \mathbf{W}_t^i \times (1 + \text{rand}(0,1))$, where α is the malicious amplification factor. According to the previous results, we directly select a moderate value for α as 1.0. As for the tolerance ratio ψ, we select $\psi = 0.50$.

We select three factors that may have influences on the performances of federated learning. Specifically, we will investigate the effect of the fraction of malicious clients $\tilde{C} \in \{0.1, 0.3, 0.6\}$, the fraction of clients for training $C \in \{0.2, 0.5, 1.0\}$, and the data distribution characteristics (IID or Non-IID). To control the variables, when investigating one parameters, the other parameters are fixed as will be detailed in the following. All experiments are conducted with 20 rounds.

The Parameter $\tilde{C}$. Here we fix $\alpha = 1.0$ and $C = 0.5$ using IID data to investigate the effect of $\tilde{C}$, which would be selected from the set $\{0.1, 0.3, 0.6\}$. According to Fig. 4, the curves are with similar trends as in Fig. 1. It shows that when with only little fraction of malicious clients, the model seems to be

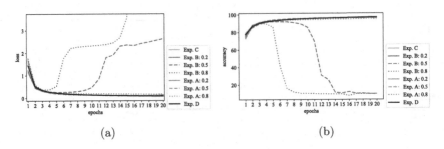

Fig. 2. Experiments with different malicious portion γ.

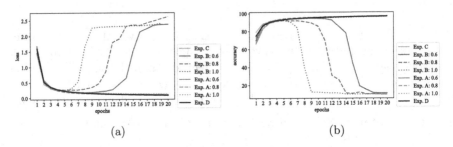

Fig. 3. Experiments with different contaminating probability β.

normally training. Generally, when training with more malicious clients, the performances tend to be worse. However, their performances would be unaffected with malicious detection scheme. In the following, we will fix $\tilde{C} = 0.3$.

The Parameter C and Data Distribution. In this part, we investigate the effect of the training fraction C and the data distribution (i.e., IID or Non-IID). We fix $\alpha = 1.0$ and $\tilde{C} = 0.3$, and select C from the set $\{0.2, 0.5, 1.0\}$. The results are as shown in Fig. 5 and 6. It is clear that training with more clients would make the training curve more stable and smooth, while the curves of those models with $C = 0.2$ actively fluctuate. From Fig. 5 with IID data, only the models with malicious clients and without detection have bad performances, while other models are with similar good performances. It is worth nothing that, in Fig. 6 with Non-IID data, those models with malicious clients and using detection scheme are with more stable curves and better performances than other models without detection scheme.

5.4   Case 3, Experiments About RoFL with the Authenticator

To simulate the cheater attackers, we apply a modification of the local weights using: $\mathbf{W}_t^i = \alpha \mathbf{W}_t^i \times (1 + \mathrm{rand}(0, 1))$, where α is the malicious amplification factor. The results are shown in Fig. 7. The malicious detection module is demonstrated to be effective: Exp. A, Exp. C and Exp. D have lower loss values and higher

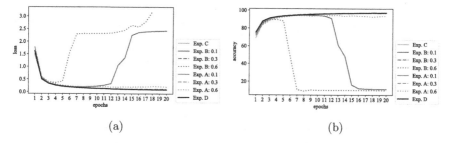

(a)   (b)

Fig. 4. Experiments with different malicious fractions $\tilde{C}$.

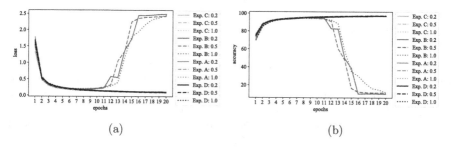

(a)   (b)

Fig. 5. Experiments with different client fractions C using IID data.

accuracy compared with Exp. B (as shown by blue lines in the figure). It can be seen that the proposed RoFL with identity authentication can protect the server from cheater attackers and has high security.

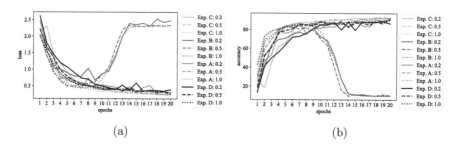

(a)   (b)

Fig. 6. Experiments with different client fractions C using Non-IID data.

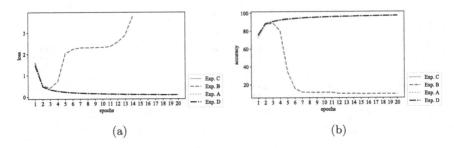

Fig. 7. Experiments with different settings in case 3. (Color figure online)

6   Conclusion

In this paper, to save the situation when the assumption that all clients randomly selected in the training process are trustworthy doesn't hold, we proposed a robust federated learning scheme (**RoFL**) which added the detection procedure to prevent malicious attacks. To validate the performance of **RoFL**, we conducted massive experiments on MNIST digit recognition dataset, and designed several experiments with different parameter settings to investigate the specific influence of each factor. We trained a CNN model targeting image classification tasks with detecting malicious attacks during the process of exchanging updated information. The experimental results were explicitly analyzed, and the principle as well as the validity were explained. The results demonstrated the effectiveness of our model when dealing with malicious clients. In our future work, we will do calculations and assign a weighting to each client after evaluating their training effects and contributions to the overall model. In this way, we can improve the accuracy and increase the training speed to a greater extent.

Acknowledgement. The research was financially supported by the National Natural Science Foundation of China (No. 61972366), the Provincial Key Research and Development Program of Hubei (No. 2020BAB105), the Foundation of Henan Key Laboratory of Network Cryptography Technology (No. LNCT2020-A01), and the Foundation of Hubei Key Laboratory of Intelligent Geo-Information Processing (No. KLIGIP-2021B06).

References

1. Aledhari, M., Razzak, R., Parizi, R.M., Saeed, F.: Federated learning: a survey on enabling technologies, protocols, and applications. IEEE Access **8**, 140699–140725 (2020)
2. Xiao, R., Ren, W., Zhu, T., Choo, K.-K.R.: A mixing scheme using a decentralized signature protocol for privacy protection in bitcoin blockchain. IEEE Trans. Dependable Secure Comput. **18**(4), 1793–1803 (2019)
3. Li, B., Liang, R., Zhou, W., Yin, H., Gao, H., Cai, K.: LBS meets blockchain: an efficient method with security preserving trust in SAGIN. IEEE Internet Things J. **9**(8), 5932–5942 (2021)

4. Liu, Y., et al.: A blockchain-based decentralized, fair and authenticated information sharing scheme in zero trust Internet-of-Things. IEEE Trans. Comput. **72**(2), 501–512 (2023)
5. McMahan, B., Moore, E., Ramage, D., Hampson, S., y Arcas, B.A.: Communication-efficient learning of deep networks from decentralized data. In: Artificial Intelligence and Statistics, pp. 1273–1282. PMLR (2017)
6. Yang, Q., Liu, Y., Chen, T., Tong, Y.: Federated machine learning: concept and applications. ACM Trans. Intell. Syst. Technol. (TIST) **10**(2), 1–19 (2019)
7. Bonawitz, K., et al.: Towards federated learning at scale: system design. arXiv preprint arXiv:1902.01046 (2019)
8. Mandal, K., Gong, G.: PrivFL: practical privacy-preserving federated regressions on high-dimensional data over mobile networks. In: Proceedings of the 2019 ACM SIGSAC Conference on Cloud Computing Security Workshop, pp. 57–68 (2019)
9. Buescher, N., Boukoros, S., Bauregger, S., Katzenbeisser, S.: Two is not enough: privacy assessment of aggregation schemes in smart metering. Proc. Priv. Enhancing Technol. **2017**(4), 198–214 (2017)
10. Bonawitz, K., et al.: Practical secure aggregation for privacy-preserving machine learning. In: Proceedings of the ACM SIGSAC Conference on Computer and Communications Security 2017, pp. 1175–1191 (2017)
11. Bhowmick, A., Duchi, J., Freudiger, J., Kapoor, G., Rogers, R.: Protection against reconstruction and its applications in private federated learning. arXiv preprint arXiv:1812.00984 (2018)
12. Agarwal, N., Suresh, A.T., Yu, F., Kumar, S., Mcmahan, H.B.: cpSGD: communication-efficient and differentially-private distributed SGD. arXiv preprint arXiv:1805.10559 (2018)
13. Xu, R., Baracaldo, N., Zhou, Y., Anwar, A., Ludwig, H.: HybridAlpha: an efficient approach for privacy-preserving federated learning. In: Proceedings of the 12th ACM Workshop on Artificial Intelligence and Security, pp. 13–23 (2019)
14. Truex, S., et al.: A hybrid approach to privacy-preserving federated learning. In: Proceedings of the 12th ACM Workshop on Artificial Intelligence and Security, pp. 1–11 (2019)
15. Li, T., Sahu, A.K., Talwalkar, A., Smith, V.: Federated learning: challenges, methods, and future directions. IEEE Sig. Process. Mag. **37**(3), 50–60 (2020)
16. Mothukuri, V., Parizi, R.M., Pouriyeh, S., Huang, Y., Dehghantanha, A., Srivastava, G.: A survey on security and privacy of federated learning. Future Gener. Comput. Syst. **115**, 619–640 (2021). https://www.sciencedirect.com/science/article/pii/S0167739X20329848
17. Mothukuri, V., Khare, P., Parizi, R.M., Pouriyeh, S., Dehghantanha, A., Srivastava, G.: Federated-learning-based anomaly detection for IoT security attacks. IEEE Internet Things J. **9**(4), 2545–2554 (2022)
18. Konečný, J., McMahan, H.B., Yu, F.X., Richtárik, P., Suresh, A.T., Bacon, D.: Federated learning: strategies for improving communication efficiency. arXiv preprint arXiv:1610.05492 (2016)
19. Deng, L.: The MNIST database of handwritten digit images for machine learning research [best of the web]. IEEE Sig. Process. Mag. **29**(6), 141–142 (2012)

FDP-LDA: Inherent Privacy Amplification of Collapsed Gibbs Sampling via Group Subsampling

Tao Huang[1,2], Hong Chen[1,2(✉)], and Suyun Zhao[1,2]

[1] Key Laboratory of Data Engineering and Knowledge Engineering of Ministry of Education, Renmin University of China, Beijing, China
{huang-tao,chong,zhaosuyun}@ruc.edu.cn
[2] School of Information, Renmin University of China, Beijing, China

Abstract. Latent Dirichlet allocation (LDA) is a widely used fundamental tool for text analysis. Collapsed Gibbs sampling (CGS), as a widely adopted algorithm for learning the parameters of LDA, has the risk of privacy leakage. In this paper, we study the inherent privacy of CGS which is exploited to preserve the privacy for latent topic updates. We propose a method, called group subsampling, and a novel centralized privacy-preserving algorithm, called Fast-Differentially-Private LDA (FDP-LDA) to amplify the inherent privacy and improve the efficiency of traditional differentially private CGS. Theoretically, the general upper bound of the amplified inherent privacy loss in each iteration of FDP-LDA is verified mathematically. To our best knowledge, this is the first work that analyzes the inherent privacy amplification of differentially private CGS. Experimentally, results on real-world datasets validate the improved performances of FDP-LDA.

Keywords: Differential privacy · Latent dirichlet allocation · Collapsed gibbs sampling

1 Introduction

LDA [2], as a popular machine learning model, is widely used to discover the latent semantic structure of text corpus [16,18,19]. CGS [4,11,12], a widely adopted algorithm, often exploits massive datasets to estimate the parameters of LDA while these datasets, e.g. clinic diagnostic records, may contain sensitive information about individuals. To alleviate the issue, some studies combine differential privacy (DP) [5] with CGS to provide CGS with guaranteed privacy [6,20,21]. Specifically, CGS could be analyzed via exponential mechanism [5] in terms of ε-DP [20,21] and thus possesses a certain degree of guaranteed privacy. The inherent privacy of CGS can be utilized to preserve the privacy for latent topic updates. However, the incurred inherent privacy loss is determined by the

---

Supported by organization x.

logarithm of large word counts and thus is often large. Besides, existing topic updates by utilizing the inherent privacy sample new topics for all words in each iteration of CGS and thus might suffer from bad efficiency when the dataset is large.

To overcome these weaknesses, we propose a method, called group subsampling, to amplify the inherent privacy of CGS. By mathematical reasoning, we give a general upper bound of the amplified inherent privacy of CGS. We then introduce group subsampling into differentially private CGS and proposed a novel algorithm called FDP-LDA which amplifies the inherent privacy and achieves better efficiency. Experiments on real-world datasets validate the improvements of our proposed method.

2   Preliminaries

In LDA, a text corpus is considered as a mixture of K latent topics, and each document d_m in a text corpus is represented by a K-dimensional distribution $\vec{\theta_m}$. Additionally, each latent topic k is represented by a V-dimensional distribution $\vec{\phi_k}$, where V represents the total number of unique words in the text corpus. The objective of the LDA training process is obtaining $\vec{\theta_m}$ and $\vec{\phi_k}$.

CGS is a widely used parameter inference method for LDA and has been demonstrated to possess a certain degree of guaranteed differential privacy [6,20,21] as CGS works in the same way as the exponential mechanism [5]. Specifically, CGS samples new topic z_i for the corresponding word w_i according to a distribution over K topics, which is denoted as $\mathbf{P} = (p_1, \ldots, p_k, \ldots, p_K)^\top$. Define a utility function $u(w, k) = \log p_k$ with sensitivity $\Delta u_k = \Delta \log p_k \neq 0$, then p_k can be written as $p_k = \exp\left(\log p_k\right) = \exp\left(\frac{\Delta u_k}{\Delta u_k} \log p_k\right) = \exp\left(\frac{\Delta u_k \cdot u(w,k)}{\Delta u_k}\right)$. According to Definition 3.4 in [5], topic sampling for a single word preserves Δu_k-DP.

3   Framework of FDP-LDA

FDP-LDA is presented in Algorithm 1. Model assumptions for privacy analysis are firstly given. Details about group subsampling are discussed before we present the guaranteed privacy for group subsampling. Finally, we briefly discuss the total privacy loss of each iteration in FDP-LDA.

3.1   Model Assumptions

We assume that the centralized data curator is trustworthy and doesn't violate raw data privacy. Suppose that there exists an adversary who monitors the whole CGS process, namely he masters topic-word counts n_k^t as well as the sampled topic k in each iteration. The neighboring dataset W' of the original private dataset W is constructed by replacing a single word t in W with t'.

3.2 Guaranteed Privacy for Group Subsampling

Group subsampling works as follows. For a word type t with M_t occurrences in document d_m, the M_t occurrences of t is partitioned into s (s is determined by q) groups uniformly. The group sizes are $\lceil M_t q \rceil$, $\lceil M_t q \rceil$, ..., $M_t - (s-1)\lceil M_t q \rceil$. Group can be regarded as 'bucket'. The capacity of each 'bucket' is determined by word count for this unique word type and group subsampling ratio q. When a 'bucket' is replete, the remaining words are assigned to the next 'bucket'.

Based on the explanation of group subsampling, we follow the idea of [20]. The inherent privacy loss of CGS ε_i is solely determined by the privacy loss ε_i^r, ε_i^t and $\varepsilon_i^{t'}$ incurred by topic updates on the replaced word w_r, related words t and t' [20]. We make detailed discussions over these items and prove that group subsampling amplifies the inherent privacy of each iteration in CGS.

Algorithm 1. FDP-LDA

Input: Document corpus W, Prior parameters τ,β, Group subsampling ratio q, Topic number K, Total iteration number $ITER$, privacy parameters α and ε_s for word count perturbation

Output: Distributions $\Phi = \{\vec{\phi_k}\}_{k=1}^K$, $\Theta = \{\vec{\theta_m}\}_{m=1}^M$

// **Group subsampling and Initialization**

for $d_m \in W$ **do**

    for each distinct $w = t \in d_m$ **do**

        Partition M_t occurrences of word type t by group subsampling ratio q into groups $g_1, g_2, \ldots, g_s$ of size $m_1, m_2, \ldots, m_s$

        Initialize topics for all groups, count n_k^t and n_m^k

    end for

end for

// **Differentially Private Collapsed Gibbs Sampling**

Set $Iter = 0$

while $Iter <= ITER$ **do**

    for $k \in [K]$ **do**

        Add noise η to each n_k^t independently: $n_k^t \leftarrow n_k^t + \eta$

    end for

    for $d_m \in W$ **do**

        for each distinct word $w = t \in d_m$ **do**

            for each group g_i for w **do**

                Let z be the topic assignment to g_i in the previous iteration

                Decrement n_k^t, n_m^k, n_k, n_m by m_i

                Compute sampling distribution $\widetilde{\mathbf{p}}$:

$$p_k \propto \frac{n_k^t + \beta}{\sum_{t=1}^V (n_k^t + \beta)} \cdot \frac{n_m^k + \tau}{\sum_{k=1}^K (n_m^k + \tau)}$$

                Sample topic via $\widetilde{\mathbf{p}}$ and update n_k^t, n_m^k, n_k, n_m

            end for

        end for

    end for

    $Iter \leftarrow Iter + 1$

end while

Output: Distributions $\Phi = \{\vec{\phi_k}\}_{k=1}^K$, $\Theta = \{\vec{\theta_m}\}_{m=1}^M$

Theorem 1. *Suppose W' is constructed from W by replacing $w_r = t \in W$ by t'. Then there exist constants $\bar{w}_k^{t'} = \max_k \left\{ n_k^{t'} + \beta \right\}$, $\bar{w}_k = \max_k \{ n_k + V\beta \}$, $\bar{t}_m^k = \max_{k,m} \left\{ n_m^k + \tau \right\}$, $\bar{l}_m = \max_m \{ n_m + K\tau \}$ and $\hat{n} \in \mathbb{N}^+$, the privacy loss ε_i^r incurred by the topic updates on the replaced word w_r in the i-th iteration can be bounded by*

$$2\{\log \frac{\bar{w}_k^{t'} - cq + \hat{n}}{\bar{w}_k^{t'}} + \log \frac{\bar{w}_k}{\bar{w}_k - cq} + \log \frac{\bar{t}_m^k}{\bar{t}_m^k - cq} + \log \frac{\bar{l}_m - cq}{\bar{l}_m}\}. \qquad (1)$$

Corollary 1. *When $q \to 0$, the privacy loss ε_i^r incurred by the topic updates on the replaced word w_r in the i-th iteration can be bounded by $2 \max_k \left\{ \left| \log \frac{n_k^{t'} + \beta}{n_k^t + \beta} \right| \right\}$.*

Theorem 2. *Suppose W' is constructed from W by replacing $w_r = t \in W$ by t'. Then there exist constants $\bar{w}_k^t = \max_k \{ n_k^t + \beta \}$, $\bar{w}_k^{t'} = \max_k \left\{ n_k^{t'} + \beta \right\}$, $\bar{w}_k = \max_k \{ n_k + V\beta \}$, $\bar{t}_m^k = \max_{k,m} \left\{ n_m^k + \tau \right\}$, and $\bar{l}_m = \max_m \{ n_m + K\tau \}$, the privacy losses ε_i^t and $\varepsilon_i^{t'}$ incurred by the topic sampling on the related words t and t' in the i-th iteration can be bounded by*

$$2\{\log \frac{\bar{w}_k^t - cq + 1}{\bar{w}_k^t} + \log \frac{\bar{w}_k}{\bar{w}_k - cq} + \log \frac{\bar{t}_m^k}{\bar{t}_m^k - cq} + \log \frac{\bar{l}_m - cq}{\bar{l}_m}\}, \qquad (2)$$

$$2\{\log \frac{\bar{w}_k^{t'} - cq + 1}{\bar{w}_k^{t'}} + \log \frac{\bar{w}_k}{\bar{w}_k - cq} + \log \frac{\bar{t}_m^k}{\bar{t}_m^k - cq} + \log \frac{\bar{l}_m - cq}{\bar{l}_m}\}, \qquad (3)$$

respectively.

Corollary 2. *When $q \to 0$, the privacy losses ε_i^t and $\varepsilon_i^{t'}$ incurred by the topic updates on the related words $w = t$ and t' in the i-th iteration can be both bounded by $2 \log \left(1 + \frac{1}{\beta} \right)$.*

Theorem 1 and Theorem 2 are generalized results about the inherent privacy loss when sampling on the replaced word and the related words because Corollary 1 and Corollary 2, which give the upper bound of the inherent privacy of conventional differentially private CGS on the replaced word and the related words when $q \to 0$, are equivalent to proposition 1 and proposition 2 proposed by [20].

Theorem 3. *The privacy losses ε_i^r, ε_i^t and $\varepsilon_i^{t'}$ are monotonically decreasing w.r.t group subsampling ratio q and monotonically increasing w.r.t topic number K.*

Thus, we can conclude that better inherent privacy amplification effects can be achieved by increasing q or decreasing K. Intuitively, a large value of group subsampling ratio q means that topic updates for this group are assigned to more words. Privacy of a single word is 'shared' among those words belonging to the same group and thus privacy leakage is mitigated.

3.3   Guaranteed Privacy for FDP-LDA

For the extrinsic injected noises ηs on word counts, the sensitivity of word count n_k^t is 1 under the word replacement assumption. When ηs are sampled from Gaussian distribution $N(0, \sigma^2)$ where $\sigma^2 = \frac{\alpha}{2\varepsilon_s}$, the perturbation on word counts in each iteration of FDP-LDA satisfies (α, ε_s)-RDP [13]. The inherent privacy loss ε_i of CGS in terms of ε-DP could be expressed in terms of RDP. Namely, the inherent privacy loss of CGS in each iteration satisfies $(\alpha, \frac{1}{2}\varepsilon_i^2\alpha)$-RDP by Proposition 3.3 in [3]. According to Proposition 1 in [13], the total privacy loss of FDP-LDA in each iteration satisfies $(\alpha, \frac{1}{2}\varepsilon_i^2\alpha + \varepsilon_s)$-RDP.

4   Experiments

We show the experiment results of FDP-LDA in this section. FDP-LDA is implemented on two real-word datasets: NIPS[1] and ENRON[2]. We divide these datasets to form the training datasets, and the remaining parts of these datasets are used as the test datasets, also called the held-out datasets. We conduct simple preprocessing on these datasets and remove all stop words and punctuations.

Perplexity [2], as a measurement of how well a probability model predicts a sample, is our evaluation metric. Conventional differentially private CGS, namely HDP-LDA [20], is our baseline. Different from HDP-LDA, FDP-LDA conducts group subsampling before topic updating. For fairness, the noise η added to word counts is generated from the Gaussian mechanism. We utilize RDP [13] to evaluate the privacy loss of injected noise η. The noise scale is kept constant to show the inherent privacy amplification of FDP-LDA. We set $\alpha = 14, \varepsilon_s = 2$. The default hyperparameters τ, β are set to 1 and 0.01, respectively. Total iteration number $ITER$ is set to 50. Comparisons between utility and efficiency are presented in the following sections.

4.1   Inherent Privacy and Model's Utility

As we keep the privacy loss of injected noise constant, perplexity is solely determined by the inherent privacy loss. To show the impacts of group subsampling ratio q, we vary q from 0.1 to 0.9 with the step being 0.2. To show the impacts of topic amount K, we conduct experiments under $K = 20, 40, 60, 80, 100$. The results are shown in Fig. 1 and Fig. 2.

From Fig. 1 and Fig. 2, it's observed that perplexity increases when we increase the value of q. Moreover, when $q = 0$, FDP-LDA degrades to HDP-LDA and we can see that HDP-LDA has the smallest perplexity. Besides, larger K results in lower perplexity.

From Theorem 3.3, it is demonstrated that a larger value of q or a smaller value of K results in low privacy loss. When $q = 0$, FDP-LDA degrades to HDP-LDA, which possesses the largest inherent privacy loss. The results of Fig. 1 and Fig. 2 demonstrate that the utility of the LDA model usually has a negative correlation with the inherent privacy loss [15, 20, 21].

---

[1] https://archive.ics.uci.edu/ml/datasets/bag+of+words.
[2] https://archive.ics.uci.edu/ml/datasets/bag+of+words.

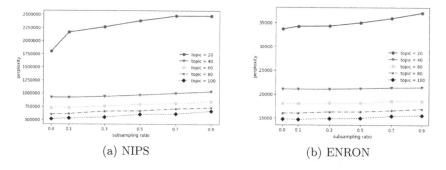

(a) NIPS   (b) ENRON

Fig. 1. Perplexity w.r.t subsampling ratio

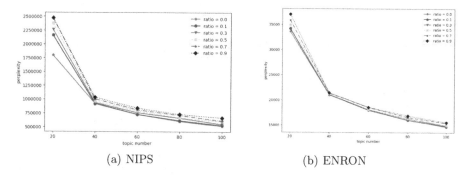

(a) NIPS   (b) ENRON

Fig. 2. Perplexity w.r.t topic number

4.2   Efficiency Improvements of FDP-LDA

Denote t_{fdp} and t_{hdp} as the average running time of each iteration in FDP-LDA and HDP-LDA, respectively. We show efficiency improvement of FDP-LDA via *speed-up ratio* $:= \frac{t_{hdp}-t_{fdp}}{t_{hdp}}$. The results are shown in Fig. 3. A positive speed-up ratio indicates improvement of efficiency and a larger speed-up ratio indicates better improvement of efficiency.

From Fig. 3, it's observed that speed-up ratios are always greater than 0, and greater q results in larger speed-up ratios. This demonstrates that FDP-LDA possesses better efficiency compared with HDP-LDA. Greater q results in better efficiency.

5   Related Work

The pioneering works [6,8] detailedly analyzes the bound of sensitivity and the guaranteed inherent privacy of Bayesian inference. When utilizing CGS, a specific Bayesian inference method, to estimate the parameters based on exponential families, studies [1,7] analyze the certain conditions of exploiting the inherent privacy of CGS while these ideal assumptions are often practically infeasible.

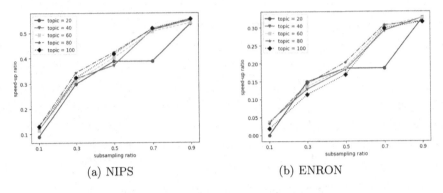

(a) NIPS   (b) ENRON

Fig. 3. Speed-up ratio w.r.t subsampling ratio

Zhao et al. [20,21] analyze the inherent privacy loss when utilizing CGS to estimate the parameters of the LDA model which is modeled by a special exponential distribution. However, the proposed inherent privacy loss is large and the guaranteed privacy is often limited.

The complexity of traditional CGS is $O(NZ)$ and is often a large number, where N and Z are the total amounts of words and latent topics. To improve the efficiency of traditional CGS, some efficient CGS algorithms [9,10,14,17,18] are proposed recently. *FastLDA* [14] reduces operations per sample to improve the efficiency of CGS. Yao et al. [17] get better efficiency of CGS by reducing the complexity $O(NZ)$ of traditional CGS to $O(N(Z_w + Z_d))$ where Z_w and Z_d are the numbers of distinct topics that are assigned to a word w and a document d respectively. Usually, $(Z_w + Z_d)$ is much smaller than Z. Li et al. [10] utilize the sparsity in the topic model and reduce the complexity from $O(NZ)$ to $O(NZ_d)$. Yuan et al. [18] proposed a compute-and-memory efficient distributed LDA implementation which is called *LightLDA*. The complexity of LightLDA is $O(N)$. Hu et al. [9] observe that topic distributions of words are skewed and only a subset of documents can approximately represent the semantics of the whole corpus. They reduce N via approximate semantics and reduce Z via skewed topic distribution.

6   Conclusion and Future Work

In this paper, we propose group subsampling, which amplifies the inherent privacy amplification of CGS. Moreover, FDP-LDA, which possesses better-guaranteed privacy and efficiency, is proposed. We make a comprehensive analysis of the upper bound of the privacy loss of the amplified inherent privacy. This is the first result of the inherent privacy amplification of CGS. We conduct extensive experiments to validate the improved performances of FDP-LDA. In future work, we plan to extend our analysis to Bayesian non-parametric models such as the Dirichlet process.

Acknowledgements. Hong Chen is the corresponding author. This work was supported by National Natural Science Foundation of China (62072460, 62076245, 62172424), Beijing Natural Science Foundation (4212022).

References

1. Bernstein, G., Sheldon, D.R.: Differentially private Bayesian inference for exponential families. In: Advances in Neural Information Processing Systems, pp. 2919–2929 (2018)
2. Blei, D.M., Ng, A.Y., Jordan, M.I.: Latent dirichlet allocation. J. Mach. Learn. Res. **3**, 993–1022 (2003)
3. Bun, M., Steinke, T.: Concentrated differential privacy: simplifications, extensions, and lower bounds. In: Hirt, M., Smith, A. (eds.) TCC 2016. LNCS, vol. 9985, pp. 635–658. Springer, Heidelberg (2016). https://doi.org/10.1007/978-3-662-53641-4_24
4. Carlo, C.M.: Markov chain Monte Carlo and Gibbs sampling. Lecture Notes for EEB 581 (2004)
5. Dwork, C., Roth, A., et al.: The algorithmic foundations of differential privacy. Found. Trends Theor. Comput. Sci. **9**(3–4), 211–407 (2014)
6. Foulds, J., Geumlek, J., Welling, M., Chaudhuri, K.: On the theory and practice of privacy-preserving Bayesian data analysis. arXiv preprint arXiv:1603.07294 (2016)
7. Ge, Y.F., Cao, J., Wang, H., Chen, Z., Zhang, Y.: Set-based adaptive distributed differential evolution for anonymity-driven database fragmentation. Data Sci. Eng. **6**(4), 380–391 (2021). https://doi.org/10.1007/s41019-021-00170-4
8. He, J., Liu, H., Zheng, Y., Tang, S., He, W., Du, X.: Bi-labeled LDA: inferring interest tags for non-famous users in social network. Data Sci. Eng. **5**(1), 27–47 (2020). https://doi.org/10.1007/s41019-019-00113-0
9. Hu, C., Cao, H., Gong, Q.: Sub-Gibbs sampling: a new strategy for inferring LDA. In: 2017 IEEE International Conference on Data Mining (ICDM), pp. 907–912. IEEE (2017)
10. Li, A.Q., Ahmed, A., Ravi, S., Smola, A.J.: Reducing the sampling complexity of topic models. In: Proceedings of the 20th ACM SIGKDD International Conference on Knowledge Discovery and Data Mining, pp. 891–900 (2014)
11. Liu, J.S.: Monte Carlo Strategies in Scientific Computing. Springer, New York (2008). https://doi.org/10.1007/978-0-387-76371-2
12. MacKay, D.J., Mac Kay, D.J.: Information Theory, Inference and Learning Algorithms. Cambridge University Press, Cambridge (2003)
13. Mironov, I.: Rényi differential privacy. In: 2017 IEEE 30th Computer Security Foundations Symposium (CSF), pp. 263–275. IEEE (2017)
14. Porteous, I., Newman, D., Ihler, A., Asuncion, A., Smyth, P., Welling, M.: Fast collapsed Gibbs sampling for latent dirichlet allocation. In: Proceedings of the 14th ACM SIGKDD International Conference on Knowledge Discovery and Data Mining, pp. 569–577 (2008)
15. Wang, Y., Tong, Y., Shi, D.: Federated latent dirichlet allocation: a local differential privacy based framework. In: AAAI, pp. 6283–6290 (2020)
16. Wang, Y., et al.: Towards topic modeling for big data. arXiv preprint arXiv:1405.4402 (2014)

17. Yao, L., Mimno, D., McCallum, A.: Efficient methods for topic model inference on streaming document collections. In: Proceedings of the 15th ACM SIGKDD International Conference on Knowledge Discovery and Data Mining, pp. 937–946 (2009)
18. Yuan, J., et al.: LightLDA: big topic models on modest computer clusters. In: Proceedings of the 24th International Conference on World Wide Web, pp. 1351–1361 (2015)
19. Yut, L., Zhang, C., Shao, Y., Cui, B.: LDA* a robust and large-scale topic modeling system. Proc. VLDB Endow. **10**(11), 1406–1417 (2017)
20. Zhao, F., Ren, X., Yang, S., Han, Q., Zhao, P., Yang, X.: Latent dirichlet allocation model training with differential privacy. IEEE Trans. Inf. Forensics Secur. **16**, 1290–1305 (2020)
21. Zhao, F., Ren, X., Yang, S., Yang, X.: On privacy protection of latent dirichlet allocation model training. arXiv preprint arXiv:1906.01178 (2019)

Multi-modal Fake News Detection Use Event-Categorizing Neural Networks

Buze Zhao$^{(\boxtimes)}$, Hai Deng, and Jie Hao

College of Computer Science and Technology, Nanjing University of Aeronautics
and Astronautics, Nanjing 210016, China
13022510390@163.com, haojie@nuaa.edu.cn

Abstract. Multi-modal fake news detection has drawn great attention
for their promising potential of preventing the spread of fake information
in social media, and it has become an important task to be addressed due
to many negative effects, such as reader misdirection, economic recession,
and political volatility. In this paper, we propose a new event-categorizing
neural networks (ECNN) framework, which combines a multi-modal feature extractor and an event categorizer to extract transferable features
of different events for fake news detection. Moreover, a residual network
is introduced to enrich the features extracted by the feature extractor.
The experimental results show that the proposed ECNN model improves
the accuracy and F1 scores compared to the baseline approach.

Keywords: Multi-modal · Event categorizer · Fake news detection ·
Early detection · Residual networks

1 Introduction

Social media platforms provide easy access to timely and comprehensive multimedia information. According to the latest data from the Pew Research Center,
while nearly half of U.S. people still obtain the news through social media in
2021 compared to 2020. The study suggests that a large part of the reason is
that social media is filled with a lot of fake news, which may intentionally provide the untrue event information [11]. Fake news may have a huge negative
influence and can even turn into a public event with harmful consequences. For
example, during the 2016 presidential election in the United States, numerous
fake news articles about the two candidates were generated and shared over 37
million times on Facebook, with the top 20 most discussed fake news items having greater influence than the actual news. Due to the serious negative impact,
it is urgent to develop an automated detection scheme to reduce the spread of
fake new.

Up to now, a number of fake news detection methods have been used to
identify fake news, including traditional learning methods [8] and deep learning-
based models [2]. However, it is challenging to detect and identify fake news on
emerging events on social media [11]. Due to the lake of corresponding prior

B. Li et al. (Eds.): APWeb-WAIM 2022, LNCS 13423, pp. 301–308, 2023.
https://doi.org/10.1007/978-3-031-25201-3_23

knowledge, it is difficult to obtain timely posts to validate such events, and they will result in the unexpected performance of the existing models. Therefore, the research target is to design an effective model: remove particular features that aren't transferable from one event to the next and retaining features that are shared between events to identify fake news.

In this paper, we investigate the multi-modal fake news detection problem and propose an event-categorizing neural networks framework. Compared with the existing results, the main contributions of this paper can be listed as:

(1) An event-categorizing neural networks model which consisting of the multi-modal feature extractor, the event categorizer and the fake news detector is proposed. In this ECNN model, the multi-modal features of the data can be extracted and then the event-specific features are eliminated to obtain the transferable features of the event, these addressed features are further utilized for fake news detection.
(2) To enhance the richness of the extracted features, the residual networks are presented for feature extraction of multi-modal data, and they provide a superior performance for fake news detection.
(3) We conduct extensive experiments on two large-scale real-world datasets to demonstrate the effectiveness of the proposed ECNN in terms of multi-modal fake news detection.

2   Related Work

In this section, we briefly review the related works on fake news detection. From the previous work in [11], fake news is defined as intentionally fabricated information that can be proven to be false. Fake news detection methods typically focus on the use of news content and social context [15]. Textual and visual aspects are the most common sources of news content components. Textual features capture specific writing styles [9] and boasting emotions [5], which are usually found in fake news content. Moreover, modeling of potential textual representations with deep neural networks [14] has good performance in detecting fake news content. Visual features are extracted from visual elements [8] to capture different features of fake news.

In the social context-based approach, someone capture the user-based features from user profiles to measure their features for detection [12]. In recent years, certain challenges in fake news detection have received extensive attention, such as user response generation, interpretable detection, and unsupervised detection. In this paper, we investigate an adversarial-learning-based method for fake news detection which aims to learn the common feature representation of different events in order to improve the performance of fake news detection.

3   Methodology

In this section, the ECNN model is based on the multi-modal feature extractor, the event categorizer and the fake news detector.

3.1   Multi-modal Feature Extractor

Text Feature Extractor. The core module of the text extractor in this model is text-CNN [3]. Text-CNN has a great capability to extract shallow text features, which is effective and widely used when focusing on intent classification in short text domains. Since our text data consists mainly of short sentences describing news information, text-CNN is a good fit, and the architecture of text-CNN is depicted in Fig. 2. It implies how to capture features for fake news detection by utilizing various filters with different window sizes.

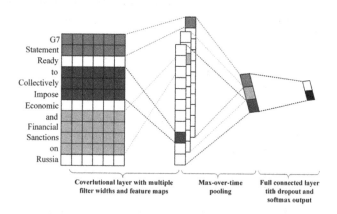

Fig. 1. The architecture of Text-CNN.

This model initializes the Embedding layer using the pre-trained word2vec for the detailed steps of the text feature extractor. The Embedding layer is then fixed, with each word described by a word embedding vector, which can be represented as $A_k \in R^j$ for the k-th word in the sentence. Then, let $A_{1:n}$ denotes a sentence containing n words, we have $A_{1:n} = A_1 \oplus A_2 \oplus \cdots \oplus A_n$, where $\oplus$ denotes the concatenation operator. Taking the i-th word as a continuous sequence of h words at the beginning, the filter is given as

$$a_i = \sigma(W_c \cdot A_{i:i+h-1}), \tag{1}$$

where h represents the window size of the filter, $\sigma(\cdot)$ and W_c denote the ReLU activation function and the weight of a_i, respectively. Further, the sentence feature vector is described as $a = [a_1, a_2, \ldots, a_{n-h+1}]$.

For each feature vector, to improve the accuracy of the extracted features, the K-Max Pooling operation is introduced. For each filter, we now have the corresponding features. This process is continued until all of the filters' features are obtained. We use different window sizes to extract text characteristics with varying granularities. For a particular window size, we have n_h different filters, there are c possible window sizes, therefore we have $c \cdot n_h$ filters. After the K-Max Pooling procedure, the text features are written as $R_{A_c} \in R^{c \cdot n_h}$. Then we employ a fully-connected layer in order to guarantee that the text feature representation ($R_A \in R^p$) and the visual feature representation have the same dimension (denoted as p): $R_A = \sigma(W_{af} \cdot R_{A_c})$, where W_{af} represents the fully connected layer weight matrix.

Visual Feature Extractor. We employ the pre-trained resnet50 [6] to capture visual feature, we also use the K-Max Pooling operation after the resnet50 network to take the values of all feature values scoring in Top-k, and then add a fully connected layer to the last layer of the network, which adjusted the visual feature dimension to p. The p dimensional visual features are represented as $R_v \in R^p$, and the last fully-connected layer can be expressed as

$$R_v = \sigma(W_{vf} \cdot R_{v_{resnet50}}), \tag{2}$$

where W_{vf} denotes the weight of the fully connected layer, $R_{v_{resnet50}}$ represents the visual feature determined by resnet50. The output of the multi-modal feature extractor can be represented as: $R_F = R_A \oplus R_V \in R^{2p}$. We denote $G_f(M; \theta_f)$ as the multi-modal feature extractor, where M is a post with textual and visual parts, which are the inputs of the event categorizer, and θ_f is the parameters in the multi-modal feature extractor.

3.2 Event Categorizer

Then, to classify different events, the event categorizer including a ReLU activation function and two fully connected layers is presented. The event categorizer is denoted as $G_e(R_F; \theta_e)$, where R_F is the vector extracted by the multi-modal feature extractor with $R_F = R_A \oplus R_V \in R^{2p}$, θ_e denotes the parameter of G_e. Denoting Y_e as event labels, and the loss function of the event categorizer can be described as $L_e(\theta_f, \theta_e) = -E_{(m,y)\sim(M,Y_e)}[\sum_{k=1}^{K} 1_{[k=y]} \log(G_e(G_f(m, \theta_f); \theta_e))]$.

The event categorizer parameters with as little loss $L_e(\theta_f, \theta_e)$ as possible can be represented as $\hat{\theta}_e = \arg\min_{\theta_e} L_e(\theta_f, \theta_e)$. To eliminate the uniqueness of each event feature, it is necessary to maximize the recognition loss $L_e(\theta_f, \hat{\theta}_e)$ by finding the optimal parameter θ_f.

3.3 Fake News Detector

To evaluate if these posts are fake or true, we apply softmax to create a fully-connected layer. Softmax is a multi-classification generalization of the binary classification function sigmoid, with the aim of presenting multi-classification results as probabilities. To avoid over-fitting, we utilize the drop out procedure in the prediction. The $G_d(R_F; \theta_d)$ is represented as the fake news detector, where R_F is the output of multi-modal feature extractor and θ_d denotes the parameters of this Module. The probability of the post is fake:

$$P_\theta(m_i) = G_d(G_f(m_i; \theta_f); \theta_d). \tag{3}$$

where m_i is i-th multimedia post. The detection loss can be expressed as

$$L_d(\theta_f, \theta_d) = -E_{(m,y)\sim(M,Y_d)}[y \log(P_\theta(m)) + (1-y)\log(1-P_\theta(m))], \quad (4)$$

where Y_d denotes the set of labels of news. To minimize $L_d(\theta_f, \theta_d)$, the optimal parameters $\hat{\theta}_f$ and $\hat{\theta}_d$ can be derived as $(\hat{\theta}_f, \hat{\theta}_d) = \arg \min_{\theta_f, \theta_d} L_d(\theta_f, \theta_d)$.

3.4 Model Integration

As shown in Fig. 1, the ECNN model presented in this paper is composed of a multi-modal feature extractor, an event categorizer and a fake news detector.

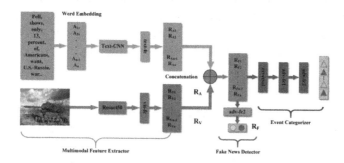

Fig. 2. Architecture of the ECNN model.

During the training phase, the multi-modal feature extractor $G_f(M; \theta_f)$ attempts to deceive $G_e(R_F; \hat{\theta}_e)$ to obtain the event commonality representation by maximizing $L_e(\theta_f, \theta_e)$. Then, the event categorizer $G_e(R_F; \theta_e)$ is presented to identify each event by minimizing the event classification loss based on a multi-modal feature representation. Meanwhile, combine the extractor $G_f(M; \theta_f)$ and the detector $G_d(R_F; \theta_d)$, the detection loss function $L_d(\theta_f, \theta_d)$ can be minimized and the performance of fake news detection is further improved. The final loss is defined as follows:

$$L_{sum}(\theta_f, \theta_d, \theta_e) = L_d(\theta_f, \theta_d) - \lambda L_e(\theta_f, \theta_e). \quad (5)$$

where λ is the weights of loss functions of both the event categorizer and the fake news detector modules. We set λ to 1, the optimization parameters of the final objective function are $(\hat{\theta}_f, \hat{\theta}_d) = \arg \min_{\theta_f, \theta_d} L_{sum}(\theta_f, \theta_d, \hat{\theta}_e)$ and $\hat{\theta}_e = \arg \min_{\theta_e} L_{sum}(\hat{\theta}_f, \theta_e)$. To find the best set of parameters, we employ stochastic gradient descent, and the parameter θ_f is updated as $\theta_f \leftarrow \theta_f - \eta(\frac{\partial L_d}{\partial \theta_f} - \lambda \frac{\partial L_e}{\partial \theta_f})$.

The gradient reversal layer (GRL) in [4] is employed here to decay the learning rate, we have $\eta' = \frac{\eta}{(1+\alpha \cdot p)^\beta}$, where η denotes the initial learning rate with a value of 0.01; α and β belong to the hyperparameters and set $\alpha = 10$ and $\beta = 0.75$; p represents the relative value of the iterative process.

4   Results

In this section, the social media dataset used for the experiments is first presented, then the performance analysis of the proposed ECNN model is given.

4.1   Datasets

We employ two widely used multi-modal fake news detection datasets, namely the Twitter dataset and the Weibo dataset. The Twitter dataset is from [1], we only use the text and image information of the tweets in the Twitter dataset, where the number of true news and fake news are 5723 and 6232, respectively, and the number of images is 682. The second dataset is the Weibo dataset from [7]. To assure the quality of the entire dataset, we first delete duplicate and low-quality images. Eventually, the number of true news and fake news are 5249 and 5265, respectively, and the number of images is 9714.

4.2   Baselines

Multi-modal Models. The multi-modal approach considers both text and image information, baselines include NeuralTalk [13], TCNN-URG [10] and att-RNN [7].

NeuralTalk [13]: NeuralTalk aims to generate captions for images. The potential features are obtained by averaging the output of the RNN for each time step and then giving features to a fully connected layer for prediction.

TCNN-URG [10]: TCNN-URG consists of two main components: a two-stage convolutional neural network and a conditional variational auto-encoder for learning representations from news articles and extracting features from user comments, respectively.

att-RNN [7]: The att-RNN uses an attention strategy that is used to extract features from textual, visual, and social contexts. In this study, the part of social context information was removed.

A Variant of the Proposed ECNN. A multi-modal feature extractor, a fake news detector, and an event categorizer are the three components that make up the entire model. However, the model with only multi-modal feature extractor and fake news detector can still perform fake news prediction. Therefore, removing the event categorizer constitutes a variant form ECNN-.

4.3   Performance Comparison

We employ the accuracy, precision, recall, and F1 metrics to evaluate the performance of the fake news detection system. Table 1 shows the average performance of the procedure after it was run five times. From the experimental results shown in Table 1, it implies that on both datasets, the ECNN model proposed in this paper performs a superior performance on accuracy and F1 scores. From Table 1, it can be obtained that att-RNN beats NeuralTalk and TCNN-URG among the

Table 1. The performance comparison on two datasets.

Dataset	Method	Accuracy	Precision	Recall	F1
Twitter	NeuralTalk	0.627	0.718	0.522	0.588
	TCNN-URG	0.675	0.732	0.623	0.663
	att-RNN	0.681	**0.732**	**0.619**	0.663
	ECNN-	0.647	0.713	0.523	0.622
	ECNN	**0.705**	0.729	0.586	**0.684**
Weibo	NeuralTalk	0.720	0.681	0.803	0.733
	TCNN-URG	0.765	0.733	0.808	0.754
	att-RNN	0.803	**0.823**	**0.827**	0.808
	ECNN-	0.795	0.774	0.787	0.788
	ECNN	**0.829**	0.814	0.815	**0.827**

multi-modal models, demonstrating that including the attention mechanism can help the model extract essential characteristics more efficiently and enhance prediction model performance.

The event categorizer is removed as a module in the proposed alternative version of the model ECNN-. As a result, not enough transferable features would be learned between different events. In contrast, the complete model improved by 3.4% in accuracy and 3.9% in F1 scores, and was able to maintain a performance similar to the best baseline in terms of precision. The experimental results demonstrate the effectiveness of the event categorizer for performance improvement. Compared with the Twitter dataset, the Weibo dataset has the feature of data balance. Hence, the experimental results are better than the Twitter dataset in general.

In terms of accuracy and F1 scores, it can be seen that our proposed model outperforms all multi-modal baselines on the dataset. It can draw the following conclusions: (1) combining multi-modal data improves the model's fake news detection performance; and (2) textual content and visual images must be modeled because they contain complementary information and can also address data imbalance.

5   Conclusions

In this paper, we propose a new neural networks framework named ECNN. The modal consists of a multi-modal feature extractor, an event categorizer and a fake news detector. The multi-modal feature extractor deceives the event categorizer and thus obtains transferable common features of different events, which can be able to detect emerging fake news. Extensive experiments on two social media datasets can demonstrate the excellent performance of the model.

References

1. Boididou, C., et al.: Verifying multimedia use at MediaEval 2015. In: MediaEval, vol. 3, no. 3, p. 7 (2015)
2. Chen, T., Li, X., Yin, H., Zhang, J.: Call attention to rumors: deep attention based recurrent neural networks for early rumor detection. In: Ganji, M., Rashidi, L., Fung, B.C.M., Wang, C. (eds.) PAKDD 2018. LNCS (LNAI), vol. 11154, pp. 40–52. Springer, Cham (2018). https://doi.org/10.1007/978-3-030-04503-6_4
3. Chen, Y.: Convolutional neural network for sentence classification. Master's thesis, University of Waterloo (2015)
4. Ganin, Y., Lempitsky, V.: Unsupervised domain adaptation by backpropagation. In: International Conference on Machine Learning, pp. 1180–1189. PMLR (2015)
5. Guo, C., Cao, J., Zhang, X., Shu, K., Yu, M.: Exploiting emotions for fake news detection on social media. arXiv preprint arXiv:1903.01728 (2019)
6. He, K., Zhang, X., Ren, S., Sun, J.: Deep residual learning for image recognition. In: Proceedings of the IEEE Conference on Computer Vision and Pattern Recognition, pp. 770–778 (2016)
7. Jin, Z., Cao, J., Guo, H., Zhang, Y., Luo, J.: Multimodal fusion with recurrent neural networks for rumor detection on microblogs. In: Proceedings of the 25th ACM International Conference on Multimedia, pp. 795–816 (2017)
8. Jin, Z., Cao, J., Zhang, Y., Zhou, J., Tian, Q.: Novel visual and statistical image features for microblogs news verification. IEEE Trans. Multimed. **19**(3), 598–608 (2016)
9. Potthast, M., Kiesel, J., Reinartz, K., Bevendorff, J., Stein, B.: A stylometric inquiry into hyperpartisan and fake news. arXiv preprint arXiv:1702.05638 (2017)
10. Qian, F., Gong, C., Sharma, K., Liu, Y.: Neural user response generator: fake news detection with collective user intelligence. In: IJCAI, vol. 18, pp. 3834–3840 (2018)
11. Shu, K., Sliva, A., Wang, S., Tang, J., Liu, H.: Fake news detection on social media: a data mining perspective. ACM SIGKDD Explor. Newsl. **19**(1), 22–36 (2017)
12. Shu, K., Zhou, X., Wang, S., Zafarani, R., Liu, H.: The role of user profiles for fake news detection. In: Proceedings of the 2019 IEEE/ACM International Conference on Advances in Social Networks Analysis and Mining, pp. 436–439 (2019)
13. Vinyals, O., Toshev, A., Bengio, S., Erhan, D.: Show and tell: a neural image caption generator. In: Proceedings of the IEEE Conference on Computer Vision and Pattern Recognition, pp. 3156–3164 (2015)
14. Zhang, A., Li, B., Wang, W., Wan, S., Chen, W.: MII: a novel text classification model combining deep active learning with BERT. Comput. Mater. Continua **63**(3), 1499–1514 (2020)
15. Zhou, X., Zafarani, R., Shu, K., Liu, H.: Fake news: fundamental theories, detection strategies and challenges. In: Proceedings of the Twelfth ACM International Conference on Web Search and Data Mining, pp. 836–837 (2019)

Unified Proof of Work: Delegating and Solving Customized Computationally Bounded Problems in a Privacy-Preserving Way

Yue Fu$^{(\boxtimes)}$, Qingqing Ye, Rong Du, and Haibo Hu

Department of Electronic and Information Engineering, The Hong Kong Polytechnic University, Kowloon, Hong Kong SAR
{yuesandy.fu,roong.du}@connect.polyu.hk, {qqing.ye,haibo.hu}@polyu.edu.hk

Abstract. Proof of work (PoW), which was initially introduced for combating junk email spam, is extremely energy-thirsty, but the computed outcomes are of very little relevance to any practical usage. Some existing works tackle this issue by replacing the useless cryptographic puzzles with some pre-determined practical problems. However, the allowable problem types are usually limited and fixed, and some problems are of no significant value. This paper presents a Unified Proof of Work (UPoW) mechanism using flexible problems as puzzles, which features a computation-trading platform where delegators post their problems and PoW executors compute them as a by-product. The proposed UPoW puts known problem types together and frames them as verifiable computation problems using Fully Homomorphic Encryptions (FHE) techniques to replace PoW puzzles. It also secures and preserves the functionality of the delegated problems, which is showcased by delegating a Private Set Intersection (PSI) problem.

Keywords: Security · Proof of work · Outsourced cloud computing

1 Introduction

Proof of work, which was initially proposed to combat junk email spam, is a widely used mechanism to guarantee a certain amount of work is indeed done for doing some task [2]. Today, it is usually employed to combat denial of service attacks where a malicious user deliberately requests expensive operations from some online service, exhausts the server's resources, and prevents legitimate users from accessing the service. Since the request can come from any device, e.g., a flash-based one [7], such a security issue stems from the fact that the user does almost no work, but the server bears a hefty workload. Therefore, the server will request the user to solve some PoW puzzles before it performs the expensive task. Since PoW forces malicious users to bear an extensive computation, it is widely known as the solution of the double-spending problem in cryptographies,

B. Li et al. (Eds.): APWeb-WAIM 2022, LNCS 13423, pp. 309–317, 2023.
https://doi.org/10.1007/978-3-031-25201-3_24

e.g., Bitcoin [6]. Unfortunately, such computations are about nothing other than guaranteeing some workload is indeed done, and the involved cryptographic puzzles are of very little relevance to any practical use.

To improve over the energy-thirsty but useless PoW, many attempts are made on utilizing the otherwise wasteful energy for computing **pre-determined** practical problems as a by-product. Such attempts, appearing to be "Proof of Something", has become a hot topic in these years, e.g., PoX [9] and proof of federated learning (PoFL) [8]. However, the allowable problem types in each "Proof of Something" are unchangeable, and users are therefore unable to customize their computational tasks. This paper aims to replace otherwise wasteful PoW puzzles with computationally bounded problems that accomplish valuable tasks. Since it is common to outsource some security demands to centralized, trusted institutes (e.g., Cloudflare) today, we consider creating a market where delegators submit problems that they would like solutions to, which are turned into some PoW puzzles that users have to find solutions. Such puzzles, a privacy-preserving way of outsourcing these computations, are efficiently verifiable and are sufficient for PoW.

The main challenge we have to address in this paper is that problems cannot be published in their raw forms, and not all instances are allowed, otherwise problem hardness will not be guaranteed even for well-studied problems—if delegators post some trivial instances as "challenges", they may not require work [1]. Even worse, a malicious delegator can collude with PoW executers and let them pass the PoW verification without actual work by posting a problem that they already know what the answer should be. To tackle this security issue, we present a general UPoW framework, in which PoW executors do not know what they are computing, but the answer for the obfuscated computational tasks still benefits delegators. We design an obfuscation protocol that can principally ally existing well-defined "Proof of Something" into unified PoW puzzles. Such puzzles are framed as verifiable computation problems using FHE, which also secures and preserves the functionality of the delegated problems.

The rest of the paper is organized as follows. Section 2 introduces some preliminaries. Section 3 presents the detailed UPoW protocol. Section 4 shows the experimental results, followed by a conclusion in Sect. 5.

2   Preliminaries

2.1   System Framework

Figure 1 is an overview of our system framework. Generally, in our proposed UPoW framework, we have the following roles: delegators who would like to outsource their problems, a trusted platform turning these practical problems posted by delegators into UPoW puzzles and selling UPoW service to its customers, customers who buy UPoW service from the platform, and users (of the customer's) who are the actual executors of UPoW. We have a problem pool in the platform where delegators post their problems. The allowable problems are some computationally bounded ones, so we can measure a **hardness score** for each problem such that correct answers for them are combined with some necessitate actual work. Then, the platform obfuscates the problems into trapdoors and builds up a trapdoor queue that provides continuous UPoW flow for its customers. Users, who are the UPoW executors, keep working on the trapdoors until the sum of hardness scores meets the current hardness threshold so as to pass the PoW verification. Finally, the platform verifies the submitted answers, decides whether the user passes the PoW verification or not, and returns the outputs to the delegators if yes.

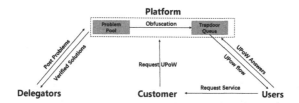

Fig. 1. An overview of the UPoW framework.

2.2   Security Model

Trust Assumptions. In our protocol, the platform is a centralized service to bridge problem delegators and customers who buy its PoW service. In practice, it can be a professional third-party cloud service vendor (e.g., Cloudflare) who collects delegator's problems, generates PoW puzzles, and sells the PoW service to the platform's customers. In this process, delegators need to trust the platform to post their input, and customers also need to trust the platform to outsource their PoW verification demands. Therefore, the platform is required to be trusted by both sides.

Functionality-Preserving Trapdoors. In the context of UPoW, if arbitrary problems are allowed from delegators, they will be able to create easy instances of that problem that does not need actual work (e.g., if we allow any instance of "evaluating an n-degree polynomial", input $x = 0$ will be an easy one). Even worse, PoW executors can conspire with malicious delegators by posting problems that PoW executors already know the answers to, and allows it to pass the PoW verification without actual workload.

Such security issue points to the fact that PoW executors should not know the problems in their initial forms. In this paper, we tackle this issue by transforming the problems into functionality-preserving trapdoors that are cryptographically indistinguishable among others and only reveal the trapdoors to UPoW executors. As such, this privacy-preserving approach guarantees both the availability of UPoW and the data privacy of problem delegators'. Details will be discussed in Sect. 3.

2.3   Cryptographic Primitives: Homomorphic Encryptions

Our construction of UPoW will employ homomorphic encryption schemes, which allow arithmetic circuits to be evaluated directly on ciphertexts. That is, the

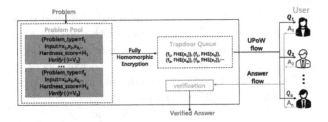

Fig. 2. The architecture of UPoW.

following property holds without knowledge of the private key: Given $Enc(x_1)$, $Enc(x_2)$ and an operation $\times$, we can compute $Enc(x_1) \times Enc(x_2)$. Without concerning efficiency, this enables outsourced cloud computing on private data and private searching [5].

Although there are already maturely designed Partial Homomorphic Encryption (PHE) schemes supporting single arithmetic, Fully Homomorphic Encryptions (FHE) that support arbitrary combinations of additions and multiplications are real concerns for practical applications. Since Gentry proposed the first FHE scheme in 2009 [4], many attempts are made for improved performance, e.g., leveled FHE that supports only circuits of a certain bounded depth. We will use this in our constructions.

3   UPoW: All Put Together

In this section, we constrain the problem to a set of pre-given computationally-bounded classes, and delegators can only choose an input for any of the determined problem classes. Without loss of generality, let f_i denote a problem class that can be either an arithmetic circuit, or simply a commonly requested function that UPoW executors are familiar with, and x denote the input delegated for that problem class to find $f_i(x)$ for. As there are only limited problem classes with extensive inputs, we expect that the sample space consisting of all possible inputs for the problem class is large enough for a specific element to hide in. Therefore, our basic idea is to employ semantically secure public-key encryptions to hide the input and preserve the homomorphism of arithmetic. Figure 2 shows the architecture of UPoW. In the problem pool, problems are packed into modules with the following four factors:

- *Problem_type*, recording the problem form denoted by f_i, either an arithmetic circuit or a function. i is the index of the problem for UPoW executors to invoke.
- *Input*, recording inputs x_i of problem f_i, which are posted from delegators.
- *Hardness_score*, recording a generally acknowledged hardness score of problem f_i.
- *Verify($\cdot$)*, recording a well-defined algorithm to verify the correctness of solutions for $f_i(x)$.

Fig. 3. A PSI protocol (Freedman et al.)

Algorithm 1. Protocol: Delegating PSI with UPoW.

Input: Alice's input is the coefficients of her polynomial $\{a_0, a_1, ...a_{k_A}\}$. Bob's input is his set $\{y_1, y_2, ..., y_{k_B}\}$

1: Bob invokes the *Problem_type* evaluating n-degree polynomials. Then he posts $\{a_0, a_1, ...a_{k_A}\}$ and $\{y_1, y_2, ..., y_{k_B}\}$ as inputs to the platform.
2: The platform chooses the secret-key parameters for FHE, encrypts all the inputs and randomly permutes them into the module where *Problem_type=n-degree polynomial evaluation*.
3: UPoW executors compute this problem as UPoW, return it to the platform for verification, and the platform returns the answer to Alice.

Delegators can choose a problem type that already exists in the problem pool and post their preferred inputs for that problem. The inputs are packed into the problem pool accordingly, and fully homomorphic encryption is performed on each of them to generate trapdoors. Trapdoors written in the form $(f_i, FHE(x_i))$ are randomly sent to the trapdoor queue to ensure all inputs are indistinguishable for UPoW executors so that they cannot tell which one was posted by a specific delegator. This process produces a continuous UPoW flow for the UPoW executors timely.

To better understand the protocol, we showcase delegating a PSI problem using UPoW. A two-party PSI mechanism enables Alice and Bob (each holds a set of inputs $\{x_1, x_2, ..., x_{k_A}\}$ and $\{y_1, y_2, ..., y_{k_B}\}$ respectively) to jointly compute the intersection of their inputs without leaking any additional knowledge. For randomly sampled value r, a basic idea for PSI problems is the fact that

$$f(x) = (x - x_1)(x - x_2)...(x - x_i)r = \begin{cases} 0, x = x_i \\ random, otherwise \end{cases} \tag{1}$$

Figure 3 shows an early protocol proposed by Freedman et al. With the scale of their sets increasing, both the polynomial degree and the evaluation times go high, and step 3 will be computationally heavy. We delegate this step using UPoW, as shown in Algorithm 1. Since the involved computation is simple, evaluating it in the ciphertext space of FHE will be practical. Assume that the evaluation of an n-degree polynomial is already well-defined as one of UPoW's *Problem_type*, Bob can delegate his polynomial evaluation to UPoW executors. To reduce the multiplication overhead, we apply the Horner's rule, which is a commonly used efficient algorithm for polynomial evaluations, on the polynomial $a_0 + a_1 x + a_2 x^2 + ... + a_{k_A} x^{k_A}$ by rewriting it into the form $a_0 + x(a_1 + x(a_2 + x(a_3 + ... + x(a_{k_A - 1} + x a_{k_A})...)))$ and evaluating the polynomial "from the inside out".

4 Experiments

We will evaluate the performance of our proposed UPoW on PSI protocols through simulation. The benchmark machine has an 8-core Intel CPU i7-10700K at 3.80 GHz and 128 GB RAM. We perform all tests using this single machine with the operating system Windows 10. For polynomial evaluations in plaintext, only the relative time is meaningful for comparison, so we run them on Matlab R2021a for simplicity. For fully homomorphic encryptions, we use the Microsoft SEAL library[1], which implements the Fan-Vercauteren scheme [3] in C++. We build the project using Visual Studio 2019 with C++ CMake Tools for Windows.

Table 1. Parameter settings for SEAL.

Name	n = poly_modulus_degree	q = coeff_modules	t = plain_modulus
SEAL2048	2048	54 (40 + 14)	4096
SEAL4096	4096	109 (36 + 36 + 37)	4096
SEAL8192	8192	218 (43 × 2 + 44 × 3)	4096
SEAL16384	16384	438 (48 × 3 + 49 × 6)	4096
SEAL32768	32768	881 (55 × 15 + 56)	4096

4.1 The Parameters for SEAL

SEAL uses power-of-2 cyclotomic rings of integers, where the plaintext space is $\mathbb{Z}_t[x]/(x^n + 1)$, and the ciphertext space is $\mathbb{Z}_q[x]/(x^n + 1)$. Here, n is a power of 2, t and q are integers. Let $R = \mathbb{Z}[x]/(x^n + 1)$, then the plaintext space and the ciphertext space can be written as $R_t = R/tR, R_q = R/qR$. For the Fan-Vercauteren scheme adopted in SEAL, it has plaintext and ciphertext spaces of this type, in which n refers to "poly_modulus_degree", q refers to parameter "coeff_modules" (which is related to "poly_modulus_degree"), and t refers to "plain_modulus." Since SEAL provides a default parameter "CoeffModulus::BFVDefault(poly_modulus_degree)" to determine q, in our simulations q will be set using this. All the encryption parameters we use for SEAL are listed in Table 1.

4.2 Computational Overhead

In this part, we evaluate the PSI polynomial in the ciphertext space according to our UPoW protocol. The performance metric is the running time, and the benchmark will be the delegator evaluating the polynomial on his machine using Horner's rule. The overhead mainly roots from evaluating the polynomial in the ciphertext space, which is usually expensive for current FHE schemes. The results can be found in Table 2, where we can see the utility of UPoW executors

---

[1] https://www.microsoft.com/en-us/research/project/microsoft-seal/.

Table 2. Running time of UPoW with different FHE parameters.

Size of Alice	Running time					
	Bob (μs)	UPoW executors (s)				
		SEAL2048	SEAL4096	SEAL8192	SEAL16384	SEAL32768
50	0.63	0.06	0.21	0.84	3.79	17.76
100	1.26	0.19	0.42	1.67	7.51	35.66
500	6.32	0.6	2.05	8.169	36.898	175.052
1000	13.02	1.18	4.10	16.46	76.88	358.61
5000	62.66	5.77	20.24	81.69	375.28	1772.21
10000	124.86	11.63	40.53	163.53	742.57	3523.09

is relatively poor, even if their computational power is supposed to be million-times of the benchmark. There are two possible ways of improvement. First, FHE is still a rapidly developing technology at its early stage. In the future, it is expectable that we can employ more efficient FHE schemes on UPoW. Second, the situation is not that bad even for the current FHE schemes. From Table 2 we can see, with n decreasing, evaluating the polynomial in the ciphertext space becomes cheaper. Although SEAL does not support $n < 2048$ (regarded as "insecure" to the security standard), it is expectable that the evaluation time will be lower in a smaller ciphertext space.

For the two points, we have an indirect methodology for verification. In special UPoW, for most practical problems f where both addition and multiplication are involved, we must encrypt them with FHE to preserve the functionality without loss of generality, although the efficiency of existing FHE algorithms is of little relevance of practice. For instance, when we use Horner's rule on polynomial evaluations, the polynomial is rewritten into a nested composition of addition and multiplication, thus using FHE is a must to preserve the operation. In the following simulation, we consider evaluating the polynomial in its initial form, i.e., the delegator encrypts all the coefficients and the variant of a polynomial with semantically secure multiplicative homomorphic encryptions (e.g., RSA), and outsources all the inter-term multiplications to UPoW executors only. In this scheme, when UPoW executors return all the results computed in the ciphertext space, the delegator is supposed to decrypt all of them and undertake the remaining intra-term additions on the plaintexts, which are computationally light. For fairness of comparison, we set the plaintext space and the ciphertext space approximately equal. Since evaluations in the ciphertext space of RSA are relatively efficient, the utility of this scheme is expected to be excellent.

Figure 4 shows the results of the running time on the simulation. In subfigure (a) and (b), there are one benchmark, which is still the delegator evaluating the polynomial on his machine using Horner's rule, and four groups for UPoW executors with different times of computational power of the benchmark. In subfigure (a), the size of Alice is fixed to 10000, and the same for subfigure (b) to Bob. Note that Horner's rule requires n times of multiplications, while this

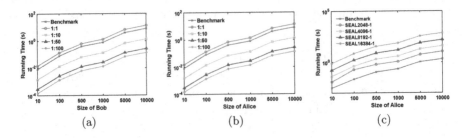

Fig. 4. Running time comparison in an RSA ciphertext space.

scheme requires $2n$ times of multiplications in the ciphertext space of RSA in total: n times for calculating $x, x^2, ...x^n$, and another n times for multiplying them with the coefficients. Therefore, the running time of the latter is expected to be longer than that of the former. From the results, we can see the utility is way better than using SEAL, even though the times of computational power are set relatively small. In subfigure (c), we compare the running time of the evaluation in the RSA ciphertext space and that in FHE ciphertext spaces of different parameters, and the benchmark is essentially the 1:1 ones in the first two subfigures. Since the running time will be too small compared to the running time of SEAL, it is amplified by 10000 times in subfigure (c), which is still the lowest one among the five curves. Hopefully, if an efficient FHE scheme comparable to RSA occurs in the future, UPoW will also be practical for all eligible functions.

5   Conclusion

This paper develops a novel protocol UPoW, replacing cryptographic puzzles of PoW with practical problems and enabling delegators to leverage the otherwise wasted energy to find answers for their problems.

Acknowledgement. This work was supported by National Natural Science Foundation of China (Grant No: 62072390, 62102334), and the Research Grants Council, Hong Kong SAR, China (Grant No: 15222118, 15218919, 15203120, 15226221, 15225921, and C2004-21GF).

References

1. Ball, M., Rosen, A., Sabin, M., Vasudevan, P.N.: Proofs of useful work. IACR Cryptol. ePrint Arch. **2017**, 203 (2017)
2. Dwork, C., Naor, M.: Pricing via processing or combatting junk mail. In: Brickell, E.F. (ed.) CRYPTO 1992. LNCS, vol. 740, pp. 139–147. Springer, Heidelberg (1993). https://doi.org/10.1007/3-540-48071-4_10
3. Fan, J., Vercauteren, F.: Somewhat practical fully homomorphic encryption. IACR Cryptol. ePrint Arch. **2012**, 144 (2012)

4. Gentry, C.: Fully homomorphic encryption using ideal lattices. In: Proceedings of the Forty-first Annual ACM Symposium on Theory of Computing, pp. 169–178 (2009)
5. Hu, H., Xu, J., Xu, X., Pei, K., Choi, B., Zhou, S.: Private search on key-value stores with hierarchical indexes. In: 2014 IEEE 30th International Conference on Data Engineering, pp. 628–639. IEEE (2014)
6. Nakamoto, S.: Re: Bitcoin P2P e-cash paper. The Cryptography Mailing List (2008)
7. On, S.T., Xu, J., Choi, B., Hu, H., He, B.: Flag commit: supporting efficient transaction recovery in flash-based DBMSs. IEEE Trans. Knowl. Data Eng. **24**(9), 1624–1639 (2011)
8. Qu, X., Wang, S., Hu, Q., Cheng, X.: Proof of federated learning: a novel energy-recycling consensus algorithm. IEEE Trans. Parallel Distrib. Syst. **32**(8), 2074–2085 (2021)
9. Shoker, A.: Sustainable blockchain through proof of exercise. In: 2017 IEEE 16th International Symposium on Network Computing and Applications (NCA), pp. 1–9. IEEE (2017)

TSD3: A Novel Time-Series-Based Solution for DDoS Attack Detection

Yifan Han[1], Yang Du[1(✉)], Shiping Chen[2(✉)], He Huang[1], and Yu-E Sun[3]

[1] School of Computer Science and Technology, Soochow University,
Suzhou, Jiangsu, China
yfhan@stu.suda.edu.cn, {duyang,huangh}@suda.edu.cn
[2] School of Optical-Electrical and Computer Engineering,
University of Shanghai for Science and Technology, Shanghai, China
chensp@usst.edu.cn
[3] School of Rail Transportation, Soochow University, Suzhou, Jiangsu, China
sunye12@suda.edu.cn

Abstract. Distributed Denial-of-Service (DDoS) attack has long been one of the biggest threats to network security. Most existing approaches collect and analyze the network traffic in a fixed window (*e.g.*, 1 min or 5 min) to detect ongoing attacks. However, they cannot track temporal information, such as the arriving moments of packets and the persistence of malicious flows in the time dimension, which inevitably harms their effectiveness. To this end, this work proposes a novel solution called Time-Series DDoS Detection (TSD3). First, we design an attention-based traffic sampling algorithm to support short-period (*e.g.*, 1 s) traffic monitoring. The proposed sampling solution can continuously track network flows with limited storage and communication resources and naturally attach the flow records with fine-grained time information, *i.e.*, slice index. Then we perform time-series analysis by encoding the flow records of successive periods to persistence distributions and training a classifier to identify the attacking or normal flows. The experimental results based on real-world network traces show that our approach significantly outperforms the state-of-the-art methods in terms of *Accuracy*, *Recall*, and *F1-score*.

Keywords: DDoS detection · Network security · Time-series analysis

1 Introduction

Distributed denial-of-service (DDoS) attack is a kind of cyber attack that attempts to disrupt legitimate access to target systems or resources by overwhelming them with a flood of Internet traffic. The increasing number of attacks and the resulting severe impacts make DDoS one of the biggest threats to network security. In the past 2020, over 50 million DDoS attacks occurred worldwide, causing considerable losses to users as well as service providers. According to the report from Neustar, the 849 companies surveyed lost $2.2 billion in

B. Li et al. (Eds.): APWeb-WAIM 2022, LNCS 13423, pp. 318–333, 2023.
https://doi.org/10.1007/978-3-031-25201-3_25

revenue responding to the attacks[1]. In general, an attacker will employ multiple compromised computers as sources of their attacking traffic to increase effectiveness. The diversity in attacking sources increases the difficulty of defense and, in turn, gives chances to attackers. Facing the growing threat caused by DDoS, it is urgent to make out an effective defending mechanism.

Many solutions have been proposed to identify and block malicious traffic [4,7–17], which can be roughly classified into two types: *signature-based* detection and *anomaly-based* detection. *Signature-based* detection can identify the specific types of attacks by matching the signatures (*e.g.*, unique combinations of packet field values) of monitored traffic with those of known attacking traffic stored in the database [16]. Despite the simplicity and efficiency in capturing known attacks, its inability facing with unknown attacks, evasion attacks, and variants of known attacks cannot be ignored. Differently, *anomaly-based* methods rely on detecting the unexpected behaviors (*i.e.*, anomaly in feature patterns) of network traffic, which allows it to maintain its robustness faced with unknown attacks [4,7–15,17]. For example, since DDoS attacks are launched by compromised sources, the spreads (the number of distinct source addresses) of attacking traffic are significantly larger than normal ones. Based on it, non-duplicate sampling (NDS) [11] treats the increase in the spread as an indication of attacks and proposes to measure the spread within a single period to identify DDoS attacks. However, the above approaches only focus on traffic of a single period. They cannot accurately identify the stealthy DDoS attacks where malicious hosts send service requests to a server with a low-profile speed, slowing down the server without making the attack obvious. Considering that stealthy attackers will persistently send requests during the attack, the k-persistent spread estimator (KPSE) [12] is designed to track network traffic across consecutive t periods to estimate k-persistent spread (*i.e.*, the number of distinct source addresses that appear at least k of them). After that, they will report the stealthy DDoS attacks with an abnormally large k-persistent spread.

A common shortage of existing detection approaches is that they can only monitor traffic at coarse-grained time granularity (*e.g.*, 1-min or 5-min measurement periods). Their period-level traffic records will lose fine-grained temporal information, such as the specific arriving moments of incoming packets. Thus, we want to improve the performance of DDoS detection with fine-grained traffic monitoring by extracting and identifying the patterns of attacks in time series (*e.g.*, persistence of malicious requests). Besides, NDS and KPSE employed a threshold mechanism to report the attacks when the monitored features exceeded a pre-set threshold. Nevertheless, the artificial threshold may lead to a high false alarm rate and a low accuracy [3].

To address the above problems, this paper presents a novel DDoS detection approach named Time-Series DDoS Detection (TSD3). First, considering that benign sources mainly produce transient requests while malicious sources (*i.e.*, attackers) will persistently send requests, we designed attention sampling to put

---

[1] Neustar: Worldwide ddos attacks and cyber insights research report. https://www. home.neustar/.

more attention on transient and persistent requests. We assign their sources with larger probabilities while lower probabilities for others (*i.e.*, noted as normal in our scenario). It can achieve higher resource efficiency than NDS and KPSE, where all sources are collected with equal probabilities. Second, to obtain fine-grained temporal information of attackers, we divide the traditional 1-minute or 5-minute measurement window into smaller equal-length periods (*e.g.*, one second) and run attention-sampling to capture traffic in shorter time slices. Sources sampled in the same period are stored together and share the same timestamp, enabling us to take a deep insight into the network patterns in the time dimension. Another advantage is that the proposed method can support continuous sampling with much smaller on-chip memory since the number of sources that appeared in a short period is smaller than that of a long period. Moreover, compared to traditional solutions that employ longer measurement periods, TSD3 can detect attacks at the end of each short period and timely report the ongoing attacks, increasing the timeliness of detection.

However, we find it still challenging to run time-series analysis for detection since the number of requests sent to different hosts varies. Besides, each source's occurrences in multiple periods is a time series, which also raises the difficulty of discrimination. To this end, we introduce a novel distribution-based representation to encode the extracted time series. Then we can employ classification algorithms for automatic detection, freeing us from the problem caused by the threshold mechanism and ensuring a higher detecting accuracy.

The proposed method is implemented in an on-chip/off-chip architecture [11]. It can guarantee both timeliness and accuracy in DDoS attack detection under high-speed networks. Before evaluation, we also explored the effect of the parameters on the sampling to balance the overhead and performance of the system. Finally, we run extensive experiments on real-world Internet traces downloaded from CAIDA [1,2] to evaluate the performance of the proposed solution. The experimental results show that our TSD3 significantly outperforms the state-of-the-art methods in *Accuracy, Recall* and *F1-score*.

In summary, this work makes the following contributions:

- We propose a novel time-series-based solution that detects DDoS attacks with high accuracy, low communication cost, and small memory usage.
- We design attention sampling, which can provide fine-grained temporal information to help track network traffic and achieve high resource efficiency by assigning larger sampling probabilities to transient and persistent sources.
- We introduce a novel distribution-based representation to encode the time-series of network traffic and apply a trained classifier for automatic detection.
- We implement the proposed solution in an on-chip/off-chip architecture and evaluate its performance by comparing it with state-of-the-art detection methods under fair settings using real-world network traces.

2   DDoS Attack Detection Problem

This work investigates the DDoS attack detection problem, whose goal is to distinguish malicious traffic from benign ones depending on the proposed time-series-based solution. Here, the network traffic is modeled as the packet stream $P = \{\mathbb{P}_1, \mathbb{P}_2, \mathbb{P}_3, ...\}$. Since our goal is to identify the malicious traffic towards targets, we may consider all packets sent to the same destination address as a flow (*i.e.*, per-destination flow). Thus, we abstract each packet $\mathbb{P} \in P$ as a pair $<f, e>$ identified by the flow label f (*i.e.*, destination address) and the element label e (*i.e.*, source address).

Based on the above definitions, we model the packet stream as a flow set $F = \{f_1, f_2, f_3, ...\}$ where each flow f_i is a substream in P, consisting all the packets carrying the same flow label f_i. We use the term *spread* to stand for the number of *distinct* elements in a flow, so the spread of a per-destination flow equals the number of distinct sources IP addresses in this flow.

Finally, we formally define the **DDoS attack detection problem** as follows: *A DDoS attack detection system monitors network traffic passing through a central router or gateway and provides judgments of whether a specific server is under a DDoS attack.*

3   Novel Time-Series-Based Solution to DDoS Detection

This section describes our novel solution to detect DDoS attacks based on time-series analysis of network flows. After introducing the motivation and system model, we show each module of the proposed system and corresponding operations in turn. Finally, we explain and verify the effectiveness of our method.

3.1   Motivation

Inspired by the two crucial features of DDoS attacking flows (*i.e.*, persistency and synchronism) [14], we find that sources (elements) in attacking flows have a large probability of appearing for long, while the appearance of elements in benign flows is relatively short. In other words, the persistent and transient elements are representative signals of attack flows and benign flows, respectively. We should distinguish them out in the process of sampling for the targeted exploration of their characteristics and improvement of system performance. Therefore, the first requirement of our approach is that the sampling mechanism should be able to hold elements' sampling/occurrence conditions in the previous cycle. Combined with the conditions of the current cycle, we can deduce the persistent state of each element, thus giving different "attention" to better capture the representative ones.

We also found that the sampling algorithms of existing methods that maintain an on-chip structure to aid sampling decisions need to stop at the end of each period to reset the structure. For instance, the on-chip structures of NDS and KPSE, which are used for duplicated filtering, need to be refreshed at the end of each period to ensure its filtering effect in the next period. Considering the high speed of networks, even a short pause in sampling will miss many incoming packets and hugely impact the sampling performance. This impact is more significant as our algorithm adopts small time slices (*e.g.*, 1-second periods). Hence, another requirement is that our sampling algorithm supports continuous execution without pausing and refreshing.

3.2 Overview of Proposed Approach

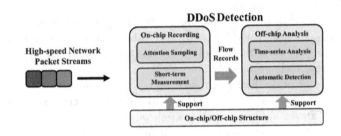

Fig. 1. System model of the Time-series DDoS Detection (TSD3).

Figure 1 depicts the structure of our approach, which is implemented in an on-chip/off-chip architecture [5,11] and placed at a central router/gateway to monitor packet streams passing through. The short-term measurement tasks with attention sampling are continuously performed on the network processing chips to leverage on-chip memory for acceleration. Although the on-chip space (SRAM) is fast enough for line-speed read/write operations, its limited capacity is insufficient to hold all records. As a result, the obtained flow records are sent to off-chip memory for more extensive storage. The subsequent analysis and detection are also performed off-chip. The analysis will extract time-series events from records and transform them into distribution-based features. Finally, these features will be fed into a trained classifier to identify the DDoS attack automatically.

3.3 Online Attention Sampling Algorithm

Attention Sampling: Considering the above two requirements, the main idea of the proposed attention sampling is: *sampling packets continuously and with the ability to give different levels of attention based on elements' attributes, for targeted sampling.* It should be noted that, unlike packet sampling that downloads every element with equal probability, our attention sampling is expected to adjust sampling probability according to the states of elements adaptively. Then representative ones deserving more attention are more likely to be captured for investigation. We believe this advantage is more suitable for DDoS detection because it is more sensitive to the differences in elements' properties between various flows and may help discriminate and identify attacks.

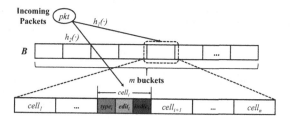

Fig. 2. Data structure of attention sampling.

Data Structure: We design an automatic upgrade and degrade mechanism to realize attention sampling. As shown in Fig. 2, the main structure of the proposed mechanism is an array noted as B, which consists of m buckets, and each bucket comprises n cells (a cell has 4 bits). Each cell includes three parts noted as $type$, $edit$ and $indic$. The first part stands for the persistent status of the corresponding element. We use -1 to stand for the short-term status, 0 for normal, and 1 for the long-term. The second part is used as a duplicated filter because we only need to record the first appearance of elements in each period. In particular, instead of only using 1 to denote the occurrence of an element, we associate the indication with the index of periods to avoid possible pauses and refreshes. Meanwhile, the second and third parts will be combined to indicate the sampling/appearing status for the elements in the last period and guide the attention assignment. Besides, attention sampling also maintains two map functions $H_1(\cdot)$ and $H_2(\cdot)$. Their output ranges are $\{0, 1, ..., m - 1\}$ and $\{0, 1, ..., n - 1\}$, respectively. All bits in B are initialized to 0 at the beginning.

Sampling Procedure: The pseudo-code of attention sampling is presented in Algorithm 1. Without loss of generality, we consider the packet with flow label f and element e (noted as $Pkt_{<f,e>}$) that arrives in period t. We first compute $H_1(f \oplus e)$ and $H_2(f \oplus e)$ to choose the bucket and cell for it. Here, notation $\oplus$ is the XOR operation. For illustration, we denote the selected one as $cell$ (line 1–3). Moreover, we compute the $status$ used for the current period t according to its parity (line 4). In order to filter the duplicated, we only take further actions when the $edit$ part of the chosen $cell$ does not equal $status$. Then, by checking $cell.type$ we can determine the type of the element as follows. Meanwhile, $edit$ and $indic$ can help assign appropriate sampling probabilities automatically.

- *Short-term (line 5–7):* If $cell.type$ is -1, it means that the type of this element is short-term and it deserves more attention. We will directly offload this packet for storage. Then we upgrade the persistent status of this element to normal by updating $cell.type$ to 0 and writing it back. It should be noted that the $cell.indic$ will be assigned as $\neg status$ once the sampled action happens. It can help in the later period to determine whether this element has been sampled in previous by judging whether $cell.edit$ equal $cell.indic$.
- *Normal (line 8–16):* Once the incoming packet is in normal status as $cell.type$ equals 0, we will check $cell.status$ and $cell.indic$ to determine whether the element should be demoted. If the result is $True$, which means this element was not present or sampled in the past period. So we can treat it as a

---

Algorithm 1: Attention sampling

Input: $Pkt_{<f,e>}$ arrived during period t

1   $index_{bucket} \leftarrow H_1(f \oplus e)$

2   $index_{cell} \leftarrow H_2(f \oplus e)$

3   $cell \leftarrow B[index_{bucket}][index_{cell}]$

4   $status \leftarrow t \mod 2$

5   **if** $cell.type = -1$ *and* $cell.edit \neq status$ **then**

6 | $samplingPkt_{<f,e>}$

7 | $cell \leftarrow [0, status, \neg status]$

8   **else if** $cell.type = 0$ *and* $cell.edit \neq status$ **then**

9 | **if** $cell.status = cell.indic$ **then**

10 | $samplingPkt_{<f,e>}$

11 | $cell \leftarrow [0, status, \neg status]$

12 | **else if** $cell.status \neq cell.indic$ **then**

13 | generate a random float $r \in [0,1]$

14 | **if** $r \leq p$ **then**

15 | $samplingPkt_{<f,e>}$

16 | $cell \leftarrow [1, status, \neg status]$

17   **else if** $cell.type = 1$ *and* $cell.edit \neq status$ **then**

18 | **if** $cell.status = cell.indic$ **then**

19 | generate a random float $r \in [0,1]$

20 | **if** $r \leq p$ **then**

21 | $samplingPkt_{<f,e>}$

22 | $cell \leftarrow [1, status, \neg status]$

23 | **else if** $cell.status \neq cell.indic$ **then**

24 | $samplingPkt_{<f,e>}$

25 | $cell \leftarrow [0, status, \neg status]$

---

short-term one and directly offload it. Otherwise, we will sample it with a pre-set sampling probability p. Here, the probability p is related to the algorithm's throughput and performance, which can be set by user-specific.

- *Persistent (line 17–25):* When $cell.type$ is 1, we treat this packet as a persistent one which is likely from attackers. We will offload and update its structure. On the contrary, if the computing result shows that this packet needs to be degraded, we will treat it as normal and sample it with probability p.

It can be seen that, by setting additional status bits, we are able to preserve the sampling status of elements in previous cycles. It helps automatically update the persistence status of elements and guides the allocation of attention in the current cycle. At the same time, we find that traditional methods only use "1" to denote elements' occurrence to filter the duplicated. It leads to an unavoidable pause at the end of each period to reset the on-chip space to "0". To achieve continuous sampling, we propose associating the notation with the index of cycles, freeing us from the refresh. So, our proposed mechanism well satisfies the two requirements mentioned above and achieves the goal of attention sampling.

Recall that the detection accuracy of existing DDoS detection methods is far from expectations due to their coarse-grained analysis. Their ignorance of fine-grained temporal information prevents them from observing some characteristics of attacks, such as persistence in the time dimension. We hope to capture more timestamps within the traditional long monitoring window. So, by continuously running short-term sampling tasks with our attention sampling, flow records are naturally integrated with the index of its arriving period, obtaining the fine-grained timestamps. Then, we are able to carry out time series analysis and have a deep view of the traffic, extracting the unique characteristic of attacks.

3.4   Time-Series-Based Solution for DDoS Detection

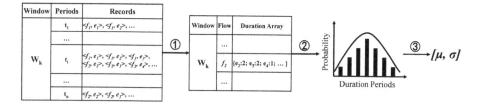

Fig. 3. Workflow of time-series analysis for DDoS detection.

Next, we will present the details of the time-series analysis and automatic detection implemented at off-chip space. Based on the obtained traffic through attention sampling, we further assemble the consecutive n periods' records into a window (noted as W_k). Compared with the longer slices adopted by traditional methods, the small partitioning of traffic enables us to respond to users' requests quickly without waiting for the end of a long period. Our analysis and detection can be carried out at the end of each short period once the user requests, thus timely reporting the ongoing attacks and reducing detection delays.

Counting: The workflow of the proposed time-series analysis is depicted in Fig. 3. The first step is to count the appearances of each element for flows in the current window, and its pseudo-code is shown in Algorithm 2. Initially, we create an empty table C to preserve the records of each flow f in W_k, where the duration information of a flow will be kept as an independent dictionary (i.e., duration array). Then, we traverse the off-chip traffic records of all n periods in W_k. For each flow f, we list all its elements and count their occurrences (i.e., duration periods) respectively. Without loss of generality, we consider the use-case in Fig. 3. We might as well assume that flow f_2 is sampled and recorded only in t_i and t_n. Then, we count the appearances for all its three elements (e_2, e_3, e_4) and update the corresponding duration array. In this scenario, e_2 and e_3 appeared in both t_i and t_n, while e_4 only appeared in t_i. Hence, we have the counting result of f_2 is $\{e_2 : 2; e_3 : 2; e_4 : 1\}$, where keys and values are elements in f_2 and their corresponding duration periods, respectively.

Encoding: Despite the obtained number of continuous periods for each element, it is still challenging to exploit them for attack detection. We need to overcome

---

Algorithm 2: Counting Duration

Input: Off-chip record of Window W_k, an empty table C

1 **for** *each period in input Window* **do**
2   **for** *each record* $< f, e >$ *in period* t_i **do**
3   **if** *flow f in C* **then**
4   **if** *element e in $C[f]$* **then**
5                 $C[f][e] \leftarrow C[f][e] + 1$
6   **else**
7                 $C[f][e] \leftarrow 1$
8   **else**
9             $C[f] \leftarrow \{\}, C[f][e] \leftarrow 1$

---

two major difficulties, *i.e.* the number of elements in various flows is different, and the occurrence of each element is a time series. Therefore, we need to model the appearances of elements and encode their time series to simplify the detection. We denote the number of all sampled elements in flow f as N_f. Obviously, for each element, the range of its appearing periods is $[0, n]$. Remember that n stands for the number of periods of a window. We use M_j to represent the number of elements that appear j periods out of total n periods. Therefore, the frequency of elements in the flow f that last for j periods is,

$$Q_{f,j} = \frac{M_j}{N_f}. \tag{1}$$

We apply the variable X to stand for the number of appearances of an element. Our goal is to study the probability of X in different flows. Given that the number of samples is sufficient, we use the frequency in (1) to approximate the probability when X equals x.

$$Pr(X = x) \approx Q_{f,x} \tag{2}$$

Accumulating the probabilities of all possible persistent periods in f, we have the sum equal to 1.

$$\sum_{j=0}^{n} Pr(X = j) = 1 \tag{3}$$

Since X is a discrete random variable, we use the probability mass function (pmf) to describe its probability distribution. For each flow, we have

$$f_X(x) = \begin{cases} Pr(X = x), & x \in [0, 1, 2, ..., n], \\ 0, & otherwise. \end{cases} \tag{4}$$

However, the discrete form of the pmf may raise difficulties for later classification. Instead, we hope to simplify the representation of the analyzing results, thereby assisting the later training and detecting process. So, we adopt the probability density function (PDF), which is used to describe the probability of a continuous random variable, to approximate the pmf of each flow in (4).

Feature Extraction: To fit into classification algorithms for automatic learning and detection, we consider using numerical features to illustrate the geometric characteristic of the PDF. At the third step, We extract out the position parameter μ and the scale parameter σ of the PDF as the final representation of flow features. Here, μ describes the position of central tendency in this distribution. In other words, the variable X has a larger probability of taking a value close to μ. The second parameter σ describes the dispersion of the distribution. It is also called the shape parameter. The larger the σ is, the flatter the PDF curve is, and the corresponding distribution is thus more scattered. On the contrary, the smaller the σ is, the thinner and taller the PDF curve is, and the corresponding distribution is thus more concentrated.

Finally, the origin traffic traces stored in off-chip memory are transformed into a two-dimensional vector (μ, σ) as the encoded representation of the time-series for each flow, which will serve as the input of the detecting module. However, it has been proved that simple threshold mechanisms are prone to incorrectly classifying legitimate flows as illegal ones and cause a high false-positive rate [3]. So, we apply a support vector machine (SVM) in the detecting module for automatic classification. The per-flow vectors will be input into a trained SVM, while the outputs are judgments of whether a specific flow is malicious.

3.5 Analysis and Explanation

We want to stress that there exists a significant difference between the traffic patterns of different streams in the time dimension. Commonly, to launch a DDoS attack, an adversary will employ enormous compromised computers as sources to flood the target. Therefore, the spread (*i.e.*, the number of distinct source addresses) of a malicious flow is higher than that of a normal one. Besides, attackers need to generate fake requests simultaneously and continuously, thus ensuring a stable volume of attacks and increasing its effectiveness. This means that the per-flow spread will maintain high throughout the attack. In other words, persistency and synchronism are two crucial features in the time series of DDoS malicious flows [14]. Hence, we draw out the unique traffic pattern of DDoS, *i.e.*, elements in attacking flows are more likely to last for a longer time. While in benign flows, the situation is just on the contrary.

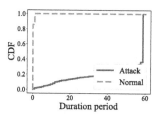

Fig. 4. The CDF of duration periods.

We also run a simple experiment to verify the persistency and synchronism of attacking flows. First, we randomly select a benign flow and an attacking flow from the applied datasets [1,2], and continuously record the appearances of their elements for 60 periods (each lasting for one second). To explore traffic characteristics in the time dimension, we only emphasize the number of appearing periods of an element instead of its exact appearances in a single period. Therefore, within each period, we only need to record the first occurrence of the element. Then, the cumulative distribution function (CDF) is applied to illustrate the distribution of occurrences of elements in a stream. As shown in Fig. 4, all elements in the benign flow appeared in less than ten periods. While for those in the attacking flow, nearly 80% of them lasted for over 50 periods.

This feature may greatly help in the distinction of attacks. However, traditional methods fail to capture such due to a lack of corresponding records. We noticed that most existing methods perform their traffic measurement and analysis within a long period. Their observed traffic is recorded at the period granularity. While some fine-grained temporal information, such as the specific arriving moments of incoming packets, are ignored. For instance, NDS [11] only records the first appearance of a packet within a long time slice while ignoring its later activities. A possible false-report case is that when faced with a sudden increase in the spread caused by ticket grabbing (*i.e.*, a large number of users visit the ticketing site at the same time), NDS will falsely report it as an attack. Then, users' legitimate behaviors will be hindered. Moreover, the lack of temporal information makes it impossible for them to observe the persistence of the attack in the time dimension. So, these approaches cannot detect stealthy DDoS attacks where malicious requests are sent continuously.

While in our solution, through periodically running short-term measurement tasks, we can continuously track flow elements and gain a deeper look at the unique patterns of attack traffic in the time dimension. It can effectively avoid false positives caused by the mistake of sudden increases of spread. It can also maintain efficiency in the face of stealthy DDoS attacks. Meanwhile, the new encoding method of time-series of network traffic also allows us to avoid the low accuracy caused by the threshold mechanism and further improve the detecting performance. To sum up, our TSD3 successfully addresses the shortcomings of existing methods.

4   Experiments and Results Analysis

4.1   Experiment Setup

Platform: We conducted our evaluations on a server equipped with two six-core Intel Xeon E5-2643 v4 3.40GHz CPU and 256 GB RAM.

Dataset: We evaluate the performance of our time-series-based solution to DDoS detection through extensive experiments on real-world Internet traffic traces. We apply one minute of benign traffic downloaded from CAIDA [2] consisting of 29,172,463 packets as benign traffic. As for attacking traffic, we use one hour of the "DDoS flooding attack 2007" dataset [1], including $302,373,723$ packets. We notice that there are $156,725$ normal per-destination flows as our positive samples, while only one malicious flow is negative. The imbalance cannot be addressed by simply dividing the later dataset into successive intervals with equal length. Hence, we apply sliding-window technology to enhance it where the window and step sizes are set to one minute and one second separately. Finally, a total of $3,540$ malicious per-destination flows were obtained.

Implementation: We compare our TSD3 with two state-of-the-art methods, *i.e.*, NDS [11] and KPSE [10]. All methods are implemented in C++. Their on-chip memory and sampling probability are all the same for fair comparisons. Particularly, for the KPSE, we divide one-minute traffic into five equal-length periods and count the number of distinct elements that appear in at least two of them (*i.e.*, per-flow 2-persistent spread). The labeled analysis results of three methods are split by $3 : 1$ for the training and evaluation of the classifier.

4.2   Effects of System Parameters

We first conduct experiments to explore the impact of parameters on system overhead, which allows us to configure the parameters flexibly according to the resources of the system. We focus on the on-chip memory size s and the sampling probability p. These two parameters will affect the packet download ratio (*i.e.*, the proportion of downloaded packets to the total packets) and the downloaded elements (*i.e.*, the number of packets downloaded per period), which are tightly related to the system overhead.

Table 1. Packet download ratio and number of downloaded elements.

On-chip size	Sampling probability	Packet download ratio	Downloaded elements
50 KB	0.1	0.004	5426
	0.2	0.008	9530
	0.5	0.02	20208
100 KB	0.1	0.004	6614
	0.2	0.008	10439
	0.5	0.02	21191

As shown in Table 1, when the sampling probability is set as 0.1, only four-thousandths of the total data packets are downloaded for time-series analysis and detection. Besides, under the same on-chip space, with the increase of sampling probability, so do the packet downloaded ratio and the number of elements downloaded per period. In addition, with the fixed sampling probability, we find that on-chip space has little effect on the number of elements downloaded. We also try to fix the sampling probability and increase the on-chip space from 50 KB to 100 KB. Experiments show that the packet downloaded ratio remains despite a slight increase in the number of downloaded elements. Therefore, the sampling probability is the main factor affecting the download overhead in our method. Considering the limited resources of communication and storage, we should try to reduce the number of packets to be sampled while ensuring detection accuracy. Hence, the value of on-chip memory size s and the sampling probability p are set as 50 KB and 0.2 in our experiments.

4.3   Performance Analysis

Table 2. Detection results.

	Accuracy	Recall	F1-score
TSD3 (t = 1 s)	**0.98**	**0.98**	**0.98**
TSD3 (t = 2 s)	**0.98**	**0.98**	**0.98**
TSD3 (t = 5 s)	0.87	0.87	0.87
NDS [11]	0.86	0.80	0.83
KPSE [10]	0.94	0.96	0.95

The comparison experiments include the following common performance metrics: *Accuracy* (the fraction of correctly classified instances), *Recall* (the class-conditional accuracy), and *F1-score* (the harmonic mean of the precision and recall). The experimental results are shown in Table 2.

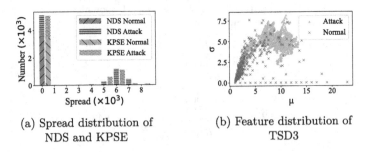

(a) Spread distribution of   (b) Feature distribution of
    NDS and KPSE   TSD3

Fig. 5. Distribution of extracted features.

For an intuitive understanding of the experimental results, we also draw the distribution of the extracted features of the three methods. Figure 5(a) shows that both NDS and KPSE can easily distinguish the abnormal flows with a spread larger than 1000 with a simple threshold mechanism. However, there still exists some malicious flows whose spreads are lower than 1000 and mixed with benign ones. It is impossible to find an exact bound between attack and normal [3]. Hence, NDS and KPSE rank third and second, respectively, in terms of *Accuracy*. KPSE performs better than NDS for its additional considerations of persistence of attacks. However, we want to stress that its adopted time slice (20-second period in our scenario) is still too coarse to achieve satisfying performance.

As depicted in Fig. 5(b), our TSD3 can clearly make the distinguishment. On the one hand, our attention sampling makes the capturing process more targeted. The rich temporal information also enables us to observe the characteristics of attacks in the time dimension. On the other hand, we use frequency distributions to encode the extracted time series, enabling automatic learning and classification using machine learning methods. Compared with artificial thresholds, our methods further improves the accuracy. Moreover, we also compare the performance of our method at different time granularities to prove the importance of fine-grained temporal information for network management. As shown in Table 2, our approach can achieve better results on minor scales (*i.e.*, 1 s or 2 s). When the time granularity is larger (*i.e.*, 5 s), the effect of attack detection will significantly reduce.

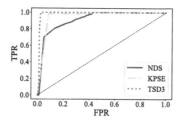

Fig. 6. ROC curve of three DDoS detection methods.

We also apply the receiver operating characteristic (ROC) curve [6] for evaluation to illustrate the diagnostic ability of a binary classifier system. With normalized units, the area under the curve (AUC) is equal to the probability that a classifier will rank a randomly chosen positive instance higher than a randomly chosen negative one. The larger the AUC, the better the classification effect of the classifier. As shown in Fig. 6, our TSD3 has a larger AUC than the two state-of-the-art methods.

5   Conclusion

This work proposes a novel DDoS attack detection approach named Time-Series DDoS Detection (TSD3). First, we design attention sampling, which can continuously track representative sources with limited storage and communication resources. Through periodically running sampling tasks, we naturally assign fine-grained timestamps to off-chip records. Thus we can observe the characteristics of attacking flows in the time dimension via time-series analysis. Moreover, we propose a novel representation to encode the above-extracted features as probability distributions and use a trained machine learning classifier for automatic learning and detecting. The experimental results based on real-world Internet traces demonstrate that our solution has a better performance in identifying DDoS attacks than the the-state-of-art approaches.

Acknowledgements. This work was supported by National Natural Science Foundation of China under Grant No. 62072322, No. 61873177, and No. U20A20182, Natural Science Foundation of Jiangsu Province under Grant No. BK20210706, and Jiangsu Planned Projects for Postdoctoral Research Funds under Grant No. 2021K165B.

References

1. CAIDA: The CAIDA UCSD DDoS Attack 2007 dataset (2007). https://www.caida.org/catalog/datasets/ddos-20070804_dataset. Accessed 16 Sept 2021
2. CAIDA: The CAIDA UCSD Anonymized Internet Traces 2016 (2016). https://www.caida.org/data/passive/passive_2016_dataset.xml. Accessed 28 July 2019
3. Chandola, V., Banerjee, A., Kumar, V.: Anomaly detection: a survey. ACM Comput. Surv. (CSUR) **41**(3), 1–58 (2009)
4. Cohen, R., Katzir, L., Yehezkel, A.: A minimal variance estimator for the cardinality of big data set intersection. In: Proceedings of the 23rd ACM SIGKDD International Conference on Knowledge Discovery and Data Mining (KDD 2017), pp. 95–103 (2017)
5. Du, Y., Huang, H., Sun, Y.E., Chen, S., Gao, G.: Self-adaptive sampling for network traffic measurement. In: IEEE INFOCOM 2021-IEEE Conference on Computer Communications, pp. 1–10. IEEE (2021)
6. Fawcett, T.: An introduction to ROC analysis. Pattern Recogn. Lett. **27**(8), 861–874 (2006)
7. Meng, S., Wang, T., Liu, L.: Monitoring continuous state violation in datacenters: exploring the time dimension. In: Proceedings of the 26th International Conference on Data Engineering (ICDE 2010), pp. 968–979. IEEE (2010)
8. Gorovits, A., Gujral, E., Papalexakis, E.E., Bogdanov, P.: LARC: learning activity-regularized overlapping communities across time. In: Proceedings of the 24th ACM SIGKDD International Conference on Knowledge Discovery and Data Mining (KDD 2018), pp. 1465–1474 (2018)
9. Herodotou, H., Ding, B., Balakrishnan, S., Outhred, G., Fitter, P.: Scalable near real-time failure localization of data center networks. In: Proceedings of the 20th ACM SIGKDD International Conference on Knowledge Discovery and Data Mining (KDD 2014), pp. 1689–1698 (2014)
10. Huang, H., et al.: You can drop but you can't hide: k-persistent spread estimation in high-speed networks. In: IEEE INFOCOM 2018-IEEE Conference on Computer Communications, pp. 1889–1897. IEEE (2018)
11. Huang, H., et al.: Spread estimation with non-duplicate sampling in high-speed networks. IEEE/ACM Trans. Networking **29**(5), 2073–2086 (2021)
12. Huang, H., et al.: An efficient k-persistent spread estimator for traffic measurement in high-speed networks. IEEE/ACM Trans. Networking **28**(4), 1463–1476 (2020)
13. Ying, X., Wu, X., Barbará, D.: Spectrum based fraud detection in social networks. In: Proceedings of the 27th International Conference on Data Engineering (ICDE 2011), pp. 912–923. IEEE (2011)

14. Li, C., Yang, J., Wang, Z., Li, F., Yang, Y.: A lightweight DDoS flooding attack detection algorithm based on synchronous long flows. In: Proceedings of the IEEE Global Communications Conference (GLOBECOM 2015), pp. 1–6. IEEE (2015)
15. Namaki, M.H., et al.: Kronos: lightweight knowledge-based event analysis in cyber-physical data streams. In: Proceedings of the 36th International Conference on Data Engineering (ICDE 2020), pp. 1766–1769. IEEE (2020)
16. Paxson, V.: Bro: a system for detecting network intruders in real-time. Comput. Netw. 31(23–24), 2435–2463 (1999)
17. Ting, D.: Approximate distinct counts for billions of datasets. In: Proceedings of the 2019 International Conference on Management of Data (SIGMOD 2019), pp. 69–86 (2019)

Executing Efficient Retrieval Over Blockchain Medical Data Based on Exponential Skip Bloom Filter

Weiliang Ke[1], Chengyue Ge[1], and Wei Song[1,2,3(✉)]

[1] School of Computer Science, Wuhan University, Wuhan, China
{keweiliang,2018302050011,songwei}@whu.edu.cn
[2] The Big Data Institute at Wuhan University, Wuhan University, Wuhan, China
[3] Intellectual Computing Laboratory For Cultural Heritage, Wuhan University, Wuhan, China

Abstract. With public enthusiasm for bitcoin rising, blockchain as the basic technology has attracted increasing attention from both industry and academia. With the particular advantages of decentralization and integrity assurance, many hospitals have moved their medical data management systems to the blockchain. However, the existing medical blockchain applications generally encounter low throughput and the absence of the corresponding index structure. So, the retrieval efficiency of the existing blockchain system is unsatisfactory for the management of the large-scale patients' healthcare records. To achieve efficient and secure medical data retrieval, we propose an efficient retrieval algorithm for blockchain medical data based on exponential skip Bloom filter. While improving the efficiency, the proposed method also can resist privacy leakage under brute-force attack. We prove the security and low overheads of the proposed scheme by theoretical analysis. The experimental results further validate the effectiveness and efficiency of this work.

Keywords: Blockchain · Bloom filter · Medical data

1 Introduction

The past several years have witnessed the surge of popularity of cryptocurrencies such as Bitcoin [14] and Ether. Behind the cryptocurrency network is the blockchain technology to provide the security and integrity for every decentralized transaction. Blockchain is a distributed database, which manages the time series data and ensures that each block is tamper-resist. Based on the distributed consensus protocol, blockchain acts as a public ledger so that every loyal node can verify all transactions without a trusted server [5] unless the majority of the system are corrupted. Nowadays, various blockchain-based applications spring up and prove the potential and scalability of blockchain technology.

W. Ke and C. Ge—Contributed equally to this work.

© The Author(s), under exclusive license to Springer Nature Switzerland AG 2023
B. Li et al. (Eds.): APWeb-WAIM 2022, LNCS 13423, pp. 334–348, 2023.
https://doi.org/10.1007/978-3-031-25201-3_26

As for now, there have been a great number of personal information management systems based on blockchain, among which the medical field contains various researches. The majority of existing blockchain applications for medical information management attach importance to the storage pattern of personal data [9], data integrity [1] and secure data sharing [19]. In these cases, either storing health records directly on the blockchain or storing the signatures on the blockchain for verification involves the query operations on the blockchain. Without an efficient data retrieval method, the blockchain may suffer from low response time due to the time overheads of queries on enormous items stored. However, the minority of related works focus on the efficiency of the secure retrieval processing like [12,16].

We believe that there are at least three obstacles to executing efficient queries on the medical data blockchain system. Firstly, it is impossible to directly change the structure of the underline blockchain because it is fixed for recording transactions fairly. Secondly, concerning complicated medical data, designing an index structure needs to trade off the query efficiency and the maintenance overheads. Thirdly, the sensitivity of medical data requires the security of the query process. Optimistically, some secure sharing methods [19] for medical data over blockchain provide solutions for the third obstacle. So it is highly desirable to design a retrieval index structure and the corresponding retrieval algorithm to satisfy both efficiency and low maintenance overheads.

Usually, patients, hospitals, and medical data consumers such as insurance companies participate in a medical blockchain system. The medical data is generated by the hospital, and is authorized to the data consumer. For example, the insurance company attempts to find Alice's health record regarding lung cancer when reviewing the qualification of claim settlement. To efficiently retrieve the medical data at the blockchain system, we designed an exponential skip Bloom filter index and corresponding retrieval algorithm for the medical information management system over the blockchain. The main contributions of our paper can be summarized as follows:

1. We design an exponential skip Bloom filter index for the huge scale of medical data over a blockchain. The maintenance overhead of the index structure is relatively low due to our way of inserting data by which we only need to update two existing Bloom filter nodes without influencing others.
2. We design an efficient retrieval algorithm based on the proposed index structure, which makes full use of the nature of the Bloom filter.
3. We conduct theoretical analysis to prove the security of the proposed retrieval scheme. We also carry out simulation experiments to validate the efficiency and security of the proposed method.

2   Related Work

Our main motivation is to develop an efficient query method to improve the medical information service. So, we reviewed the related work, especially for search index and algorithm over encrypted data on the blockchain and related scenarios containing current blockchain applications for medical data management.

2.1   Blockchain Applications for Medical Data Management Service

Many scholars focus on blockchain technology to solve access control, secure sharing, storage pattern, and traceability of Electronic Health Record (EHR) in the current healthcare industry. Query over the blockchain is an indispensable step for these methods.

For storing data in the blockchain, LevelDB was used as the underlying data storage system for most blockchain systems [17], which was more convenient to write data instead of reading it. He et al. [9] designed the pattern to store healthcare bill information and defined the access control policies with Smart Contract. The system used Patient ID, Chart ID, and Service ID to query bill information and EHR. To achieve traceability and integrity of EHR, Huang et al. [11] created a proof-chain to store all users' manipulation on EHR. The restriction, that each query request should be accepted by willing patients, increases response time. Aujla et al. [1] connects blockchain with medical IoT data to verify data integrity. In this method, each block consists of electronic health records of the specific patient only so that query operation will be executed on much more blocks. As for secure sharing, Zhang et al. [19] introduced a deniably authenticated searchable encryption scheme (DASES) to support medical image data sharing on the blockchain. This method needs to search for target medical image data over blockchain before executing the proposed decryption step.

Even though the methods mentioned above are effective in their main focus, it is difficult to obtain further improvement without an efficient and secure retrieval algorithm.

2.2   Search Algorithm over Encrypted Data on Blockchain

In the field of information retrieval with unstructured big data, there are two main indexes, i.e., forward index and inverted index. With a forward index, it takes a heavy cost to match the input keyword with each index because the number of search operations grows linearly as the files increase [13]. While with an inverted index, the way of firstly searching keywords index, which links to several relative documents, improves efficiency greatly. Hu et al. made use of smart contract to construct a fair privacy-preserving search scheme where each party, in the multi-user setting particularly, is fairly treated and incentivized to do correct computations [10]. Chen et al. proposed a scheme that utilizes the complex Boolean expression to extract the electric health records to construct the index [4]. Pei et al. [15] realized an efficient query on blockchain by proposing a new semantic keyword extraction algorithm and constructing an on-chain inverted index structure called Merkle Semantic Trie. Unfortunately, maintaining the security of an inverted index structure needs too much storage for creating different keys for each data owner, let alone the complexity of the inverted index itself will sacrifice more storage overhead. Peng et al. [12] proposed Searchain, a peer-to-peer oblivious keyword search mechanism over blockchain. The core design of Searchain was to broadcast query requests with authorization (OKSA), which protects the privacy of data users. However, the latency will be increased if some peers are offline. Shlomi et al. [16] used Hadoop's MapReduce infrastructure to provide some SQL queries over blockchain. But to insert a new item into B+ Tree-based index is complicated. The query restriction is that the user need to know the range of the targeted block number in advance.

To realize higher efficiency compared to an inverted index with lower maintaining overhead and index security, we chose Bloom filter as a priority. Bloom filter [2] has a great advantage in terms of space efficiency and query time. Besides, it queries elements without saving the elements themselves, which ensures the security of the index. In Bitcoin, the Bloom filter can filter out UTXO that does not belong to the target wallet [6]. Ethereum virtual machine(EVM) adopts Bloom filter to save log data for efficient query [7]. Nonetheless, although search phase can be finished quickly if the user knows the approximate location, it will start from the latest block if user knows nothing about the target log. Ding et al. [8] proposed a multi-keyword searching scheme by establishing a tree-shaped Bloom filter index. However, the space efficiency and maintenance cost of this scheme are not ideal.

3 Problem Definition

3.1 System Model

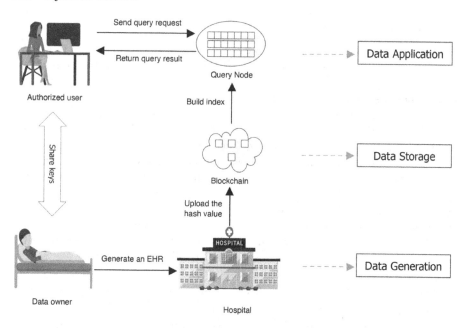

Fig. 1. Structure of medical blockchain system

In this subsection, we will introduce the architecture of our system, which includes three layers, i.e., a blockchain layer, a query layer maintaining exponential skip Bloom filter index, and an application layer providing service. Figure 1 shows the system model, which includes 4 actors, i.e., data owners(e.g. patients), data consumers(e.g. doctors or insurance companies), servers (institutions such as hospitals), and query nodes.

Blockchain Layer. In the blockchain, data owners and data consumers are light nodes, which only store the block header while the hospitals and query nodes represent full nodes, which store the entire blockchain data. Concretely, when a data owner receives EHRs from a doctor, he can use some encryption methods such as Attribute-Based Encryption (ABE) to encrypt the plain text first. Then the server transforms encrypted EHRs or signatures into a hash value, and uploads the hash value to the blockchain after the new block is verified and accepted by every node.

Query Layer. In the query layer, a query node is designed to establish an exponential skip Bloom filter index structure, which stores the keyword of health records on the blockchain to execute the efficient query. The index structure and algorithm will be introduced later.

Application Layer. When data consumers want to query the specific EHR information or the signature over the blockchain, they send a query request to the query node, which implements an efficient query algorithm on the blockchain according to the index structure it constructs. After receiving the corresponding result, the data consumers can decrypt the information finally.

3.2   Design Goals

Nowadays, related works focus on storage patterns, verification of data integrity, and secure sharing of medical data over the blockchain. Faced with the massive data over blockchain, appropriate and detailed retrieval methods are insufficient. In some business cases regarding healthcare such as insurance claims, the insurance company needs to search for the applicant's electronic health record or signature for verification. When the patient comes for a follow-up visit, the doctor needs to review his history. For the sake of privacy preservation, some secure access control methods can be adapted to process EHRs, which has attracted many scholars' attention but is not the main focus of our scheme. When executing queries on the blockchain, classic information retrieval indexes and algorithms encounter with several weaknesses such as higher storage overhead of index structure or additional consideration for security, which has been introduced in Sect. 2. In conclusion, we intend to design an index structure and search algorithm with Bloom filter over blockchain, which satisfies the following features:

Fast Query Speed. Confronted with huge amounts of medical data, our search algorithm can filter plenty of irrelevant health records with the help of the Bloom filter. What's more, we designed an exponential skim Bloom filter index structure to accelerate search speed.

Lower Maintainence. The Bloom filter is a bit vector with very high space utilization, which means a huge amount of data will be hashed as one bit in the Bloom filter. On the other hand, with our exponential skim Bloom filter index structure, the update will not influence the entire index structure but only two Bloom filter indexes after simple computation.

Security of Index. Without saving the elements themselves, the Bloom filter ensures the security of the index. We also perform security analysis on our index structure and find that it is hard to guess the exact information in the Bloom filter index node.

4 Preliminaries

4.1 Bloom Filter

Bloom filter [2] was first proposed by Burton Howard Bloom in 1970, which can determine whether an element is in a set or not. Composed of a long binary vector (bit vector) and a series of random hash functions, the Bloom filter stands out among general algorithms in terms of space efficiency and query time. Initially, there are m bits in the binary vector which are initialized to 0. To add an element of a set S = $\{s_1, s_2, \ldots, s_n\}$ to the Bloom filter, an array of hash function [3] $\{h_0, h_1, h_2, \ldots, h_{k-1}\}$ where k represents the number of hash function should be applied to get an array of binary vector in which k bits are 1. In Bloom filter, hash collision is evitable if hash functions are well designed. To determine whether an element is included in the set S, k hash functions map the data and then check whether each $h_i(x)(1 \leq i \leq k)$ is 1. To achieve the best performance, given the n elements and error rate err, the bit width of Bloom filter m and the number of hash functions k should satisfy the following two equations.

$$m = -\frac{nln(err)}{(ln2)^2} \tag{1}$$

$$k = \frac{m}{n}ln2 \tag{2}$$

4.2 Blockchain Technology

Blockchain is a distributed database that manages the time series data and ensures that each block is tamper-resist. With cryptography knowledge, blockchain successfully achieves secure storage of simple data and reliable verification in the meantime. More specifically, each block consists of a block header and body. A block header includes the hash value of the previous block, the difficulty of mining, timestamp, root number of Merkle tree and so on while the block body stores all transactions in this block. In our scenario, both data consumers and data owners act as light nodes that only stores the block header while hospitals and query node function as the full node to process all transactions. Compared with traditional maintenance on a single server, the blockchain database is maintained and backed up by all nodes over blockchain. In this case, tampering transaction on one node is invalid unless the whole system has verified the integrity of the transaction.

5   Efficient Retrieval Algorithm for Blockchain Data

5.1   Construction of Exponential Skip Bloom Filter Index Structure

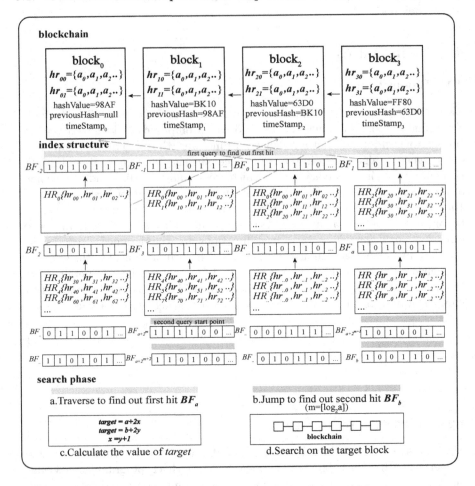

Fig. 2. Structure of exponential skip Bloom filter index.

Assuming that the hospital packs up a series of health records that contain some attributes such as { "*stomach tumor*", "*malignant*", "*three months*"}, we generate an index for each health record to insert into Bloom filter. For each block, the set HR_i contains all the encrypted health records of patients in this block where i indicates the index of block. The whole structure of both blockchain data and the Bloom filter index is shown in Fig. 2. A notation list is shown in Table.1 to explain each terminology.

To achieve an efficient query on the blockchain, we construct a set of Bloom filters $\{BF_0, BF_1, ..., BF_i, ..., BF_n\}$ where i indicates the index of the block. BF_i contains health record set HR of k blocks with the rules we set as follows:

1. For each Bloom filter BF_i, it extends two kinds of link: block link and index link. The block link extends to $block_i$ while the index link connects itself with the set HR of some assigned blocks.
2. The index link extends to some health record sets. BF_i contains the Bloom filter set $\{HR_{i+2^0}, HR_{i+2^1}, HR_{i+2^2}, ...HR_{i+2^k}\}$ where k is the upper bound and $HR_i = \{hr_{i0}, hr_{i1}, ..., hr_{in}\}$. The hash function set $\{h_0, h_1, ..., h_n\}$ is used to map the health record index to the Bloom filter BF.
3. The total number of blocks' health records mapped to BF_i depends on the number of blocks. What's more, the error rate decides the value of bit width of BF and the number of hash function n.

To solve the problem that the first two blocks can not be searched, we add two Bloom filters BF_{-2}, BF_{-1} at the beginning where BF_{-2} contains health records of $block_0$ and BF_{-1} contains health records of $block_0$ and $block_1$. Then all the blocks can be queried according to our algorithm introduced in 5.2.

Table 1. Notations used in this paper.

Symbol	Descriptions
a	Attribute of electronic health record
hr_{ij}	The jth health record on the ith block
HR_i	A set of health records on the ith block
BF_i	Bloom filter lined with the ith block
$block_i$	The ith block

5.2 The Principle of the Efficient Query Algorithm over Blockchain

When the query node performs a query on the blockchain, it needs to generate query Bloom filter bf_q according to the query index. As shown in Fig. 2, our method to perform efficient queries over the blockchain is mainly based on Bloom filter to filter some invalid blocks. Like the principle of Bloom filter, finding out the potential block is the same as selecting out the BF nodes whose positions of "1" bit cover all that are in bf_q. According to the index structure we mentioned above, the target block index t must satisfy the Eq. 3 where non-negative p is the index of block and non-negative q is the exponent. Firstly, the query node searches from BF_{-2}. Assuming that BF_a hits the result firstly, which means the potential block index is in the set $A = \{a + 2^0, a + 2^1, a + 2^2...\}$, the index a is recorded. Intuitively, the search process will start again from BF_{a+1}. Assuming that the BF_b ($b > a$) hits the result for the second time, we can construct Eq. 4 where both x and y is a non-negative number. Since we are searching from back to end, the relationship of x and y has only one kind of pattern. The first hit means the value of a is the smallest among all values of p which satisfies the Eq. 3 while the value of x is the biggest among all values of q. In the same way, because BF_b hits the result secondly, the value of b is the second smallest and the value of y is the second biggest which indicates that x and y satisfy the equation shown in Eq. 5. In summary, our algorithm mainly focuses on the first two hits to get the value of a and b. Then with the three equations, we can easily figure out the value of x and y. Finally, the index of target block t can be easily calculated.

$$t = p + 2^q \tag{3}$$

$$a + 2^x = b + 2^y \tag{4}$$

$$x = y + 1 \tag{5}$$

5.3   Further Optimization of the Algorithm

After the first hit, the value of a is recorded. However, if the process to find out the value of b is to start query from a, the efficiency of the algorithm will downgrade with the increase of x or y. To shorten this overhead, we figure out the closest start for the second query with Theorem 1.

Theorem 1. *For the first hit index a, if the target index is $a + 2^x$, then $a < 2^x$.*

Proof. Assume that there is a Bloom filter BF_a hitting the result for the first time but it satisfies $a > 2^x$. Then the value of a can be decomposed to $a - 2^x + 2^x$, so the target index $a + 2^x$ can be decomposed to $a - 2^x + 2^{x+1}$. With the relationship shown in Eq. 3, when querying the $a - 2^x$, it must have hit the target. But it contradicts the assumption that BF_a hits the target for the first time. Hence, $a < 2^x$ can be proved.

With Theorem 1, some Bloom filter nodes $block_{a+2^x}$, where x satisfies $2^x < a$, can be excluded. From the three equations mentioned above, the relationship of a and b can be calculated as Eq. 6. To find out the next query starting point which is closest to b but not bigger than b, we firstly divide a to the combination $n + 2^m$ in which m is the biggest exponent and n is natural number as shown in Eq. 7, which satisfies $n < 2^m$. With Theorem 1, we can confirm that $a = n + 2^m < 2^x$, which indicates $2^m < 2^x$ and $2^m \leq 2^{x-1}$, so the relationship of m and x can be concluded as $a + 2^m \leq a + 2^{x-1}$. That is to say, the max value m which is $\lfloor log_2 a \rfloor$ satisfies inequality $a + 2^m \leq b$. Therefore, the next starting point BF_{a+2^m} and continues by the increasing of m. For example, when BF_7 first hits the result, then the query node will query BF_{7+4}, BF_{7+8} and so on until the second hit takes place.

$$b = a + 2^{x-1} \tag{6}$$

$$a = n + 2^m (m = \lfloor log_2 a \rfloor) \tag{7}$$

5.4   A Constant Value to Stop Endless Query

Assuming the situation that having searched for many Bloom filters, the first hit still does not take place, the time overhead will be huge if any measure is not taken. Hence, we give a constant value to stop searching to the last Bloom filter.

Given that the total number of blocks is n, the search can be stopped directly if the BF_{end} does not hit for the first time.n and end satisfies Eq. 8 and Eq. 9 respectively. The reasons why the value of end is max $(2^{t-1} - 1, n - 2^t)$ are as follows. Firstly, if the target index is $block_n$, the first hit will be $n - 2^t$. Secondly, if the target index belongs to interval $[2^t, n)$, every index in this interval can be transformed into $2^t + h, 0 \leq h < n - 2^t$. In that case, if any number in this interval is the target number, it must be hit before $n - 2^t$. Thirdly, if the target index belongs to interval $[2^{t-1}, 2^t)$, every index in this interval can be transformed into $2^{t-1} + h, 0 \leq h < 2^{t-1}$. Hence, the first hit must happen before searching $2^{t-1} - 1$th Bloom filter. At last, considering that the target number falls on $[0, 2^{t-1})$, the first hit must also take place before searching $2^{t-1} - 1$th Bloom filter. In conclusion, we set the finishing point as $max(2^{t-1} - 1, n - 2^t)$.

$$2^t \leq n < 2^{t+1} \tag{8}$$

$$end = max(2^{t-1} - 1, n - 2^t) \tag{9}$$

Algorithm 1. Retrieval Algorithm over Blockchain

Input: bf_q, query Bloom filter ; n, number of blocks ;
Output: block number of the target health record
1: $t = \lfloor log_2 a \rfloor$;
2: $imax = max(2^{t-1} - 1, n - 2^t)$;
3: **for** int $i = 0$; $i \leq imax$; $i{+}{+}$ **do**
4:     **if** BF_i hits bf_q **then**
5:         $m = \lfloor log_2 i \rfloor$
6:         **while** $i + 2^m < n$ **do**
7:             **if** BF_{i+2^m} hits bf_q **then return** $i + 2^{m+1}$
8:             **else**
9:                 $m{+}{+}$
10:            **end if**
11:        **end while**
12:    **end if**
13: **end for**

Assuming the health record is in $block_{10000}$, the search phase can be concluded as shown in yellow line and blue line of Fig. 3. Firstly, search phase will start from beginning and hit BF_{1808} because $1808 + 2^{13} = 10000$ (step a). Next searching will start from BF_{2832} because $1808 + 2^{10} = 2832$. It continues from BF_{3856} and hits BF_{5904} for the second time because $5904 + 2^{12} = 10000$ (step b). So the target block can be calculated by $1808 + 2^{log(5904-1808)+1} = 10000$ (step c). The last step is to look up $block_{10000}$ for the health record (step d).

5.5   Maintenance Mechanism of Index

To maintain our exponential jump type Bloom filter index structure, we will perform three kinds of operations on it which are **Initialization, Insert** and **Update**.

Initialization: Before obtaining the health records of all patients on the blockchain, the query node initializes certain quantities of Bloom filter nodes according to the number of blocks. Each Bloom filter node has two states: "active" and "inactive". The state "active" means that the node has an index link with a certain health record while the "inactive" state means that the node has no links.

Insert: After finishing initialization, the query node starts to insert encrypted heath records into the Bloom filter for the first time. Firstly, according to the number of health records and the error rate, the bit width and the number of the hash function can be calculated. Then the query node creates index links under the rule of exponential increase to connect the BF with health records set HR and also creates block links extended to the corresponding block. The last step is to update the state of some nodes from inactive to active.

Update: When a new block is created, the health records set of this block will be inserted into only two previous Bloom filters, which are BF_a and BF_b in 5.23. For example, if the newly created block is $block_{13}$, then the health records of it will be inserted into BF_{13-2^3} and BF_{13-2^2}. The detail of algorithm is shown in Algorithm 2.

Algorithm 2. Update Bloom Filter

Input: HR, encrypted heath records set ; d, index number of the new block ; BF, Bloom filter index

Output: BF_{d-2^i}; $BF_{d-2^{(i-1)}}$

  1: $i = log_2 d$;

  2: calculate the bit width of BF_{d-2^i} and $BF_{d-2^{(i-1)}}$;

  3: map HR into BF_{d-2^i} and $BF_{d-2^{(i-1)}}$;

  4: **return** BF_{d-2^i}, $BF_{d-2^{(i-1)}}$;

---

6   Performance and Security Analysis

6.1   Performance Analysis

When processing the search request containing the query Bloom filter, the query node needs to find out the first two hits and calculate the potential result. The cost of query processes is $a + log(b - a) + m$ where a represents the index of Bloom filter hit firstly, b represents the index of Bloom filter hit secondly and m

represents the max value which satisfies the inequality $a + 2^m \leq b$. If the target block is t, a could be calculated as $BF_{t-2^{log_2 t}}$. When inserting a new block into the index structure, there is no need to scan the entire index structure. With the update method mentioned in Sect. 5, the time complexity is $O(1)$, which is efficient for maintaining large-scale data.

6.2   Storage Overhead

When it comes to Bloom filter index, there are three types of data: Bloom filter node, index link and block link. If the average bit of Bloom filter is m, then there are 2^m Bloom filter nodes in total and $(m + 1) \times 2^m$ bit space to store because each Bloom filter node has m bit width and 1 bit to indicate state. Secondly, considering N blocks in blockchain, each Bloom filter node i connects $\lfloor log_2(N-i) \rfloor + 1$ Bloom filters through index link. Assuming the storage occupied by an index link is 1 byte, total space of index links is $(\sum_{i=0}^{N-1} \lfloor log_2(N-i) \rfloor + 1) + 3$ bytes in consideration of the first two Bloom filters BF_{-2} and BF_{-1}. Lastly, the space to store block links is related to the number of blocks. So, if the block link needs 1 byte to store, the total space will be N bytes.

In conclusion, we can find that with the addition of health records, the storage of Bloom filters and block links will be mainly influenced. The former is indispensable for quick query while the latter costs very little. Hence, we can achieve high efficiency with appropriate storage overhead.

6.3   Secure Analysis

Our method mainly involves three types of information which include the encrypted health record attributes hr on the blockchain, the Bloom filter stored at the query node, and the query requests from the data user. hr is encrypted by the data owner through some methods, such as ABE, which can ensure data security on the blockchain.

Under the KPA (Known Plaintext Attack) model, we assume that it exists an adversary A masters the Bloom filter and the user's query request and tries to break the security of the proposed scheme. In the threaten model, A is able to act as an authorized user to ask the query node to execute the query over the index and attempts to guess the exact information in the Bloom filter index node with a non-negligible probability. After A executes QH queries over the index, since the hash function is a type of secure pseudo-random function, he can guess the word in an index node with the probability roughly $QH/2^m$, where m is the width of the Bloom filter. Consider the parameters $QH = 300,000$ (the query amount for one month and 10,000 queries every day) and $m=150$, the probability of A exactly guessing the word in Bloom filter is no more than 2.10×10^{-40} in a new query, which is negligible. Therefore, through the analysis, we can draw a conclusion that the user's data privacy is guaranteed by our scheme under the KPA model.

7   Performance Evaluation

Through creating different scales of blockchain system, uploading health record attributes to blockchain, creating Bloom filter set according to the rules we set before, and then executing query algorithm, we conduct our experiment to test the efficiency of our method in three dimensions which is the influence of Bloom filter error rate on efficiency, the influence of blockchain scale on efficiency and maintenance overhead. Our test dataset (https://archive.ics.uci.edu/ml) has 5 million health records including treatment process, disease description, and physical information of patients such as { "$patient_nbr$" : "8222157"}. The experimental environment involves one machine running macOS 12.1 with 16 GB and an Apple M1 Pro core.

7.1   Query Efficiency of Search Method

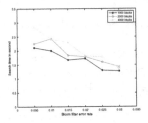 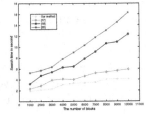 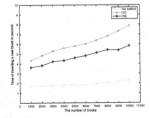

Fig. 3. Influence of error rate on query efficiency   **Fig. 4.** Time cost in different blockchain system   **Fig. 5.** The cost of maintenance

Due to the fact that the time overhead in our method is related to the number of hash functions in Bloom filter and the number of hash functions is decided by the error rate of Bloom filter, we experimented with the influence of error rate on query efficiency. At the same time, we created different scales of blockchain with 1000, 2000, and 4000 blocks to test the search time. Figure 3 shows that the increase in error rate will not significantly affect the search time. What's more, we found that if the error rate can be defined appropriately, the amount of data does not influence the search time seriously. For example, when the error rate is 0.02 and 0.03, the search time is very close to each other.

To test efficiency in different scales of blockchain system, we compared query efficiency in four methods. Figure 4 shows that the search time of our method will not increase sharply with the enlarging of blockchain. We chose the traditional query method on Ethereum[1] and two tree-shaped methods [8,18]. Ethereum[1] searches Bloom filter from start to end. Both [8,18] use Bloom filters to build a tree-shaped index structure and search layer by layer. Although these tree-shaped methods scan fewer Bloom filter nodes, they will spend more search time at the first few nodes. The root Bloom filter has a huge number of bits so they need more hash functions. The space cost of [8] is greater than that of

---

[1] https://www.ethereum.org.

[18]. In our method, the amount of data included in each Bloom filter node is average, so search time spent on each node is lower than [8,18].

7.2 Maintenance Overhead of Our Method

To test the maintenance overhead of our scheme, we compared the time of inserting pieces of health records from a new block into the index structure between our method and [8,18] in different scales of the blockchain system. The experimental performance is shown in Fig. 5. It seems that the time of inserting a new block is quite low and stable no matter how large the whole blockchain system is. To insert a new query index into the Bloom filter, the inserting time of our method is remarkably lower than [8,18], which is $O(1)$. When inserting a new block in the Bloom filter index, we just need to find out two previous Bloom filter nodes. [8,18] need to scan the entire tree to update some Bloom filter nodes. Especially, the root Bloom filter contains all heath records of blockchain in [8], its maintenance overhead will escalate with the increase of the number of blocks.

8 Conclusion

In our paper, we proposed an efficient retrieval algorithm for blockchain data based on exponential skip Bloom filter. The combination of Bloom filter and blockchain accelerates speed of figuring out the certain health record on large -scale data. The main contribution is to use mathematic theory and create a special index structure to filter irrelevant blocks in order to speed up the query process. Moreover, the maintenance overhead and storage overhead of the proposed index structure is rather small so that the whole method can satisfy application requirements in healthcare. In future work, we will focus on the optimization of authorization and further improvement in efficiency.

Acknowledgements. This work was partially supported by the National Natural Science Foundation of China (No. 62072349), National Key Research and Development Project of China (No. 2020YFC1522602), and the Science and Technology Major Project of Hubei Province, China (No. 2021BEE057).

References

1. Aujla, G.S., Jindal, A.: A decoupled blockchain approach for edge-envisioned IoT-based healthcare monitoring. IEEE J. Sel. Areas Commun. **39**(2), 491–499 (2020)
2. Bloom, B.H.: Space/time trade-offs in hash coding with allowable errors. Commun. ACM **13**(7), 422–426 (1970)
3. Carter, J.L., Wegman, M.N.: Universal classes of hash functions. J. Comput. Syst. Sci. **18**(2), 143–154 (1979)
4. Chen, L., Lee, W.K., Chang, C.C., Choo, K.K.R., Zhang, N.: Blockchain based searchable encryption for electronic health record sharing. Futur. Gener. Comput. Syst. **95**, 420–429 (2019)

5. Crosby, M., Pattanayak, P., Verma, S., Kalyanaraman, V., et al.: Blockchain technology: beyond bitcoin. Appl. Innov. **2**(6–10), 71 (2016)
6. Dai, X., et al.: LVQ: a lightweight verifiable query approach for transaction history in bitcoin. In: 2020 IEEE 40th International Conference on Distributed Computing Systems (ICDCS), pp. 1020–1030. IEEE (2020)
7. Dannen, C.: Introducing Ethereum and Solidity, Vol. 1. Springer, Cham (2017). https://doi.org/10.1007/978-1-4842-2535-6
8. Ding, Y., Song, W., Shen, Y.: Enabling efficient multi-keyword search over fine-grained authorized healthcare blockchain system. In: Wang, X., Zhang, R., Lee, Y.-K., Sun, L., Moon, Y.-S. (eds.) APWeb-WAIM 2020. LNCS, vol. 12318, pp. 27–41. Springer, Cham (2020). https://doi.org/10.1007/978-3-030-60290-1_3
9. He, X., Alqahtani, S., Gamble, R.: Toward privacy-assured health insurance claims. In: 2018 IEEE International Conference on Internet of Things (iThings) and IEEE Green Computing and Communications (GreenCom) and IEEE Cyber, Physical and Social Computing (CPSCom) and IEEE Smart Data (SmartData). pp. 1634–1641. IEEE (2018)
10. Hu, S., Cai, C., Wang, Q., Wang, C., Luo, X., Ren, K.: Searching an encrypted cloud meets blockchain: a decentralized, reliable and fair realization. In: IEEE INFOCOM 2018-IEEE Conference on Computer Communications, pp. 792–800. IEEE (2018)
11. Huang, H., Sun, X., Xiao, F., Zhu, P., Wang, W.: Blockchain-based eHealth system for auditable EHRs manipulation in cloud environments. J. Parallel Distrib Comput. **148**, 46–57 (2021)
12. Jiang, P., Guo, F., Liang, K., Lai, J., Wen, Q.: Searchain: Blockchain-based private keyword search in decentralized storage. Futur. Gener. Comput. Syst. **107**, 781–792 (2020)
13. Magdy, S., Abouelseoud, Y., Mikhail, M.: Privacy preserving search index for image databases based on surf and order preserving encryption. IET Image Proc. **14**(5), 874–881 (2020)
14. Nakamoto, S.: Bitcoin: a peer-to-peer electronic cash system. Decentral. Bus. Rev. p. 21260 (2008)
15. Pei, Q., Zhou, E., Xiao, Y., Zhang, D., Zhao, D.: An efficient query scheme for hybrid storage blockchains based on Merkle semantic trie. In: 2020 International Symposium on Reliable Distributed Systems (SRDS), pp. 51–60. IEEE (2020)
16. Song, X., Li, B., Dai, T., Tian, J.: A trust management-based route planning scheme in LBS network. In: International Conference on Advanced Data Mining and Applications, ADMA 2022. LNCS, vol. 13087, pp. 307–322. Springer, Cham (2022).https://doi.org/10.1007/978-3-030-95405-5_22
17. Wang, Q., He, P., Nie, T., Shen, D., Yu, G.: Survey of data storage and query techniques in blockchain systems. Comput. Sci. **45**(12), 12–18 (2018)
18. Yoon, M.K., Son, J., Shin, S.H.: Bloom tree: a search tree based on bloom filters for multiple-set membership testing. In: IEEE INFOCOM 2014-IEEE Conference on Computer Communications, pp. 1429–1437. IEEE (2014)
19. Zhang, Y.l., Wen, L., Zhang, Y.j., Wang, C.f.: Deniably authenticated searchable encryption scheme based on blockchain for medical image data sharing. Multim. Tools Appli. **79**(37), 27075–27090 (2020)

IMPGA: An Effective and Imperceptible Black-Box Attack Against Automatic Speech Recognition Systems

Luopu Liang[1], Bowen Guo[2], Zhichao Lian[2(✉)], Qianmu Li[1], and Huiyun Jing[3]

[1] School of Computer Science and Engineering, Nanjing University of Science and Technology, Nanjing 210094, China
[2] School of Cyberspace Security, Nanjing University of Science and Technology, Wuxi 214443, China
lzcts@163.com
[3] China Academy of Information and Communications Technology, Beijing 100191, China

Abstract. Machine learning systems are ubiquitous in our lives, so it is necessary to study their vulnerabilities to improve the reliability and security of the systems. In recent years, adversarial example attacks have attracted considerable attention with remarkable success in fooling machine learning systems, especially in computer vision. For automatic speech recognition (ASR) models, the current state-of-the-art attack mainly focuses on white-box methods, which assume that the adversary has full access to the details inside the model. However, this assumption is often incorrect in practice. The existing black-box attack methods have the disadvantages of low attack success rate, perceptible adversarial examples and long computation time. Constructing black-box adversarial examples for ASR systems remains a very challenging problem. In this paper, we explore the effectiveness of adversarial attacks against ASR systems. Inspired by the idea of psychoacoustic models, we design a method called Imperceptible Genetic Algorithm (IMPGA) attack based on the psychoacoustic principle of auditory masking, which is combined with genetic algorithms to address this problem. In addition, an adaptive coefficient for auditory masking is proposed in the method to balance the attack success rate with the imperceptibility of the generated adversarial samples, while it is applied to the fitness function of the genetic algorithm. Experimental results indicate that our method achieves a 38% targeted attack success rate, while maintaining 92.73% audio file similarity and reducing the required computational time. We also demonstrate the effectiveness of each improvement through ablation experiments.

Keywords: ASR system · Adversarial example · Genetic algorithm · Psychoacoustic model

1 Introduction

With the rapid development of machine learning technologies, deep neural networks (DNNs) have been widely applied in a lot of fields, e.g., autonomous driving [1], image classification [2], speech recognition [3] and so forth. However, a series of recent studies [4–7] have demonstrated that neural networks are vulnerable to adversarial attacks, where tiny perturbations are designed to be added to legitimate examples to successfully fool neural networks, leading to harmful or irreversible consequences. Adversarial examples are inputs that have been specifically designed by an adversary to cause a machine learning system to produce a misclassification [8]. Adversarial attacks aim to fool a machine learning model by perturbing the inputs in such a way to affect its decision while not altering that of a human [2].

To date, current research on adversarial attacks against machine learning models has mainly focused on image classification models with many exciting attempts. A range of methods have achieved admirable experimental results, e.g., Fast Gradient Sign Method (FGSM) [5], iterative-FGSM (I-FGSM) [7], MI-FGSM [9], Deepfool [10], and C&W attack [6], etc. Nonetheless, adversarial attacks on automatic speech recognition (ASR) systems have not received enough attention until recently. Adversarial attacks against ASR systems refer that the adversary constructs an audio adversarial example in the ASR system, which sounds very similar to the original audio but can be recognized as target text, as is shown in Fig. 1. Similar to image adversarial attack methods, adversarial attack methods have also been explored in the field of speech, such as Houdini [11], CommanderSong [12], the C&W attack borrowed from image attack method [13], and the psychoacoustic model-based design speech adversarial attack methods [14, 15], etc.

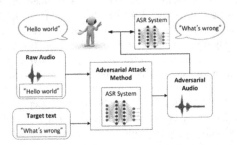

Fig. 1. An illustration of the adversarial attack against ASR systems.

Depending on how much information the adversary has access to the model, adversarial attacks on ASR systems can be divided into two categories, (1) white-box attack: The adversary knows all the details of the ASR system. (2) black-box attack: The adversary does not know the details of the ASR system before attacking the model. Existing adversarial attacks against ASR systems mainly focus on white-box attack methods. The most advanced white-box attack method [15] has developed effective and imperceptible audio adversarial examples with a success rate of 100% and robust audio adversarial examples with a success rate of 64.6%. However, the complex structure of the ASR system is not conducive to the development of white-box attacks. Meanwhile, white-box attacks that require full access to system details (such as network architecture, weights, related parameters, etc.) make unrealistic assumptions and thus fail to meet the requirements of real-world scenarios. Furthermore, audio adversarial examples are poorly transferable, making it difficult to generate audio adversarial examples on alternative models to attack the target ASR model [12]. These shortcomings greatly limit their usability in practical applications in ASR system attacks.

Compared with the white-box attack, the black-box attack only needs to obtain the output probability of the neural network, which is more in line with the actual situation. However, designing efficient black-box attacks on ASR systems still faces many challenges. Alzantot et al. [16] used a genetic algorithm-based method to iteratively add adversarial perturbations to the original speech for the first time, and finally successfully attacked the target model. However, this method only achieves significant attack results on smaller datasets and lightweight speech recognition models. Taori et al. [17] combined genetic algorithm and gradient estimation, only got a 35% attack success rate with a maximum of 3000 iterations per target. Furthermore, the target text is limited to two words, which prevents the adversary from expressing its attack intent clearly. Chen et al. devised a method called Devil's Whisper [18]. The core idea of this attack is to achieve a black-box attack on the transferability of ASR systems through an alternative model, while the number of queries contains about 1500 times. In addition, Wang et al. proposed a Selective Gradient Estimation Attack (SGEA) method [19], which successfully attacked the DeepSpeech ASR model [20]. The method combines mini-batch gradient estimation and coordinate selection strategy, which greatly reduces the number of queries, but the designed adversarial examples are not robust.

Our contributions are as follows:

- In this paper, we improve the construction method of adversarial examples on ASR systems and propose a new black-box attack method against ASR systems, called Imperceptible Genetic Algorithm (IMPGA) attack.
- In order to design more imperceptible adversarial examples, we make use of the psychoacoustic principle of auditory masking, and only add the adversarial perturbation to regions of the original audio which will not be heard by a human, even if this perturbation is not "quiet" in terms of the energy.
- To accelerate the convergence and balance attack success rate and audio similarity, we propose an adaptive auditory masking threshold and redesign the fitness score function in the genetic algorithm.

Compared with previous methods, our black-box attack method improves attack success rate and audio similarity while reducing the number of queries, which is a step closer to the practical black-box attack method.

2 Background and Related Work

2.1 ASR Systems

Compared to image classification models, ASR systems are much more complex. Typically, image classification models take pixels directly as input and a class label as output, while ASR systems take features extracted from raw audio as input to a neural network, and finally get transcribed text corresponding to multiple labels from the network as output. ASR systems can be divided into two categories: traditional ASR systems and end-to-end ASR systems. Compared with traditional ASR systems, end-to-end ASR systems are becoming more and more popular due to their relatively simple structure and superior performance.

Baidu's DeepSpeech model [20] is a state-of-the-art end-to-end ASR system based on DNNs, which has been extensively explored in previous studies [13, 17, 21]. The details of DeepSpeech can be referred to [20].

2.2 Adversarial Attacks on ASR Systems

In the real world, speech recognition models have been widely used in various fields, but existing research has proved that ASR models based on deep neural networks are vulnerable to attacks from speech adversarial examples.

Most existing attack methods against ASR systems focus on white-box methods. Houdini [11] proposed a general adversarial attack framework for pose estimation, semantic segmentation and speech recognition. However, this method generates adversarial audio spectrograms instead of adversarial audio samples. Carlini and Wanger [13] solved this problem by refactoring the feature extraction method and proposed an iterative optimization algorithm to generate audio adversarial examples, which reaches 100% attack success rate against DeepSpeech model. In order to make noise inaudible, Schönherr et al. [14] proposed an optimization method based on psychoacoustic Hiding. Qin et al. [15] spread the C&W attack [13], making imperceptible and robust audio adversarial examples by leveraging the psychoacoustic principle of auditory masking and realistic simulated environmental distortions.

The reason why white-box attack methods always achieve high attack success rates is that these methods use the full information of the recognition model, which often does not match the actual situation. Therefore, it is necessary to study black-box attack methods that are more suitable for the real situation. Alzantot et al. [16] proposed a targeted black-box attack on a speech command classification model by applying the genetic algorithm, which achieves a success rate of 87%. However, only a single English word can be successfully attacked in this method. Taori et al. [17] extended this work by combining genetic algorithms and gradient estimation to conduct a targeted black-box attack on the ASR system, successfully generating adversarial examples that are consistent with the target sentence. While under 300,000 queries, the success rate is 35%, and the target phrase is only two words. Furthermore, Khare et al. [22] proposed a multi-objective genetic algorithm to attack ASR systems. However, the average word error rate (WER) between the transcribed text and the target text is high in this method. In addition, the performance of this attack may not be objectively evaluated because it does not take into account many important metrics, such as the number of queries and the attack success rate.

Overall, currently existing solutions have limitations such as the low success rate, tight constraints on target words and high time cost. This present situation motivates us to propose a new adversarial attack against ASR systems so as to attain a higher attack success rate while achieving a more imperceptible attack. The proposed method is expected to handle more complicated target texts.

3 Imperceptible Genetic Algorithm (IMPGA) Attack Method

Adversarial attacks can be generated given various information about the neural network, such as loss function or model structure. However, in practical situations, the neural networks behind such ASR systems are usually not released publicly, so the adversary only has limited access to the ASR. Given this limitation, we adopt the same threat model as the previous attack method [16, 17], using the open-source Mozilla DeepSpeech as a black-box system, which assumes that the adversary does not know the parameters and network architecture of the ASR system. Meanwhile, the attack process does not refer to any information about how the transcription is done, and only the output probabilities of the model can be accessed.

3.1 Threat Model

Given a target ASR system M and an original input x, adversarial attacks can produce the audio adversarial example x' that is very close to the legitimate input audio x by perturbing x while the model has different transcription results for x and x', i.e.,

$$M(x) \neq M\left(x'\right).$$
(1)

A targeted adversarial attack is a more powerful attack method: not only must the classification of x and x' be different, but the network must assign a specific label (chosen by the adversary) to instance x'. In other words, the adversarial example x' generated by the targeted attack can be transcribed into the desired target text t. i.e.,

$$M(x) \neq M\left(x'\right) \text{ and } M\left(x'\right) = t.$$
(2)

Given the original input x and the target t, we perform a black-box target attack on the model M via adding perturbations to x, which aims to get the perturbed audio sample x', such that $M\left(x'\right) = t$. In order to let humans can not notice the target and minimize the added audible noise in the input, we maximize the cross-correlation between x and x'. A sufficient value of the perturbation size is determined by using our proposed black-box attack method, so we do not need access to details such as the gradient of M to execute the attack. The purpose of this paper is to demonstrate that it is possible to design targeted adversarial examples with only slight distortion.

3.2 IMPGA Attack Method

We present an overview of our attack before elaborating the detailed design. The basic premise of our proposed algorithm is that it takes raw audio samples and adds noise to the samples through trial and error, so that the perturbed adversarial audio resembles original input, but is decoded as a target by the ASR system, as is shown in Fig. 2. Since we have no knowledge of the network architecture and parameters of the model, we resort to the gradient-free optimization method to address this problem. Genetic algorithm, which has been applied to previous attack methods [16, 17], usually goes through three steps:

selection, crossover and mutation. Genetic algorithm also refers to the fitness function and iteratively makes the population evolve towards a better solution to simulate natural evolutionary processes.

Genetic algorithm can handle problems of this nature well with black-box adversarial attacks on speech-to-text systems because it is completely independent of the gradients of the model. The method proposed in [16] used a limited dataset consisting of audio samples with only one word and classification with a predefined number of categories. The target text in the method of [17] contains only two words. In order to extend the algorithm that is able to achieve target texts with more vocabulary while increasing the imperceptibility of the attack, we try to modify the genetic algorithm and introduce the idea of psychoacoustic hiding.

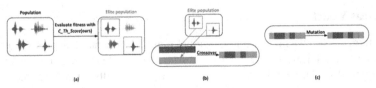

Fig. 2. Diagram of our genetic algorithm. In this figure, picture (a) represents the selection step to select an elite group from the audio population using our proposed C_Th_Score. Picture (b) represents the crossover strategy, where the elite samples are crossed as parents to get the offspring samples. Picture (c) represents the mutation step, in our method we use a momentum mutation strategy to add noise to the audio.

Fitness Function Based on the Psychoacoustic Model. A good understanding and explanation of the human auditory mechanism is essential for building imperceptible adversarial examples. Therefore, in this work, we rely on the extensive work done in the audio field to capture the characteristics of human perception of audio. In this attack method, we use frequency masking, which refers to the phenomenon that a louder signal (the "masker") can make other signals at nearby frequencies (the "maskees") imperceptible [23, 24]. In simple terms, a masker can be thought of as creating a "masking threshold" in the frequency domain of the audio. Any signal below this threshold is practically imperceptible.

Audio signals usually change rapidly over time. Since the masking threshold of an audio is measured in the frequency domain, we first perform a preprocessing step on the original audio, computing the short-time Fourier transform of the original audio signal to obtain the frequency spectrum of the overlapping portion of the signal (called a "window"). The window size N is set to 2048 samples, which are extracted in a batch size of 512 samples. We denote the i th box of the spectrum of the x th frame by $s_x(i)$. Next, we compute the log-magnitude power spectral density (PSD) $p_x(i)$ using Eq. 3:

$$p_x(i) = 10\log_{10}\left|\frac{1}{N}s_x(i)\right|^2. \tag{3}$$

The normalized PSD estimate $\overline{p_x}(i)$ is defined by [24]:

$$\overline{p_x}(i) = 96 - \max_{i}\{p_x(i)\} + p_x(i). \tag{4}$$

Given an audio input, in order to compute its masking threshold, first we need to identify the maskers. Then, each masker's masking threshold can be approximated using the simple two-slope spread function, which is derived to mimic the excitation patterns of maskers. Finally, the global masking threshold $\theta_x(i)$ is a combination of the individual masking threshold as well as the threshold in quiet via addition. We refer interested readers to [24] for details on computing the masking threshold.

When we add a perturbation δ to the audio input x, if the normalized PSD estimated value of the perturbation $p_\delta(i)$ is below the frequency masking threshold of the original audio $\theta_x(i)$, the perturbation will be masked by the original audio which is inaudible relative to humans. The normalized PSD estimated of the perturbation $\overline{p_\delta}(i)$ can be calculated via:.

$$\overline{p_\delta}(i) = 96 - \max_i\{p_x(i)\} + p_\delta(i), \tag{5}$$

where $p_x(i)$ and $p_\delta(i)$ can be brought into formula 3 for calculation, $p_x(i)$ and $p_\delta(i)$ are the PSD estimation of the raw and the perturbation audio input.

In summary, given an audio example x and a target text y, we address the problem of constructing an adversarial example $x' = x + \delta$ by using the fitness function we apply in the genetic algorithm:

$$C\_Th\_Score(x, y, \delta) = l_{net}(f(x+\delta), y) + \alpha \cdot l_\theta(x, \delta), \tag{6}$$

where l_{net} requires that the adversarial examples fool the audio recognition system into making a targeted prediction y. The optimization goal of the term l_{net} is to make $f(x+\delta) = y$ (i.e., x' is recognized as y by the ASR model). In the DeepSpeech ASR model, the term l_{net} is Connectionist Temporal Classification (CTC) loss function [25]. The l_θ constrains the normalized PSD estimated value of the perturbation $\overline{p_\delta}(i)$ to be under the frequency masking threshold of the original audio $\theta_x(i)$. The hinge loss is used here to compute the loss for masking threshold:

$$l_\theta(x, \delta) = \frac{1}{\lfloor\frac{N}{2}\rfloor + 1} \sum_{i=0}^{\lfloor\frac{N}{2}\rfloor} \max\{\overline{p_\delta}(i) - \theta_x(i), 0\}, \tag{7}$$

where N is a predefined window size, and $\lfloor\frac{N}{2}\rfloor$ outputs the largest integer not larger than $\frac{N}{2}$. The adaptive parameter α that balances the relative importance of the network loss $l_{net}(f(x+\delta), y)$ and the imperceptibility loss $l_\theta(x, \delta)$ is adaptively updated according to the performance of the current attack situation. Specifically, every 200 iterations, if the current adversarial example successfully fools the ASR system, then α is increased to attempt to make the adversarial example less perceptible. Correspondingly, every 500 iterations, if the current adversarial example fails to make the targeted prediction, we decrease α. The effect of parameter α is discussed in the ablation experiment.

Genetic Algorithm. The genetic algorithm works by improving population on each iteration, or generation, through evolutionary operations such as Selection, Crossover and Mutation [26].

Selection. We calculate the score for each example in the population at each iteration with reference to the fitness function to determine which samples are the best performers in the generation. The fitness function is significant for the speed and effectiveness of the algorithm. Our fitness function is the *C_Th_Score*, which as mentioned in Eq. 6, is used to determine the similarity between an input audio sequence and a given phrase. Next, we select the samples with the best scores from the current generation of the population, and these samples are our elite population. Elite population contains samples of desirable traits that we want to pass on to future generations.

Crossover. We then select the parent samples of the next generation from the elite group and perform a crossover operation to create a child sample by extracting approximately half of the elements from *parent*1 and *parent*2 respectively. Our criterion for selecting a sample as a parental sample is the value of the fitness function of the sample.

Mutation. In this step, we add a mutation with some probability to our new child. The mutation step is the only source of noise in the algorithm. In the mutation step, we randomly add noise to the sample with a certain probability. Random mutation is critical because it can lead to traits that are beneficial to the population and jumping out from local optimum, which are then passed on to offspring through crossover. Without the mutation, very similar samples would emerge across generations. Therefore, the way to get rid of this local maximum is to push it in different directions by mutation to get a higher score. In this method, we update the mutation probability according to the momentum update of Eq. 8 and proceed to the next iteration.

$$p_{new} = \beta \cdot p_{old} + \frac{\gamma}{|currScore - prevScore|}, \qquad (8)$$

where p_{old} is the mutation probability calculated in the previous iteration, *currScore* is the highest fitness score in the currently calculated audio population, *prevScore* is the highest fitness score in the audio population calculated in the previous iteration, β and γ are the balance parameters for mutation probability and fitness score weights.

The population will continue to improve over time as only the best traits of the previous generations as well as the best mutations will remain. Eventually, either the algorithm will reach the max number of iterations, or one of the samples is exactly decoded as the target, and the best sample is returned. We use the Levenshtein Distance during the attack to observe whether the attack is successful. The Levenshtein Distance calculates the minimum number of letter-level operations required to convert transcribed text to target text, including substitutions, deletions, and insertions. We consider the attack to be successful when the distance between the decoding of the original audio and the currently generated adversarial sample is zero.

3.3   Gradient Estimation

Genetic algorithms work well when the solution target space is large and a relatively large number of mutation directions may be beneficial to the problem solving. Genetic algorithms have the advantage of being able to efficiently search large spaces. However, this advantage of genetic algorithms becomes a shortcoming when an adversarial sample approaches its target perturbation. When the targeted attack method is close to success, the adversarial audio samples only need a few perturbations in some key regions to obtain correct decoding. In such cases, gradient estimation techniques tend to be more effective.

In a gradient estimation attack, the adversary adds perturbations proportional to the estimated gradient, instead of using the true gradient of ASR models as in a white-box attack method [6, 8]. Gradient estimation attacks do not require the adversary to train local models, which is another advantage. Since training such local models may require more queries based on the training data, this program will be a complicated and expensive process for real world situations.

Therefore, in our proposed targeted attack algorithm, we use the gradient estimation method when the edit distance between the current decoding result and the target decoding result falls below a certain threshold. To approximate the gradient of a black-box system $FD_x(g(x), \delta)$, we can use the technique proposed by [27]:

$$FD_x(g(x), \delta) = \begin{bmatrix} (g(x + \delta_1) - g(x))/2\delta \\ \vdots \\ (g(x + \delta_n) - g(x))/2\delta \end{bmatrix}. \tag{9}$$

In the above formula, x refers to the vector of inputs representing the audio file. δ_i refers to a vector that the value of the i th position is a small δ, and the rest of the positions are all zero. $g(\cdot)$ represents the fitness function, which in our case is C_Th_Score. Essentially, we perform a small perturbation on each index and individually see what the difference is in C_Th_Score, which allows us to compute gradient estimates with respect to the input audio x. If the gradient of the function g exists, then $\lim_{\delta \to 0} FD_x(g(x), \delta) = \nabla_x g(x)$. $\nabla_x g(x)$ is the true gradient of the function g. The gradient estimate method is beneficial for a black-box adversary aiming to approximate a gradient based attack, since the gradient can be directly estimated with access to only the function values.

However, performing this calculation totally would be unacceptable, because only one gradient estimate for a simple 3-s audio clip which is sampled at 16 kHz requires up to nearly 50,000 queries to the model. Therefore, we only randomly sample 100 indices to perturb each generation when using the gradient estimate method. When adversarial examples are already close to the target, gradient estimation takes into account more informed perturbations in exchange for higher computational speed. Our complete algorithm is provided in Algorithm 1, and is composed of a genetic algorithm step and followed by a gradient estimation step.

Algorithm 1 IMPGA attack for generating adversarial audio example

---

Input: Original input x , Target text t
Output: An adversarial audio example x'
1: population = $[x] \times populationSize$
2: **while** $itr < maxItr$ and $Decode(x')! = t$
3:    $scores = -C\_Th\_Score$(population, t)
4:    x' = population[$Argmax(scores)$]
5:    **if** $LevenshteinDistance(t, Decode(x')) > 2$ **then**
        //do genetic algorithm
6:        **while** $populationSize$ children have not been made **do**
7:            Select $parent1$ from $topk$(population) according to $softmax$(populationScore)
8:            Select $parent2$ from $topk$(population) according to $softmax$(populationScore)
9:            child = $Mutate(CrossOver(parent1, parent2), p)$
10:       **end while**
11:       newScores = $-C\_Th\_Score$(newPopulation, t)
12:       $p = MomentumUpdate(p, newScores, scores)$
13:    **else**
        //do gradient estimation
14:       topElement = top(population)
15:       gradPop = n copies of topElement, each mutated slightly at one index
16:       grad = $(-C\_Th\_Score$ (gradPop) $- scores) / \delta_{mutation}$
17:       population = topElement + grad
18:    **end if**
19: **end while**
20: **return** x'

---

4   Evaluation

4.1   Datasets and Evaluation Metrics

Datasets. In this paper, we use the speech commands dataset [28] that consists of 65,000 audio files. In this crowd-sourced dataset, each audio file is a one second audio clip of single words like: "yes", "down", "right", etc. We randomly select 100 audio files with 10 different labels as the original audio files in our experiments.

We also choose the LibriSpeech dataset [29] in our evaluation, which is a corpus of 16 kHz English speech derived from audiobooks and is suitable for training or evaluating ASR systems. LibriSpeech contains 1,000 h of free speech for download, so it is appropriate for validating the effectiveness of adversarial attacks against ASR systems. We randomly select 100 audio examples with durations less than 3 s in the dev-clean set of the LibriSpeech, and treat their labels as the target text. These audio files contain a minimum of two words and a maximum of five words, and sentences with long words like wonderful, excellent, illustration, etc.

Metrics. We evaluate the performance of our algorithm in three primary ways. The first standard is the attack success rate, which is defined as:

$$Suc\_rate = N_s/N_o \times 100\%, \tag{10}$$

where N_s is the number of examples that the generated adversarial examples are able to transcribe to the target by the ASR system, and N_o is the total number of attempted attacks, which is 100 in our experiment.

We also calculated the generally acknowledged metric of the cross-correlation coefficient between the original audio sample and the adversarial example, which is defined as:

$$corr = Cov(X, Y)/\sigma_X \sigma_Y, \tag{11}$$

where $Cov(X, Y)$ is the covariance of X and Y, σ_X and σ_Y are the standard deviations of X and Y. This metric is applied for determining the similarity between the two audio samples.

Another evaluation metric is the average number of required iterations for successful multi-step attacks. We denote it as:

$$itr = \sum_{i=1}^{n} itr\left(x_i'\right)/n, \tag{12}$$

where $\sum_{i=1}^{n} itr\left(x_i'\right)$ refers to the sum of the required iterations for all successful multi-step attacks and n is the number of audio adversarial samples in the same situation. The significance of introducing this indicator is that it can be used as an evaluation basis for the length of time spent in an attack. Similarly, we define $itr_{<1000} = \sum_{i=1}^{k} itr_{<1000}\left(x_i'\right)/k$ as the number of required iterations below 1,000 for successful multi-step attacks.

4.2 Evaluation Results

Table 1 shows a complete comparison of Genetic Algorithm Attacks [16], Genetic Algorithms Combined with Gradient Estimation [17] and our method. Experiments demonstrate that our algorithm possesses higher attack success rate, higher audio similarity and fewer required iterations for successful attacks.

Among the audio samples generated by our proposed algorithm, we achieve an average correlation similarity of 92.73% between the final adversarial samples and the original samples using cross-correlation. After the attack process with the maximum number of iterations set to 3,000, 38% adversarial samples achieved accurate decoding in less than 3,000 generations, and even 12 adversarial examples achieve accurate decoding in less than 1,000 generations. From the point of view of the time spent in the attack, the average number of iterations spent by our proposed method in the case of a successful attack is 1,375, which is improved 19.1% at least compared to the previous algorithms [16, 17]. The reason is that the addition of the psychoacoustic model improves the imperceptibility of the audio, that is, the similarity between the original audio and its corresponding adversarial examples. The proposed adaptive coefficient α balances the weight between finding adversarial examples that can be accurately decoded as targets and improving the audio similarity during the attack process, which helps to reduce the computational time for generating adversarial samples while improving the attack success rate.

Table 1. Comparison with prior work.

Method	Suc_rate	corr	itr	itr$_{<1000}$
[16]	30%	84.48%	1,914	3%
[17]	34%	87.82%	1,700	8%
Our method	**38%**	**92.73%**	**1,375**	**12%**

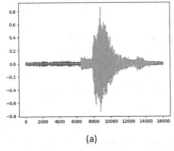
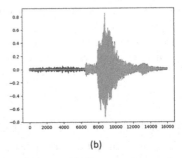

(a)   (b)

Fig. 3. Overlapping of adversarial (blue) and original (orange) audio example waveforms. Picture (a) shows the difference between Taori et al.'s [17] adversarial waveform and the original waveform. Picture (b) illustrates the difference between our adversarial waveform and the original waveform.

Figure 3 presents a helpful visualization of the similarity between an original audio sample and adversarial audio samples produced by different attack methods. By overlapping the waveforms of the original audio sample and its corresponding adversarial audio samples, we can see almost no major changes in the audio, and our method produces less noise. As the visualization shows, the variation that the adversarial example makes from the original example is to apply relatively low noise evenly around the original audio sample. Using the audio adversarial example produced by our proposed method, the sample appears to still be transcribed as the label corresponding to the original speech when heard by a human, but decoded by the DeepSpeech model as the target adversarial phrase.

4.3 Ablation Study

In ablation experiments, we explore the effect of psychoacoustic model and adaptive parameter α in Eq. 6 on attack effectiveness. As a baseline, we use a simplified version of the algorithm, forgoing both the masking thresholds and the adaptive parameter α

module. In the second scenario, we involve the proposed masking thresholds. This minimizes the amount of added noise but reduces the attack success rate to some extent. In the final situation, we add the adaptive parameter step, which results in the full version of the proposed algorithm, with significantly improved scores for all evaluation metrics we use. For the ablation experiments, the maximum number of iterations is still set to 3,000. The complete ablation study results are shown in Table 2.

Table 2. Comparisons of the proposed attack with different experiment settings.

The experiment settings	Suc_rate	corr	itr
baseline	34%	87.82%	1,700
$(\alpha_{up}, \alpha_{down}) = (1.0, 1.0)$	32%	88.39%	1,988
$(\alpha_{up}, \alpha_{down}) = (1.0, 0.8)$	36%	88.74%	1,700
$(\alpha_{up}, \alpha_{down}) = (1.2, 0.8)$	**39%**	90.63%	1,601
$(\alpha_{up}, \alpha_{down}) = (1.5, 0.8)$	38%	**92.73%**	**1,375**

In Table 2, α_{up} means that for every 200 iterations, if the current adversarial example successfully fools the ASR system, α will be updated to $\alpha \times \alpha_{up}$. α_{down} means that for every 500 iterations, if the current adversarial example fails to make the target prediction, α will be updated to $\alpha \times \alpha_{down}$. $(\alpha_{up}, \alpha_{down}) = (1.0, 1.0)$ refers that we do not use adaptive parameter α strategy. As is shown in the first two rows of Table 2, we compare the method incorporating the psychoacoustic model with a baseline approach for solely validating the effectiveness of the psychoacoustic model strategy. Since the psychoacoustic model strategy can make the attack more imperceptible, this strategy significantly improves the average similarity score.

Furthermore, the adaptive parameter tuning strategy is proposed to accelerate the convergence speed and balance attack success rate and audio similarity. In order to examine the effectiveness of the adaptive parameter strategy, we evaluate the effects of the attack method before and after integrating the adaptive parameter α with the psychoacoustic model. By comparing rows 2 and 3, 4, 5 in Table 2, we can notice that the adaptive parameter α is effective in terms of accelerating convergence. Meanwhile, the larger value of α_{up}, the greater weight of the psychoacoustic model during optimization, resulting in higher similarity of the generated adversarial examples. Therefore, the results of the ablation experiments demonstrate that both the psychoacoustic model part and the adaptive parameter tuning strategy in our proposed method are effective.

5 Conclusion

In combining genetic algorithms and the psychoacoustic principle of auditory masking, we are able to generate a black-box audio adversarial attack model for the DeepSpeech ASR model that produces better adversarial samples than using genetic algorithms alone. By initially using a genetic algorithm and a more guided gradient estimation search, while introducing the psychoacoustic model and applying adaptive parameter tuning, our method improves the attack success rate and reduces the time it takes to generate adversarial examples compared to previous work. Meanwhile, we are able to achieve this while maintaining a high degree of similarity. By comparing with existing methods, we experimentally demonstrate the feasibility and advantages of our proposed method and expand breadth for future research.

Note that we focus on scenarios in this work where audio adversarial examples are constructed that can be directly fed into the ASR system, rather than real-world physical attacks where the audio is played by speakers and recorded by microphones. We believe that future work still needs to be done and will leave this important and challenging problem to our future work.

Acknowledgments. This work is supported by the National Key R&D Program of China (2021YFF0602104-2).

References

1. Chen, C., Seff, A., Kornhauser, A., Xiao, J.: Deepdriving: Learning affordance for direct perception in autonomous driving. In: Proceedings of the IEEE International Conference on Computer Vision, pp. 2722–2730 (2015)
2. Krizhevsky, A., Sutskever, I., Hinton, G.E.: ImageNet classification with deep convolutional neural networks. In: Advances in Neural Information Processing Systems, vol. 25 (2012)
3. Wang, K., Guan, D., Li, B.: Deep group residual convolutional ctc networks for speech recognition. In: Gan, G., Li, B., Li, X., Wang, S. (eds.) ADMA 2018. LNCS (LNAI), vol. 11323, pp. 318–328. Springer, Cham (2018). https://doi.org/10.1007/978-3-030-05090-0_27
4. Szegedy, C., et al.: Intriguing properties of neural networks. In: Proceedings of ICLR, pp. 1–5 (2014)
5. Goodfellow, I.J., Shlens, J., Szegedy, C.: Explaining and harnessing adversarial examples. In: Proceedings of ICLR, pp. 1–11 (2015)
6. Carlini, N., Wagner, D.: Towards evaluating the robustness of neural networks. In: Proceedings of IEEE Symposium on Security and Privacy (SP), pp. 39–57 (2017)
7. Kurakin, A., Goodfellow, I.J., Bengio, S.: Adversarial examples in the physical world. In: Proceedings of ICLR, pp. 1–9 (2017)
8. Biggio, B., et al.: Evasion attacks against machine learning at test time. In: Blockeel, H., Kersting, K., Nijssen, S., Železný, F. (eds.) ECML PKDD 2013. LNCS, vol. 8190, pp. 387–402. Springer, Heidelberg (2013). https://doi.org/10.1007/978-3-642-40994-3_25
9. Dong, Y., et al.: Boosting adversarial attacks with momentum. In: Proceedings of the IEEE Conference on Computer Vision and Pattern Recognition, pp. 9185–9193 (2018)
10. Moosavi-Dezfooli, S.M., Fawzi, A., Frossard, P., DeepFool: a simple and accurate method to fool deep neural networks. In: Proceedings of the CVPR, pp. 2574–2582 (2016)
11. Cisse, M.M., Adi, Y., Neverova, N., Keshet, J.: Houdini: fooling deep structured visual and speech recognition models with adversarial examples. In: Advances in Neural Information Processing Systems 30, Long Beach, CA, USA, pp. 6980–6990 (2017)
12. Yuan, X., et al.: CommanderSong: a systematic approach for practical adversarial voice recognition. In: 27th USENIX security symposium (USENIX Security 2018), pp. 49–64 (2018)
13. Carlini, N., Wagner, D.: Audio adversarial examples: targeted attacks on speech-to-text. In: IEEE Security and Privacy Workshops (SPW), pp. 1–7. IEEE (2018)

14. Schönherr, L., Kohls, K., Zeiler, S., Holz, T., Kolossa, D.: Adversarial attacks against automatic speech recognition systems via psychoacoustic hiding. The Internet Society (2019)
15. Qin, Y., Carlini, N., Cottrell, G., Goodfellow, I., Raffel, C.: Imperceptible, robust, and targeted adversarial examples for automatic speech recognition. In: International Conference on Machine Learning, pp. 5231–5240. PMLR (2019)
16. Alzantot, M., Balaji, B., Srivastava, M.: Did you hear that? Adversarial examples against automatic speech recognition. arXiv:1801.00554 (2018)
17. Taori, R., Kamsetty, A., Chu, B., Vemuri, N.: Targeted adversarial examples for black box audio systems. In: Proceedings of IEEE SPW, pp. 15–20 (2019)
18. Chen, Y., et al.: Devil's whisper: a general approach for physical adversarial attacks against commercial black-box speech recognition devices. In: USENIX Security 2020, pp. 2667–2684 (2020)
19. Wang, Q., Zheng, B., Li, Q., Shen, C., Ba, Z.: Towards query-efficient adversarial attacks against automatic speech recognition systems. IEEE Trans. Inf. Forensics Secur. **16**, 896–908 (2020)
20. Hannun, A., et al.: Deep speech: Scaling up end-to-end speech recognition. ArXiv preprint arXiv:1412.5567 (2014)
21. Sriram, A., Jun, H., Gaur, Y., Satheesh, S.: Robust speech recognition using generative adversarial networks. In: ICASSP, pp. 5639–5643 (2018)
22. Khare, S., Aralikatte, R., Mani, S.: Adversarial black-box attacks on automatic speech recognition systems using multi-objective evolutionary optimization. arXiv:1811.01312 (2018)
23. Mitchell, J.L.: Introduction to digital audio coding and standards. J. Electron. Imaging **13**(2), 399 (2004)
24. Lin, Y., Abdulla, W.H.: Principles of psychoacoustics. In: Lin, Y., Abdulla, W.H. (eds.) Audio Watermark, pp. 15–49. Springer, Cham (2015). https://doi.org/10.1007/978-3-319-07974-5
25. Graves, A., Fernández, S., Gomez, F., Schmidhuber, J.: Connectionist temporal classification: labelling unsegmented sequence data with recurrent neural networks. In: Proceedings of the 23rd International Conference on Machine Learning, pp. 369–376 (2006)
26. Holland, J.H.: Genetic algorithms. Sci. Am. **28**, 77–80 (1992)
27. Bhagoji, A.N., He, W., Li, B., Song, D.: Exploring the space of black-box attacks on deep neural networks. arXiv preprint arXiv:1712.09491 (2017)
28. Warden, P.: Speech commands: a dataset for limited-vocabulary speech recognition. arXiv preprint arXiv:1804.03209 (2018)
29. Panayotov, V., Chen, G., Povey, D., Khudanpur, S.: LibriSpeech: an ASR corpus based on public domain audio books. In: ICASSP, pp. 5206–5210. IEEE (2015)

FD-Leaks: Membership Inference Attacks Against Federated Distillation Learning

Zilu Yang[1] , Yanchao Zhao[1]([⊠]) , and Jiale Zhang[2]

[1] College of Computer Science and Technology, Nanjing University of Aeronautics and Astronautics, Nanjing 211106, China
yczhao@nuaa.edu.cn
[2] School of Information Engineering, Yangzhou University, Yangzhou 225009, China

Abstract. With the enhanced sensing and computing power of mobile devices, emerging technologies and applications generate massive amounts of data at the edge network, assigning new requirements for improving the privacy and security. Federated learning signifies the advancement in protecting the privacy of intelligent IoT applications. However, recent researches reveal the inherent vulnerabilities of the federated learning that the adversary could obtain the private training data from the publicly shared gradients. Although model distillation is considered a state-of-the-art technique to solve the gradient leakage by hiding gradients, our experiments show that it still has the risk of membership information leakage at the client level. In this paper, we propose a novel client-based membership inference attack in federated distillation learning. Specifically, we first comprehensively analyze model distillation in defensive capabilities for deep gradient leakage. Then, by exploiting that the adversary can learn other participants' model structure and behavior during federated distillation, we design membership inference attacks against other participants based on private models of malicious clients. Experimental results demonstrate the superiority of our proposed method in terms of efficiency in resolving deep leakage of gradients and high accuracy of membership inference attack in federated distillation learning.

Keywords: Federated learning · Model distillation · Membership inference attack

1 Introduction

With the advancements in machine-learning technologies, internet of things (IoT) system [5] requires collecting and processing large amounts of data from different IoT devices, which is crucial for implementing high-quality deep learning models [2]. Security and privacy issues (e.g., data leakage) in IoT system are significant obstacles for data providers to share data in edge networks [13], and leakage of private data can lead to severe problems [8]. *Federated learning* [7,11,14] is considered as a promising distributed learning framework to solve the above problems. Unlike other collaborative learning frameworks, federated learning updates the global model by aggregating all of the participants' local parameters without requiring the participants to share local training datasets, so that the federated learning protocol can solve the privacy leakage challenge by distributing the training work to local domains.

Although the federated learning can provide privacy guarantees for participants' local training data, even generalized great models in the federated learning framework are still highly vulnerable to membership inference attacks. A major reason why membership inference attacks succeed in the federated learning framework is that gradient parameters are susceptible to various inference attacks as they remember information about their training data. As a result, the gradient of transmission between server and client are distinguishable for members and non-members of the training datasets. Meanwhile there are other privacy leakage issues in the federated learning. Ligeng Zhu [15] demonstrates that it is possible to obtain private training data from publicly shared gradients, in which a malicious adversary is able to reconstruct the participant's local private data from the gradients, which poses a considerable challenge to privacy preservation in federated learning.

There are two types of state-of-the-art defenses: 1) Ensure data information leakage in gradient parameters is mitigated utilizing *gradient perturbation*. 2) Use the *federated distillation* method to transfer the gradients to *logits* while reducing the communication overhead between clients and server and mechanically address gradient leakage problems [6]. The federated distillation method exploits the private data insensitivity and low dimensionality of the logits parameters transmitted during model distillation process, so that it can prevent malicious adversaries from distinguishing membership and non-membership information of the target dataset by observing changes in the logits parameters, and the attacker's goal of reconstructing the participants' local training datasets from the logits parameters.

However, Our research discovers that while model distillation methods can mechanically mitigate membership inference attacks against federated learning, by designing elaborate membership attack classifiers of malicious clients, the membership information of the private data of the global participants is still at risk of leakage. We name our scheme as the client-based membership inference in federated learning. The reason why we call *client-based* is that in the process of model distillation, the malicious attacker acts as a participant in the federated distillation, and its local model learns information about the structure and local data of the private models of other participants in the process of knowledge distillation. In fact, the client based attack is more accessible and risky than the server based ones. The adversary's membership attack classifier designed based on its own local model is still capable of infer membership and non-membership information of other participants' local training datasets.

Our contributions can be summarized in the following three points:

- **Resolving deep leakage of gradients:** We discovered that the federated learning framework incorporating the model distillation algorithm can not only reduce the communication overhead of IoT systems, but also mechanistically address the threat of gradient privacy leakage of federated learning.

- **Client-based membership Inference:** We first demonstrate that IoT systems using federated distillation are fundamentally vulnerable to membership inference attacks launched by malicious participants. In addition, we present details on how to design logits-based membership inference attacks against federated learning.
- **Excellent performances in experiment:** In experiments, we set two main metrics, the degree of leakage of logits and the accuracy of member inference attacks. Experiments demonstrate that logits can significantly hinder private information leakage, but still cannot prevent member inference attacks.

2 Related Work

2.1 Membership Inference

Reza Shokri et al. [12] first proposed membership inference attacks against machine learning models, where a malicious adversary mainly uses the overfitting properties of the model to train an attack on the binary classifier region to subdivide the differences in the confidence score vectors of the target model. To perform membership inference against a target model, Reza Shokri make adversarial use of machine learning and train our own inference model to recognize differences in the target model's predictions on the inputs that it trained on versus the inputs that it did not train on. Later, Milad Nasr [10] exploits the privacy vulnerability of the stochastic gradient descent algorithm used for federated learning, where each training data sample leaves a distinguishable footprint on the gradient of the loss function of the model parameters, and designs a white-box membership inference attack against the participant's neural network model. Recently, Luca Melis [9] demonstrates that in federated learning, adversary can infer properties that hold only for a subset of the training data and are independent of the properties that the joint model aims to capture.

2.2 Model Distillation

Knowledge distillation (KD) is a method of model compression proposed by Hinton [3], which refers to the use of a more complex Teacher model that has been trained to guide the training of a lighter Student model, thus reducing the model size and computational resources while maintaining the accuracy of the original Teacher model as much as possible. Later, Daliang Li [6] developed a federated learning framework, *FEDMD*, using *transfer Learning* and *knowledge distillation*. This framework allows different clients to design different network structures according to their computational capabilities, and the clients perform model updates through the average logits sent from the server. Because the update of the model relies on logits transferred between the server and the client, FEDMD can alleviate the gradient privacy leakage problem raised by

Ligeng Zhu [15], and it can reduce the amount of data transferred throughout the framework, which is ideal for adaptation in lightweight federated learning such as IoT system.

3   Federated Distillation

Federated distillation learning is essentially a federated learning process that involves an online knowledge distillation operation [4]. In the basic operational process of FD, each device treats itself as a student and considers the average model output of all other devices as the output of its teacher. Each model output is a set of logit values normalized by a softmax function, hereafter denoted as a logit vector whose size is given by the number of labels. The teacher-student output difference is measured periodically using cross-entropy [1], which becomes the student's loss regularizer, called the distillation regularizer, thus gaining knowledge of the other devices during distributed training.

For the deep leakage of gradients in federated learning, federated distillation learning protocol transformed the data between the server and client from the gradient to the logits output of the model. This adequate modification essentially solves the privacy leakage problem of the gradient in the universal federated learning framework from the mechanism logits. The dimension of the output is the same as the label, which is more undersized than the dimension of the gradient parameters in the universal federated learning framework and retains less private data information of the participants. Moreover, in our enhanced FD framework, the process of exchanging logits among clients to learn the knowledge of other models is based on the public dataset provided by the server, which primarily solves the problem of privacy leakage of participants' private data.

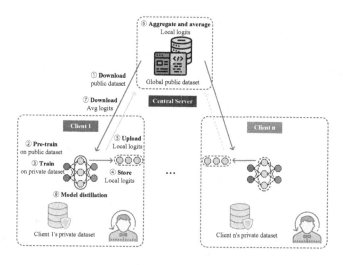

Fig. 1. Overview of federated distillation (FD).

Figure 1 illustrates the fundamental layout of our enhanced federated distillation collaborative learning framework. There are m participants in the FD process. Each has a minimal labeled dataset $D_k := \{(x_i^k, y_i)\}_{i=1}^{N_k}$, which can be IID or non-IID. In addition, the FD server saves a sizeable public dataset $D_{public} := \{(x_i^{public}, y_i^{public})\}_{i=1}^{N_{public}}$ collected by the server from a public source and accessed by each client. Each participant independently designs their model f_k to achieve a classification task. In addition, instead of sharing model parameters among participants, clients share their model outputs corresponding to the standard training dataset D_{public}.

The revised federated distillation learning protocol is shown in Algorithm 1. In the federated distillation process, firstly, each client trains the model on the public training dataset D_{public} using transfer learning and then on its own private dataset D_k. In per round of FD communication, each client computes the logits vector $logits_k(x_i^{public})$ on the public dataset D_{public} collected and distributed by the server, and uploads $logits_k(x_i^{public})$ output of the model to the server. After receiving a certain number K of logits uploaded by clients, the server aggregates and averages all the corresponding logits to generate $logits_{avg}(x_i^{public})$ and sends it to each client. Finally, each client performs model training on the received logits $logits_{avg}(x_i^{public})$ employing knowledge distillation method until all clients' private model f_k have reached the state of convergence.

The federated distillation protocol can primarily alleviate the problem of deep gradient leakage in the federated learning framework. Specifically, Ligeng Zhu [15] demonstrated that it is possible to obtain private training data from publicly shared gradients. They named this leakage as deep leakage from gradients and experiment verified its effectiveness on computer vision and natural language processing tasks. To recover the data from gradients, they first randomly initialize a dummy input x' and label input y'. They then feed these "dummy dat" into models and get "dummy gradients".

$$\nabla W' = \frac{\partial \ell(F(x', W), y')}{\partial W} \tag{1}$$

Optimizing the dummy gradients close as to original also makes the dummy data close to the real training data. Given gradients at a certain step, they obtain the training data by minimizing the following objective.

$$
\begin{aligned}
x'^*, y'^* &= \arg\min_{x', y'} \|\nabla W' - \nabla W\|^2 \\
&= \arg\min_{x', y'} \|\frac{\partial \ell(F(x', W), y')}{\partial W} - \nabla W\|^2
\end{aligned} \tag{2}
$$

While the distance $\|\nabla W' - \nabla W\|$ are optimized, the dummy data will eventually get closer to the real data. In the federated distillation protocol, we will use an the similar method to illustrate the extent of logits leakage. To recover the data from logits, we Similarly initialize a dummy input x' and label input y', and then put these "dummy dat" into models and get "dummy logit". Optimizing the dummy logits close as to original logits downloaded from server, the dummy data will approach the real data in the same way. However, due to the specificity of the communication carrier logits of the federated distillation protocol, the degree of information leakage is very low, so we can assume that federated distillation can mechanically prevent gradient deep leakage, the relevant experimental demonstration will be shown in section V.

Algorithm 1. The FD framework that relies on logits for communication

---

Input: Public dataset D_{public}, private dataset D_k, client model $f_k, k = 1...m$.
Output: Trained model f_k
 1: Pre-training: Each client trains f_k to convergence on the public dataset D_{public}.
 2: Transfer learning: Each client trains f_k on the private dataset D_k.
 3: **while** f_k not converged **do**
 4:    **for** N rounds **do**
 5:       **Communicate:** Each client compute the logits vector $logits_k(x_i^{public})$ on the public dataset D_{public}, and transmits the result to the cloud server.
 6:       The number of clients uploading the logits verctor is k.
 7:       **if** $k \geq K$ **then**
 8:          **Aggregate:** The server compute the updated average logits:
 9:             $\overline{logits}(x_i^{public}) = \frac{1}{m}\sum_k logits_k(x_i^{public})$.
10:          **Distribute:** Sever send the updated average logits $\overline{logits}(x_i^{public})$ to each client.
11:          **Distillation:** Each client trains its model f_k to approach the updated average logits $\overline{logits}(x_i^{public})$.
12:          **Revist:** Each client trains its model f_k on its private data for some epochs.
13:       **end if**
14:    **end for**
15: **end while**

---

4   Client-Based Membership Inference

4.1   Overview of the Attack

The model distillation method converts the gradients passing model parameters in federated learning into logits. Logits are used in the learning process of model knowledge during federated distillation. They do not contain as much information about the global and local models as the gradient parameters, so applying logits to federated learning as a vehicle for passing model parameters is especially defensible against gradient membership inference attacks. However, considering that there is a malicious adversary among the participants and that the hostile adversary distills the knowledge of the other participants' models after obtaining the logits that have been aggregated and averaged by the server, the model trained by the malicious participant will still remember some details of the private data of the other participants.

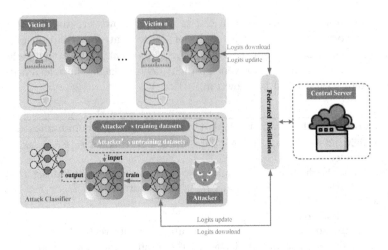

Fig. 2. Membership inference attack in federated distillation learning.

Figure 2 illustrates our membership inference attack in federated distillation learning. Our membership inference attack exploits the observation that while model distillation can address the gradient leakage problem in federated learning, malicious participants can still infer about private data's member and non-member information against other victims participating in federated learning by designing a malicious attack classifier. The main reason why models trained by knowledge distillation method contains information about the private data of their teachers' models. Malicious attacks make their own private model the target model of the attack. The attacker's objective is to construct an attack model that can recognize such differences in the target model's behavior and use them to distinguish members from non-members of the other models' training dataset based solely on the target model's output.

To train our attack model, We do not have to actively train the shadow model [10] because the attacker sees its own private model as the target model and the malicious adversary is aware of the ground truth of its own private model. Therefore, we can use supervised training on the inputs and the corresponding outputs (each labeled "in" or "out") of the adversary's private models to teach the attack model how to distinguish the target models' outputs on members of its training dataset from its outputs on non-members.

Formally, let the malicious client's private model $f_{adv}()$ be the target model $f_{target}()$. Then D_{target}^{train} becomes the training set for that target model, and D_{target}^{train} contains $(x^{\{i\}}, y^{\{i\}})_{target}$ labeled record data. The record data $x_{target}^{\{i\}}$ is the input to the target model, and $y_{target}^{\{i\}}$ is the actual label of the record data with label classification size c_{target}. The output of the target model is a vector of confidence scores of dimension c_{target}. All vectors of this output are in $[0,1]$ and their sum is 1.

Let $f_{attack}()$ be the attack model. Its input x_{attack} and the target model's score vector are of the same dimension, both c_{target}. The purpose of $f_{attack}()$ is to implement a membership inference attack, so the attack classifier is a binary classifier with a two-dimensional output, "in" and "out".

Figure 3 illustrates our end-to-end attack process. Given a target record data (x, y) for our attack, the record is first fed into the target model $f_{target}()$ to obtain the model's prediction vector y= $f_{target}()$. The distribution of y is related mainly to the actual classification label of the target record data x. So we input the label y of the target record data (x, y) and the prediction vector y of the target model $f_{target}()$ into the attack model $f_{attack}()$, and the attack model calculate the membership probability $Pr\{(x, y) \in D_{target}^{train}\}$, i.e., the probability that the true label y and the prediction vector y belong to the target training set, which is also the probability that the record input to attack model is classified as "in" class. The essential prerequisite for our attack to be realized is the training set of the attack model and the internal parameters of the target model. The malicious adversary already has knowledge of the internal structure of its own private model (i.e., the target model) and the training set D_{target}^{train}. Then the attacker's collection of training data for the attack binary classifier and the acquisition of prediction vectors for the target attack data can be achieved by accessing its own private model that has participated in federated distillation.

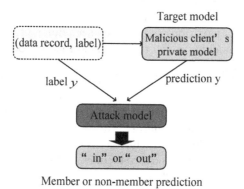

Fig. 3. End-to-end attack process.

4.2   Training the Attack Model

The main idea behind our client-based membership inference attack technique is that the local models of malicious adversary obtained by machine learning training through the federated distillation framework behave similarly to the private models of other benign participants involved in federated distillation. This observation is empirically borne out by our experiments in the rest of this paper. Our results show that learning how to infer membership in adversary's private models' training dataset (for which attacker know the ground truth and can easily compute the cost function during supervised training) produces an attack model that successfully infers membership in the target model's training

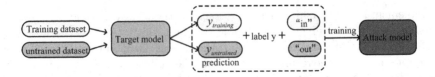

Fig. 4. Process of training the attack model.

dataset, too. We query the target model with the malicious adversary's private model's own training dataset and a disjoint test set of the same size. The data contained in the training dataset is marked as "in", and the rest of the data not involved in the federated distillation training is marked as "out". The attacker now has a record dataset, the corresponding output of its own private model and the in/out labels. The attack model aims to infer the labels from the records and the corresponding outputs.

Figure 4 shows how to train the attack model. For $(x, y) \in D_{target}^{train}$, the prediction vector $y = f_{target}(x)$ is computed, and the records (y, y, in) are added to the attack training set D_{attack}^{train}. Let $D_{target}^{untrained}$ be a set of records that do not intersect with the training set of the malicious adversary's private model. Then, the prediction vector y is computed based on $(x, y) \in D_{target}^{untrained}$, and the records (y, y, out) are added to the attack training set D_{attack}^{train}. Finally, the D_{attack}^{train} is partitioned into c_{target} partitions associated with a different class label. For each label y, a separate model is trained that predicts the "in" or "out" membership state of $(x, y) \in D_{target}$.

The reason why our malicious client can perform membership inference attacks lies in the overfitting problem in the machine learning process. So the malicious adversary identifies the subtle differences between the prediction scores obtained after feeding the trained dataset and the untrained dataset into the target model by training a dichotomous classifier that enables the attacking dichotomous classifier to achieve the purpose of membership and non-membership identification of the private data of all participants who participated in the federated distillation training.

5   Experimental Evaluation

5.1   Datasets and Experimental Setup

Datasets. In our experimental evaluation, we implement a federated distillation learning protocol on four benchmark datasets: MNIST, Extended MNIST (EMNIST), CIFAR-10 and CIFAR-100. MNIST and CIfAR-10 are used as the public dataset, EMNIST and CIFAR-100 are used as the private dataset for each client. Details of these datasets are described in the Table 1.

- **MNIST:** This dataset includes ten classes of handwritten digits from '0' to '9'. It is commonly used in training various image processing models. Total of 70,000 images are divided as the training set (60,000 images) and the testing set (10,000 images).

Table 1. Summary of datasets used in our experiments

Collaborative task	Public dataset	Private classes	Number of private data samples per class per client
EMNIST/MNIST I.I.D	MNIST	Letters [a - f] classes	6
EMNIST/MNIST Non I.I.D	MNIST	Letters from one writers	3
CIFAR I.I.D	CIFAR10	CIFAR100 random 6 subclasses	6
CIFAR Non I.I.D.	CIFAR10	CIFAR100 superclasses [0 - 5]	3

- **EMNIST:** The EMNIST dataset is a set of handwritten character digits derived from the NIST Special Database 19 and converted to a 28×28 pixel image format and dataset structure that directly matches the MNIST dataset.
- **CIFAR-10:** It consists of a training set of 60,000 images and a testing set of 10,000 images with 32×32 pixies in ten classes. These images are mainly cats, dogs, horses, etc.
- **CIFAR-100:** The CIFAR-100 dataset is a subset of the Tiny Images dataset and consists of 60000 32×32 color images. The 100 classes in the CIFAR-100 are grouped into 20 superclasses. There are 600 images per class. There are 500 training images and 100 testing images per class.

Experimental Settings. We implement our experiments within a federated distillation learning prototype by using the Tensorflow2.0 and Keras framework. All experiments are done on a RHEL7.5 server with NVIDIA Quadro P4000 GPU with 32GB RAM and Ubuntu 16.04LTS OS platform.

In terms of dataset processing, we set up ten participants, one of whom is assumed as the adversary. We construct MNIST and EMNIST as a set of datasets, CIFAR-10 and CIFAR-100 as a set of datasets, respectively. MNIST is set as the public dataset required for model distillation that the server can provide, and EMNIST is set as the private dataset of the client in an independent and non-independent homogeneous distribution, respectively. Correspondingly, CIFAR-10 is used as a public dataset and CIFAR-100 as a private dataset.

Table 2. Clients' models for MNIST/EMNIST and CIFAR

Model	MNIST/EMNIST			CIFAR-10/CIFAR-100		
	1st conv layer filters (n1)	2nd conv layer filters (n2)	3rd conv layer filters (n3)	1st conv layer filters (n1)	2nd conv layer filters (n2)	3rd conv layer filters (n3)
0	128	256	None	128	256	None
1	128	384	None	128	128	192
2	128	512	None	64	64	64
3	256	256	None	128	64	64
4	256	512	None	64	64	128
5	64	512	256	64	128	256
6	64	128	192	64	128	192
7	128	192	256	128	192	256
8	128	128	128	128	128	128
9	128	128	192	64	64	64

- **Model settings:** We list the architectures of the models used by each participant in the MNIST/EMNIST and CIFAR datasets in Table 2. Considering the superiority of federated distillation compared to federated learning in terms of model heterogeneity, we set the parameters of the models for the 10 participants to be different.
- **Training configurations:** For both MNIST and CIFAR datasets of the federated distillation learning method, we set the number of participants to 10 and federated learning rounds is set as $N = 20$. Before executing the federated distillation learning protocol, each participant train for $E = 20$ local epochs with the initial learning rate $\eta = 0.05$, and then execute logits match process for $M = 3$ rounds.

5.2   Impact of Model Distillation

Performance of Federated Distillation Learning. In our federated distillation learning experiments, we take into consideration the effectiveness of the federated distillation learning. Here, we mainly evaluate the mean accuracy on centralized-Machine Learning, federated learning and federated distillation learning under different data distribution settings. Figure 5 shows the accuracy of centralized-Machine Learning (centralized ML), federated learning protocol (FL), and federated distillation learning (FD) for 20 communication rounds. As shown in Fig. 5(a), the accuracy of the federated distillation learning can be up to 82% as the federated learning's highest accuracy is 85%. That demonstrates that the model distillation method can maintain very good performance in a federated learning environment. As depicted in Fig. 5(b), the difference in performance between federated learning and federated distillation learning is within 5%. In Fig. 5(c) and Fig. 5(d), in the setting of CIFAR as dataset, both the I.I.D. and Non I.I.D. cases, federated distillation learning performs only a little worse than federated learning. The experiment can demonstrate that federated distillation and federated learning performance is very close under independent homogeneous and non-independent homogeneous distribution settings for MNIST and CIFAR datasets.

Effect of Model Distillation on Gradient Leakage. Figure 6 shows the deep leakage of federated learning and logits leakage of federated distillation learning on images from CIFAR-100. In gradient leakage part in federated learning, we start with Gaussian noise (first column) and try to match the gradients produced bu the dummy data and real ones. When the optimization finishes, the recover result are almost identical to ground truth image, despite few negligible artifact pixels. In our federated distillation learning protocol, we investigated the degree of leakage of logits utilizing the same method, the effect of image recovery is far from gradient leakage. The comparison demonstrates that the logits model distillation method of the federated distillation protocol can primarily alleviate the gradient leakage problem in federated learning.

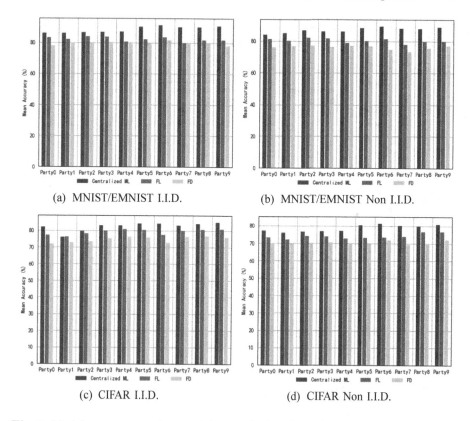

Fig. 5. Model accuracy on three training methods, where the dataset setting changes and Data Distribution is different.

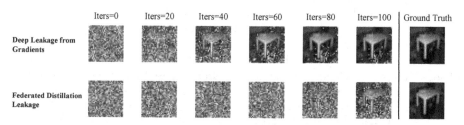

Fig. 6. The visualization showing the deep leakage from federated learning and logits leakage in federated distillation learning on images from CIFAR.

5.3 Effectiveness of Client-Based Membership Inference Attack

We evaluate this client-based membership inference attack by randomly reconstituting records from the target models' training and test datasets. In our attack evaluation, we use sets of the same size (i.e., with the same number of members and non-members) to maximize the uncertainty of inference, hence a baseline accuracy of 0.5. Here, we mainly evaluate the attack classification accuracy on membership inference task under different dataset settings.

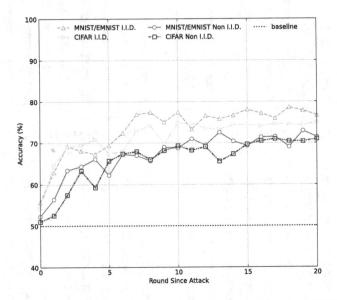

Fig. 7. Performance of membership inference attacks for MNIST and CIFAR datasets under different data distribution settings.

As shown in Fig. 7, the accuracy of membership inference task showed an obvious upward trend (up to 78.62%) as the communication round goes on. This is because the malicious client's model learns the knowledge of other participants' models using model distillation method in the federated distillation process, so that the parameters and structure of the malicious model record valid information about other participants' private data, and by downloading updates of logtis parameters and training its own model based on them, adversary is able to achieve a fairly good knowledge of the membership and non-membership information of the private data set not used by itself. As depicted in Fig. 7, accuracy of membership inference task based on MNIST/EMNIST I.I.D dataset (red line) reach a high level of 78.62% and the result of membership inference attack against CIFAR I.I.D model (purple line) performs a little worse than I.I.D. data setting, but still able to reach 75.02%. In the case of non-independent data distribution, the effectiveness of the membership inference attack decreases somewhat compared to that of the independent identically distributed condition but still reaches 70%. Based on the illustrations of the effectiveness of the attack task in the above four scenarios with different data settings, we demonstrate the vulnerability of the federated distillation protocol to membership inference attacks on malicious client.

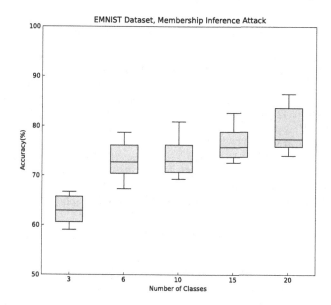

Fig. 8. Accuracy of the membership inference attack against EMNIST dataset classification models trained on the federated distillation framework. The boxplots show the distribution of accuracy over 20 communication rounds (with a different number of classes).

5.4 Evaluation of the Number of Classes

We trained all client models by setting their private datasets as EMNIIST with 2, 6, 10, 15, 20 classes. Models with fewer classes leak less information about their training inputs, this result is similar to the conclusions in centralized machine learning and federated learning. As the number of classes increases, the model needs to extract more distinctive features from data to be able to classify inputs with high accuracy. Informally, models with more output classes need to remember more about their training data, thus they leak more information. Figure 8 shows the relationship between the number of classes and the accuracy of membership inference.

6   Summary and Future Work

In this paper, we evaluate the performance of federated distillation protocol and the task of malicious client-based membership inference attacks with four benchmark datasets, illustrating the feasibility of the federated distillation learning for solving the deep leakage problem of gradients and the vulnerability of client-based membership inference attacks. In future work, we plan to investigate a more robust federated distillation framework to address the membership information leakage that still exists on the client side by means of logits perturbation.

References

1. De Boer, P.T., Kroese, D.P., Mannor, S., Rubinstein, R.Y.: A tutorial on the cross-entropy method. Ann. Oper. Res. **134**(1), 19–67 (2005)
2. Hao, M., Li, H., Luo, X., Xu, G., Yang, H., Liu, S.: Efficient and privacy-enhanced federated learning for industrial artificial intelligence. IEEE Trans. Industr. Inf. **16**(10), 6532–6542 (2019)
3. Hinton, G., Vinyals, O., Dean, J.: Distilling the knowledge in a neural network. arXiv preprint arXiv:1503.02531 (2015)
4. Jeong, E., Oh, S., Kim, H., Park, J., Bennis, M., Kim, S.L.: Communication-efficient on-device machine learning: federated distillation and augmentation under non-IID private data. arXiv preprint arXiv:1811.11479 (2018)
5. Khan, M.A., Salah, K.: IoT security: review, blockchain solutions, and open challenges. Futur. Gener. Comput. Syst. **82**, 395–411 (2018)
6. Li, D., Wang, J.: FedMD: heterogenous federated learning via model distillation. arXiv preprint arXiv:1910.03581 (2019)
7. Li, P., et al.: Multi-key privacy-preserving deep learning in cloud computing. Futur. Gener. Comput. Syst. **74**, 76–85 (2017)
8. Lu, Y., Huang, X., Dai, Y., Maharjan, S., Zhang, Y.: Blockchain and federated learning for privacy-preserved data sharing in industrial iot. IEEE Trans. Industr. Inf. **16**(6), 4177–4186 (2019)
9. Melis, L., Song, C., De Cristofaro, E., Shmatikov, V.: Inference attacks against collaborative learning. arXiv preprint arXiv:1805.04049 (2018)
10. Nasr, M., Shokri, R., Houmansadr, A.: Comprehensive privacy analysis of deep learning: passive and active white-box inference attacks against centralized and federated learning. In: 2019 IEEE Symposium on Security and Privacy (SP), pp. 739–753. IEEE (2019)
11. Shokri, R., Shmatikov, V.: Privacy-preserving deep learning. In: Proceedings of the 22nd ACM SIGSAC Conference on Computer and Communications Security, pp. 1310–1321 (2015)
12. Shokri, R., Stronati, M., Song, C., Shmatikov, V.: Membership inference attacks against machine learning models. In: 2017 IEEE Symposium on Security and Privacy (SP), pp. 3–18. IEEE (2017)
13. Wang, J., Cao, B., Yu, P., Sun, L., Bao, W., Zhu, X.: Deep learning towards mobile applications. In: 2018 IEEE 38th International Conference on Distributed Computing Systems (ICDCS), pp. 1385–1393. IEEE (2018)
14. Yang, Q., Liu, Y., Chen, T., Tong, Y.: Federated machine learning: concept and applications. ACM Trans. Intell. Syst. Technol. (TIST) **10**(2), 1–19 (2019)
15. Zhu, L., Han, S.: Deep leakage from gradients. In: Yang, Q., Fan, L., Yu, H. (eds.) Federated Learning. LNCS (LNAI), vol. 12500, pp. 17–31. Springer, Cham (2020). https://doi.org/10.1007/978-3-030-63076-8_2

Spatial and Multi-media Data

When Self-attention and Topological Structure Make a Difference: Trajectory Modeling in Road Networks

Guoying Zhu[1], Yu Sang[2], Wei Chen[1(✉)], and Lei Zhao[1]

[1] School of Computer Science and Technology, Soochow University, Suzhou, China
1909401156@stu.suda.edu.cn, {robertchen,zhaol}@suda.edu.cn
[2] School of Artificial Intelligence and Computer Science, Jiangnan University,
Wuxi, China
7213107001@stu.jiangnan.edu.cn

Abstract. The ubiquitous GPS-enabled devices (e.g., vehicles and mobile phones) have led to the unexpected growth in trajectory data that can be well utilized for intelligent city management, such as traffic monitoring and diversion. As a building block of the smart-mobility initiative, trajectory modeling has received increasing attention recently. Despite the great contributions made by existing studies, they still suffer from the following problems. (1) The topological structure of a road network is underutilized. (2) The existing methods cannot characterize the stopping probability of a trajectory. To this end, we develop a novel model entitled TMRN (**T**rajectory **M**odeling in **R**oad **N**etworks), which is composed of the following three modules. (1) Road2Vec: the module is developed to learn the representations of road segments by fully utilizing the topology information of a road network. (2) LWA: the lightweight attention-based module is designed to capture the long-term regularity of trajectories. (3) MOP: a novel matching operation is proposed to calculate the transition probability of the next segment for the current path. The extensive experiments conducted on two real-world datasets demonstrate the superiority of TMRN compared with state-of-the-art methods.

Keywords: Self-attention · Topological structure · Trajectory modeling

1 Introduction

The massive trajectory data in road networks underlie a wide range of location-aware applications, such as urban traffic analysis for government, targeted advertising based on destination prediction for merchants, and popular routes recommendation for drivers. As an essential building block of the above applications, trajectory modeling has received increasing attention in existing studies [18,25], inspired by the high performance of the Markov chain [9,14] and deep learning methods [1] in modeling sequence information.

Uniquely, different from the routes without fixed structure and generated by animals [2,20], hurricanes [24], and flights [7,22], trajectories collected from road networks contain abundant topological structure information and usually can be denoted as sequences of road segments [6,25]. The purpose of trajectory modeling [25] is to model the likelihood of a given trajectory $\tau = (e_1, e_2, \cdots, e_n)$,

$$P(\tau) = P(e_1) \prod_{i=1}^{n-1} P(e_{i+1} \mid e_{1:i}) \tag{1}$$

where e_i denotes a road segment and $P(e_{i+1} \mid e_{1:i})$ is the transition probability which captures the probability that the trajectory τ takes the next road segment e_{i+1}. The existing methods for trajectory modeling can be divided into the following two categories. (1) In traditional work, the Markov chain has been widely explored [21,26]. By way of illustration, [21] develops a probabilistic inference model to understand human mobility during disasters, where the Markov Decision Process (MDP) [17] and Inverse Reinforcement Learning (IRL) [31] are utilized for training parameters of the learned evacuation graph. (2) In recent advances, researchers have applied deep learning techniques to trajectory modeling, such as Recurrent Neural Network (RNN) and Long Short-Term Memory (LSTM). Precisely, to capture the variable-length sequence and meanwhile to address the constraint of topological structure on trajectory modeling, two RNN-based models, namely CSSRNN and LPIRNN, are proposed [25]. This work is the first attempt to adopt deep learning techniques to model road network constrained trajectories. Following this study, an LSTM-based model, namely DND, is proposed to model vehicle trajectories by simultaneously considering the distant and neighboring dependencies for destination prediction [18].

Existing studies have made significant contributions to trajectory modeling, but there remains great scope for further improvement. Firstly, although the self-attention mechanism [23] based deep learning methods have proven more powerful than traditional models (e.g., RNN and LSTM) in capturing long-range dependence, there has been no work utilizing this mechanism for trajectory modeling in road networks. Secondly, although the topological structure of a road network has been considered to a certain extent [25], incorporating this factor into the self-attention mechanism for more effective trajectory modeling has not been explored. Thirdly, although we can model the probability $P(e_{i+1} \mid e_{1:i})$ for current path $e_{1:i}$ based on Eq. (1), i.e., answer the question *"Which road segment is the next segment of $e_{1:i}$?"*, the closely related question *"When will the trajectory stop?"* cannot be answered with Eq. (1), since the probability related to the stopping segment e_n has not been modeled, and this is the main drawback of the existing work.

Having observed the shortcomings of existing studies, we propose a novel model called TMRN and redefine the trajectory modeling as:

$$P(\tau) = P(e_1) \left(\prod_{i=1}^{n-1} P(e_{i+1} \mid e_{1:i}) \right) P(e_v \mid e_{1:n})$$

where e_v is an artificial and nonexistent road segment that means e_n is the stopping segment of τ. In a real application, if the predicted next segment for the current path is e_v, the moving object will stop in the current segment. Moreover, the proposed model TMRN consists of the following three components. Firstly, a novel module, namely Road2Vec, containing two binary classifiers, is developed to learn the coarse-grained representation of each road segment. Secondly, a lightweight attention-based module called LWA is designed to assign different weights for hidden vectors and capture the long-term regularity of trajectories. Thirdly, a matching operation called MOP is proposed to calculate the transition probability of the next road segment for the current path.

To sum up, we make the following major contributions in this study.

- To the best of our knowledge, we are the first to redefine the trajectory modeling by tackling the next and stopping road segment prediction simultaneously.
- We develop a novel model entitled TMRN, and the model contains three main components, i.e., road segment representation learning, lightweight self-attention mechanism, and a matching operation.
- We conduct extensive experiments on two real-world datasets, and the experimental results demonstrate the high performance of TMRN. The source code is available at *https://github.com/goodapweb-2022/tmrn*.

The rest of this paper is organized as follows. We present the related work in Sect. 2 and formulate the problem in Sect. 3. The proposed model TMRN is introduced in Sect. 4, followed by the experiments in Sect. 5. The paper is concluded in Sect. 6.

2   Related Work

Markov Chain Based Trajectory Modeling. Understanding the mobility patterns of moving objects by trajectory modeling underlies a wide range of applications such as route planning [3,31], next location prediction [4], and destination prediction [26,27]. Specifically, the method PROCAB is developed for probabilistically reasoning from observed context-aware behaviors by assigning as much probability as possible to drivers' preferred routes [31]. Following this work, Song et al. [21] mine the short-term and long-term evacuation behaviors for individuals, where a probabilistic model is developed based on the Markov decision process and inverse reinforcement learning. To model the heterogeneous routing decisions in trajectories, a framework based on Bayesian inverse reinforcement learning is proposed [30]. To infer movement trajectories when detailed tracking information is not available due to privacy and data collection constraints, Li et al. [11] present a complete and computationally tractable model based on a first-order Markov. The problem of next location prediction is investigated with three basic models in [4], where the Global Markov Model, Personal Markov Model, and Region Markov Model are integrated with linear regression in different ways. Pecher et al. [16] study the trajectory prediction in the context of dynamic data-driven applications systems based on the T-Drive trajectory dataset [29]. Qiao et al. [19] focus on improving the prediction accuracy of non-Gaussian mobility data by constructing a hybrid Markov-based model.

Deep Learning Based Trajectory Modeling. Having witnessed the high performance of deep learning models in modeling sequence information [10,28], many researchers have applied these techniques to trajectory modeling [13,25]. As the first attempt at adopting deep learning techniques to model road network constrained trajectories, Wu et al. [25] propose two models, CSSRNN and LPIRNN. Inspired by the idea of [12], topology information is manually fed into the neural network in the way of state-constrained softmax function in CSSRNN. The second model, LPIRNN, contains two components, i.e., a shared task layer that produces intermediate information and trajectory prediction by multiple individual tasks. Following this, O'Keeffe et al. [15] explore whether RNN can be used to generate realistic mobility patterns for taxis and personal cars. Specifically, the distribution of street popularities is defined as the relative number of times in the i-th street traversed by a fleet of taxis moving on the street network during a referenced period. To model vehicle trajectory for destination prediction, the distant and neighboring dependencies are considered [18]. These two dependencies are investigated based on the macro characteristics (e.g., travel purpose, driving preferences, and road traffic conditions) and local road connectivity. After revisiting RNN-based approaches for trajectory modeling, Liang et al. [13] expose two common drawbacks, i.e., these methods are awkward to learn trajectories with continuous-time dynamics and vulnerable to noise sample locations. A novel model called TrajODE is devised to address the problems, where the continuous-time characteristics of neural ordinary differential equations and the robustness of stochastic latent space are combined.

Despite the significant contributions of the studies above, there remains great scope for achieving higher performance trajectory modeling. Consequently, we develop a novel model called TMRN by taking full advantage of the self-attention mechanism and a road network's topological structure in a unified manner.

3   Problem Definition

Definition 1. *Road Network. A road network is represented as a directed graph $G = (V, E)$, where V is a set of vertices (i.e., intersections), and E is a set of edges (i.e., road segments). Each edge e of E represents a road segment from an intersection v ($v \in V$) to its adjacent intersection v' ($v' \in V$).*

Definition 2. *Adjacent Set. Given a road network $G = (V, E)$, we use adjacent set, which is defined as $\mathcal{A}$, to record the adjacent road segments, where each element $\mathcal{A}_e$ of $\mathcal{A}$ denotes the set of adjacent road segments of e ($e \in E$).*

Definition 3. *Trajectory [25]. A trajectory in a road network is a sequence of road segments in the form of $\tau = (e_1, e_2, \cdots, e_n)$.*

Definition 4. *Finished Trajectory. Given a historical trajectory $\tau = (e_1, e_2, \cdots, e_n)$ in a road network, the corresponding finished trajectory is defined as $\tau' = (e_1, e_2, \cdots, , e_n, e_v)$, where e_v is a virtual and nonexistent road segment, and e_v indicates that the trajectory τ has stopped in e_n.*

Definition 5. *Prefix Trajectory. Given a finished trajectory* $\tau' = (e_1, \cdots, e_n, e_v)$ *in a road network, the i-th prefix trajectory of* τ' *is denoted as* $\tau_p^i = (e_1, \cdots, e_i)$, *where* $1 \leq i \leq n+1$ *and* e_{n+1} *is* e_v.

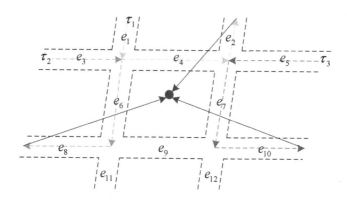

Fig. 1. An example of finished and prefix trajectory

Considering the example in Fig. 1, there are three historical trajectories $\tau_1 = (e_1, e_4, e_2)$, $\tau_2 = (e_3, e_6, e_8)$, and $\tau_3 = (e_5, e_7, e_{10})$. Firstly, before training the proposed model TMRN, we add the virtual road segment e_v (i.e., the red line in Fig. 1) to the end of each trajectory and obtain finished trajectories $\tau_1' = (e_1, e_4, e_2, e_v)$, $\tau_2' = (e_3, e_6, e_8, e_v)$, and $\tau_3' = (e_5, e_7, e_{10}, e_v)$. Then, we can randomly select some prefix trajectories from $\{\tau_1', \tau_2', \tau_3'\}$, such as (e_1, e_4), (e_3, e_6, e_8, e_v), and (e_5, e_7, e_{10}) to optimize the parameters of TMRN.

Problem Formulation. Given a road network G, a set of historical trajectories $S = \{\tau_1, \tau_2, \cdots, \tau_n\}$, and a new prefix trajectory $\tau_p^i = \{e_1, e_2, \cdots, e_i\}$, this work aims to train a model that can be used to answer the following question:

– Which road segment is the most likely next segment for current path τ_p^i?

Obviously, there are two possibilities for this question. By way of illustration, given a new prefix trajectory $\tau_p^i = \{e_1, e_2, \cdots, e_i\}$ and assuming the model TMRN has been optimized (1) if the predicted result is e_v, it means the moving object will stop in the current segment e_i, (2) if the predicted result is not e_v, it means the object will continue to move along the predicted road segment.

4   Proposed Model TMRN

4.1   Overview of TMRN

Observed from Fig. 2, the proposed model TMRN contains the following three modules. (1) Road2Vec: the module is introduced to learn the representation of each road segment. Intuitively, road segments included by historical trajectories usually have a relatively small average frequency in a large-scale road network with many segments, which brings a great challenge for effective trajectory modeling. To address the problem, we take the contextual information of road segments into account, i.e., exploring whether two road segments occur in the same trajectory and are adjacent. (2) LWA: a lightweight module based on the self-attention mechanism is developed to give the hidden layer better representations by assigning hidden vectors with different weights. Specifically, a hidden vector with a higher weight indicates that it is more critical for describing the product in multi-head self-attention. The self-attention mechanism can model trajectories more effectively than the recurrent network as it can capture the long-range dependence better. (3) MOP: a novel matching operation is designed to calculate the transition probability of the next segment or e_v for the current path.

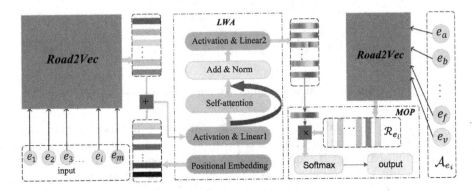

Fig. 2. Overview of the proposed model TMRN.

4.2 Module Road2Vec

As presented in Fig. 3, the trainer of Road2Vec contains two binary classifiers developed based on the idea of [6], which takes two road segments e_x and e_y as inputs and produces the probability $P_s(e_x, e_y)$ and $P_a(e_x, e_y)$ for e_x and e_y to indicate whether they occur in the same trajectory and are adjacent, respectively.

$$P_s(e_x, e_y) = \sigma\left(l_s\left(f\left(U_{e_x}\right) \circ f\left(U_{e_y}\right)\right)\right)$$

$$P_a(e_x, e_y) = \sigma\left(l_a\left(f\left(U_{e_x}\right) \circ f\left(U_{e_y}\right)\right)\right)$$

$$\sigma(x) = \frac{1}{1 + e^{-x}}$$

where $U\ (|U| = p \times q)$ is a matrix, each row U_{e_x} represents the embedding vector of a road segment e_x, p is the number of road segments, and q denotes the embedding size of a vector. The function $f(\cdot)$ denotes the normalization, l_s and l_a denote the linear transformation. To train the module Road2Vec, we extract a set of segment pairs from the given trajectory set S, where each pair contains two labels $L_s(e_x, e_y)$ and $L_a(e_x, e_y)$. Specifically, $L_s(e_x, e_y)$ is set to 1 if e_x and e_y occur in the same trajectory. Otherwise, it is set to 0. Similarity, we set $L_a(e_x, e_y)$ to 1 if e_x and e_y are adjacent and to 0 if they are nonadjacent. The loss functions are defined as:

$$Loss_s = L_s(e_x, e_y) \log P_s(e_x, e_y) + (1 - L_s(e_x, e_y)) \log(1 - P_s(e_x, e_y))$$

$$Loss_a = L_a(e_x, e_y) \log P_a(e_x, e_y) + (1 - L_a(e_x, e_y)) \log(1 - P_a(e_x, e_y))$$

$$Loss = \beta \times Loss_s + (1 - \beta) \times Loss_a$$

The parameter β measures the importance of $Loss_s$ and $Loss_a$ for Road2Vec. Note that the purpose of building Road2Vec is to fully utilize the topological structure of a road network for more effective trajectory modeling, and obtain a coarse-grained representation, which will be finally optimized in Sect. 4.4, of each road segment.

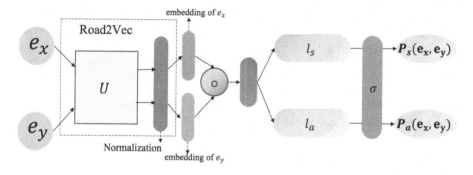

Fig. 3. Trainer of the module Road2vec.

4.3 Module LWA

LWA is a lightweight self-attention-based module proposed to get the expected representation of the predicted road segment. It mainly contains positional embedding, self-attention layer, and activation & linear layer. Positional embedding adds information about the order of road segments in a trajectory to the input. The self-attention layer assigns hidden vectors with different weights to represent the predicted road segment better. The linear layers in LWA aim to adjust the size of the hidden vector.

Positional Embedding. The order of road segments in a trajectory provides important information for trajectory modeling. For example, the closer the segment is to the next segment of the current path, the more influence it has on the predicted result. Consequently, the position embedding [5] is considered in this part, and it is given as:

$$POS = \begin{cases} P(t, 2i) = & \sin(t/10000^{2i/q}) \\ P(t, 2i+1) = & \cos(t/10000^{2i/q}) \end{cases}$$

where t is the position of a road segment in a trajectory, i and q denote the index and embedding size of the road segment, respectively.

Given a prefix trajectory $\tau = (e_1, e_2, \cdots, e_i, e_{i+1})$, we mask e_{i+1} with e_m and get a new sequence $x = (e_1, e_2, \cdots, e_i, e_m)$. Based on Road2Vec and Positional Embedding, we can get the matrix $E \in R^{m \times q}$, the j-th row of which is the representation vector $Road2Vec(e_j) + POS(e_j)$ of the road segment e_j in x. m denotes the sequence length.

Activation & Linear Layer1. The activation function used in this work is GELU, which is defined as:

$$GELU(y) = yP(Y \leq y) = y \cdot \Phi(y)$$

where Y is a Gaussian random variable with zero-mean and unit variance. The linear layers in LWA are used to adjust the size of the hidden representation of a road segment to make the model more flexible, where the first Activation & Linear Layer is defined as:

$$X = (GELU(E))W_1 + b_1$$

where $W_1 \in R^{q \times r}$ is the parameter matrix and b_1 is the bias. $X \in R^{m \times r}$ is the input of the self-attention layer.

Self-attention Layer. In the self-attention layer, three fully connected networks $W_q, W_k, W_v \in R^{r \times r}$ are used to map the input $X \in R^{m \times r}$ into three matrices, i.e., the query matrix $Q^{m \times r}$, the key-value pair matrices $K^{m \times r}$ and $V^{m \times r}$, where r is the embedding size of a road segment.

$$Q, K, V = X[W_q, W_k, W_v] + b$$

where b is the bias. Then, the attention is defined as:

$$ATT(Q, K, V) = softmax(\frac{QK^\top}{\sqrt{r}})V$$

This paper applies the multi-head scheme to compute the attention scores and obtain information from different representation subspaces. Each head focuses on one part of the hidden representation of the road segment. Mathematically, the calculation is given as:

$$Head_j(Q, K, V) = ATT(QW_j^Q, KW_j^K, VW_j^V)$$

$$MultiHead(Q, K, V) = Concat(Head_1, ..., Head_h)W$$

where $W \in R^{r \times r}$ is the parameter matrix, and h is the number of heads. $W_j^Q, W_j^K, W_j^V \in R^{r \times r/h}$ are parameter matrices on the j-th head.

Add & Norm Layer. As discussed in [23], Add & Norm Layer is beneficial to speed up the training and improve the generalization performance by preventing the numerical variation. $f(\cdot)$ represents the normalization in Eq. (2),

$$Add\&Norm = f(X + MultiHead(Q, K, V)) \tag{2}$$

Activation & Linear Layer2. The second Activation & Linear Layer transforms the size of the output of the self-attention layer to match the size of the vectors in Road2Vec,

$$(\mathbf{h}_{e_1}, \mathbf{h}_{e_2}, \cdots, \mathbf{h}_{e_i}, \mathbf{h}_{e_m}) = Add\&Norm \times W_2 + b_2 \tag{3}$$

where $W_2 \in R^{r \times q}$ and b_2 is the bias.

4.4   Module MOP

For the current path $\tau = (e_1, e_2, \cdots, e_i)$, we use $A_{e_i} = \{e_a, e_b, \cdots, e_f, e_v\}$ to denote the set of adjacent segments of e_i, where each segment in $\{e_a, e_b, \cdots, e_f\}$ is adjacent to e_i based on the topological structure of the given road network. The current path may stop in e_i, thus e_v is also added into A_{e_i}. Next, we calculate the transition probability $P(e_k|e_{1:i})$ for each segment e_k in A_{e_i}. Detailedly, we construct a matrix $\mathcal{R}_{e_i} \in R^{|A_{e_i}| \times q}$ for A_{e_i}, where the j-th row of $\mathcal{R}_{e_i}$ is the embedding $Road2Vec(e_j)$ of the j-th segment in A_{e_i}. Based on the expected representation $\mathbf{h}_{e_m}$ of the segment e_{i+1} generated by Eq. (3) and the matrix $\mathcal{R}_{e_i}$, we calculate the transition probabilities with following method:

$$\{P(e_a \mid e_{1:i}), P(e_b \mid e_{1:i}), \cdots, P(e_f \mid e_{1:i}), P(e_v \mid e_{1:i})\} = softmax(f(\mathbf{h}_{e_m})\mathcal{R}_{e_i}^{\top})$$

where $f(\cdot)$ is the normalization. The segment e_k with the maximum transition probability $P(e_k|e_{1:i})$ is the predicted next road segment. The final loss function is defined as follows.

$$Loss = -\sum_{j=1}^{|A_{e_i}|} y_j \log P_j \tag{4}$$

where P_j is the j-th probability in $\{P(e_a \mid e_{1:i}), P(e_b \mid e_{1:i}), \cdots, P(e_f \mid e_{1:i}), P(e_v \mid e_{1:i})\}$. y_j is set to 1 if the j-th segment in A_{e_i} is the next segment of e_{i+1} in training dataset, otherwise, y_j is set to 0.

Training Details. (1) The module Road2vec obtains the coarse-grained representations of road segments, and the obtained embedding vectors are fine-tuned during the training of TMRN. Specifically, we use the coarse-grained embeddings as initial inputs and update them to optimize the loss function in Eq. (4). (2) In order to ensure parallelism, we set a max length, which is larger than that of each trajectory in the given dataset, and use the operation *padding*, i.e., adding 0 to trajectories with a length less than the max length. For example, the sequence $(e_1, e_2, \cdots, e_i)$ is set to $(e_1, e_2, \cdots, e_i, e_m, 0, \cdots, 0)$ if the length of it is less than the max length. (3) All parameters of TMRN are adjusted based on the Adam algorithm during the training.

5   Experiment

5.1   Experimental Datasets

BJ(Beijing dataset): The road network, which contains 20225 road segments, is collected from the city Beijing. The dataset has 69448 trajectories generated by 12447 taxis from 2012-10-1 to 2012-10-7. To optimize the parameters of TRMN, we choose an 80-20 split on the dataset. The numbers of road segment pairs used for training Road2vec, prefix trajectories in training TMRN, and prefix trajectories in testing TMRN are 683626, 419421, and 56603.

PT(Porto dataset): PT is an open dataset, generated by 442 taxis from 2013-01-07 to 2014-06-30, contains 441019 trajectories, and the corresponding road network includes 2274 road segments. To optimize the parameters of TRMN, we choose a 80-20 split on the dataset. The numbers of road segment pairs used for training Road2vec, prefix trajectories in training TMRN, and prefix trajectories in testing TMRN are 50238, 1292413, and 430805.

5.2   Experimental Setup

The parameters of TMRN are presented in Table 1. All algorithms are implemented on an Intel Xeon 2.20 GHz 64 processors machine with 2 TB memory.

Table 1. Parameters setting

Dataset	Number of heads	Learning rate	Dropout	Embedding size
BJ	6	0.0001	0.06	72
PT	9	0.00005	0.06	216

5.3   Evaluation Metric

Given a set of test trajectories, we use the metrics ACC and NLL, which have been widely used in existing work [19,25] and are defined as follows, to evaluate the performance of TMRN. Additionally, the metric $ACCofe_v$ is proposed for modeling the stopping segment of a trajectory.

$$ACC = \frac{1}{\sum_{i=1}^{N} k_i} \sum_{j=1}^{k_i-1} \mathbf{1}\{argmax_{e \in A_{e_j}} P(e \mid e_{1:j}) = e_{j+1}\} \qquad (5)$$

$$NLL = -\frac{1}{N} \sum_{i=1}^{N} \sum_{j=1}^{k_i-1} \log P(e_{j+1}|e_{1:j}) \qquad (6)$$

$$ACCofe_v = \frac{number\ of\ correct\ predictions\ of\ e_v}{number\ of\ all\ predictions\ of\ e_v} \qquad (7)$$

where N is the number of the finished trajectories and k_i is the length of the finished trajectory τ_i. Although the compared methods don't consider e_v initially, we can adjust them by adding e_v to the datasets without changing models' structures, so that they can predict e_v.

5.4   Compared Methods

The compared methods can be divided into the following two categories. Firstly, Markov chain-based methods: (1) MarkCF [11], the method is developed based on the first-order Markov chain; (2) MCST, the algorithm is a variant of [8] and it matches the partial trajectory with full trajectories in the training set that contain the last k road segments of the partial trajectory; (3) HMP, the approach is a variant of [19], where locations are replaced with road segments. Secondly, Deep learning-based methods: (1) CSSRNN [25], the method is an extension of traditional RNN and addresses the issue of topological constraints; (2) TODE [13], the model is designed based on Neural Ordinary Differential Equations; (3) TM-RNN and TM-LSTM, these two approaches are developed by replacing the module LWA with the traditional RNN and LSTM. Notably, modules Road2vec and MOP are contained by these methods.

The ACC, NLL, and $ACCofe_v$ of all methods are presented in Table 2. Without surprise, the hybrid Markov chain-based method HMP has better performance than MarkCF and MCST, as more information is considered in HMP. Deep learning-based methods have higher ACC and lower NLL than Markov chain-based approaches due to their senior performance in modeling sequence information. This advantage of the deep learning-based methods is even more pronounced in the denser dataset PT. The methods TM-RNN and TM-LSTM have better performance than other baselines, and the reason is twofold. Firstly, we train a module, Road2vec, before trajectory modeling to learn the representation of each road segment. Secondly, the module MOP takes full advantage of the topological structure of a road network. Thirdly, the self-attention mechanism is beneficial for processing long sentences. Our proposed model TMRN has the highest prediction accuracy. The result is expected since it fully utilizes the advantage of the lightweight self-attention and the topological of a road network.

Table 2. Prediction results of compared methods

	Method	MarkCF	MCST	HMP	CSSRNN	TODE	TM-RNN	TM-LSTM	TMRN
BJ	ACC	37.51%	61.85%	64.32%	64.47%	65.18%	67.17%	68.82%	**72.83%**
	NLL	96.08	46.13	42.16	42.01	40.92	37.10	36.78	**30.10**
	$ACCofe_v$	2.41%	6.72%	11.52%	41.36%	42.88%	42.72%	48.14%	**49.86%**
PT	ACC	27.94%	42.58%	44.52%	76.91%	76.97%	78.02%	79.21%	**81.08%**
	NLL	117.63	85.37	81.01	33.42	34.51	32.09	30.17	**27.49**
	$ACCofe_v$	4.78%	6.96%	10.32%	80.37%	82.76%	84.05%	85.37%	**88.60%**

Table 3. Performance of baseline methods

	Method	TMRN-LM	TMRN-LSTM	TMRN-RL	TMRN
BJ	ACC	65.81%	68.82%	58.26%	**72.83%**
	NLL	40.79	36.78	56.94	**30.10**
	$ACCofe_v$	47.76 %	48.14%	37.37%	**49.86%**
PT	ACC	80.17%	79.17%	77.81%	**81.08%**
	NLL	28.78	36.78	35.31	**27.49**
	$ACCofe_v$	87.95%	85.37%	82.73%	**88.60%**

5.5 Impact of Different Modules

To further validate the benefits brought by Road2vec, LWA, and MOP, we develop following baseline methods: TMRN-LM (replace Road2Vec with Word2Vec), TMRN-LSTM (replace LWA with LSTM), and TMRN-RL (replace MOP with Mask method in CSSRNN).

From the results in Table 3, we have the following observations. (1) The three modules Road2vec, LWA, and MOP are significant for our proposed model, as TMRN has much better performance than all baseline algorithms. (2) The module MOP, which is developed based on the structure of a road network, is the most critical component for our proposed model TMRN, as the method TMRN-RL without MOP has the lowest prediction accuracy compared with TMRN-LM and TMRN-RM.

5.6 Analysis of Parameters

Varying Number of Heads. From the results in Table 4, we observe that the performance of TMRN is sensitive to the number of heads of lightweight self-attention. A too-small number is not appropriate for TMRN, as the multi-head attention allows a model to jointly attend to information from different representation subspaces at different positions, and this will be inhibited with a single attention head [23]. Additionally, a too-large value of the number may lead to information distraction. To achieve the best performance of TMRN, we set the number of heads to 6 for BJ and 9 for PT.

Table 4. Varying number of heads

	Heads number	1	3	6	9	12
BJ	ACC	69.83%	70.19%	**72.83%**	69.55%	70.32%
	NLL	32.41	33.07	**30.10**	35.95	36.35
	$ACCofe_v$	46.92%	47.36%	**49.86%**	47.88%	45.78%
PT	ACC	80.31 %	80.51%	80.62%	**81.08 %**	70.32%
	NLL	29.12	28.39	28.12	**27.49**	28.41
	$ACCofe_v$	86.71%	88.51%	87.73%	**88.60%**	87.61%

Table 5. Varying learning rate

		Learning rate	0.00005	0.000075	0.0001	0.000125	0.00015
BJ	ACC		71.31%	71.67%	**72.83%**	71.98%	70.98%
	NLL		31.14	34.31	**30.10**	30.42	33.42
	$ACCofe_v$		47.36%	43.57%	**49.86%**	46.21%	45.14%
		Learning rate	0.00001	0.000025	0.00005	0.000075	0.0001
PT	ACC		80.72%	80.84%	**81.08%**	81.07%	80.39%
	NLL		30.07	29.83	**27.49**	27.59	28.34
	$ACCofe_v$		87.47%	87.19%	88.60%	**89.91%**	87.17%

Varying Learning Rate. The learning rate is another critical parameter of TMRN, and we present the prediction accuracy of TMRN by varying the parameter from 0.00001 to 0.0002 for BJ and from 0.00001 to 0.0001 for PT in Table 5. Observed from which, the increase of learning rate firstly leads to the increase of prediction accuracy since TMRN avoids the local optimum in this progress. However, given a too large learning rate, it is difficult for TMRN to obtain the best performance, as the global optimum may be missed during the iteration. Consequently, the learning rate is set to 0.0001 for BJ and 0.00005 for PT in this study.

Varying Dropout. The dropout is varied from 0.04 to 0.1, and the corresponding results are presented in Table 6. Observed from which, a too small or large drop is not optimum for TMRN. This is because a small dropout may lead to the overfitting of a deep learning model. Meanwhile a large dropout may lead to the loss of information. Consequently, we set the dropout to 0.06 for both BJ and PT to achieve the highest prediction accuracy.

Table 6. Varying dropout

		Dropout	0.02	0.04	0.06	0.08	0.1
BJ	ACC		70.15%	70.26%	**72.83%**	70.23%	70.41%
	NLL		35.13	34.47	**30.10**	35.36	35.10
	$ACCofe_v$		49.36%	49.38%	**49.86%**	47.63%	47.27%
PT	ACC		80.79%	80.48%	**81.08%**	80.76%	80.70%
	NLL		27.99	30.30	**27.49**	27.93	27.98
	$ACCofe_v$		85.39%	86.15%	**88.60%**	88.54%	85.39%

Table 7. Varying embedding size

		36	54	72	90	108
	Embedding size	36	54	72	90	108
BJ	ACC	69.06%	70.83%	**72.83%**	70.45%	69.69%
	NLL	34.75	33.53	**30.10**	36.13	37.29
	$ACCofe_v$	47.38%	49.78%	**49.86%**	48.57%	41.36%
	Embedding size	144	180	216	252	288
PT	ACC	80.89%	70.83%	**81.08%**	81.03%	80.73%
	NLL	27.70	33.53	**27.49**	28.51	28.54
	$ACCofe_v$	86.45%	86.73%	**88.60%**	86.63%	86.73%

Varying Embedding Size. The results for varying embedding size are presented in Table 7. As expected, the increase of the embedding size initially leads to the increase of prediction accuracy as more information is captured by learned vectors in this progress. Meanwhile, a too-large vector may bring some noisy information and decrease prediction accuracy. As a result, the embedding size is set to 72 for BJ and 216 for PT.

6　Conclusion and Future Work

Understanding the mobility pattern of moving objects can benefit intelligent traffic management. As a building block of it, trajectory modeling has received increasing attention. To further improve the effectiveness of existing studies, we develop a novel model called TMRN that is composed of three components. Firstly, the module Road2vec is introduced to learn representations of road segments. Then, a lightweight self-attention-based module is designed to assign different weights for hidden vectors. Finally, a matching operation is proposed to predict the next road segment for the current path. The extensive experiments conducted on two real-world datasets demonstrate the higher performance of TMRN compared with that of state-of-the-art algorithms. In future work, the distance information between different road segments can be utilized to model trajectories more effectively.

Acknowledgement. This work is supported by the National Natural Science Foundation of China under Grant No. 61902270, and the Major Program of the Natural Science Foundation of Jiangsu Higher Education Institutions of China under Grant No. 19KJA610002.

References

1. Alahi, A., Goel, K., Ramanathan, V., Robicquet, A., Fei-Fei, L., Savarese, S.: Social LSTM: human trajectory prediction in crowded spaces. In: CVPR, pp. 961–971 (2016)

2. Ardakani, I., Hashimoto, K., Yoda, K.: Understanding animal behavior using their trajectories - a case study of gender specific trajectory trends. In: Streitz, N.A., Konomi, S. (eds.) HCI, pp. 3–22 (2018)

3. Banovic, N., Buzali, T., Chevalier, F., Mankoff, J., Dey, A.K.: Modeling and understanding human routine behavior. In: CHI, pp. 248–260 (2016)

4. Chen, M., Yu, X., Liu, Y.: Mining moving patterns for predicting next location. Inf. Syst. **54**, 156–168 (2015)

5. Devlin, J., Chang, M., Lee, K., Toutanova, K.: BERT: pre-training of deep bidirectional transformers for language understanding. In: NAACL-HLT, pp. 4171–4186 (2019)

6. Fu, T., Lee, W.: Trembr: exploring road networks for trajectory representation learning. Trans. Intell. Syst. Technol. **11**(1), 1–25 (2020)

7. Georgiou, H.V., Pelekis, N., Sideridis, S., Scarlatti, D., Theodoridis, Y.: Semantic-aware aircraft trajectory prediction using flight plans. Int. J. Data Sci. Anal. **9**(2), 215–228 (2020)

8. Groves, W., Nunes, E., Gini, M.L.: A framework for predicting trajectories using global and local information. In: Computing Frontiers Conference, pp. 37:1–37:10 (2014)

9. Kafsi, M., Grossglauser, M., Thiran, P.: Traveling salesman in reverse: conditional Markov entropy for trajectory segmentation. In: ICDM, pp. 201–210 (2015)

10. Li, B., et al.: T-PORP: a trusted parallel route planning model on dynamic road networks. IEEE Trans. Intell. Transp. Syst. (2022)

11. Li, M., Ahmed, A., Smola, A.J.: Inferring movement trajectories from GPS snippets. In: WSDM, pp. 325–334 (2015)

12. Liang, C., Berant, J., Le, Q.V., Forbus, K.D., Lao, N.: Neural symbolic machines: learning semantic parsers on freebase with weak supervision. In: ACL, pp. 23–33 (2017)

13. Liang, Y., Ouyang, K., Yan, H., Wang, Y., Tong, Z., Zimmermann, R.: Modeling trajectories with neural ordinary differential equations. In: IJCAI, pp. 1498–1504 (2021)

14. Nascimento, J.C., Figueiredo, M.A.T., Marques, J.S.: Trajectory classification using switched dynamical hidden Markov models. Trans. Image Process. **19**(5), 1338–1348 (2010)

15. O'Keeffe, K., Santi, P., Ratti, C.: Modeling vehicular mobility patterns using recurrent neural networks. CoRR abs/1910.11851 (2019)

16. Pecher, P., Hunter, M., Fujimoto, R.: Data-driven vehicle trajectory prediction. In: SIGSIM-PADS, pp. 13–22 (2016)

17. Puterman, M.L.: Markov Decision Processes: Discrete Stochastic Dynamic Programming. Wiley Series in Probability and Statistics. Wiley, Hoboken (1994)

18. Qian, C., Jiang, R., Long, Y., Zhang, Q., Li, M., Zhang, L.: Vehicle trajectory modelling with consideration of distant neighbouring dependencies for destination prediction. Int. J. Geogr. Inf. Sci. **33**(10), 2011–2032 (2019)

19. Qiao, Y., Si, Z., Zhang, Y., Abdesslem, F.B., Zhang, X., Yang, J.: A hybrid Markov-based model for human mobility prediction. Neurocomputing **278**, 99–109 (2018)

20. Sakuma, T., et al.: Efficient learning algorithm for sparse subsequence pattern-based classification and applications to comparative animal trajectory data analysis. Adv. Robot. **33**(3–4), 134–152 (2019)

21. Song, X., Zhang, Q., Sekimoto, Y., Horanont, T., Ueyama, S., Shibasaki, R.: Modeling and probabilistic reasoning of population evacuation during large-scale disaster. In: KDD, pp. 1231–1239 (2013)

22. Stecz, W., Gromada, K.: Determining UAV flight trajectory for target recognition using EO/IR and SAR. Sensors **20**(19), 5712 (2020)
23. Vaswani, A., et al.: Attention is all you need. In: NIPS, pp. 5998–6008 (2017)
24. Wang, D., Tan, P.: JOHAN: a joint online hurricane trajectory and intensity forecasting framework. In: KDD, pp. 1677–1685 (2021)
25. Wu, H., Chen, Z., Sun, W., Zheng, B., Wang, W.: Modeling trajectories with recurrent neural networks. In: IJCAI, pp. 3083–3090 (2017)
26. Xue, A.Y., Zhang, R., Zheng, Y., Xie, X., Huang, J., Xu, Z.: Destination prediction by sub-trajectory synthesis and privacy protection against such prediction. In: ICDE, pp. 254–265 (2013)
27. Xue, A.Y., Zhang, R., Zheng, Y., Xie, X., Yu, J., Tang, Y.: Desteller: a system for destination prediction based on trajectories with privacy protection. Proc. VLDB Endow. **6**(12), 1198–1201 (2013)
28. Yang, D., Fankhauser, B., Rosso, P., Cudré-Mauroux, P.: Location prediction over sparse user mobility traces using RNNs: flashback in hidden states. In: Bessiere, C. (ed.) IJCAI, pp. 2184–2190 (2020)
29. Yuan, J., et al.: T-drive: driving directions based on taxi trajectories. In: GIS, pp. 99–108 (2010)
30. Zheng, J., Ni, L.M.: Modeling heterogeneous routing decisions in trajectories for driving experience learning. In: UbiComp, pp. 951–961 (2014)
31. Ziebart, B.D., Maas, A.L., Dey, A.K., Bagnell, J.A.: Navigate like a cabbie: probabilistic reasoning from observed context-aware behavior. In: UbiComps, vol. 344, pp. 322–331 (2008)

Block-Join: A Partition-Based Method for Processing Spatio-Temporal Joins

Ting Li and Jianqiu Xu[✉]

Nanjing University of Aeronautics and Astronautics, Nanjing, China
{liting830,jianqiu}@nuaa.edu.cn

Abstract. Spatio-temporal joins are important operations in spatio-temporal databases. The rapid increase in the amount of spatio-temporal objects makes the cost of spatio-temporal joins expensive and requires an efficient method for spatio-temporal joins. In this paper, we propose a block-join method for spatio-temporal joins by partitioned blocks. We first partition the entire spatio-temporal data space of two trajectory datasets into equal-sized blocks. Spatio-temporal objects with similar spatio-temporal attributes will be split into the same block. To achieve a uniform distribution of trajectories inside a block called block trajectories, we merge two blocks into one block called unequal-sized blocks. Then, we evaluate block trajectories in the same and adjacent blocks to get pairs satisfying the spatio-temporal join conditions. The pairs are sorted and removed duplicated pairs to get precise results. Using both real and synthetic datasets, we carry out comprehensive experiments in a prototype database system to evaluate the efficiency of our methods. The experimental results show that our approach outperforms alternative methods in the system by a factor of 2-10x on large datasets.

Keywords: Spatio-temporal join · Trajectory data · Partition

1 Introduction

With the widespread use of GPS technology, a large amount of spatio-temporal data is collected [3,4,21]. Typically, spatio-temporal data are unevenly distributed, which brings great challenges to spatio-temporal operations. Spatio-temporal data can be described as a moving object containing spatial and temporal information which is stored in a spatio-temporal database. In spatio-temporal databases, spatio-temporal joins are important and expensive operations.

In this paper, we focus on the spatio-temporal join that combines two trajectory data sets according to the spatial and temporal predicates. Assuming that spatio-temporal objects move only in the x-direction over time, then we can display the trajectories in two-dimensional axes, which is shown in Fig. 1.

© The Author(s), under exclusive license to Springer Nature Switzerland AG 2023
B. Li et al. (Eds.): APWeb-WAIM 2022, LNCS 13423, pp. 397–411, 2023.
https://doi.org/10.1007/978-3-031-25201-3_30

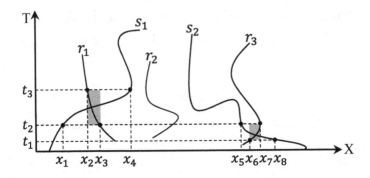

Fig. 1. Example of spatio-temporal join.

As an example in Fig. 1, given the relations $R = \{r_1, r_2, r_3\}$ and $S = \{s_1, s_2\}$ in which stores the trajectories of two trajectory data sets, respectively. For each pair of (r_i, s_j) returned as spatio-temporal join results needs to satisfy the following conditions: (i) $\exists$ a time interval $[t_i, t_j](i < j)$ that the time period of r_i and the time period of s_j overlap at $[t_i, t_j]$;(ii) the line $[x_m, x_n](m < n)$ of each trajectory r_i and s_j mapped to the two-dimensional space within $[t_i, t_j]$ intersects. Therefore, the pairs of (r_1, s_1) and (r_3, s_2) satisfy the spatio-temporal join in Fig. 1. For applications, we can find people who are close contacts to COVID-19 patients by analysing whether their activity time and movement trajectories intersect. Based on the close contacts, secondary contacts can be also determined by their spatio-temporal movement. According to the results of close contacts and secondary contacts, the Centers for Disease Control and Prevention can carry out relevant epidemic prevention work.

The straightforward method for answering spatio-temporal join is to iteratively evaluate each pair of trajectories from two data sets. However, the straightforward method is time-consuming and has high time complexity. Sorting trajectories according to their start time point is a good way to improve the efficiency of the straightforward method. If the start time point of $s_1 \in S$ does not intersect with the end time point of $r_1 \in R$, then $\{s_2, \ldots, s_n\} \subseteq S$ will not be compared to r_1, which will reduce the number of comparisons. Although sorting trajectories can organize trajectories by time dimension and improves the efficiency of spatio-temporal join, how to organize spatio-temporal objects according to the spatio-temporal similarity is a great challenge for spatio-temporal join. Only comparing the trajectories organized by the spatio-temporal similarity will efficiently perform the spatio-temporal join.

To this end, this paper makes several contributions. First, we propose the partitioning methods of equal-sized blocks and unequal-sized blocks. All the trajectories within a block are split into so-called block trajectories. Second, the block-join algorithms based on the partitioned blocks are proposed. Third, we conduct experiments on a synthetic dataset and a real dataset. The experimental evaluation shows that our proposed methods are efficient.

The rest of this paper is organized as follows. We briefly review related work in Sect. 2. In Sect. 3, we introduce the definitions of the problems studied. Then, we introduce our framework in Sect. 4. Our partitioning methods are described in Sect. 5 and block-join algorithms are proposed in Sect. 6. Experimental results are presented in Sect. 7, and we conclude in Sect. 8.

2 Related Work

Join operations in spatio-temporal databases can be categorized into three main categories: (i) *spatial joins*, (ii) *temporal joins* and (iii) *spatio-temporal joins*.

Spatial joins compute all pairs of spatial objects in two datasets satisfying a distance constraint. The paper [15] studied the following six algorithms that can perform spatial joins: S3 [10], SGrid [12], TOUCH [16], PBSM [17], PS [18], and NL [14] and finds that the data being real or synthetic, the data distribution, and the join order can influence the performance of algorithms. However, the partitioning methods in spatial joins only consider the data distribution in the spatial dimension without the time dimension.

For temporal joins, overlap joins determining pairs with overlapping intervals between two valid-time relations are common. OIP [6] is proposed to improve the efficiency of partitioning and indexing schemes for overlap joins. Besides overlap joins, an index-based Hilbert-TJ [19] is proposed for mapping temporal data into Hilbert curve space, which is clustered for a predictive spatio-temporal interval join. However, temporal joins are also only consider the time dimension.

Spatio-temporal joins take both spatial and temporal data space into account and return pairs satisfying spatial and temporal predicates as results. ST-kNN join [11] is proposed for considering both spatial closeness and temporal concurrency based on kNN. Spatio-temporal joins on historical indoor tracking data are proposed in [13]. TRSTJ [2] is proposed to find groups of moving objects that have followed similar movements in different times. The equal-sized partitioning to organize spatio-temporal objects based on both spatial and temporal data space has been studied in spatio-temporal joins. However, according to the distribution of the spatio-temporal objects, the unequal-sized partitioning based on three-dimensional space needs to be studied.

Different application scenarios of spatio-temporal joins are as follows. A method for spatio-temporal join which joins time-series grid data and time-series point data is proposed in [9], which is applied to estimate damage situations following a natural disaster. The paper [5] studied the trajectory similarity join aiming to find similar trajectory pairs from two large collections of trajectories. A model for spatio-temporal join selectivity estimation based on rigorous probabilistic analysis is developed in [20], which reveals factors that affect selectivity.

3 Problem Definition

We utilize a relational framework to define spatio-temporal data in which a spatio-temporal data type *mpoint* [8] is embedded as an attribute to represent spatio-temporal objects. That is, spatio-temporal objects are stored in a relation with schema $R(oId : int, Trip : mpoint)$. Each tuple in R states a spatio-temporal object identified by an attribute *Trip* which is typically defined by the data type *mpoint*. The *mpoint* is defined by a triple $(\langle t_i, t_j \rangle \ \langle x_i, y_i \rangle \ \langle x_j, y_j \rangle)$, where t_i refers to the start time point of a trajectory and t_j refers to the end time point of a trajectory. Then, $\langle x_i, y_i \rangle$ refers to the location at t_i and $\langle x_j, y_j \rangle$ refers to the location at t_j. We give an example for a relation R in Table 1.

Table 1. A relation R stores spatio-temporal objects.

oId: int	Trip: mpoint
1	$(\langle t_1, t_2 \rangle \ \langle x_1, y_1 \rangle \ \langle x_2, y_2 \rangle)$
2	$(\langle t_3, t_4 \rangle \ \langle x_3, y_3 \rangle \ \langle x_4, y_4 \rangle)$
3	$(\langle t_5, t_6 \rangle \ \langle x_5, y_5 \rangle \ \langle x_6, y_6 \rangle)$

A spatio-temporal join operator is performed on two relations storing spatio-temporal objects to pairs only those tuples that match in spatio-temporal attributes. That is, a tuple r from R and a tuple s from S are returned as a result pair only if r and s satisfy conditions of spatio-temporal join on attribute *Trip*.

Based on a relational framework, relational operators can be leveraged to formulate expression queries. The time period of a spatio-temporal object is $T(r.Trip)$ and the line mapped to the two-dimensional space of a spatio-temporal object is defined as $L(r.Trip)$. If two trajectories overlap within a time interval $t = [t_i, t_j](i < j)$, their lines mapped to the two-dimensional space within $[t_i, t_j]$ are defined as $L(r.Trip, t)$. Then, the spatio-temporal join over relations R and S is defined in Definition 1.

Definition 1. *Spatio-Temporal Join.* *Given two relations R and S, $R \bowtie S = \{(r, s) \mid r \in R \land s \in S \land T(r.Trip) \cap T(s.Trip) = t \land L(r.Trip, t) \cap L(s.Trip, t)\}$.*

If $T(r.Trip)$ and $T(s.Trip)$ of two spatio-temporal objects overlap within t, but their $L(r.Trip, t)$ and $L(s.Trip, t)$ do not intersect, this situation does not satisfy the spatio-temporal join. Only when two spatio-temporal objects satisfying both spatial and temporal conditions is spatio-temporal join.

4   The Framework

We outline the framework for processing spatio-temporal joins based on equal-sized blocks in Fig. 2. The procedure consists of three steps: partitioning the three-dimensional data space into blocks, producing block trajectories and performing the block-join algorithm. For further optimization, we merge equal-sized blocks which have few block trajectories inside into unequal-sized blocks to achieve a uniform distribution of block trajectories.

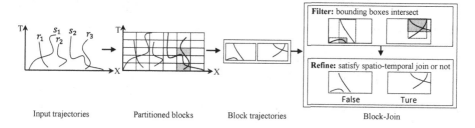

Fig. 2. The query process of equal-sized blocks.

Our spatio-temporal join framework supports processing both equal-sized blocks and unequal-sized blocks. The query procedure for unequal-sized blocks follows the same steps as processing equal-sized blocks but involves an extra step of merging equal-sized blocks into unequal-sized blocks.

We list frequently used notations throughout this paper in Table 2.

Table 2. Notations.

Notation	Description
R, S	Relations of trajectories
$T(r.Trip)$	The time period of a trajectory
$L(r.Trip)$	The line mapped to the two-dimensional space of a trajectory
$Box3d$	The entire 3-D minimum box of two trajectory relations
$numX$, $numY$, $numT$	Numbers of blocks on x, y and t axes
$blockNum$	Total numbers of blocks
B	The relation of blocks
TR, TS	Relations of block trajectories
sTR, sTS	Sorted relations in ascending order by bId of block trajectories
$bIdP$	The relation storing pairs of bId for the blocks to be compared

5   Partitioned Blocks

In order to organize trajectories from two trajectory data sets by spatio-temporal similarity, we partition the three-dimensional data space of two trajectory data sets into 3-D blocks. Each block represents a similar spatio-temporal data space. Only trajectories inside the same or adjacent blocks will be performed spatio-temporal join operation. Partitioned blocks can be classified into the following two types: (i) equal-sized blocks and (ii) unequal-sized blocks.

5.1   Equal-Sized Blocks

For each trajectory, we can get its corresponding 3-D minimum bounding box. We combine all the 3-D minimum bounding boxes of trajectories in R and S to get the entire 3-D boxes occupied by R and S, respectively. Then, the $Box3d$ used to be partitioned is combined with the 3-D boxes of R and S. If we only partition the 3-D box of one trajectory relation, there may exist trajectories of another trajectory relation that are out of the 3-D box. What is more, the results of the spatio-temporal join will be incomplete.

Then, we partition the $Box3d$ into 3-D equal-sized blocks. In three-dimensional space, the first block starts from the minimum spatio-temporal point $(minX, minY, minT)$ of $Box3d$. The next block will add a partition interval based on the values of $numX$, $numY$ and $numT$, which are the numbers of blocks on each axis. The partition interval of the x-axis is as follows.

$$xInterval = \lceil \frac{Box3d.maxX - Box3d.minX}{numX} \rceil \tag{1}$$

In the same way, we can get the partition intervals of y and t axes. Let B be a relation of blocks. Each $b \in B$ is denoted by $b(bId, block)$ in which $b.bId = \{1, \ldots, numX \times numY \times numT\}$. The numbering order of blocks defined as bId is traversing from the x-axis direction, then the y-axis direction, and finally the t-axis. Then, we give the definition of $block$ in Definition 2.

Definition 2. Block. *The block identified by an attribute block is defined as* $block = (\langle minX, minY, minT \rangle, \langle maxX, maxY, maxT \rangle)$ *which refers to the 3-D box occupied by a block.*

The trajectories from two trajectory data sets in Fig. 1 can be partitioned into equal-sized blocks in two-dimensional space, which is shown in Fig. 3.

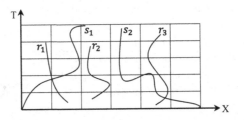

Fig. 3. Example of equal-sized blocks in two-dimensional space.

For the three-dimensional space, we give an example of $2 \times 2 \times 2$ blocks in Fig. 4(a) and an example of a block in Fig. 4(b).

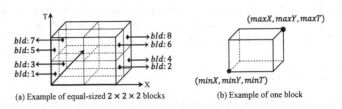

(a) Example of equal-sized $2 \times 2 \times 2$ blocks   (b) Example of one block

Fig. 4. Example of equal-sized blocks in three-dimensional space.

5.2 Block Trajectories

After partitioning the *Box3d* into equal-sized blocks, we split trajectories into block trajectories. Let *TR* be a relation of block trajectories for *R*. Each $tr \in TR$ is denoted by $tr(Id, bId, subTrip)$ in which $tr.Id$ refers to $r.Id$ of a trajectory and $tr.bId$ refers to bId of a block where tr is located in. The $tr.subTrip$ refers to the block trajectory which is given the definition in Definition 3.

Definition 3. Block Trajectory. *The block trajectory identified by an attribute subTrip which is defined by mpoint refers to the segments of a trajectory inside a block. The block trajectory needs to satisfy the following conditions: (i) $tr.subTrip \subseteq r.Trip$ (ii) $T(tr.subTrip) \subseteq T(r.Trip)$.*

We traverse all the tuples in relations *B* and *R* and the block trajectories satisfying that *r.Trip* has segments inside *b.block* will be returned. The segments of a trajectory outside a block will not be returned as a block trajectory. All the block trajectories from *R* are stored in a relation named *TR*. Similarly, we can get *TS* from *B* and *S*. Since we partition the 3-D box of two trajectory relations, no one trajectory will be missed. To sum up, we split a trajectory if and only if its 3-D minimum bounding box intersects with a block, which is considered in two cases shown in Fig. 5.

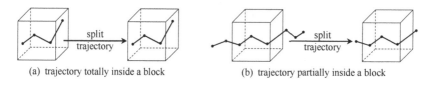

(a) trajectory totally inside a block   (b) trajectory partially inside a block

Fig. 5. Two cases of splitting trajectory.

For the first case in Fig. 5(a), the trajectory is totally inside a block, and all the segments of the trajectory are returned as block trajectory within the block. As for the second case in Fig. 5(b), the trajectory is partially inside a block, and segments of trajectory except totally outside the block are returned as segments of block trajectory. In order to ensure the accuracy of results for spatio-temporal join, we do not split the segments. Therefore, if a segment is partially inside the block, it is also returned as a segment of the block trajectory.

For partitioned blocks, we first partition the entire 3-D minimum bounding box of two relations into equal-sized blocks. In order to achieve a uniform distribution of block trajectories, we merge the blocks having few numbers of block trajectories on the basis of equal-sized blocks and get unequal-sized blocks.

5.3 Unequal-Sized Blocks

Based on equal-sized blocks, we propose the method of unequal-sized blocks to achieve a uniform distribution of block trajectories, which not only improves the running time of bad equal-sized blocks but also further improves the performance of good equal-sized blocks.

Instead of decomposing one block, we use the way of merging two blocks into one block to achieve a uniform distribution of block trajectories. Decomposing one block will increase the total number of blocks, which will increase the number of comparisons for spatio-temporal join and storage costs. Merging blocks is performed in the x-axis direction in this paper. In addition, merging blocks on the y-axis is also feasible. Since partitioning on the time dimension has a more important effect on spatio-temporal join than on the spatial dimension, we do not merge blocks on t-axis and maintain the partitioning granularity on t-axis.

During the process of partitioning, we count the number of block trajectories in each equal-sized block. Then, we determine the minimum number of block trajectories within a block. Traversing all the blocks, if the number of block trajectories within a block is less than or equal to the minimum number, this block will be merged into the next block and then deleted. For a block located at the right edge of the x-axis, it will be merged into its previous block. In addition, merging blocks can be performed multiple times until achieving a uniform distribution of block trajectories. An example in two-dimensional space shown in Fig. 6 is unequal-sized blocks of trajectories in Fig. 1.

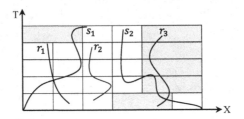

Fig. 6. Example of unequal-sized blocks.

For three-dimensional space, we show the process of merging blocks for $4 \times 2 \times 2$ blocks in Fig. 7. The adjacent blocks in the pair painted grey in Fig. 7(a) are two blocks that will be merged into one block.

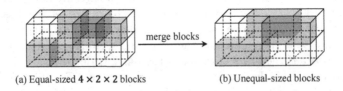

(a) Equal-sized **4 × 2 × 2** blocks   (b) Unequal-sized blocks

Fig. 7. Example of merging blocks in three-dimensional space.

5.4   Adjacent Blocks

For spatio-temporal join, block trajectories mostly satisfy the spatio-temporal join inside the same block. However, there are some block trajectories satisfying the spatio-temporal join at a common vertex, edge, or face of a block. To avoid missing results, we consider not only the spatio-temporal join of the same blocks but also adjacent blocks. For $3 \times 3 \times 3$ blocks, there are 26 adjacent blocks for the center block ($xId = 2, yId = 2, tId = 2$). The 26 adjacent blocks intersect with the center block at 8 vertices, 12 edges and 6 faces.

Therefore, before performing the spatio-temporal join, we have to know all the adjacent blocks of each block. We store the pairs of ($bId1$, $bId2$) into a relation $bIdP$ in which $bId1$ refers to the bId of a block and $bId2$ refers to the bId of its adjacent block. In order to perform the spatio-temporal join conveniently, we also pair the bId of a block and itself into $bIdP$ which can be represented as ($bId1$, $bId1$). An example of tuples in relation $bIdP$ which represents the pairs of $2 \times 2 \times 2$ blocks in Fig. 4 is given in Table 3.

Table 3. Example of a relation $bIdP$ for $2 \times 2 \times 2$ blocks.

$bId1$: int	$bId2$: int
1	$(1,1), (1,2), (1,3), (1,4), (1,5), (1,6), (1,7), (1,8)$
2	$(2,2), (2,3), (2,4), (2,5), (2,6), (2,7), (2,8)$
...	...
8	$(8,8)$

To avoid duplicate pairs of adjacent blocks, we store a pair of ($bId1$, $bId2$) if and only if $bId2$ is greater than $bId1$. The duplicate pairs, such as $(1,3)$ and $(3,1)$, will cause repeat comparisons during the process of spatio-temporal join.

6   Block-Join

We propose a block-join method to answer spatio-temporal join. There are two phases: *filter* and *refine*. Given two relations of block trajectories TR and TS, we compare all the pairs of block trajectories from two relations which are in the same and adjacent blocks. The pairs of block trajectories satisfying the conditions of spatio-temporal join will be removed duplicate pairs and returned as results.

6.1   Filter and Refine Phase

Using the 3-D minimum bounding box of each trajectory can roughly determine whether two trajectories intersect or not, which is called the filter phase. If the bounding boxes do not intersect, the trajectories can not intersect. However, the bounding boxes intersecting does not mean that the trajectories intersect. Therefore, we need the refine phase to do a precise comparison. The refine phase is performed by two parts which are spatio-temporal joins of same blocks and adjacent blocks. We compare all the segments from two block trajectories one by one, if two segments satisfy the conditions of spatio-temporal join, the pairs of Id for two block trajectories are returned as preliminary results. In order to get final results, we sort preliminary results and remove duplicate pairs.

6.2 Algorithms

We give an overview of the block-join algorithm in Algorithm 1.

Algorithm 1: Block-Join

---

    input : R, S: relations of trajectories
    output: W: results relation

1 Combine two 3-D boxes of R and S into $Box3d$
2 Partition $Box3d$ into equal-sized blocks
3 **if** *the distribution of block trajectories is non-uniform* **then**
4 | merge equal-sized blocks into unequal-sized blocks

5 Split trajectories into relations of block trajectories TR and TS
6 Locate *tupleId* of the range for block trajectories to be compared
7 Join TR and TS
8 **return** R;

Algorithm 2: Locating

---

    input : TR, TS: relations of block trajectories for R and S
    output: U, V: relations of representing the *tupleId* of bId for TR and TS

1 Sort TR and TS in ascending order by bId to get sTR and sTS
2 Initialize U, V and $bIdLocation$
3 **for** $i \leftarrow 1$ **to** $blockNum$ **do**
4 | **for** *all sorted block trajectories str in sTR* **do**
5 | | **if** $str.bId = i$ **then**
6 | | | $bIdLocation$++;
7 | $U \leftarrow U \cup U(bIdLocation)$;
8 repeat lines 3-7 for sTS and get V
9 **return** (U, V);

To reduce the number of comparisons for block-join, we sort relations TR and TS in ascending order by bId and record the *tupleId* corresponding to each bId in relations U and V, which is summarized in Algorithm 2. Therefore, the block-join does not have to compare all the block trajectories one by one. With the *tupleId* and sorted relations TR and TS, the block-join can directly locate the range of block trajectories that will be compared.

By the pairs of bId and sorted block trajectories sTR and sTS, we perform the spatio-temporal join operation on the same blocks and adjacent blocks which is summarized in Algorithm 3. The 3-D minimum bounding boxes are used to filter out pairs of block trajectories that are unlikely to intersect. If the pairs of block trajectories satisfy the conditions of spatio-temporal join, the pairs of $(tr.Id, ts.Id)$ corresponding to each block trajectory will be stored in the result relation. However, there are duplicate pairs in the results, such as block trajectories with $tr.Id = 3$ may intersect the block trajectories with $ts.Id = 5$ in two or more blocks, pairs of $(3, 5)$ may be stored in the result relation more than once. Therefore, we sort the result relation in ascending order by both $tr.Id$ and $ts.Id$. Then, we remove duplicate pairs and get the final results.

Algorithm 3: Join

---

    input : U, V: relations of $tupleId$
           sTR, sTS: sorted relations of block trajectories
           $bIdP$: relation of pairs ($bId1$, $bId2$)
           $blockNum$: total numbers of blocks
    output: W: result relation of pairs ($tr.Id$, $ts.Id$)

1 **for** *all pairs (bId1, bId2) of $bIdP_i$ in bIdP* **do**
2   **for** $j \leftarrow (U[bIdP_i.bId1 - 1] + 1)$ **to** $U[bIdP_i.bId1]$ **do**
3   **for** $k \leftarrow (V[bIdP_i.bId2 - 1] + 1)$ **to** $V[bIdP_i.bId2]$ **do**
4   **if** *the bounding boxes of tr and ts intersect* **then**
5   **if** $T(tr.subTrip) \cap T(ts.subTrip) = t$ and
                $L(tr.subTrip, t) \cap L(ts.subTrip, t)$ **then**
6                $W \leftarrow W \cup W((tr.Id, ts.Id))$;

7 Sort W and remove duplicate pairs in W;
8 **return** W;

---

7 Experimental Evaluation

7.1 Setup

We implement our methods in C++ and perform the evaluation in an extensible database system SECONDO [7]. A desktop PC (Intel(R) Xeon(R) E5-2609CPU, 1.9 GHz, 32 GB memory, 1 TB hard disk) running Ubuntu 20.04 (64 bits, kernel version 5.11.0) is used. We use a synthetic dataset named MyTrains and a real dataset including GPS records of taxis from Beijing [1] named Taxi to conduct our experiments. The overall running time is used as a performance metric.

The dataset Trains is a relation describing trips of underground trains in Berlin. MyTrains is enlarged from Trains by translating 20 times. MyTrains consists of 11,240 trajectories and 1,030,880 GPS record points. The trajectory datasets used for the spatio-temporal join are TrainsA and TrainsB, and their number of trajectories is half of the MyTrains. The dataset Taxi consists of 10,357 trajectories and the total number of GPS record points is about 15 million. The trajectory datasets Taxi1 and Taxi2 which are used to perform the spatio-temporal join contain half trajectories of Taxi, respectively.

Given two trajectory relations $R = \{r_1, \ldots, r_m\}$ and $S = \{s_1, \ldots, s_n\}$, which means that there are m tuples in R and n tuples in S. In an extensible database system SECONDO, spatio-temporal join is implemented by the operator *symmjoin* which is used to compare with our block-join methods. The operator *symmjoin* implements a symmetric, non-blocking, nested loop join technique and its complexity is $m \cdot n$ to perform the spatio-temporal join for R and S. Both the symmetric join and block-join return same results within a same dataset, which verifies the correctness of our methods.

We fix the value of $numT$ at 1, and change the values of $numX$ and $numY$ to correspond to the spatial partition. Then, we fix the values of $numX$ and $numY$ at 1, and change the value of $numT$, corresponding to the temporal partition. After determining the optimal partition value of the t-axis, change the values of $numX$ and $numY$ to implement the spatio-temporal partition.

7.2   Performance of Spatial-Temporal Join

Equal-Sized Blocks. According to the size of entire $Box3d$ for two trajectory datasets, we set different partition values of $numX$, $numY$, and $numT$ and perform the spatio-temporal join operation on the synthetic dataset. Therefore, we vary three parameters: (i) $numX = \{1,2,3,4,5\}$, (ii) $numY = \{1,2,3,4,5\}$ and (iii) $numT = \{9,18,27,36\}$.

First, we fix the values of $numX$ and $numY$ at 1 and vary the value of $numT$ from $\{9,18,27,36\}$. As shown in Fig. 8(a), when the value of $numT$ is equal to 18, the spatio-temporal join uses less overall running time. Then, we fix the value of $numT$ at 18 and vary the values of $numX \times numY$ from 1 to 25. The least overall running time is the value of $numX \times numY = 5$ in Fig. 8(b).

As shown in Fig. 8(a) and Fig. 8(b), the experimental results of only partitioning the space dimension or the time dimension are worse than the spatio-temporal partition for the synthetic dataset.

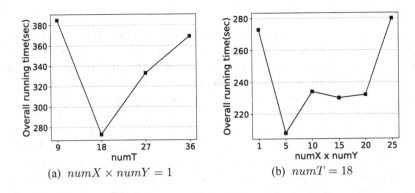

(a) $numX \times numY = 1$ (b) $numT = 18$

Fig. 8. Equal-sized blocks on the synthetic dataset.

Similarly, we set different partition values of $numX$, $numY$, and $numT$ of the real dataset. Then, we vary three parameters: (i) $numX = \{1,2,3,4,5\}$, (ii) $numY = \{1,2,3,4,5\}$ and (iii) $numT = \{5,10,15,20\}$ and perform the spatio-temporal join operation.

First, we fix the values of $numX$ and $numY$ at 1 and vary the value of $numT$ from $\{5,10,15,20\}$. As shown in Fig. 9(a), when the value of $numT$ is equal to 15, the spatio-temporal join uses less overall running time. Then, we fix the value of $numT$ at 15 and vary the values of $numX \times numY$ from 1 to 25. The least overall running time is the value of $numX \times numY = 10$ in Fig. 9(b).

As shown in Fig. 9(a) and Fig. 9(b), the experimental results of the spatio-temporal partition are better than only partitioning the space dimension or the time dimension for the real dataset.

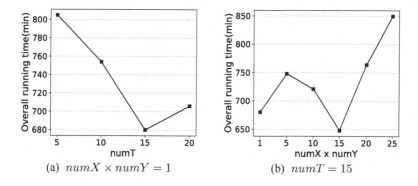

(a) $numX \times numY = 1$ (b) $numT = 15$

Fig. 9. Equal-sized blocks on the real dataset.

According to the experimental results, the spatial-temporal partition is better than a single spatial partition or temporal partition for the spatio-temporal join.

Unequal-Sized Blocks. We merge equal-sized blocks in Fig. 8(b) into unequal-sized blocks on the synthetic dataset. Performing the spatio-temporal join based on unequal-sized blocks and the overall running time is shown in Fig. 10(a). After merging blocks to achieve a uniform distribution of block trajectories, the efficiency of all equal-sized blocks is shorter than equal-sized blocks. The overall running time between the symmetric join and our block-join is shown in Fig. 10(b). Our block-join method based on partitioned blocks is better than the symmetric join.

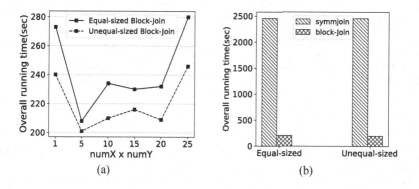

(a)   (b)

Fig. 10. Comparisons on the synthetic dataset.

As for the real dataset, we also perform the spatio-temporal join using both equal-sized blocks and unequal-sized blocks and unequal-sized blocks perform better. The comparisons between equal-sized blocks and unequal-sized blocks are shown in Fig. 11(a). Our block-join algorithms outperform the symmetric join by a factor of 2-3x on the real dataset which is shown in Fig. 11(b).

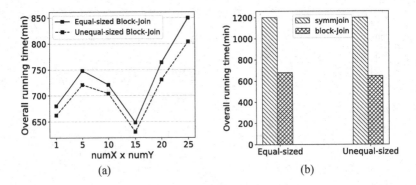

Fig. 11. Comparisons on the real dataset.

8   Conclusions

We studied spatio-temporal joins of spatio-temporal data. A spatio-temporal join method based on partitioned blocks called block-join is proposed to efficiently perform the spatio-temporal join. Partitioning the entire spatio-temporal data space of two spatio-temporal data sets into equal-sized blocks is to organize the spatio-temporal data within a spatio-temporal similarity. In order to achieve a uniform distribution of block trajectories, we merge two equal-sized blocks into one block called unequal-sized blocks. We develop block-join algorithms for block trajectories in the same and adjacent blocks which are paired if and only if two block trajectories satisfy conditions of spatio-temporal joins. We sort pairs and remove duplicate pairs to get results. Experimental results demonstrated that our method outperforms alternative methods. The future work is to consider the automatic setting of partition values for partitioned blocks.

Acknowledgements. This work is supported by National Natural Science Foundation of China (61972198), Natural Science Foundation of Jiangsu Province of China (BK20191273).

References

1. https://www.microsoft.com/en-us/research/publication/t-drive-trajectory-data-sample/
2. Bakalov, P., Hadjieleftheriou, M., Tsotras, V.J.: Time relaxed spatiotemporal trajectory joins. In: Proceedings of the 13th Annual ACM International Workshop on Geographic Information Systems, pp. 182–191 (2005)

3. Chen, H., Du, J., Zhang, W., Li, B.: An iterative end point fitting based trend segmentation representation of time series and its distance measure. Multimedia Tools Appl. 13481–13499 (2020). https://doi.org/10.1007/s11042-019-08440-0
4. Dai, T., Li, B., Yu, Z., Tong, X., Chen, M., Chen, G.: PARP: a parallel traffic condition driven route planning model on dynamic road networks. ACM Trans. Intell. Syst. Technol. (TIST) 12(6), 1–24 (2021)
5. Dan, T., Luo, C., Li, Y., Zheng, B., Li, G.: Spatial temporal trajectory similarity join. In: Asia-Pacific Web (APWeb) and Web-Age Information Management (WAIM) Joint International Conference on Web and Big Data, pp. 251–259 (2019)
6. Dignös, A., Böhlen, M.H., Gamper, J.: Overlap interval partition join. In: Proceedings of the 2014 ACM SIGMOD International Conference on Management of Data, pp. 1459–1470 (2014)
7. Güting, R.H., Behr, T., Düntgen, C.: SECONDO: a platform for moving objects database research and for publishing and integrating research implementations. IEEE Data Eng. Bull. 33(2), 56–63 (2010)
8. Güting, R.H., et al.: A foundation for representing and querying moving objects. ACM Trans. Database Syst. (TODS) 25(1), 1–42 (2000)
9. Hayashi, H., Asahara, A., Sugaya, N., Ogawa, Y., Tomita, H.: Spatio-temporal join technique for disaster estimation in large-scale natural disaster. In: Proceedings of the 6th ACM SIGSPATIAL, pp. 49–58 (2015)
10. Koudas, N., Sevcik, K.C.: Size separation spatial join. In: Proceedings of the 1997 ACM SIGMOD, pp. 324–335 (1997)
11. Li, R., et al.: Distributed spatio-temporal k nearest neighbors join. In: Proceedings of the 29th International Conference on Advances in Geographic Information Systems, pp. 435–445 (2021)
12. Lo, M.L., Ravishankar, C.V.: Spatial hash-joins. In: Proceedings of the 1996 ACM SIGMOD International Conference on Management of Data, pp. 247–258 (1996)
13. Lu, H., Yang, B., Jensen, C.S.: Spatio-temporal joins on symbolic indoor tracking data. In: 2011 IEEE 27th International Conference on Data Engineering, pp. 816–827 (2011)
14. Mishra, P., Eich, M.H.: Join processing in relational databases. ACM Comput. Surv. (CSUR) 24(1), 63–113 (1992)
15. Nobari, S., Qu, Q., Jensen, C.S.: In-memory spatial join: the data matters! In: 20th International Conference on Extending Database Technology: EDBT/ICDT 2017 Joint Conference (2017)
16. Nobari, S., Tauheed, F., Heinis, T., Karras, P., Bressan, S., Ailamaki, A.: Touch: in-memory spatial join by hierarchical data-oriented partitioning. In: Proceedings of the 2013 ACM SIGMOD International Conference on Management of Data, pp. 701–712 (2013)
17. Patel, J.M., DeWitt, D.J.: Partition based spatial-merge join. ACM SIGMOD Rec. 25(2), 259–270 (1996)
18. Preparata, F.P., Shamos, M.I.: Computational geometry: an introduction (2012)
19. Raigoza, J., Sun, J.: Temporal join processing with hilbert curve space mapping. In: Proceedings of the 29th Annual ACM Symposium on Applied Computing, pp. 839–844 (2014)
20. Sun, J., Tao, Y., Papadias, D., Kollios, G.: Spatio-temporal join selectivity. Inf. Syst. 31(8), 793–813 (2006)
21. Ulm, G., Smith, S., Nilsson, A., Gustavsson, E., Jirstrand, M.: OODIDA: on-board/off-board distributed real-time data analytics for connected vehicles. Data Sci. Eng. 6(1), 102–117 (2021)

RN-Cluster: A Novel Density-Based Clustering Approach for Road Network Partition

Yingying Ding and Jianqiu Xu[✉]

Nanjing University of Aeronautics and Astronautics, Nanjing, China
{dyy_97,jianqiu}@nuaa.edu.cn

Abstract. As an indispensable part of urban road management, road network partition is essential for the design and planning of road systems. In this paper, we focus on the problem of urban road network partitioning and propose a density-based clustering method RN-Cluster. Specifically, by taking road junctions as vertices, the relationships between road junctions and road sections as edges to construct the road knowledge graph, our method can effectively infer the implicit information of each road or junction. In addition, we propose a measurement of attenuation of neighborhood connectivity (ANC) as the density measurement to select core vertices. Our experiments show that the proposed method can efficiently divide the urban road network into different reasonable areas while achieving state-of-the-art performance on multiple metrics.

Keywords: Clustering method · Spatial partition · Road graph

1 Introduction

With the acceleration of urbanization, urban road networks are becoming increasingly complex. The proper functioning of road networks is fundamental for development of daily socio-economic activities and the sustainability of modern communities [17]. As an important step of urban traffic management, road network partitioning is closely related to the problems of urban traffic congestion [11,15,19].

In intelligent transportation [24], clustering is often used to partition road networks. However, at present, the partitioning of urban road networks poses three significant challenges. Firstly, the partition of a road network is not only the mining of vertex set knowledge. When clustering road networks, spatial adjacency constraints should also be considered to aggregate road networks into spatially connected homogeneous subareas. Secondly, the algorithm need to be robust and can adapt to different scale road networks and ensure efficiency in a large-scale road network. Thirdly, the generated output partitions need to be effectively adapted to the regional relationship of road networks with different shapes.

At present, some scholars have divided the road network in combination with the knowledge of graph theory. The graph partition methods [3–5,12] mainly focus on connection relationships without considering the geographical locations of junctions and length of roads. This could lead to a massive lack of information utilization if only the connection relationship is considered.

In this paper, we propose a novel density-based road network partition method. Specially, we use road junctions as vertices and road sections as edges to build an urban road graph. Then, we can extract core vertices according to spatial information by using the proposed attenuation of neighborhood connectivity (ANC) as the measure of density. Finally, we can obtain a set of subgraphs through further expansion and aggregation, which can be organized into final road network partitions. The contributions of this paper are as follows:

- We propose a density-based clustering approach, i.e., RN-Cluster, to partition the road network. Experimental results show that our method can efficiently divide the urban road network into reasonable, evenly distributed, and spatially compact areas and infer the implicit knowledge of general road network data.
- We propose the Attenuation of Neighborhood Connectivity (ANC) as the density measurement in RN-Cluster. This factor captures the change of road density ratio in a specific area within a certain distance change, which can effectively distinguish core vertices from road graphs.

2   Related Work

In traffic control management, it is widely accepted that potential strategies to alleviate traffic congestion should be designed based on the sub-partition of the urban road network [7]. Some scholars use clustering algorithms to study road network partition. For example, Yagoda et al. [22] put forward 'intersection coupling index', which considered the link volume and link length between adjacent intersections. Feng et al. [8] proposed a sub-area merging model based on a two-dimensional graph theory clustering algorithm for road network merging and used F-test to determine the optimal merging results. Pascale et al. [16] proposed a homogeneous sub-area detection method based on the spatial-temporal clustering method. Lopez et al. [14] proposed density-based spatial and data point clustering post-processing methods by using Normalised cut, DBSCAN, and Growing Nerve Gas (GNG). [13,21] proposed road network partition methods based on Kmeans clustering and improved FCM algorithm and compared their advantages and disadvantages through actual road network analysis. However, the k value of the Kmeans algorithm is preset, which is difficult to estimate and does not have universality. Yu et al. [23] proposed a road network partition method based on community detection. They combined the travel speed correlation between road sections and the fast unfolding method to divide the urban road network into sub-partitions of densely correlated road sections.

Unlike other road network partitioning methods, RN-Cluster aims to make up for the shortcomings of traditional clustering algorithms that do not consider the connectivity of related road sections, which has more practical implications for overall structural analysis.

3 Proposed Method

3.1 Framework

Figure 1 shows the general flow of our approach. We first build a road graph $\mathcal{G}$ from the standard road network $\mathcal{N}$ to achieve lower coupling. Then, we use the proposed density-based approach to cluster the road junctions and sections. Finally, the road graph $\mathcal{G}$ is divided to obtain the final subgraphs $\{\mathcal{G}_1, \mathcal{G}_2, \ldots, \mathcal{G}_K\}$.

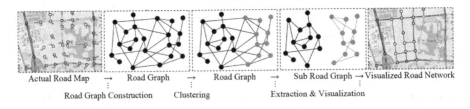

Actual Road Map   →   Road Graph   →   Road Graph   →   Sub Road Graph → Visualized Road Network

Road Graph Construction   Clustering   Extraction & Visualization

Fig. 1. Framework sketch of our proposed RN-Cluster.

3.2 Density-Based Clustering of Road Graph

Algorithm 1 shows the main clustering process. First, we initialize the attributes of vertices (Algorithm 1, line 1), then we start to traverse each vertex. If the vertex is not accessed, we obtain its neighbor vertex set, as shown in Eq. 1 and calculate the neighborhood connectivity (ANC) of the vertex. If the ANC is greater than threshold ρ, we set this vertex as a core vertex, then add it to a new cluster and use the *expandCluster* () function to expand this cluster. If the above conditions are not met, we classify the vertex as a noise (Algorithm 1, lines 2–12). Function *expandCluster* () is the iterative recursive process of Algorithm 1, lines 2–12, we just don't create a new cluster when expanding. At last, Algorithm 1 returns the clustering results set $\mathcal{C}$.

The latitude and longitude describe the geographical location on a vast scale, these values change subtly within a city. We using meters to measure the distance to calculate the actual distance between junctions.

We use the A^* algorithm to calculate the shortest path. In addition, we add an *R-Tree* index to the input data. When calculating the distance, we only calculate the data within the bounding box range instead of traversing all vertices which can significantly reducing the time overhead.

3.3 The Attenuation of Neighborhood Connectivity (ANC)

In Algorithm 1, line 6, we use the Attenuation of Neighborhood Connectivity (ANC) as the density measurement to select core vertices. The ANC factor uses the degree of the shrinking neighborhood and expanding neighborhood as the

---

Algorithm 1: Road Graph Clustering

Input: Road graph $\mathcal{G}$, which contains set of vertices $\mathcal{V}$, set of relations $\mathcal{E}$.
Distance parameter ϵ. Neighborhood density threshold ρ.
Output: Clustering results $\mathcal{C}$

1 Initial $v.clusterNo = -1, v.visited = false, v.isCore = false$ for every $v \in \mathcal{V}$;
2 **for** v *in* $\mathcal{V}$ **do**
3   **if** $v.visited$ *is False* **then**
4           $v.visited = True$;
5           $\mathcal{V}_{nbr} = F_{nbr}(v, \epsilon)$;
6   **if** $ANC(v) \geq \rho$ **then**
7               $v.isCore = true$;
8   Create a cluster $\mathcal{C}_i$, add v to $\mathcal{C}_i$;
9               $expandCluster(\mathcal{C}_i, \mathcal{V}_{nbr})$;
10   **end**
11   **end**
12 **end**

---

basis and captures the change of road density ratio in a specific area within a certain distance change. The ANC is defined as follows:

$$F_{nbr}(v_i, \epsilon) = \{v | d(v, v_i) < \epsilon, v \in \mathcal{V}\} \tag{1}$$

$$ANC(v_i) = \frac{\sum_{v \in F_{nbr}(v_i, \epsilon)} deg^+(v)}{\sum_{v \in F_{nbr}(v_i, \epsilon + \triangle\epsilon)} deg^+(v)} \tag{2}$$

where $\mathcal{V}$ represents the vertex set in road graph $\mathcal{G}$, $d(v, v_i)$ represents the reachable distance between vertices, $\epsilon + \triangle\epsilon$ means the scale of the expanded area. In our experiments, we set $\triangle\epsilon = 10^3$.

4 Experimental Results

In this paper, we use an open-source, object-oriented database system SECONDO [10] as our experimental platform. We use Nanjing road network data which is downloaded from the open-source online map data platform OpenStreetMap (OSM) [20]. As the second-largest city in the East China region, Nanjing can be significantly typical and representative of road traffic analysis. Since undefined roads or small paths can have a negative impact on the results, we selected several road types from all roads and conducted our experiments on these roads. The selected road types are "Motorway", "Trunk", "Primary", "Secondary", "Tertiary", "Residential" and "Road", contains 18,451 roads in total.

4.1 Evaluation Metrics

We use different quantitative metrics to evaluate the proposed methodology's performance.

Considering the cluster connectivity between road junctions in the road network, we use the modularity [1] metric, which is defined as follows:

$$Q = \frac{1}{2m} * \sum_{i \neq j} \left(A_{ij} - \frac{k_i * k_j}{2m} \right) \delta \left(\mathcal{C}_i, \mathcal{C}_j \right) \tag{3}$$

where A_{ij} represents any element of the adjacency matrix corresponding to the whole network. m represents the total number of edges of the network and k_i represents the degree of vertex i. $\delta \left(\mathcal{C}_i, \mathcal{C}_j \right)$ is the *Kronecker* function.

We use Complementary cumulative distribution function (CCDF) [9] as a measurement of partition uniformity. Let the number of vertices be a random variable N, we can define CCDF as follows:

$$CCDF(n) = \tilde{F}_N \left(n \right) = 1 - F_N \left(n \right) = P \left(N > n \right) \tag{4}$$

Silhouette score [18] is another classic metric used to evaluate the quality of clusters in terms of how well samples are clustered with other samples that are similar to each other. The specific calculation of Silhouette score is shown below:

$$Score = \frac{b - a}{max \left\{ a, b \right\}} \tag{5}$$

where a and b represent mean intra-cluster distance and mean nearest-cluster distance, respectively. We use reachable distance instead of Euclidean distance to calculate the distance between junctions.

4.2   Experimental Evaluation

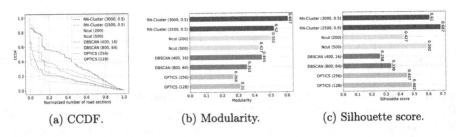

(a) CCDF.   (b) Modularity.   (c) Silhouette score.

Fig. 2. Performance of baseline algorithms. The values in parentheses represent the parameters of the method, i.e., RN-Cluster (ϵ, ρ), Ncut(α), DBSCAN ($\epsilon, minPts$) and OPTICS (n).

In this subsection, we show the performance of our method by evaluating the metrics listed in the Subsect. 4.1. We compare our RN-Cluster with a group of representative clustering approaches, i.e., Ncut [12], DBSCAN [6] and OPTICS [2] to comparatively demonstrate the effectiveness of our method. After a wide range of parameter adjustments, we chose the best two groups of parameters for an objective comparison.

Figure 2(a) shows the CCDF of the normalized number of road sections in partitions. As described above, the CCDF metric quantifies the overall partitioning uniformity. Suppose the curve is closer to a linear decline. In that case, it indicates that the road sections are more evenly distributed in network partitions. Figure 2 (b) and (c) shows the modularity and silhouette score of methods respectively. These figures clearly illustrate that our method can achieve better clustering performance than others while enjoying a higher-level partition uniformity.

4.3   Visualization of Partition Results

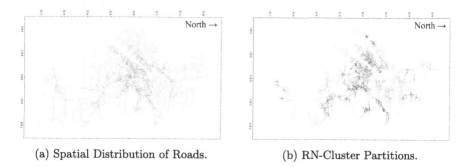

(a) Spatial Distribution of Roads.   (b) RN-Cluster Partitions.

Fig. 3. Visualization of results. The x-axis and y-axis represent the latitude and longitude respectively.

Figure 3 shows a visualization of Nanjing road network partition results. Figure 3(a) shows the road network structure of Nanjing city, which is the input of our proposed framework. We can observe that the Nanjing road network has a large number of roads and junctions, and its road network structure is complicated. Figure 3(b) demonstrates one of the results of our method. Specifically, we draw clusters with different colors and set noise roads to dotted lines.

As shown in Fig. 4, the partitioned region shape can change arbitrarily and semantically with the expansion of density. For example, (a) in Fig. 4 basically corresponds to the old urban area of Nanjing city, (d) corresponds to traffic arteries centered on the Qilin highway hinge, (e) corresponds to the urban or residential area of Lishui District and (f) corresponds to the passenger transport area centered on Lishui Railway Station. Furthermore, with ANC's effective core detection performance, the untidy and unimportant path within the cluster can be automatically detected and filtered as noise. In other words, unlike other network partition methods, RN-Cluster reduces the importance of the old hutongs within the urban road network, which is of more practical significance for overall structure analysis.

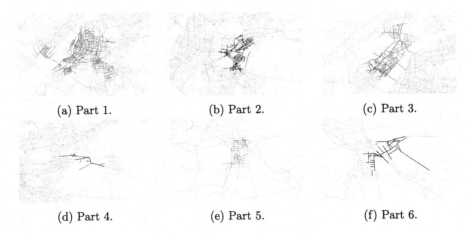

(a) Part 1.   (b) Part 2.   (c) Part 3.

(d) Part 4.   (e) Part 5.   (f) Part 6.

Fig. 4. Example of partitions. Each subfigure represents a partitioned road network region.

5   Conclusion

In this paper, we propose a novel density-based method RN-Cluster for large-scale road network partition and use a new factor to measure the attenuation of neighborhood connectivity. Our experiments show that our proposed method can efficiently divide the urban road network into different reasonable areas and infer the implicit knowledge of each road or junction and the relationships between them. RN-Cluster can generate evenly distributed and spatially compact partitions while performing well on multiple metrics, which is of great value and potential in practical application.

Acknowledgements. This work is supported by National Natural Science Foundation of China (61972198), Natural Science Foundation of Jiangsu Province of China (BK20191273).

References

1. Aaron, C., Newman, M., Cristopher, M.: Finding community structure in very large networks. Phys. Rev. E **70**(6 Pt 2), 066111 (2004)
2. Ankerst, M., Breunig, M.M., Kriegel, H.P., Sander, J.: Optics: ordering points to identify the clustering structure. ACM SIGMOD Rec. **28**(2), 49–60 (1999)
3. Anwar, T., Liu, C., Vu, H.L., Leckie, C. Partitioning road networks using density peak graphs: efficiency vs. accuracy. Inf. Syst. **64**, 22–40 (2017)
4. Blondel, V.D., Guillaume, J.L., Lambiotte, R., Lefebvre, E.: Fast unfolding of communities in large networks. J. Stat. Mech.: Theory Exp. **10**, P10008 (2008)
5. Dai, T., et al.: Parp: a parallel traffic condition driven route planning model on dynamic road networks. ACM Trans. Intell. Syst. Technol. (TIST) **12**(6), 1–24 (2021)

6. Ester, M., Kriegel, H.P., Sander, J., Xu, X., et al.: A density-based algorithm for discovering clusters in large spatial databases with noise. In: KDD, vol. 96, pp. 226–231 (1996)
7. Etemadnia, H., Abdelghany, K., Hassan, A.: A network partitioning methodology for distributed traffic management applications. Transportmet. A Transp. Sci. 10(6), 518–532 (2014)
8. Feng, S., Ma, D.: Two-dimensional graphic theory on clustering method of small traffic zones. J Harbin Insti. Technol.(2015)
9. Gutenberg, P.: Complementary cumulative distribution function (2007)
10. Güting, R.H., et al.: SECONDO: an extensible DBMS platform for research prototyping and teaching. In: ICDE, pp. 1115–1116 (2005)
11. Islam, M.R., Jenny, I.J., Nayon, M., Islam, M.R., Amiruzzaman, M., Abdullah-Al-Wadud, M.: Clustering algorithms to analyze the road traffic crashes. CoRR abs/2108.03490 (2021)
12. Ji, Y., Geroliminis, N.: On the spatial partitioning of urban transportation networks. Transp. Res. Part B Methodol. 46(10), 1639–1656 (2012)
13. Lin, X., Xu, J.: Road network partitioning method based on canopy-kmeans clustering algorithm. Arch. Transp. 54(2), 95–106 (2020)
14. Lopez, C., Krishnakumari, P., Leclercq, L., Chiabaut, N., van Lint, H.: Spatiotemporal partitioning of transportation network using travel time data. Transp. Res. Rec. 2623(1), 98–107 (2017)
15. Mahajan, R., Mansotra, V.: Predicting geolocation of tweets: using combination of CNN and BILSTM. Data Sci. Eng. 6(4), 402–410 (2021)
16. Pascale, A., Mavroeidis, D., Lam, H.T.: Spatiotemporal clustering of urban networks: Real case scenario in London. Transp. Res. Rec. 2491(1), 81–89 (2015)
17. Rivera-Royero, D., Galindo, G., Jaller, M., Betancourt Reyes, J.: Road network performance: a review on relevant concepts. Comput. Ind. Eng. 165, 107927 (2022)
18. Rousseeuw, P.J.: Silhouettes: a graphical aid to the interpretation and validation of cluster analysis. J. Comput. Appl. Math. 20, 53–65 (1987)
19. Song, X., Li, B., Dai, T., Tian, J.: A trust management-based route planning scheme in LBS network. In: International Conference on Advanced Data Mining and Applications, ADMA 2022. LNCS, vol . 13087. Springer, Champp. 307–322. Springer, Cham (2022). https://doi.org/10.1007/978-3-030-95405-5_22
20. Vargas-Munoz, J.E., Srivastava, S., Tuia, D., Falcão, A.X.: Openstreetmap: challenges and opportunities in machine learning and remote sensing. IEEE Geosci. Remote Sens Mag. 9(1), 184–199 (2021)
21. Wang, X.X.: The partition of urban traffic network and classification of traffic status based on clustering. Ph.D. thesis, Beijing Jiaotong University (2017)
22. Yagoda, H.N., Principe, E.H., Vick, C.E., Leonard, B.D.: Subdivision of signal systems into control areas (1973)
23. Yu, Q., Li, W., Yang, D., Zhang, H.: Partitioning urban road network based on travel speed correlation. Int. J. Transp. Sci. Technol. 10(2), 97–109 (2021)
24. Yuan, H., Li, G.: A survey of traffic prediction: from spatio-temporal data to intelligent transportation. Data Sci. Eng. 6(1), 63–85 (2021)

Fine-Grained Urban Flow Inferring via Conditional Generative Adversarial Networks

Xv Zhang, Yuanbo Xu(✉), Ying Li, and Yongjian Yang

Mobile Computing Intelligent (MIC) Laboratory, Key Laboratory of Symbolic Computation and Knowledge Engineering for the Ministry of Education, Jilin University, Changchun, China
xuz20@mails.jlu.edu.cn, {yuanbox,liying,yyj}@jlu.edu.cn

Abstract. Urban flow super-resolution (UFSR) can deduce fine-grained urban flow heatmap (UFH) based on coarse-grained observations and plays an essential role in urban planning (traffic prediction, public facility deployment, for instance). However, existing methods fail to capture the internal structural features of sparse UFHs and the external factors that lead to a significant waste of urban resources. To this end, we propose an enhanced super-resolution framework (Urban Flow-aware Super Resolution - Generative Adversarial Network, UrbanSG) to deduce fine-grained UFH for urban resource allocation. Specifically, we employ a conditional-GAN as the backbone, considering external factors as the specified condition. To capture the implicit urban structural correlation, we integrate the flow self-attention mechanism into our model, which focuses on urban grids with active traffic volumes. The evaluations of extensive experiments on two real-world datasets demonstrate the superiority of our framework. Especially when dealing with a sparse dataset, our method reduces error by 15.02% to the state-of-the-art baselines.

Keywords: Urban computing · Super resolution · Deep learning · Attention mechanism

1 Introduction

With the rapid development of intelligent terminals and wireless communication, an unprecedented amount of data springs up, improving numerous essential applications in the city, such as intelligent traffic management and public facility deployment. To sufficiently obtain fine-grained data, an intelligent urban system requires many sensing devices to cover the entire urban landform, which simultaneously imposes significant O&M (operation and maintenance) overhead and becomes a prohibitive factor for the global intelligence without prompt resource alignment strategies and proper allocation method. Hence, reconstructing the fine-grained urban situation through available coarse-grained observation becomes an urgent issue.

Urban flow super-resolution (UFSR) aims at inferring fine-grained traffic flows with an available coarse-grained observation. UFSR has brought significant urban planning and traffic management improvements as a variant and application of image SR for

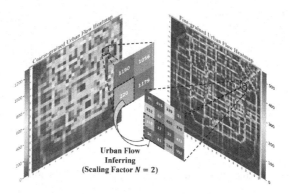

Fig. 1. UFSR: coarse-grained to fine-grained.

transportation. For example, recovering fine urban traffic flow volume with coarse-grained observed data captured from limited traffic monitoring devices reduces redundant resource allocation. Although UFSR is closely related to image SR, it confronts a unique structural constraint. Specifically, the sum of the fine-grained traffic volumes of a super-resolution region is strictly equal to that of the corresponding coarse-grained super-resolution region, as shown in Fig. 1. In other words, we consider UFSR as a mapping problem, which maps low information entropy data to that of high information entropy [11].

Nevertheless, in the complex context of the smart city, UFSR still suffers from the following challenges:

1. External Factors: Urban flow distribution is generally affected by external factors, such as time and weather. For example, office area usually has a higher density than attractions on weekdays, while the opposite is true on weekends; People prefer to be at home on stormy days and outside on sunny days. Hence, the same functional area shows various flow distributions under varying external factors. Without adequate consideration of external factors, the model will not yield a well-detailed fit.
2. Flow distribution sparsity: The high sparsity of UFH in the real world is often insufficient to support existing FSR models. Figure 2(a) shows that the base stations (represented by red dots) cannot be assigned to all urban grids due to the limited quantity and uneven distribution, which leads to data sparsity. Figure 2(b) reflects the sparsity difference under different time slices. The model without sparse data processing capability will degrade performance when facing real-world urban data.

To this end, we devise an enhanced super-resolution framework that can deduce fine-grained UFH from a coarser one. To the best of our knowledge, UrbanSG is the first attempt to exploit CGAN to complete the works above. Notably, the contributions of our work lie in the following aspects:

1. To jointly consider the city's internal relations and external factors, we present UrbanSG, a novel deep neural network model. It employs a CGAN as a backbone that considers external factors as the city's specific condition.

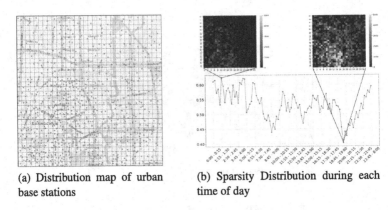

(a) Distribution map of urban base stations

(b) Sparsity Distribution during each time of day

Fig. 2. Mobile signaling data in Changchun.

2. We integrate the self-attention mechanism into our model, which conducts UrbanSG focus on the active urban grids and effectively alleviates the sparsity problem.
3. We process, analyze, and experiment in two urban scenarios. Our experimental results verify the significant advantages of UrbanSG in both effectiveness and efficiency. Moreover, the experiments from multiple perspectives validate the rationale for different components of our model.

2   Related Work

2.1   Image Super Resolution

Image super-resolution (ISR) aims to recover low-resolution images to the original high-resolution images, which has emerged as a popular research topic in CV for decades and has a promising application in image recognition [5, 14], image augmentation [2, 8], and urban computing [11]. Over the years, researchers have successfully proposed many ISR algorithms, mainly categorized into super-resolution recovery [17, 18] and super-resolution reconstruction [1]. To fully utilize the hierarchical features on residual branches, Liu et al. [12]proposed a novel residual feature aggregation framework for more efficient feature extraction. Similarly, for optimizing the residual networks, [19] introduced the global context module to streamline the residual network effectively and [22] proposed a connection block group to prevent an excess of residual network parameters by distillation technique and attention mechanism. Yang et al. [18] proposed a novel texture transformer network consisting of four closely-related modules for texture transfer and synthesis. Compared with the non-local sparse attention pattern in [16]. Yan et al. [17] proposed a graph attention network with recurrent feature mapping blocks that take advantage of the internal patch-recurrence in a natural image.

2.2   Urban Flow Analysis

Urban flow forecasting plays a significant role in several aspects, such as traffic management [11] and risk assessment [4]. Gu et al. [7] developed an interpretable bicycle flow forecasting method, which first divided the entire city into zones based on flow density to provide effective bicycle flow forecasting and interpretable flow patterns. Pan et al. [15] proposed a spatial-temporal relationship network, which accurately predicts fine-grained traffic volume at a specific time by modeling coarse-grained traffic data. Zhang et al. [20] proposed a novel graph neural network that integrates local spatial semantics into a global pattern representation of the whole city through a city's custom graph diffusion paradigm representation. Gong et al. [6] proposed three spatial-temporal models to solve the above problem to explore the dependence between time and space in urban computing and improve the prediction accuracy. In addition to the applications mentioned above, we aim to address a novel problem regarding urban transportation in this study.

3   Formulation

In this section, we declare four specific concepts and then define the problem of UFSR.

Definition 1 (Urban Grid). We divide the urban sampling area into I J square regions based on latitude and longitude. As shown in Fig. 1, each square region denotes an urban grid. For an area of interest with a given size, smaller urban grids (i.e., using more significant I, J) suggest that we can obtain a lower granularity of signal data to build a finer-grained UFH.

Definition 2 (Urban Flow Heatmap). Let $M \in \mathbb{R}^{I \times J}$ represents an urban flow heatmap, where each element $m^{i,j} \in N_+$ denotes the traffic volume of the urban grid (I, J). We define 'Urban flow' in traffic volume and people flow, counting the number of vehicles (taxis, for example) passed and the number of check-in records for mobile users in each urban grid separately.

Definition 3 (Super-Resolution Region). We define adjacent $N \times N$ urban grids as a super-resolution region given a scaling factor N. Figure 1 illustrates an example consisting of 128×128 high granularity urban grids. We can obtain 32×32 non-overlapping super-resolution regions, which finally transform into the corresponding coarse-grained UFH with the guide of urban structural constraint.

Definition 4 (Urban Structural Constraint). Similar to [11], let $m_C^{i,j} \in M_C$ be the traffic volume of the urban grid in coarse-grained UFH and $m_F^{i',j'} \in M_F$ be the flows in the corresponding super-resolution region. The structural constraint between $m_C^{i,j}$ and $m_F^{i',j'}$ defines as follows:

$$m_C^{i,j} = \sum_{i',j'} m_F^{i',j'} \qquad s.t. \lfloor \frac{i'}{N} \rfloor = i, \lfloor \frac{j'}{N} \rfloor = j, \tag{1}$$

where N denotes the scaling factor of our UFSR problem, $i = 1, 2, ..., I$ and $j = 1, 2, ..., J$.

Problem Statement (Urban Flow Super-Resolution). Given a coarse-grained UFH M_C, a scaling factor N, and the external factor E(holiday, weather, for example), the UFSR task is to learn a model G, which maps $M_C \in \mathbb{R}^{I \times J}$ into a fine-grained UFH $M_F \in \mathbb{R}^{NI \times NJ}$:

$$M_F = G(M_C \mid E, N; \theta), \tag{2}$$

where θ denotes all trainable parameters of model G.

4   Methodology

Figure 3 depicts the framework of UrbanSG, which inferences a high-resolution, super-resolved UFH M_F from a low-resolution input UFH M_C. In this section, we first provide a detailed description of our framework's workflow. Then we present our strategy for dealing with the adverse effects of sparse urban flow and complex external factors separately.

4.1   Adversarial Network Architecture

We employ a Generative Adversarial Network as the backbone to conduct the structurally constrained impact with an adversarial training process. Following the traditional GANs, UrbanSG consists of two parts: a generator G_{θ_G} and a discriminator G_{θ_G}. We optimize them together in an alternating manner to solve the adversarial min-max problem:

$$\min_{\theta_G} \max_{\theta_D} (\mathbb{E}_{M_F \sim p_{train}(M_F)}[log D_{\theta_D}(M_F)] \\ + \mathbb{E}_{M_C \sim p_{train}(M_C)}[log(1 - D_{\theta_D}(G_{\theta_G}(M_C)))]), \tag{3}$$

which aims to learn a generator G to fool the discriminator D, which is trained to distinguish between super-resolution UFHs $G_{\theta_G}(M_C)$ and honest UFHs M_F. The generator G can learn the structural constraints and the distribution associations within super-resolution regions.

First, inspired by [9], we employ B residual blocks with the same layout as the deep core of our generator network G. To make the model better suited to the UFSR task, we modify the convolution kernel filter size, the number of input channels, and the activation function of the stacked residual blocks separately. Then, we employ a upsample block to double the size of the feature map (i.e., $M_C' \in \mathbb{R}^{2I \times 2J}$). Next, we present the flow self-attention module, which conducts the model focus on active urban grids from both inter-channel and intra-channel. Considering the urban structural constraint, we present the sum restriction module that corrects all urban grids' values in UFH according to the volume distribution inside each super-resolution region. Finally, with the workflow of G, we obtain the fine-grained UFH $G_{\theta_G}(M_C) \in \mathbb{R}^{NI \times NJ}$ inferred from coarse-grained UFH M_C.

To discriminate real fine-grained UFH M_F from generated super-resolution samples, we train a discriminator network that solves the min-max problem in Eq. 2. Inspired by the VGG network, we adopt a similar model architecture that stacks H convolutional blocks with an increasing number of convolutional kernels in each block, simple but efficient.

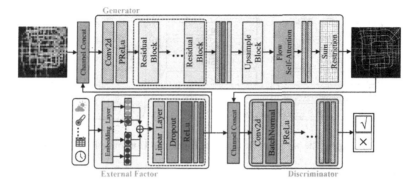

Fig. 3. The framework of UrbanSG with 4 up-scaling ($N = 4$).

Feature Extraction Module. In this section, we specify the feature extraction process and the implementation details of the residual block. We first extract the feature representation of external factors and treat it as an additional channel of the input UFH $M_C \sim P_{train}(M_C)$ for further feature extraction. Before entering the residual network, we apply a single convolution layer (filter size 9×9, channel C) to extract the low-level features. Then the extracted latent features serve as input of the following B Residual Blocks. As shown in Fig. 3, the layout of the residual block follows the guideline in [9], which employs two convolutional layers, a Batch Normalization and a PReLu function. Notably, to improve the model performance, we follow the N^2-normalization operation proposed in [11], which calculates the distribution of each urban grid; this allows the convolution process to focus further on the distribution within each super-resolution region. The detailed process is as follows:

$$
\begin{aligned}
Sum_{\tilde{M}}^{C} &= SumPooling(\tilde{M}, D), \\
Sum_{\tilde{M}}^{F} &= NNupsampling(Sum_{\tilde{M}}, D), \\
\tilde{M}' &= \tilde{M} \oslash Sum_{\tilde{M}}^{F},
\end{aligned}
\tag{4}
$$

where $\tilde{M}$ represents a feature map in the convolution process, D is the upscale factor of our UFSR task, and $\oslash$ represents element-wise division.

Finally, considering that the output flow distribution can demonstrate inter-regional dependence on the original coarse-grained UFH M_C and achieve efficient gradient back-propagation, we apply a skip connection which creates an information highway and brings an identity mapping into the feature extraction process.

$$
m_C^{i,j} = \sum_{i',j'} \lambda_{i',j'} m_F^{i',j'}
$$

$$
s.t. \sum_{i',j'} \lambda_{i',j'} = 1, \lambda \in N \quad \lfloor \frac{i'}{N} \rfloor = i, \lfloor \frac{j'}{N} \rfloor = j,
\tag{5}
$$

where λ denotes the current urban grid traffic volume distribution in the corresponding super-resolution region (Fig. 4).

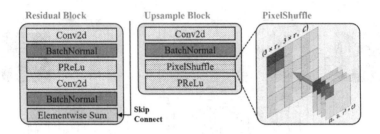

Fig. 4. The workflow of Residual Block and Upsample Block.

4.2 External Factor Integration

Significantly, it is insufficient to consider merely historical traffic data for our UFSR task. The existence of external factors can also make a vital and complicated impact on the urban grid's traffic volume. The actual impact on traffic flow becomes even implicit and non-negligible when different external factors are entwined. To this end, we build our model based on CGAN [13]: this variant of the original GAN allows generator G to produce condition-specific data by providing the critical condition (holiday, weather, for example) as an additional input to G along with a random noise input (Coarse-grained UFH in our UFSR task). Thus, we reformulate Eq. 3 as follows:

$$\min_{\theta_G}\max_{\theta_D}(\mathbb{E}_{M_F \sim p_{train}(M_F)}[logD_{\theta_D}(M_F \mid \mathbf{c})]$$
$$+\mathbb{E}_{M_C \sim p_{train}(M_C)}[log(1 - D_{\theta_D}(G_{\theta_G}(M_C \mid \mathbf{c})))], \tag{6}$$

where $\mathbf{c}$ corresponds to a conditional vector, such as the one-hot vector for a specific class label ("a rainy Saturday in May at 2 pm," for example).

Specifically, we first divide the available external factors into two feature sets based on the continuity of their values: categorical features and continuous features. We directly concatenated the continuous features (wind speed, temperature, for example) into one vector, $\mathbf{f}_{c}on$. We followed the guideline of [10], which transforms the available categorical features (weather, time of the day, and day of the week, for example) into a low-dimensional vector $\mathbf{f}_{c}at$ by applying different embedding methods to them separately. Then we concatenate the two embeddings as the ultimate latent feature representation of external factors.

$$\mathbf{f}_{ext} = [\mathbf{f}_{con} \oplus \mathbf{f}_{cat}], \tag{7}$$

where $\oplus$ denotes concatenation operation.

After getting the concatenated vector $\mathbf{f}_{e}xt$, we feed it into the Linear Layer and employ the $ReLu$ activation function to introduce nonlinear correlation. We employ the Dropout operation to randomly inactive 20% neural connections to avoid over-fitting. Then we employ two linear layers, which enable mapping the low-dimensional external factor into the high-dimensional space to explore the implicit features.

4.3 Flow Self-attention Module

Most UFSR models [11,21,22] are built with convolutional layers, which inevitably lead to drawbacks: 1) Convolution normally focuses on the local neighbors of feature maps, which is computationally inefficient to model long-range dependencies. It

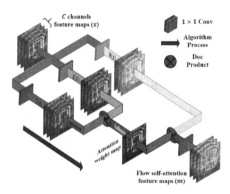

Fig. 5. The workflow of the flow Self-Attention Module.

is unreliable if we only consider the flow effects of adjacent F^2(filter size F) urban when dealing with UFH because even if two urban grids are far apart, there can be a strong correlation between their traffic flows. For example, suppose there are two railway stations far apart in a city. If one of them breaks down, the urban grid nearby the other will experience a dramatic increase in traffic. 2) Due to the limits of observation, UFH typically contains a considerable amount of static data, which aggravates the data sparsity and hinders exploring the internal data dependency; this seriously disrupts the performance of the UFSR task.

Figure 5 shows the structure of the proposed flow self-attention module. First, to calculate the attention weights, we transform the feature maps obtained from the previous hidden layer $x \in \mathbb{R}^{C \times N}$(C channels, N feature locations from the previous hidden layer) into two feature spaces, g and f by 1×1 convolution, which satisfies the following equation:

$$g\left(x\right) = W_g x, \ f\left(x\right) = W_f x, \tag{8}$$

$$\beta_{ij} = \frac{\exp\left(t_{ij}\right)}{\sum_{i=1}^{N} \exp\left(t_{ij}\right)}, \qquad t_{ij} = g\left(x_i\right)^T f\left(x_j\right), \tag{9}$$

where β_{ij} indicates how much attention the model pays to the ith grid when synthesizing the jth grid. After a 1×1 convolution with the element product of the feature map under another feature space h, we obtain the output of the attention layer: $o_j = \left(o_1, o_2, ..., o_j, ..., o_N\right) \in \mathbb{R}^{C \times N}$.

$$o_j = v\left(\sum_{i=1}^{N} \beta_{ij} h\left(x_i\right)\right), \ h\left(x_i\right) = W_h x_i, \ v\left(x_i\right) = W_v x_i, \tag{10}$$

In the Eqs. 8 to 10, respectively, $W_f \in \mathbb{R}^{C \times C}, W_g \in \mathbb{R}^{C \times C}, W_v \in \mathbb{R}^{C \times \bar{C}}$ and $W_h \in \mathbb{R}^{\bar{C} \times C}$ are learned trainable weight matrices parameters which are implemented with 1×1 convolution layers. In order to minimize the computational complexity, we attempt multiple sets of channel number $\bar{C}$, which is reduced by C/k, $where\, k \in \{1, 2, 4, 8\}$. Since no significant performance decrease occurs, we uniformly set $k = 8$ in subsequent experiments.

After deriving the output of the attention layer, we do the product with a scaling parameter and sum it with the original input feature map x. Therefore, the final output is obtained from the following equation:

$$y_i = \gamma o_i + x_i, \tag{11}$$

where γ is a trainable scalar with an initial value of 0. This way allows the model to be more dependent on the relevance of local neighbors in the early training period and then gradually assign more weights to non-local neighbors' relevance as the training epoch increases.

5 Experiment

We conduct extensive experiments on two real-world datasets to validate generalization and inference abilities under the sparse urban flow context. Then, we perform a series of ablation experiments to verify the importance of each module for the UFSR task and the inter-constraint between the modules.

5.1 Dataset

We employ two real-world datasets in our experiments to better examine our model's capabilities, namely TaxiBj and CCMobile. Table 1 details both datasets, containing two sub-datasets: urban flows and external factors.

Table 1. Detailed composition of the datasets

Dataset	TaxiBJ	CCMobile
Time Span	P1:2013.7.1–2013.10.31 P2:2014.2.1–2014.6.30 P3:2015.3.1–2015.6.30 P4:2015.11.1–2016.3.31	2017.7.5–2017.10.15
Time interval	30 min	15 min
Coarse-grained size	32×32	32×32
Fine-grained size	128×128	64×64
Up-scaling factor	4	2
External factors (time, event and meteorology)		
# Weathers	16 types	10 types
# Holidays	41	13
Temperature/C	[−24.6, 41.0]	[10.4, 34.1]
Wind Speed/mph	[0, 48.6]	/

1. **CCMobile**: We collect and aggregate check-in records from CCMobile terminal check-in data in each first week of six months. Figure 2(a) shows the experimental area, divided into 32×32 urban grids with red points indicating signal stations. Then we statistic the signal station capacity in each grid and obtain coarse/fine-grained resolution heatmaps within time slices of 15 min.

2. **TaxiBJ**: A taxi GPS dataset in Beijing, including driving trajectory and external factors. While the difference between taxiBI and CCMobile lies in: 1) TaxiBJ possesses complete external factor information. 2) TaxiBJ's data coverage is extensive, which divides the 12-month time into four different period sub-datasets. 3) TaxiBJ has a more significant up-scaling factor $N = 4$. In addition, we split both datasets without overlapping, following the ratio of 7:2:1 (training, validation, and test).

5.2 Experimental Settings

Evaluation Metrics. Following the traditional guideline, we employ three standard urban traffic data metrics (RMSE, MAE, and MAPE) to evaluate performance in different aspects, which are separately defined as:

$$
\begin{aligned}
RMSE &= \sqrt{\frac{1}{n} \sum_{i=1}^{n} \left\| X_f^i - \hat{X}_f^i \right\|_F^2}, \\
MAE &= \frac{1}{n} \sum_{i=1}^{n} \left\| X_f^i - \hat{X}_f^i \right\|_F, \\
MAPE &= \frac{1}{n} \sum_{i=1}^{n} \left\| \left(X_f^i - \hat{X}_f^i \right) \oslash X_f^i \right\|_F,
\end{aligned}
\tag{12}
$$

where n denotes the samples' total number, X_f^i and $\hat{X}_f^i$ denote the ground truth and the corresponding ith inferred value separately. RMSE and MAE are commonly used to measure the absolute magnitude of ground truth from the predicted value, while MAPE measures the relative magnitude (i.e., percentage). MAE and MAPE are less susceptible to extreme values and focus more on smooth outcomes, while RMSE employs square calculations to amplify prediction errors and is more sensitive to outlier data; thus, it favors spiky distribution.

Baselines. We compare UrbanSG with the following 6 baselines, which can be categorized into: (1) heuristic, (2)image super-resolution, (3)urban super-resolution as follows:

- **Mean Partition (Mean)**: We distribute the traffic volume evenly within an N^2 super-resolution, where N denotes the up-scaling factor.
- **Historical Average (HA)**: HA calculates the weights of each urban grid in the super-resolution region from the training data and assigns them fixedly.
- **SRCNN** [3]: SRCNN makes the first attempt to incorporate Convolutional Neural Network into ISR by scaling a coarse-grained image into high-resolution space.
- **SRResNet** [9]: SRResNet introduces the residual block, enabling the model to stack more layers and providing the theoretical basis for subsequent ISR works.

- **UrbanFM** [11]: UrbanFM innovatively incorporates the external factor into urban super-resolution and deploys the distributional up-sampling module.
- **UrbanODE** [21]: UrbanODE employs neural-ODEs and a pyramid attention mechanism for learning spatial correlations, the SOTA in the UFSR task.

5.3 Result on TaxiBJ

Model Comparison. In this subsection, we perform a comprehensive comparison test with 6 baselines and 3 variants. We report the result with $C - F$(Coarse-Fine Grained) as $16 - 64$ as our default setting, $N = 4$.

Table 2. Performance comparison among different models on TaxiBJ.

Method	Upscale	P1				P2				P3				P4			
		RMSE	MAE	MAPE	Δ	RMSE	MAE	MAPE	Δ	RMSE	MAE	MAPE	Δ	RMSE	MAE	MAPE	Δ
MEAN	4	20.918	12.019	4.469	–	20.466	13.91	5.364	–	27.442	16.029	5.6112	–	19.049	11.07	4.192	–
HA	4	4.741	2.251	0.336	−7.35%	5.381	2.551	0.334	−9.87%	5.594	2.674	0.328	−3.47%	4.125	2.203	0.313	−1.95%
SRCNN	4	4.297	2.491	0.714	−128.12%	4.612	2.681	0.689	−126.64%	4.815	2.829	0.727	−129.34%	3.838	2.289	0.667	−117.23%
SRResNet	4	4.164	2.457	0.713	−127.8%	4.524	2.66	0.688	−126.32%	4.69	2.775	0.717	−126.18%	3.667	2.189	0.637	−107.49%
UrbanFM	4	3.991	2.036	0.331	−5.75%	4.374	2.256	0.322	−5.92%	4.539	2.348	0.323	−1.89%	3.526	1.831	0.31	−0.98%
UrbanODE	4	3.86	1.963	0.313	0	4.391	2.213	0.304	0	4.479	2.287	0.317	0	3.439	1.81	0.307	0
UrbanSG	4	**3.782**	**1.891**	**0.287**	**+8.31%**	**4.129**	**2.114**	**0.279**	**+8.22%**	**4.285**	**2.173**	**0.293**	**+7.57%**	**3.107**	**1.624**	**0.288**	**+6.19%**
UrbanSG-attn	4	3.784	1.891	0.291	+7.03%	4.136	2.12	0.286	+5.92%	4.311	2.189	0.295	+6.94%	3.192	1.688	0.29	+5.54
UrbanSG-attn-ext	4	3.841	1.959	0.302	+3.51%	4.307	2.193	0.297	+2.3%	4.391	2.247	0.308	+2.84%	3.355	1.729	0.301	+1.95%
UrbanSG-attn-disc	4	4.016	2.144	0.343	−9.58%	4.397	2.268	0.33	−8.55%	4.604	2.363	0.339	−6.94%	3.606	1.839	0.317	−3.26%

Table 2 summarizes the results of the comparison experiments on TaxiBJ, where -attn, -ext, and -disc denote the deletion of the self-attention module, the external factor integration module, and the generator from UrbanSG, respectively. According to Table 2, We have the following observations: (1) UrbanSG and most of its variants outperform all baselines over all time periods in all validation metrics. Moreover, the advance of UrbanSG-attn-ext over all baselines suggests that employing generative adversarial networks as the backbone plays a leading role in improving the inference performance. (2) The performance of UrbanSG-attn shows no significant decline on TaxiBJ. We argue that it is attributed to the high data coverability of TaxiBJ limits the function of the self-attention module. Figure 6 illustrates a set of experiments for upscaling factor $N = 2, 4, 6, 8$. We can above that: (1) UrbanSG outperforms all other baselines for all Ns. (2) UrbanSG performs superior in the high-resolution UFSR task.

Study on External Factor Integration. It is insufficient to extrapolate from historical traffic volume in the complex urban computing context. Thus, we experiment on the effectiveness of external factors, where we randomly subsample the external factor from the original training set in four ratios: 100%, 50%, 30%, and 10%.

To verify the effectiveness of the external factor integration module, we compare UrbanSG with UrbanSG-ext and UrbanSG-up, as shown in Fig. 7. When we reduce the scale of external factors in training data, the distance between UrbanSG and UrbanSG-ext becomes more remarkable, which indicates that external factor integration plays the role of a meta learner and provides prior knowledge to the model training. It is the same as UrbanSG and UrbanSG-up. Our model may recover some external impacts when training data (external factor) is enough.

5.4 Result on CCMobile

Model Comparison. Table 3 illustrates the experiment results on CCMobile, which leads to the following observations: (1) Data sparsity reduces the model performance on CCMobile. However, UrbanSG still outperforms all the baselines in all three metrics. (2) A more significant up-scaling factor decreases the prediction performance. Take UrbanODE as an example; when the up-scaling factor N increases from 2 to 4, its MAE increases from 3.578 to 4.332. (3) Our model outperforms all other baselines in adapting to sparse data. For instance, when increasing N from 2 to 4, UrbanSG's performance improvement on MAE improves from 13.84% to 15.02%. By comparing the performance of UrbanSG and UrbanSG-attn, the adaption ability mainly comes from the flow self-attention module.

Table 3. Performance Comparison among different models on CCMobile.

Method	#Params/M	N = 2						N = 4					
		MAE	Δ	RMSE	Δ	MAPE	Δ	MAE	Δ	RMSE	Δ	MAPE	Δ
HA	X	10.622	−196.87%	16.408	−87.05%	0.927	−106%	11.032	−154.66%	23.015	−119.04%	0.918	−98.27%
SRCNN	7.4	4.502	−25.82%	10.006	−14.07%	0.51	−13.4%	5.232	−20.77%	12.148	−15.62%	0.544	−17.58%
SRResNet	5.5	4.403	−23.06%	9.667	−10.2%	0.508	−12.89%	5.123	−18.26%	11.81	−12.4%	0.536	−15.72%
UrbanFM	6.2	4.027	−12.55%	9.196	−4.83%	0.488	−8.44%	4.749	−9.62%	10.925	−3.98%	0.505	−9.05%
UrbanODE	**2.1**	3.578	0	8.772	0	0.45	0	4.332	0	10.507	0	0.463	0
UrbanSG	2.8	**3.083**	**+13.84%**	**8.017**	**+8.61%**	**0.395**	**+12.19%**	**3.681**	**+15.02%**	**9.395**	**+10.58%**	**0.399**	**+13.87%**
UrbanSG-attn	2.4	3.417	+4.51%	8.603	+1.93%	0.377	+2.8%	4.111	+5.11%	10.483	+2.3%	0.441	+4.84%

Study on Parameter Size. Table 3 illustrates the parameter settings for all baselines (heuristics methods without statistics of parameters). In addition, we test multiple sets of hyper-parameters (residual block, B and base channel, C) and argue that larger B and C could effectively improve model performance and bring a training time delay and increase memory space. Therefore, considering the trade-off between model performance and training cost, we set the default B-C as 16–256.

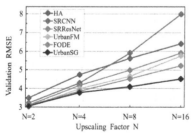
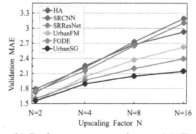

(a) Performance comparison on RMSE   (b) Performance comparison on MAE

Fig. 6. Performance comparison over various up-scaling factors.

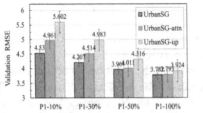

(a) Performance comparison on RMSE

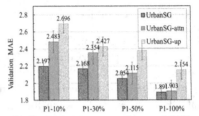

(b) Performance comparison on MAE

Fig. 7. Performance comparison over four-level external factor training sets.

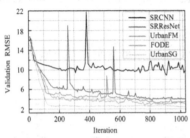

(a) Convergence efficiency comparison
between UrbanSG and baselines

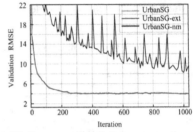

(b) Convergence efficiency comparison
between UrbanSG and variants

Fig. 8. Convergence efficiency comparison in 1000 iterations.

Study on Efficiency. We conduct a set of comparison experiments on P1, as shown in Fig. 8, which shows that UrbanSG converges faster and smoother than its variants and all baselines. UrbanSG reaches convergence at 300 iterations and can maintain a relatively stable convergence trend. Additionally, the similar performance of UrbanSG and UrbanSG-ext indicates that the external factor integration module only slightly affects the model's convergence but significantly improves performance.

Study on Flow Self-attention Module. We sum up the attention values of all instances during training and normalize them for visualization. Figure 9 shows the heatmap of attention weights and the corresponding physical urban flow heatmap. We observe that: (1) As we can see from regions 1 and 2, the flow self-attention module enables the model to focus on active urban grids and weights the inactive urban grid as 0, hence alleviating or ignoring the interference that sparsity brings to UFSR task. (2) Learning from region 3, the model avoids assigning too much attention weight to urban grids with dense traffic because if each urban grid within a super-resolution region has a large and similar traffic volume, they will share all the attention weights equally, resulting in each being small, which is consistent with common sense.

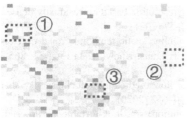

(a) Attention heatmap in CCMobile

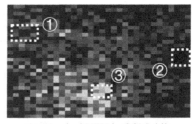

(b) Physical UFH in CCMobile

Fig. 9. Heatmap comparison between attention weights and urban flow.

6 Conclusion

This work formalizes the UFSR problem and presents a CGAN-based framework (UrbanSG) to solve it. UrbanSG has addressed the two challenges specific to this problem, i.e., flow distribution sparsity and the complexity of external factors, by leveraging the flow self-attention module and external factor fusion sub-net. Experiment shows that our approach advances baselines by at least 4.5%, 17.0% and 54.1% on TaxiBJ and 3.5%, 7.8% and 22% on CCMobile in terms of three metrics. Various empirical studies and visualizations have confirmed the advantages of UrbanSG in both efficiency and effectiveness.

Acknowledgement. This work was supported by the General Project of National Natural Science Foundation of China under Grant 62072209, the National Natural Science Foundation of authority Youth Fund under Grant 62002123, the Key project of Science and technology development Plan of Jilin Province Grant 20210201082GX, the Jilin Provincial Development and Reform Commission Project Grant 2020C017-2, the Science and technology project of Education Department of Jilin Province under Grand JJKH20221010KJ, and the CCF-Baidu Open Fund under Grant 2021PP15002000.

References

1. Cai, J., Meng, Z., Ho, C.M.: Residual channel attention generative adversarial network for image super-resolution and noise reduction. In: Proceedings of the IEEE/CVF Conference on Computer Vision and Pattern Recognition (CVPR) Workshops (2020)
2. Chen, W., Long, G., Yao, L., Sheng, Q.Z.: AMRNN: attended multi-task recurrent neural networks for dynamic illness severity prediction. In: World Wide Web Conference , pp. 2753–2770 (2020)
3. Chen, W., Wang, S., Long, G., Yao, L., Sheng, Q.Z., Li, X.: Dynamic illness severity prediction via multi-task RNNs for intensive care unit. In: 2018 IEEE International Conference on Data Mining (ICDM), pp. 917–922. IEEE (2018)
4. Chen, W., Yue, L., Li, B., Wang, C., Sheng, Q.Z.: DAMTRNN: a delta attention-based multi-task RNN for intention recognition. In: Li, J., Wang, S., Qin, S., Li, X., Wang, S. (eds.) ADMA 2019. LNCS (LNAI), vol. 11888, pp. 373–388. Springer, Cham (2019). https://doi.org/10.1007/978-3-030-35231-8_27

5. Dun, Y., Da, Z., Yang, S., Qian, X.: Image super-resolution based on residually dense distilled attention network. Neurocomputing **443**, 47–57 (2021)
6. Gong, Y., Li, Z., Zhang, J., Liu, W., Zheng, Y.: Online spatio-temporal crowd flow distribution prediction for complex metro system. IEEE Trans. Knowl. Data Eng. **34**, 865–880 (2020)
7. Gu, J., et al.: Exploiting interpretable patterns for flow prediction in dockless bike sharing systems. IEEE Trans. Knowl. Data Eng. **34**(2), 640–652 (2020)
8. Jo, Y., Kim, S.J.: Practical single-image super-resolution using look-up table. In: Proceedings of the IEEE/CVF Conference on Computer Vision and Pattern Recognition (CVPR), pp. 691–700 (2021)
9. Ledig, C., et al.: Photo-realistic single image super-resolution using a generative adversarial network. In: 2017 IEEE Conference on Computer Vision and Pattern Recognition (CVPR), pp. 105–114 (2017)
10. Liang, Y., Ke, S., Zhang, J., Yi, X., Zheng, Y.: Geoman: multi-level attention networks for geo-sensory time series prediction. In: Proceedings of the Twenty-Seventh International Joint Conference on Artificial Intelligence, IJCAI-18, pp. 3428–3434 (2018)
11. Liang, Y., et al.: Urbanfm: inferring fine-grained urban flows. In: Proceedings of the 25th ACM SIGKDD International Conference on Knowledge Discovery, pp. 3132–3142 (2019)
12. Liu, J., Zhang, W., Tang, Y., Tang, J., Wu, G.: Residual feature aggregation network for image super-resolution. In: Proceedings of the IEEE/CVF Conference on Computer Vision and Pattern Recognition (CVPR), pp. 2359–2368 (June 2020)
13. Mirza, M., Osindero, S.: Conditional generative adversarial nets. Comput. Sci. pp. 2672–2680 (2014)
14. Noor, D.F., Li, Y., Li, Z., Bhattacharyya, S., York, G.: Gradient image super-resolution for low-resolution image recognition. In: ICASSP 2019 IEEE International Conference on Acoustics, Speech and Signal Processing (ICASSP), pp. 2332–2336 (2019)
15. Pan, Z., et al.: Spatio-temporal meta learning for urban traffic prediction. IEEE Trans. Knowl. Data Eng. **34**(3), 1462–1476 (2022). https://doi.org/10.1109/TKDE.2020.2995855
16. Wang, R., Lei, T., Zhou, W., Wang, Q., Meng, H., Nandi, A.K.: Lightweight non-local network for image super-resolution. In: ICASSP 2021–2021 IEEE International Conference on Acoustics, Speech and Signal Processing (ICASSP), pp. 1625–1629 (2021)
17. Yan, Y., Ren, W., Hu, X., Li, K., Shen, H., Cao, X.: Srgat: Single image super-resolution with graph attention network. IEEE Trans. Image Process. **30**, 4905–4918 (2021)
18. Yang, F., Yang, H., Fu, J., Lu, H., Guo, B.: Learning texture transformer network for image super-resolution. In: Proceedings of the IEEE/CVF Conference on Computer Vision and Pattern Recognition (CVPR), pp. 5791–5800 (2020)
19. Yue, L., Tian, D., Chen, W., Han, X., Yin, M.: Deep learning for heterogeneous medical data analysis. World Wide Web **23**(5), 2715–2737 (2020)
20. Zhang, X., et al.: Traffic flow forecasting with spatial-temporal graph diffusion network. In: In Proceedings of the AAAI Conference. vol. 35, pp. 15008–15015 (2021)
21. Zhou, F., Jing, X., Li, L., Zhong, T.: Inferring high-resolutional urban flow with internet of mobile things. In: ICASSP 2021–2021 IEEE International Conference on Acoustics, Speech and Signal Processing (ICASSP), pp. 7948–7952 (2021)
22. Zhou, F., Li, L., Zhong, T., Trajcevski, G., Zhang, K., Wang, J.: Enhancing urban flow maps via neural odes. In: Proceedings of the Twenty-Ninth International Joint Conference on Artificial Intelligence, IJCAI-20, pp. 1295–1302 (2020)

TDCT: Transport Destination Calibration Based on Waybill Trajectories of Trucks

KaiXuan Zhu, Tao Wu, Wenyi Shen, Jiali Mao$^{(\boxtimes)}$, and Yue Shi

School of Data Science and Engineering, East China Normal University,
Shanghai, China
{51215903072,52195100007,51215903053,71205901048}@stu.ecnu.edu.cn,
jlmao@dase.ecnu.edu.cn

Abstract. Accurate transport destination is significant for improving efficiency in bulk logistics. But actually, the address information and the coordinate of latitude and longitude of transport destination are often incorrect due to manual input errors or vague address. Constantly generated logistics data including trucks' trajectories and waybills allows for the possibility of calibrating transport destination by analyzing staying behaviors of trucks. In this paper, we propose a transport destination calibration framework, called TDCT. Through clustering stay points into stay areas and then merging nearest stay areas based on *turn-off location*, *stay hotspot* can be ensured properly located. Further, a binary classification method by combining behavior features and area features is present to distinguish transport destination from other types of *stay hotspots*. Finally, a demo system is built for a steel logistics company to showcase the effectiveness of TDCT.

Keywords: Transport destination calibration · Turn-off location based merging · Behavior feature · Stay-area feature

1 Introduction

In bulk logistics field, transport destination plays a significant role in the services such as transport route planning and cargo accounts settlement. But actually, transport destination is not always correct due to typing mistakes of the operators or ambiguous destination information. As illustrated in Fig. 1, in a transport destination library of steel logistics, each record contains the address and the coordinate of latitude and longitude of one destination. Address in the second record is incomplete because it does not contain any company name or provide detailed street address. Also, from the third record to the sixth, although all of them represent a same company, their address and coordinates are inconsistent. Imagine if truck drivers transport the cargoes according to incorrect destination, they will unavoidably delay transporting. Thus, it necessitates an appropriate method to locate and annotate transport destination precisely to ensure updating of destination library.

© The Author(s), under exclusive license to Springer Nature Switzerland AG 2023
B. Li et al. (Eds.): APWeb-WAIM 2022, LNCS 13423, pp. 435–440, 2023.
https://doi.org/10.1007/978-3-031-25201-3_33

With extensive applications of positioning equipments, massive trajectories of trucks are gathered continuously. These trajectories, together with waybills provide us an opportunity to calibrate transport destinations. Based upon these data, a straightforward way is to mine stay points nearby the destination location of waybill to locate real transport destination. But destination location in transport destination library may be incorrect. Besides, in long-distance transporting, trucks may stop at other areas such as rest area and gas station. As a result, stay points nearby destination location of waybill cannot be used to identify real transport destination. Recently, few researches have devoted to delivery location correction issue for delivery service [3,4]. They recognize real delivery location based on extracting delivery caused stay points in terms of delivery time of each waybill annotated by couriers. While for bulk logistics, transport arrival time of each waybill is determined by truck driver's clicking "Complete" button, as shown in Fig. 2(a). Actually most of clicks occur considerably later than the time of transport task completing due to truck driver's habits. For instance, they don't click until they finish cargo unloading and park down trucks in the rest areas. That is, the location of driver's clicking may be far from real transport destination. So the correction method proposed in [3,4] cannot be used to calibrate transport destination for bulk logistics.

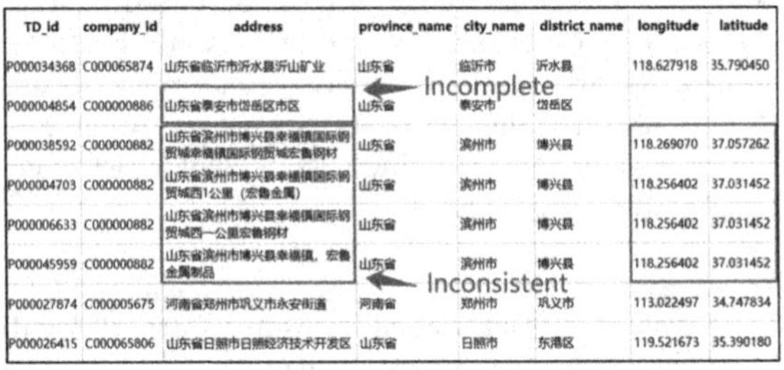

TD_id	company_id	address	province_name	city_name	district_name	longitude	latitude
P000034368	C000065874	山东省临沂市沂水县沂山矿业	山东省	临沂市	沂水县	118.627918	35.790450
P000004854	C000000886	山东省泰安市岱岳区市区	山东省	泰安市	岱岳区		
P000038592	C000000882	山东省滨州市博兴县幸福镇国际钢贸城幸福镇国际钢贸城卖钢材	山东省	滨州市	博兴县	118.269070	37.057262
P000004703	C000000882	山东省滨州市博兴县幸福镇国际钢贸城西2公里 (宏鲁金属)	山东省	滨州市	博兴县	118.256402	37.031452
P000006633	C000000882	山东省滨州市博兴县幸福镇国际钢贸城西一公里宏鲁钢材	山东省	滨州市	博兴县	118.256402	37.031452
P000045959	C000000882	山东省滨州市博兴县幸福镇, 宏鲁金属制品	山东省	滨州市	博兴县	118.256402	37.031452
P000027874	C000005675	河南省郑州市巩义市永安街道	河南省	郑州市	巩义市	113.022497	34.747834
P000026415	C000065806	山东省日照市日照经济技术开发区	山东省	日照市	东港区	119.521673	35.390180

Fig. 1. Illustration of transport destination library

However, locating and annotating transport destination have faced challenges. First, according to the observations in Fig. 2(b), transport stay hotspots (e.g., C and D) have their respective stay areas of varying sizes and many of them are close to each other. It is difficult to distinguish the locations of transport destinations from those of neighboring *stay hotspots*. Second, other types of *stay hotspots* have similar stay behaviors with transport destination, e.g., having same stay duration. It also brings challenge for annotating real transport destination. To tackle the above challenges, we present a transport destination calibration framework using waybill trajectories of trucks, called TDCT. As illustrated in Fig. 3, TDCT consists of data pre-processing, stay hotspot extraction and transport destination correction.

2 System Overview

Data Pre-processing. Initially, we obtain *waybill trajectories* by matching trucks' trajectories with each waybill in terms of the transport start and completion timestamps recorded in the waybill. To precisely identify stay points from *waybill trajectories*, we delete trajectory point that is far from most of its neighboring points, and then the point that is obviously different from most of its neighboring points in direction. Additionally, we remove the point that has abnormally high speed as compared to its neighbors. To locate each *stay hotspot*, we perform map matching [2] on each *waybill trajectory* having zero-speed points, to identify its nearest road (denoted as r_n). Meanwhile, we obtain the last matched point (denoted as p_{i-1}) and unmatched ones (denoted as $\{p_i, p_{i+1}, ..., p_j\}$) relative to r_n in *waybill trajectory*.

(a) Waybill   (b) Transport Destination

Fig. 2. Illustration of waybill and transport destination

Stay Hotspot Extraction. We regard the points that keep zero speed for a while from unmatched trajectory points as stay points. Specifically, we keep account of time gap between a zero-speed point (p_k) and its subsequent non-zero speed one in a same *waybill trajectory*, denoted as *stay duration*. When *stay duration* is beyond a preset time threshold $th_t d$ (here $th_t d$ is set as 8 min due to most unloading operation takes more than 8 min), p_k is viewed as a stay point. Its timestamp is viewed as *start time* of one stay behavior of truck. For *waybill trajectory* wt_l that p_k belongs to, the last matched point p_{i-1} in wt_l relative to the road r_n is regarded as *turn-off point* which truck drives away from r_n.

Next, we cluster all stay points using *DBSCAN* method to identify stay areas, and calculate means of the coordinates of stay points in stay areas separately to obtain their centers, denoted as $\{stay_1, stay_2, ..., stay_m\}$, as shown in Fig. 2(b). To avoid grouping stay points on both sides of the road into a same cluster, the parameter *eps* of *DBSCAN* is set as 5 m, and then the parameter min_{sample} is empirically set as 5. The reason for using *DBSCAN* is that it does not require a pre-specifying the cluster number and can find clusters (i.e. stay areas) of any shapes. Consider that a truck may unload cargoes at several warehouses inside a factory or logistics park, a *stay hotspot* contains multiple stay areas. To judge whether several stay areas belong to the same *stay hotspot*, we put forward a *turn-off location* based stay area merging method. It is observed that when trucks drive to a same *stay hotspot*, they usually turn off the road at almost the similar location, i.e. their road turn-off points are near to each other, as illustrated in Fig. 2(b). Inspired by this, the corresponding road turn-off points of stay points in different stay areas are clustered using *Meanshift* method to extract the centers of clusters with maximum density respectively. Such centers are regarded as *turn-off location* correspond to stay areas separately, and the centers' directions are obtained by calculating average directions of *turn-off points* in their belonged clusters. Further, the adjacent *turn-off locations* with almost the same direction (i.e. directions difference between *turn-off locations* are less than 15 degrees) are merged into a cluster, and then corresponding stay areas of *turn-off location* are merged to derive a *stay hotspot*. The location of *stay hotspot* is derived by calculate means of the coordinates of latitude and longitude of stay areas' centers.

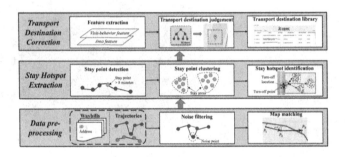

Fig. 3. Overview of transport destination calibration framework

Transport Destination Correction. We first extract *behavior feature* and *area feature* of *stay hotspots*. Actually, various transport destinations have respective limited types of cargoes and different working time constraints. Besides, as compared to the other types of *stay hotspots*, stay duration of transport destination is significantly distinct due to long-time unloading operation, and stop frequency of transport destination is more. So *behavior feature* includes cargo type transported (denoted as F_{type}), start time of one stay behavior (denoted as F_{time}), time duration of one stay behavior (denoted as F_{dura}) and stop frequency at a stay hotspot in a *waybill trajectory* (denoted as F_{stop}). Also, we consider the type of neighboring road (denoted as F_{road}), distribution of nearby *POI* type (denoted as F_{poi}) and area of *stay hotspot* (denoted as F_{area}) as *area feature*. Compared with other types of *stay hotspots*, transport destination is larger and usually locates beside the roads like provincial highway, and is near some factories and companies. To obtain its area, we extract the convex hull covered by stay points, represent by a sequence of vertices $\{(x_1,y_1)(x_2,y_2)...(x_h,y_h)\}$, and get $\frac{1}{2}[(\sum_{i=1}^{h-1} x_iy_{i+1} + x_hy_1) - (\sum_{i=1}^{h-1} y_ix_{i+1} + y_hx_1)]$. Then *XGBoost* [1] is applied to train a binary classification model for annotating transport destination. For the updated transport destinations, we obtain their address via reverse geocoding services[1]. On the basis of that, we update the coordinate of latitude and longitude and address information of related records in destination library.

3 Demonstration

The experimental datasets are from a steel logistics technology company named *JCZH*: (1) **Trucks' trajectories**. It consists of 1,376,246 GPS points generated by 2,293 trucks in Shandong Province from Jan 1th to Apri 1th, 2021, in which each trajectory contains truck ID and a series of GPS points. (2) **Waybills**. Each record contains truck ID, name and weight of cargo, start and completion timestamp of transport, and transport destination ID. It contains 71,821 waybills involving 53 original transport destinations. (3) **Road network**. The road network covered by trucks' trajectories is from OSM with 1,630,544 vertices and 652,603 edges. (4) **Ground truth**. 86 transport destination and 236 other types of stay hotspots are annotated manually by driver volunteers.

First, we compare TDCT with the method proposed in [4] (or *DTInf* for short). *DTInf* identifies real delivery location by clustering stay points nearby destination location of the waybill. We implement both methods to calibrate 53 transport destinations respectively, and then calculate the distance between calibrated location and actual location of transport destination to evaluate the performance of both methods. As shown in Fig. 4(a), TDCT significantly outperforms *DTInf*. TDCT can calibrate about 92.45% of real transport destinations, but *DTInf* only calibrate 54.72% of real ones. In addition, the location of transport destination calibrated by TDCT is closer to actual location.

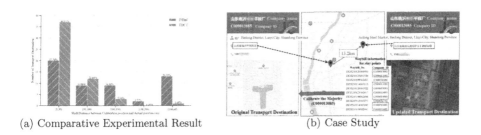

(a) Comparative Experimental Result   (b) Case Study

Fig. 4. Illustration of transport destination calibrating

Moreover, we give a case study to show the effectiveness of TDCT. As illustrated in Fig. 4(b), stay point set (marked by green dots that enclosed by red box) is identified as a transport destination. We obtain its address information through address query interface as "Anfeng Steel Market, Hedong District, Linyi

---

[1] https://lbs.amap.com/api/webservice/guide/api/georegeo.

City, Shandong Province". Then we recognize its company name by waybill ID, and we update its address and location in the library. From Fig. 4(b), we can observe its original address is "Hedong District, Linyi City, Shandong Province", and its mapped location is about 13.2 km away from our calibrated location. In 2021, TDCT has been deployed to the transport destination calibration module on *JCZH*'s logistics platform, which is used to calibrate the transport destination library regularly. Up to now, a total of 1,575 transport destinations have been corrected, and 853 redundant transport destinations have been merged.

Acknowledgments. This research was supported by NSFC (Nos.62072180, U1911203 and U1811264).

References

1. Chen, T., Guestrin, C.: Xgboost: a scalable tree boosting system. In: SIGKDD, pp. 785–794 (2016)
2. Newson, P., Krumm, J.: Hidden Markova map matching through noise and sparseness. In: SIGSPATIA,. pp. 336–343 (2009)
3. Ruan, S., et al.: Filling delivery time automatically based on couriers' trajectories. IEEE Trans. Knowl. Data Eng. **35**, 528–1540 (2021)
4. Ruan, S., et al.: Doing in one go: Delivery time inference based on couriers' trajectories. In: SIGKDD, pp. 2813–2821 (2020)

OESCPM: An Online Extended Spatial Co-location Pattern Mining System

Jinpeng Zhang[1,2], Lizhen Wang[1,3(✉)], Wenlu Lou[2], and Vanha Tran[4]

[1] Yunnan University, Kunming, Yunnan, China
zjp@ynufe.edu.cn, lzhwang@ynu.edu.cn
[2] Yunnan University of Finance and Economics, Kunming, Yunnan, China
wenlu_lou@163.com
[3] Dianchi College of Yunnan University , Kunming, Yunnan, China
[4] FPT University, Hanoi, Vietnam
hatv14@fe.edu.vn

Abstract. In spatial data mining, co-location pattern mining is intended to discover the sets of spatial features whose instances occur frequently in nearby geographic areas. Co-location pattern mining is an important task in spatial data mining and has been applied in many applications. Although many spatial co-location pattern mining methods applied to point objects have been proposed, few mining approaches are designed for extended spatial objects in practical scenes. Unlike existing works, we develop OESCPM (Online Extended Spatial Co-location Pattern Mining System) to mine co-location patterns over extended objects. OESCPM decomposes extended instances into cells and counts the number of cells belonging to different features. With simple interaction on OESCPM, users can view co-location patterns over interested extended spatial datasets online.

Keywords: Spatial data mining · Co-location pattern · Extended spatial objects · Online mining

1 Introduction

According to Tobler's theory, things that are near each other are more closely related than things that are distant from each other. Co-location pattern mining (CPM) extracts the subsets of spatial features whose instances are located together frequently in geography. For example, the set {Hospital, Pharmacy} is a co-location pattern means they are frequently located near each other. Co-location patterns have a vital impact in various fields, like geoscience, public transportation, environmental studies [1], and so on.

Although there are many CPM approaches that have been successful in reducing redundancy[1], maximal co-locations [2, 3], and dominant relationships [4], they are mostly designed for point spatial instances (objects). But in the real-world, spatial instances are generally displayed within the form of extended instances, such as parks' dissemination is polygonal and streams are lines. The CPM methods designed for point spatial instances are not available for extended spatial objects, they do not correctly reflect the neighbor relationships between objects.

In this paper, we develop an Online Extended Spatial Co-location Pattern Mining System (OESCPM) that allows users to extract co-location patterns over extended objects online, which discovers co-locations by decomposing extended instances into cells and calculating the cells' feature transaction. With OESCPM, users can easily access the prevalent co-location patterns for user-specified extended datasets.

2 System Overview

OESCPM is designed for extended spatial objects in practical scenes. Unlike a point instance is expressed as a point in space, an extended instance expresses the instance such as lines, polygons, clusters, and so on. As shown in Fig. 1, the OESCPM contains three main modules. (1) Pre-processing module pre-processes the spatial data. (2) Visualization module provides users with a visual image display of spatial datasets. (3) Mining module mining co-location patterns.

First, users specify datasets or region by their interests. Preprocess module decomposes extended spatial objects of various shapes at the macro level into cell-level. The example is given in Fig. 2, the red is the instance's area, by decomposing into cells, the irregular polygon instance is displayed as six cells. This process is similar to representing an image in pixels. Then visualization module collects spatial features' location information and plot the scatter distribution of spatial features in a canvas to form image and provides users with intuitive visual to demonstrate features' distribution. In co-location mining module, system calculates buffers (sphere of influence) for each instance, generates neighbor relations between instances at cell-level, merges cells' feature transactions, mining candidate co-location patterns, evaluates co-locations' prevalence and feedback to users on prevalent co-location patterns.

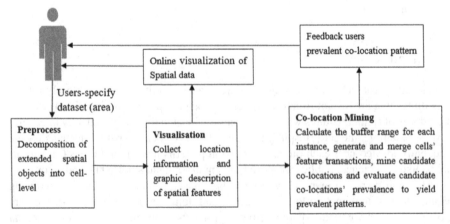

Fig. 1. Description of OESCPM operating mechanism

Fig. 2. Extended spatial instance expressed at cell-level (Color figure online)

3 Techniques

In this section, we mainly focus on the implementation of the co-location pattern mining technology with extended instances.

Mining Method. In our definition, the buffers of each instance represent the effect scope of the instance. The buffer is shown as the range wrapped by the dotted line. If the buffers of instances overlap, the instances are neighbors. The buffer intersection area of instances of multiple features expresses the co-occurrence information of these features. We count the number of cells in the intersection of instance buffers to obtain feature transactions, and use transaction mining methods such as FP-growth to obtain co-location patterns.

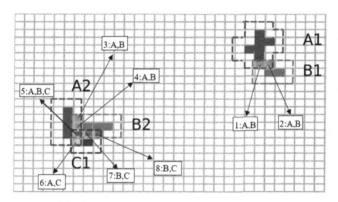

Fig. 3. Example of mining co-location patterns

For example, as described in Fig. 3, for instances A1 and B1, their buffers overlap on cells 1 and 2. For instances A2, B2 and C1, the buffers of A2 and B2 overlap on cell 3 and 4. A2, B2 and C1 overlap on cell 5, A2, C1 overlap on cell 6, B2 and C1 on cell 7 and 8. In total, features A, B have four cells overlapping, A, C have one cell overlapping, B, C have two, and A, B, C have one. The formal expression is {A, B} = 4, {A, C} = 1, {B, C} = 2, {A, B, C} = 1. By FP-growth, the candidate co-location pattern result is {A, B} = 5, {A, C} = 2, {B, C} = 3, {A, B, C} = 1. Finally, the prevalence of the candidate patterns are calculated and the prevalent patterns are obtained.

4 Demonstration Scenarios

With a flat design interface, OESCPM is encapsulated well with a friendly interface. Users can access the mining results for a specified dataset through simple interaction. In this demonstration, the data from 2010 China land cover type data (chinaLC 2010) [5] are used to show the demonstration of OESCPM. The dataset has 19 land cover types (spatial features). We use an uppercase letter to represent a land cover feature, such as J for grassland, N for Cropland, S for Rainfed Croplands, and so on.

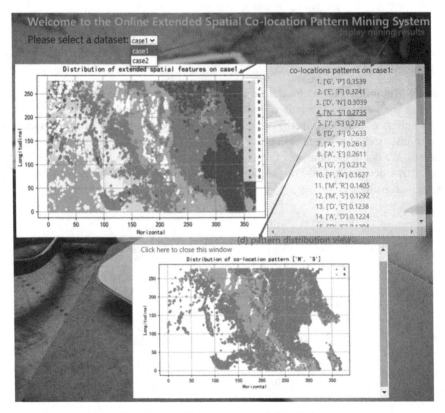

Fig. 4. User interface of OESCPM

Figure 4 shows the user interface of OESCPM. At first, the users select the dataset (area) they care about on chinaLC 2010. Figure 4(a) shows a drop-down list containing options for each region of the chinaLC, allowing the users to specify the region of interest. For example, datasets case1 and case2. Then, the system transforms the users-specify dataset into a graphical scatter plot, providing users with an intuitive way to display the distribution of spatial features. The display is shown in Fig. 4(b). Next, OESCPM mines co-location patterns over the specify dataset and displays the results in the output box in Fig. 4(c). Co-location patterns are descending ranked by co-location patterns' prevalence. In the output box, the serial number indicates the ranking of the pattern, the content of the pattern is given in square brackets, and the final decimal represents the prevalence metric of the pattern. If the patterns' number mined by the system is high, the users can scroll through the scroll bar of the output box to view additional results. Moreover, users can learn about the detailed distribution of a co-location pattern by clicking on the link for that pattern. For instance, users click on the link to {N, S} in Fig. 4(c), the link pops up a hover box showing the distribution view of pattern {N, S} on case1 shown in Fig. 4(d). After viewing, visitors can close the co-location distribution view and continue to view the distribution of other patterns.

Overall, users can access the mining results with simple clicks throughout the whole spatial co-location pattern mining process. With OESCPM, system visitors can use co-locations to guide production and living practices. For example, in the mining results on case1, the most prevalent pattern {G, P}, i.e., Open Shrublands and Barren often occur together, indicating that Open Shrublands are not amenable for reclamation into Rainfed Croplands on case1 region. The system can be applied in other scenarios by replacing the spatial database, such as species distribution analysis, cancer contamination distribution analysis and so on.

5 Conclusion

In the existing co-location pattern mining applications, the methods are designed for point objects, but they are not suitable for practical applications, so we develop the OESCPM system to discover co-location patterns over extended spatial objects. The online user interface does not require extra user actions. The demonstration scenarios showed better usability and efficiency of OESCPM. The co-location patterns of OESCPM reports can be used to guide agricultural production, species distribution analysis and other domains.

Acknowledgements. This work is supported by the National Natural Science Foundation of China (61966036); Yunnan Fundamental Research Projects (202201AS070015); and the Project of Innovative Research Team of Yunnan Province (2018HC019). We also acknowledge for the data support from "National Earth System Science Data Center, National Science & Technology Infrastructure of China" (http://www.geodata.cn). National Natural Science Foundation of China, 62276227.

References

1. Wang, L., Bao, X., Zhou, L.: Redundancy reduction for prevalent co-location patterns. IEEE Trans. Knowl. Data Eng. **30**(1), 142–155 (2017). https://doi.org/10.1109/TKDE.2017.2759110
2. Tran, V., Wang, L., Chen, H., Xiao, Q.: MCHT: a maximal clique and hash table-based maximal prevalent co-location pattern mining algorithm. Expert Syst. Appl. **175**, 114830 (2021). https://doi.org/10.1016/j.eswa.2021.114830
3. Wang, L., Zhou, L., Lu, J., Yip, J.: An order-clique-based approach for mining maximal co-locations. Inf. Sci. **179**, 3370–3382 (2009). https://doi.org/10.1016/j.ins.2009.05.023
4. Fang, Y., Wang, L., Hu, T., Wang, X.: DFCPM: a dominant feature co-location pattern miner. In: Cai, Y., Ishikawa, Y., Xu, J. (eds.) APWeb-WAIM 2018. LNCS, vol. 10987, pp. 456–460. Springer, Cham (2018). https://doi.org/10.1007/978-3-319-96890-2_38
5. National Earth System Science Data Center. National Science & Technology Infrastructure of China. http://www.geodata.cn

gTop: An Efficient SPARQL Query Engine

Yuqi Zhou[1]([✉]), Lei Zou[1], and Gang Cao[2]

[1] Peking University, Beijing, China
{zhouyuqi,zoulei}@pku.edu.cn
[2] Beijing Academy of Artificial Intelligence, Beijing, China
caogang@baai.ac.cn

Abstract. In this demonstration, we present **gTop**, a top-k query engine based on gStore which supports SPARQL queries over RDF databases. gTop can answer top-k queries with high efficiency and scalability. We use the DP-B algorithm for top-k queries and the DP-any algorithm for any-k queries. We break cyclic queries into pieces of tree queries and use DP-any to solve queries generated before assembling the results to retrieve the origin answers. Experiments show the efficiency of gTop. We provide a demonstration website to show the usage of gTop, where users can query YAGO2 with top-k SPARQL queries.

Keywords: SPARQL · Top-k · Query engine

1 Introduction

Top-k queries can be expressed as queries with the 'LIMIT' and 'ORDER BY' clauses in SPARQL language. The 'ORDER BY' clause specifies the ranking criterion and the 'LIMIT' clause tells the number of top-ranked results users want. Figure 1 shows an example of the SPARQL top-k query, which queries the top-5 cities having the largest GDP and is the birthplace of a musician. Top-k queries are a powerful tool to mine data information and are widely used by analysts.

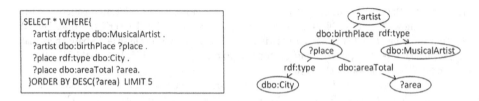

Fig. 1. An example SPARQL query

In this paper, we propose gTop, a new query engine specially designed for SPARQL top-k queries. gTop is built upon the gStore database [5], which is an open-source graph database using subgraph matching to answer SPARQL queries.

B. Li et al. (Eds.): APWeb-WAIM 2022, LNCS 13423, pp. 446–450, 2023.
https://doi.org/10.1007/978-3-031-25201-3_35

2   System Structure

The overview structure of gTop is depicted in Fig. 2. gTop uses different strategies to answer acyclic and cyclic top-k queries. Acyclic queries are directly forward into the top-k tree query solver, which uses the DP-B algorithm [2]. For cyclic queries, gTop first breaks them into tree queries through the query splitter and each sub-query will be evaluated by the any-k tree query solver. A controller takes charge of the enumeration process by dynamically deciding which sub-query to enumerate, and how many results should be outputted from the sub-query. The sub-results are assembled and then transferred to the users.

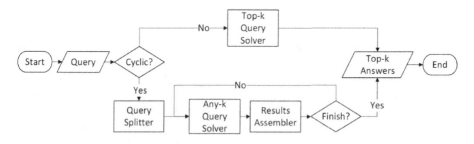

Fig. 2. The working process of gTop

Top-k Tree Query Solver. The basic structure of a SPARQL query can be expressed as a graph, as shown in Fig. 1. If there is no cycle in the query graph, we refer to it as an acyclic query, otherwise, we call it a cyclic query. For acyclic top-k queries, the DP-B algorithm is the first linear algorithm proposed [2], which is used and adapted in gTop. The top-k solver completes this task in two steps. The solver first filters each variable's candidates from top to bottom and then enumerates the top-ranked result recursively from bottom to top.

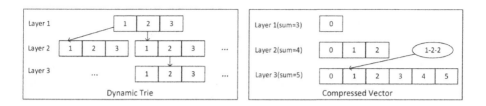

Fig. 3. An example to illustrate dynamic trie and compressed vector

DP-B uses a dynamic trie to solve the **FQ-iterator Problem**. The problem focuses on enumerating top-ranked results from m ranked sub-lists, where each combination of elements from the m sub-lists makes a valid result and the score is the sum of each element. For example, we want the min-3 results assembled from 3 ranked lists. We use a sequence of integers to identify a combination. Sequence $2 - 2 - 1$ refers to the result made of the two 2nd elements from the first two lists, and the 1st element of the last list. We observe that $1 - 2 - 1$ and $2 - 1 - 1$ are always no greater than $2 - 2 - 1$ and we call $1 - 2 - 1$ and $2 - 1 - 1$ parents of $2 - 2 - 1$.

In general, for sequences $x_1 - x_2 - \dots - x_m$ and $y_1 - y_2 - \dots - y_m$, if $\exists 1 \leq i \leq m, x_i = y_i - 1$, and $\forall 1 \leq j \leq m, j \neq i, x_j = y_j$, we call $x_1 - x_2 - \dots - x_m$ is the parent of $y_1 - y_2 - \dots - y_m$. Easy to see, a sequence is the potential to be the next result only when all its parents (if exists) have been enumerated. DP-B uses a Dynamic Trie to record whether all parents of a sequence have been enumerated. gTop adapts the DP-B algorithm for better performance in SPARQL queries, for example, it contains auxiliary indexes to support predicate variables of triples.

Any-k Tree Query Solver. Solving the cyclic top-k query is an NP-hard problem. Some algorithms break the queries into tree sub-queries and assemble the sub-results back [1,4], where the top-k tree query algorithms may not work well. For example, after getting the top-10 results, top-k algorithms may perform a whole search to the top-20 results, making no use of the previous effort. Our any-k tree solver treats k as dynamic, assuming no limit for the number of results users want and maintaining a scalable memory structure when running.

As mentioned before, the DP-B algorithm uses a dynamic trie that uses multi-layers of pointer arrays. The array size is fixed to k, limiting its ability to answer any-k queries. For an example shown in Fig. 3, to query sequence $2 - 2 - 1$, the dynamic trie first visits the 2nd pointer in the first layer which points to address $addr1$, and then the 2nd pointer in $addr1$, which maps to $addr2$, the information of the sequence in the first slot in $addr2$.

Our any-k tree queries solver uses another dynamic structure we proposed, Compressed Vector, to avoid the limitation. A compressed vector groups the sequence by the sum of the numbers in sequence. For example, in terms of 3 sub-lists, $1 - 2 - 2$ and $1 - 1 - 3$ are classified into one group Layer 3, which is composed of sequences summing to 5. A compressed vector maps each sequence into one non-negative number, that is, the lexicographical order of the sequence in the group.

Compressed Vector gets rid of the requirement to know k, avoids allocating too many memory pieces, and reduces the time to access RAM. The translation from sequences to integers (and the opposite direction also) is of low cost due to the efficient translation algorithm we implement. Such a mechanism allows the solver to continuously generate top-ranked results and is scalable to the data size and k.

Query Splitter and Results Assembler. gTop treats SPARQL queries as subgraph isomorphism problems by treating the subject and object of a triple as nodes in a graph as in Fig. 1. In SPARQL, a user can define scoring functions in literal values, which can only be objects of triples. According to this feature, we design a heuristic strategy to split queries. Once we find a cycle in the graph, we choose the edge with the worst selectivity to cut off. We keep cutting off edges until the query becomes a tree query. Then we run the DP-any algorithm to solve tree queries, once a result is enumerated, the results assembler checks whether the cut edges can be matched in the newly enumerated result, if not, ignore this result. The process stops if the any-k algorithm exhausts the results or we have already produced k valid results.

The correctness lies in the feature of SPARQL queries. The score of a result only comes from nodes, so ignoring some edges will not change the relative order of the origin top-k answers, but will bring more intermediate results. The query splitter chooses the edge with the worst selectivity to cut off to avoid overflooding intermediate results. The results assembler assures each output result is the correct match of the cyclic query.

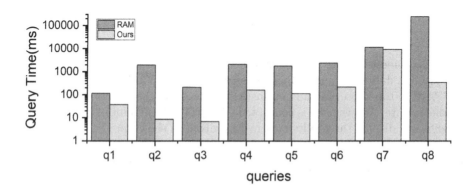

Fig. 4. The answering process of gTop

3 Experiments and Demonstration

We run our experiment on a server with an Intel(R) Xeon(R) CPU E5-2640 v3 2.60 GHzCPU and 125 GB memory running CentOS Linux. We evaluate all the algorithms by a real-world knowledge graph YAGO2 [3]. YAGO2 is a high-quality knowledge graph extracted from Wikipedia, which is widely used in many real-world applications and is widely accepted as a criterion for performance in graph algorithms.

We test the algorithms with 8 queries and the results are shown in Fig. 4. In the figure, 'RAM' refers to the naive algorithm, which ranks and gets the top-k results after finding all the results by subgraph matching. The experiment shows that our strategy has a significant improvement over the naive implementation.

The queries have 2 to 10 query triples each and the results are ranked by the linear combination of up to 4 numeric literal variables.

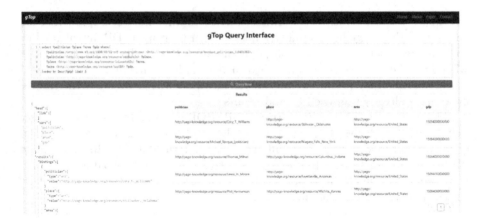

Fig. 5. The answering process of gTop

In the demonstration, we present a website for users to use gTop. We prebuild the YAGO2 databases and gTop processes query over YAGO2. Users can type in SPARQL top-k queries and get answers by pressing the 'Query Now' button on the website. The website is connected to an HTTP endpoint of gTop and the website will display the results returned by gTop, as shown in Fig. 5.

Acknowledgements. This work was partially supported by National Key R&D Program of China (2020AAA0105200).

References

1. Cheng, J., Zeng, X., Yu, J.X.: Top-k graph pattern matching over large graphs. In: Jensen, C.S., Jermaine, C.M., Zhou, X. (eds.) 29th IEEE International Conference on Data Engineering, ICDE 2013, Brisbane, Australia, 8–12 April 2013, pp. 1033–1044. IEEE Computer Society (2013). https://doi.org/10.1109/ICDE.2013.6544895
2. Gou, G., Chirkova, R.: Efficient algorithms for exact ranked twig-pattern matching over graphs. In: Proceedings of the 2008 ACM SIGMOD International Conference on Management of Data, pp. 581–594 (2008)
3. Hoffart, J., Suchanek, F.M., Berberich, K., Weikum, G.: YAGO2: A spatially and temporally enhanced knowledge base from Wikipedia. Artif. Intell. **194**, 28–61 (2013). https://doi.org/10.1016/j.artint.2012.06.001
4. Zeng, X., Cheng, J., Yu, J.X., Feng, S.: Top-K graph pattern matching: a twig query approach. In: Gao, H., Lim, L., Wang, W., Li, C., Chen, L. (eds.) WAIM 2012. LNCS, vol. 7418, pp. 284–295. Springer, Heidelberg (2012). https://doi.org/10.1007/978-3-642-32281-5_28
5. Zou, L., Mo, J., Chen, L., Özsu, M.T., Zhao, D.: gStore: answering SPARQL queries via subgraph matching. Proc. VLDB Endow. **4**(8), 482–493 (2011). https://doi.org/10.14778/2002974.2002976, http://www.vldb.org/pvldb/vol4/p482-zou.pdf

Multi-SQL: An Automatic Multi-model Data Management System

Yu Yan, Hongzhi Wang$^{(\boxtimes)}$, Yutong Wang, Zhixin Qi, Jian Ma, Chang Liu, Meng Gao, Hao Yan, Haoran Zhang, and Ziming Shen

Harbin Institute of Technology, Harbin, China
{yuyan,wangzh}@hit.edu.cn

Abstract. Nowadays, data in applications become diverse and large in scale. In order to meet the increasing demand for multi-model data management, multi-model databases have evolved into huge systems with many knobs. However, such a large system brings great challenges for users to efficiently tune. In extreme cases, it is difficult even for experienced DBAs to be the masters in multi-source engines and provide effective solutions for multi-model data. Meeting this challenge, we firstly propose an automatic multi-model data management system (Multi-SQL) based on machine learning. We design an intelligent middleware in which some artificial intelligence(AI) technologies are embedded, including automatic multi-model storage selection and automatic multi-model index recommendation methods. In this paper, we clarify the architecture, core techniques and key scenarios of Multi-SQL.

Keywords: Multi-model · Data management · Artificial intelligence

1 Introduction

Nowadays, data-centric applications increase rapidly. Such applications usually contain data in multiple models, such as relation, key-value, graph and document. For instance, in the social commerce scenarios [3], some basic information like user's basic information is suitable for storing in the relational model; the data that needs to be frequently updated and simple in structure, are proper to store with the key-value model; the extremely complicated relationships like social network would be stored as the graph model. For the data in various models, multi-model data management is in demand.

In order to deal with the complex and diverse big data scenarios, researchers have proposed some multi-model database systems. At present, there are two kinds of methods to process and manipulate the multi-model data. On the one hand, **polyglot persistence** [1] [4] manage the multi-model data by developing a middleware over the multiple storage engines. For instance, RHEEM [1] provides a general cross-platform data processing system which integrates DBMS, NoSQL, etc. However, polyglot persistence methods with the integration of

B. Li et al. (Eds.): APWeb-WAIM 2022, LNCS 13423, pp. 451–455, 2023.
https://doi.org/10.1007/978-3-031-25201-3_36

numerous independent databases also bring the tons of options and configuration knobs while enabling the database to handle the more data models [4]. On the other hand, **unified multi-model database** supports the management of multiple data models by mapping the data in various models to a unified underlying data model. For example, ArangoDB [2] uses the document model as the unified model. On this basis, it supports the storage of the key-value model and the graph model by integrating more functions.

We observe that multi-model databases have evolved into more complex systems with many knobs in order to deal with increasing demand of multi-model data.

Since the ability of human is always limited, it is impossible for a single person to be proficient with large-scale database knobs. In the increasingly complex DBMS, we can no longer rely on the database administrators for manual tuning. Thus, there exist significant demands for the automated multi-model database solutions. Unfortunately, there is no systems or approaches to study the automatic management on multi-model data, and all existing multi-model methods have to rely on DBAs to manage multi-model database.

In this paper, we develop the automatic multi-model database, Multi-SQL, which is specially designed for the automatic management of multi-model data, including automatic multi-model storage and index selection. Multi-SQL focuses on four popular data models, i.e. relational model, key-value model, document model and graph model.

Our system has the following highlights:

User-Friendly: The primary goal of Multi-SQL is to achieve user-friendliness, so that non-expert users can efficiently implement complex data management without complicated database tuning. In order to achieve this goal, we build a simple interactive interface with only a few configurations.

Automatic Multi-model Storage Selection: Given the multi-model data and workload, Multi-SQL automatically obtains proper storage structure by utilizing artificial intelligent methods embedded in Multi-SQL, which could improve the efficiency of multi-model database management.

Automatic Multi-model Index Tuning: Multi-SQL could achieve automatic index selection for multi-model data. Given the data, workload and corresponding storage structure, Multi-SQL automatically completes index establishment in multi-model data without human tuning.

2   System Architecture and Implementation

In this section, we will introduce the overall architecture of multi-SQL. We design an automatic middleware manage multi-source storages, containing the real-time storage selection and optimal index tuning. Specifically, our middleware totally combines the multiple stores management and AI technologies to save the efforts of DBA. As we can see in Fig. 1, *User Interaction* delivers the files and configurations provided by users to *Controller*. *Controller* is responsible for managing the

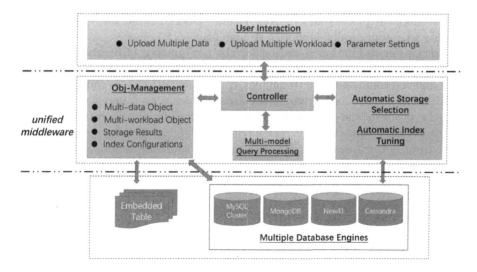

Fig. 1. System architecture

direct and indirect transmission of information in Multi-SQL. *Controller* controls all the information flow of the entire system. In order to efficiently achieve automatic management, we regard all the data, workload, storage, and index of Multi-SQL as objects. And *Obj-Management* uses an *Embedded Table* to record the mapping from these operating objects to the persistent storage. *Automatic Storage* is able to provide the optimal storage structure and index plan for the data set and workload uploaded by users. *Automatic Storage* builds in a variety of AI4DB technologies, which can realize automatic multi-model data storage and indexing. *Multi-model Query Processing* supports a unified and flexible multi-model query interface for multi-model data. Our system integrates four open source databases as the underlying storage engines, including MongoDB, Cassandra, New4J, and MySQL Cluster. They are expert in storing document data, key-value data, graph data and relational data, respectively.

3   Key Technologies

In this section, we present the key automatic techniques of Multi-SQL, including automatic storage selection, index tuning methods for multi-model data.

Storage structure is vital to improve the efficiency of workload execution. In this paper, we employ two AI techniques, ASH [6] and DUALSTORE [5], for automatic multi-model storage selection. ASH obtains the optimal storage design by comparing the cost of different storage structures through a machine-learning-based estimation model which consider some important factors for the storage structure, such as database configurations, workload and runtime features. DUALSTORE transforms the graph storage determination into a dynamic

Markov decision process and designs a tuner DOTIL based on reinforcement learning to automatically support a storage plan.

Index recommendation is crucial for the performance of databases. Multi-SQL achieves fully automated index establishment for multi-model data by automatic index recommendation techniques, containing DRLISA [8] and CNNIS [7]. CNNIS converts the index recommendation problem into a fine-grained multi-category problem and utilize the multi-core convolutional neural network to achieve index recommendation. The input is the original data set and workload, and output is the index plan. DRLISA uses the database as the learning environment and establish a deep reinforcement learning model based on the duel network to implement index tuning.

4   Demonstration Scenario

We utilize the typical multi-model scenario, social commerce [3] as the basic of demonstration, which contains various data models including relational data like user information table, graphs like social network, documents like posts, and key-value pairs like user behavior records. Next, we show three important multi-model management functions of Multi-SQL.

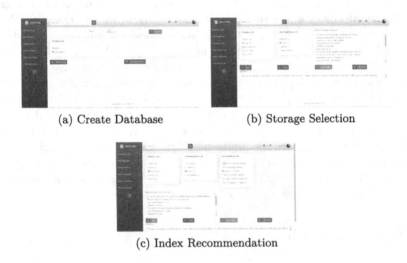

(a) Create Database   (b) Storage Selection

(c) Index Recommendation

Fig. 2. System demonstration

Upload Source Data. As shown in Fig. 2a, Multi-SQL allows users to create their own multi-model database, and then create a variety of data objects in this library. After that, users can flexibly choose their own existing objects to operate, and constantly create new objects in the process.

Storage Selection. Multi-SQL supports the storage selection function for multiple data models. As depicted in Fig. 2b, users select certain data objects and workload object, such as user information, posts, social net, etc. Multi-SQL will automatically select a storage structure for checked objects.

Index Recommendation. Multi-SQL could automatically provide the optimal indexes for the current storage and workload. As Fig. 2c presents, Multi-SQL gives more detailed configurations for the 'User_behavior' object such as "uid: Hash (fillfactor = 55)".

More detailed demonstrations and the source code of Multi-SQL are a available at the github[1].

Acknowledgement. Acknowledgements: This paper was supported by The National Key Research and Development Program of China (2020YFB1006104) and NSFC grant (U1866602).

References

1. Agrawal, D., Ouzzani, M., Papotti, P., Quiané-Ruiz, J.A., Mansour, E.: RHEEM: enabling cross-platform data processing: may the big data be with you! Proc. VLDB Endow. **11**(11), 1414–1427 (2018)
2. ArangoDB: Arangodb v3.3 documentation-data models and modeling (2017). https://docs.arangodb.com/3.3/Manual/DataModeling/
3. Busalim, A.H., Hussin, A.R.C.: Understanding social commerce: a systematic literature review and directions for further research. Int. J. Inf. Manage. **36**(6pt.A), 1075–1088 (2016)
4. Duggan, J., Zdonik, S., Elmore, A.J., Stonebraker, M., Balazinska, M.: The Big-DAWG polystore system. ACM SIGMOD Rec. **44**(2), 11–16 (2015)
5. Qi, Z., Wang, H., Zhang, H.: A dual-store structure for knowledge graphs. In: TKDE (2021)
6. Wang, H., Wei, Y., Yan, H.: Automatic storage structure selection for hybrid workload. CoRR abs/2008.06640 (2020). https://arxiv.org/abs/2008.06640
7. Yan, Yu., Wang, H., Zou, J., Wang, Y.: Automatic document data storage system based on machine learning. In: Wang, X., Zhang, R., Lee, Y.-K., Sun, L., Moon, Y.-S. (eds.) APWeb-WAIM 2020. LNCS, vol. 12318, pp. 551–555. Springer, Cham (2020). https://doi.org/10.1007/978-3-030-60290-1_45
8. Yan, Y., Yao, S., Wang, H., Gao, M.: Index selection for NoSQL database with deep reinforcement learning. Inf. Sci. **561**, 20–30 (2021). https://doi.org/10.1016/j.ins.2021.01.003, https://www.sciencedirect.com/science/article/pii/S0020025521000049

---

[1] https://github.com/HIT-MultiSQL/Multi-SQL.

Demonstration on Unblocking Checkpoint for Fault-Tolerance in Pregel-Like Systems

Zhenhua Yang[1,2], Yi Yang[1,2], and Chen Xu[1,2(✉)]

[1] East China Normal University, Shanghai, China
{zhyang,yiyang}@stu.ecnu.edu.cn, cxu@dase.ecnu.edu.cn
[2] Shanghai Engineering Research Center of Big Data Management, Shanghai, China

Abstract. Pregel-like systems are developed to execute iterative applications on massive graph data, which often leads to a long runtime. These systems are usually deployed on a cluster of commodity servers, where failures are common. Hence, fault-tolerance is crucial for them. A typical fault-tolerance technique is checkpointing, which can be achieved in a blocking or an unblocking manner. Blocking checkpointing incurs notable overhead as it pauses iterative computation. Unblocking checkpointing decrease the overhead by parallelizing checkpointing and iterative computation. However, it introduces resource contention due to parallel checkpointing tasks, which may prolong overall execution time. The *queuing strategy* and the *staleness/tardiness-aware skipping policy* can effectively improve unblocking checkpointing by alleviating the resource contention and selecting an optimal checkpoint from the queued checkpoints, respectively. In this demonstration, we showcase their internal mechanisms based on Apache Giraph.

Keywords: Graph processing · Fault tolerance · Checkpoint

1 Introduction

Large-scale graph processing plays an increasingly vital role in the world of big data, since it is used across various important application domains, ranging from social computation to online recommendation. Aiming at effectively and efficiently handling graphs involved in these areas, which are usually massive in size, the data management community has developed Pregel [3] as well as many other Pregel-like systems (e.g., Giraph [2] and Sedge [7]). These systems adopt the vertex-centric computation model and organize a whole graph processing job as a sequence of iterative *supersteps*. The supersteps are executed repeatedly until a termination condition is satisfied, which often takes a long time.

Typically, Pregel-like systems are deployed on a cluster of commodity servers, where failures are common. Therefore, fault-tolerance mechanisms are necessary. A widely employed scheme of fault-tolerance is to write checkpoints to stable storage (e.g., a distributed file system) periodically during iterative computation

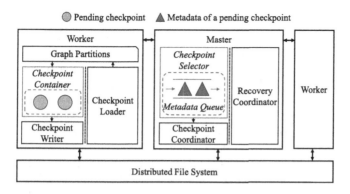

○ Pending checkpoint   ▲ Metadata of a pending checkpoint

Fig. 1. The architecture of Pregel-like systems with improved unblocking checkpointing

and retrieve the latest available checkpoint for a rollback recovery upon failures. Pregel-like systems usually store checkpoints in a blocking manner, which incurs notable overhead since it pauses iterative computation. Although unblocking checkpointing [6] can be adopted to reduce the overhead, it introduces resource contention due to parallel checkpointing tasks, which may prolong overall execution time under both failure and failure-free cases. Our prior work [8] proposed a *queuing strategy* to alleviate the resource contention by storing incoming checkpoints in a queue and restricting the maximum concurrency of checkpointing tasks. Further, a *staleness/tardiness-aware skipping policy* was proposed to select an optimal checkpoint from the queue.

In the rest of the paper, Sect. 2 revisits the above two fault-tolerance techniques. Section 3 then demonstrates the internal mechanisms of them based on Giraph, with an interactive graphical user interface (GUI). Additionally, we provide an accompanying screencast video[1] with the paper.

2   Background

In this section, we review our prior work [8] that improves unblocking checkpointing for Pregel-like systems. We focus more on the implementation and system integration of the proposed techniques.

Queuing Strategy. The rationale for unblocking checkpointing is to allow parallel execution of checkpointing tasks and iterative computation. Multiple checkpointing tasks may also be executed in parallel if the checkpointing completion time is longer than the time interval between checkpoint issuing. In this case, resource contention caused by parallel checkpointing may become significant and unignorable. The resource contention prolongs the completion time of both iterative computation and checkpointing tasks. It is worth noting that the prolonged completion time of checkpointing tasks may result in a longer recovery time. The reason behind this is that an uncompleted checkpoint requires the system to recompute from an earlier superstep upon failure.

---

[1] https://bilibili.com/BV1yB4y1D7tB.

To alleviate resource contention and reduce overall execution time, a *queuing strategy* is proposed for unblocking checkpointing. More specifically, a checkpoint container (in workers) and a checkpoint selector (in the master) are implemented to achieve this. As depicted in Fig. 1, the checkpoint container periodically takes graph snapshots as pending checkpoints in memory and reports some meta-information to the checkpoint selector. The meta-information includes the size of the snapshot, the corresponding superstep, etc. The checkpoint selector combines meta-information from all the workers to create a metadata object, which is then inserted into the metadata queue. The queue stores metadata objects following a First-In-First-Out (FIFO) policy, until the checkpoint coordinator requests more metadata objects to launch a checkpointing task. The checkpoint writers in workers then write corresponding pending checkpoints. The checkpoint coordinator is modified to restrict the maximum concurrency of checkpointing tasks to k, which is a user-defined parameter. The optimal choice of k is determined by system configuration.

Skipping Policy. The default FIFO policy employed by the queue is oblivious to the characteristics of checkpoints, which leaves optimization opportunities. Specifically, a completed checkpoint at a later superstep brings more benefits than the one at an earlier superstep, as the former leads to less recomputation time during recovery. This characteristic is denoted as *checkpoint staleness*. It can be formalized as $S(C_i) = 1/i$, where $S(C_i)$ means the staleness S of the checkpoint C at superstep i. Clearly, checkpoints with a lower $S(C_i)$ are preferred. Besides, checkpoints at different supersteps vary in size, since the amount of message to save differs on different supersteps. In general, the completion time of a checkpoint is proportional to its size. Checkpoints with less completion time are available for a recovery sooner. This characteristic is named *checkpoint tardiness* and defined as $T(C_i) = Size(C_i)$, where $T(C_i)$ means the tardiness T of the checkpoint C_i and $Size(C_i)$ means its size. It is clear that checkpoints with a lower $T(C_i)$ are preferred.

In order to leverage the characteristics of unblocking checkpoints to further decrease recovery time, a *staleness/tardiness-aware skipping policy* is designed. The skipping policy replaces the default FIFO policy used by the checkpoint selector in Fig. 1. Clearly, once a new checkpoint is completed, it can further decrease recovery time. Hence, the tardiness is preferable to the staleness. In particular, the checkpoint selector with the staleness/tardiness-aware skipping policy maintains the order of queued metadata objects by a composite key on the tardiness and the staleness. Notably, the checkpoint selector automatically drops metadata objects which have higher staleness than a completed checkpoint to decrease memory usage. Meanwhile, corresponding pending checkpoints in the workers are also dropped.

Except for the characteristics of checkpoints, some work [1,5] takes the cost of recomputation during recovery into consideration and tries to find an optimal checkpoint interval. Their work relies heavily on failure prediction techniques to trade off between the checkpointing overhead and the recomputation cost. Unfortunately, existing failure prediction models are still far from providing satisfactory results [4]. Therefore, these techniques are left as future work to be integrated into our demonstration system.

The above two improvements do not affect the recovery procedure. Upon failure, the recovery coordinator in Fig. 1 determines the latest available checkpoint and notifies the workers. Then, the workers apply a rollback recovery using the checkpoint loader module to reconstruct their respective graph partitions.

3   Demonstration

For the demonstration, we implement above fault-tolerance techniques in Giraph, as well as a web-based GUI. Our system is deployed on a cluster of three machines, which can be accessed by the GUI from a laptop. Additionally, we provide attendees with two algorithms, i.e., *Connected Components* (*CC*) and Single Source *Shortest Path* (*SP*), and the Wikipedia link (de)[2] dataset.

Queuing Strategy. Figure 2 shows the initial GUI for demonstrating the queuing strategy. Attendees specify the parameter k and the checkpoint interval. Meanwhile, they select an algorithm to execute. Upon clicking the "RUN" button, attendees see the real-time information of the queuing strategy.

As an example, Fig. 2 depicts a scenario that attendees set both the checkpoint interval and k to 1, and choose to run the *CC* algorithm. According to the left panel, there are twelve checkpoints in the checkpoint queue at superstep 18, but only the checkpoint at superstep 6 is in writing. The line chart on the right illustrates the change history of the queue length. Attendees observe what causes a change by hovering the mouse pointer over a change point on the line of the change history. For example, at superstep 14, a checkpoint is added into the checkpoint queue, accounting for the increase on the line of the change history.

Skipping Policy. Attendees investigate our staleness/tardiness-skipping policy in detail by toggling the policy radio button. The settings that attendees can specify here do not change, but the result part offers additional information (e.g., why a checkpoint is picked for writing or skipped otherwise).

By leaving other settings unchanged and clicking the "RUN" button, as depicted in Fig. 3, attendees see that the queue length is much shorter compared to Fig. 2. The "Details" panel shows an example of how the skipping mechanism works. Among the checkpoints at superstep 16, 17 and 18, the last one has the lowest staleness and tardiness, so it is chosen for writing. After the pick of the checkpoint at superstep 18, the remaining two checkpoints are skipped since they cannot further improve the recovery time if failures happen.

---

[2] https://networkrepository.com/web-wikipedia-link-de.php.

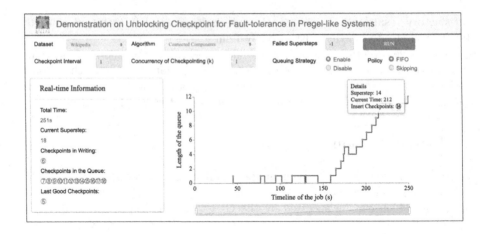

Fig. 2. Demonstrating the queuing strategy.

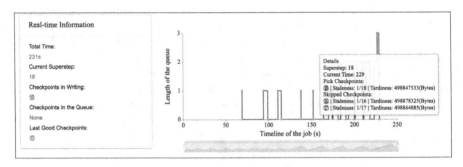

Fig. 3. Demonstrating the skipping policy.

Acknowledgments. This work was supported by the National Natural Science Foundation of China (No. 61902128).

References

1. Benoit, A., Cavelan, A., Fèvre, V.L., Robert, Y., Sun, H.: Towards optimal multi-level checkpointing. IEEE Trans. Comput. **66**(7), 1212–1226 (2017)
2. Giraph. https://giraph.apache.org
3. Malewicz, G., et al.: Pregel: a system for large-scale graph processing. In: SIGMOD, pp. 135–146 (2010)
4. Natella, R., Cotroneo, D., Madeira, H.: Assessing dependability with software fault injection: a survey. ACM Comput. Surv. **48**(3), 44:1–44:55 (2016)
5. Sigdel, P., Yuan, X., Tzeng, N.: Realizing best checkpointing control in computing systems. IEEE Trans. Parallel Distributed Syst. **32**(2), 315–329 (2021)
6. Xu, C., Holzemer, M., Kaul, M., Markl, V.: Efficient fault-tolerance for iterative graph processing on distributed dataflow systems. In: ICDE, pp. 613–624 (2016)
7. Yang, S., Yan, X., Zong, B., Khan, A.: Towards effective partition management for large graphs. In: SIGMOD, pp. 517–528 (2012)
8. Yang, Y., Yang, Z., Xu, C.: Exploiting unblocking checkpoint for fault-tolerance in pregel-like systems. In: WISE, pp. 71–86 (2021)

Author Index

Printed in the United States
by Baker & Taylor Publisher Services